Electrical Engineering Materials

About the Authors

R K Shukla is Associate Professor in the Physics Department at Harcourt Butler Technological Institute, Kanpur; a premier and autonomous institute affiliated to Gautam Buddh Technical University, Lucknow. He has nearly 19 years of teaching experience at both the undergraduate and postgraduate levels. During this period he has taught many courses like Engineering Physics, Materials Science, Engineering Materials, Solid State Physics, Electronics, etc.

Dr Shukla is actively engaged in research work, has published more than 40 research papers in international journals, and has presented 35 papers in national and international conferences. His area of research is Condensed Matter Physics. He has guided many a PhD students in their theses work and has been on the review panel of PhD theses of many universities. He is a member of the Board of Studies at Bundelkhand University, Jhansi, and has worked in various research projects of UGC and CSIR. He is also a reviewer of reputed journals like *Philosophical Magazine*.

Archana Singh is Assistant Professor in the Electrical Engineering Department at Harcourt Butler Technological Institute, Kanpur. She has completed more than seven years in teaching and has been teaching subjects like Electrical Engineering, Basic System Analysis, Network Analysis and Synthesis, and Electrical and Electronics Engineering Materials. She completed her BE in Electrical Engineering from Motilal Nehru National Institute of Technology, Allahabad. She has worked for more than 2 years in Tata Technologies Limited, a software company, in Pimpri, Pune. She also has experience in designing, development and customizing of various Oracle application modules.

Professor Singh has published several papers in journals of international repute (papers in ISRTE, ICPS-09 and NCERU-09) and various conferences. She is a member of Department of Purchase Committee, Proctorial Board, at the institute level. She is also a life member of the Indian Society for Technical Education (ISTE) and Institution of Engineers (IE). Professor Singh is currently pursuing PhD.

Electrical Engineering Materials

R K Shukla
Associate Professor
Department of Physics
Harcourt Butler Technological Institute (HBTI)
Kanpur, Uttar Pradesh

Archana Singh
Assistant Professor
Department of Electrical Engineering
Harcourt Butler Technological Institute (HBTI)
Kanpur, Uttar Pradesh

Tata McGraw Hill Education Private Limited
NEW DELHI

McGraw-Hill Offices

New Delhi New York St Louis San Francisco Auckland Bogotá Caracas
Kuala Lumpur Lisbon London Madrid Mexico City Milan Montreal
San Juan Santiago Singapore Sydney Tokyo Toronto

Tata McGraw-Hill

Published by the Tata McGraw Hill Education Private Limited,
7 West Patel Nagar, New Delhi 110 008.

Electrical Engineering Materials

ISBN (13): 978-1-25-902974-5
ISBN (10): 1-25-902974-3

Vice President and Managing Director—MHE: *Ajay Shukla*

Head, Publishing and Marketing—Higher Education: *Vibha Mahajan*

Publishing Manager (SEM & Tech. Ed.): *Shalini Jha*
Editorial Executive: *Koyel Ghosh*
Executive—Editorial Services: *Sohini Mukherjee*
Senior Production Manager: *Satinder S Baveja*
Production Executive: *Anuj K Shriwastava*

Marketing Manager—Higher Education: *Vijay Sarathi*
Senior Product Specialist (SEM & Tech. Ed.): *Tina Jajoriya*
Graphic Designer (Cover): *Meenu Raghav*

General Manager—Production: *Rajender P Ghansela*
Manager—Production: *Reji Kumar*

Typeset at Bukprint India, B-180A, Guru Nanak Pura, Laxmi Nagar-110 092 and printed at Ram Book Binding, C-114, Okhla Industrial Area, Phase-I, New Delhi-110 020

Cover Printer: SDR

RCLYCRAZDDDLD

The McGraw·Hill Companies

Contents

Preface

Overview

It gives us immense pleasure to present the first edition of *Electrical Engineering Materials*. In this book, the adopted philosophy is to take the readers from the realm of the unknown to the ambit of the known. Thus, our efforts have been to write the book in a manner such that every new concept follows from and is pre-linked to a concept that the student has been familiarised with.

The book presents concepts by providing an adequate review of a pre-requisite well-known theory. This way of presentation has a deep and everlasting impact on the minds of students. The engineering materials and their formation and characteristics, more or less, have a close dependance on the physics of the materials. The science of engineering materials cannot be understood in the absence of physics. The overall theme shows how the microscopic properties of matter are responsible for the utility of the material that they are a component of.

Objective of this Book

The major aim of the book is to discuss aspects of different materials in a very simple and friendly style, making even the toughest topic very enjoyable to read. To achieve this, we have taken the help of neat diagrams, graphs and examples and have stressed on providing a simple style of communication of the topics. Focus is given to the conceptual outlook rather than complex mathematical calculations. The book is organised in a way that offers great flexibility to meet the needs of the instructors, along with the reader. We believe that students will definitely like the classroom lecture-like presentation.

We have tried to use the internationally accepted symbols throughout the book. It is not expected that students find a formal solution to all the complicated equations, since it is unnecessary and in most cases, probably beyond their mathematical capacities. Therefore, the solutions are given and the only requirement is to verify and to learn the basic concept that leads to the application.

Target Audience

Many additional topics (chapters) like Band Theory of Solids, Modern Techniques of Material Studies and Special-Purpose Materials have been added to give the book a wider span and make it useful at the university level where courses related to engineering materials exist. This adds to the usefulness of the book at the university level for several courses such as BTech, BSc, MSc, etc. We feel that the book will be very useful to students and will cater to the needs of those who have similar courses in their curriculum.

Salient Features

- Simplistic coverage of electrical engineering materials highlighting characteristics suitable for their use in practical systems
- Extensive coverage of materials including glass, thermoplastics, organic semiconductors, etc.
- Coverage of high- and low-resistivity conducting materials and their application
- Examination-oriented pedagogical features with numerous illustrations:
  - Diagrams: 308
  - Solved Examples: 130
  - Objective-Type Questions: 250
  - Numerical Problems: 99
  - Exercises: 150
  - Fill in the Blanks: 100
  - Self-Quiz: 110

Chapter Organisation

The present book consists of eleven chapters.

Chapter 1 covers the basics of crystallography, different kinds of crystals, methods to establish the nature of the crystalline and amorphous materials, like XRD, and its underlying principle. Crystal defects, imperfections and their influence on electrical properties has also been discussed.

Chapter 2 describes the quantum theory of material formation, i.e. band theory of solids, and the difference between metals, insulators and semiconductors on the basis of band theory. The concept of Brillouin zones, and allowed and forbidden bands are also discussed. The contribution of holes to the current in a semiconductor is also beautifully described using simple examples.

Chapter 3 basically deals with popular magnetic properties of materials and their role in selection of magnetic materials for various appliances. Theories related with diamagnetic, paramagnetic and ferromagnetic materials are also described. The application of magnetic materials for memory storage devices, computer hard disks, floppy disks, credit cards and non-volatile memories is also explained. Besides the above, the materials useful for various electric machines like transformers, rotating machines, aircraft generators are described in detail. Some new promising materials like non-oriented and grain oriented steels, and hardened steel used in heavy-duty machines are also discussed.

Chapter 4 deals with dielectric materials and their behaviour in ac and dc fields along with various kinds of polarisation mechanisms. Theories related to polar and non-polar dielectrics (Langevin's theory, etc.) have been described. The concept of macroscopic and microscopic electric fields is also provided. Various kinds of breakdown phenomena in dielectric materials have been detailed. Dielectric loss, loss tangent and its significance in electrical engineering is explained in a simple and lucid style.

Chapter 5 presents conductivity of metals and superconducting materials. Various factors and their influence on thermal and electrical conducting have been described. Free electron theory and quantum mechanical theory of conducting metals is discussed in detail and the concept of Fermi energy is dealt with an extremely simple approach.

Various kinds of thermoelectric phenomena are also described. Superconductivity has also been covered along with its interpretation and application. Changes in thermodynamic properties at superconducting state and development of high-temperature superconductors are also explained. The application of this new class of materials is nicely described.

Chapter 6 gives an overview of popular conducting and insulating materials suitable for electrical engineering purposes. The properties of high-resistivity and low-resistivity materials are described in detail. Materials required for high load and heavy load contact are also explained. Besides the above, insulating gases, liquids and solids are discussed along with their characteristic properties. Comparison between materials and their suitability to various temperature ranges is given in a simple manner. Different kinds of capacitor materials and selection of insulating materials on the basis of some electrical properties are also presented.

Chapter 7 describes some electronic devices (unipolar and bipolar) along with their preparation (fabrication) and purification methods. Various kinds of breakdown mechanism in *p-n* junctions and IC chip fabrication is also discussed. The chapter also explains photonic devices like solar cell, photodiode, phototransistor, LEDS and their applications.

Chapter 8 provides the conduction mechanism in semiconductor materials. This chapter describes elemental, compound and multi-component semiconductors. How band-gap variation can be achieved depending upon the need is also explained in a very interesting way. Various kinds of growth techniques and doping methods are also discussed. Temperature dependence of carrier concentration and measurement of some semiconductor parameters is dealt with in a simplified way.

Chapter 9 covers mechanical properties of materials with the point of view of their electrical applications.

Chapter 10 presents optical properties of materials and the materials used in various photonic devices which can detect the incidence of radiation or can emit radiation of desired colour. LASER is also described along with fundamentals of fibre optics, the backbone of modern communication systems.

Chapter 11 provides a glimpse of some modern and special-purpose materials with the point of view of their application in electrical engineering. Some modern techniques of material studies like DSC, SEM, TEM and optical absorption spectroscopy with the underlying principles and operation are also covered.

Web Supplements

The Solution Manual for instructors is available on the Website
http://www.mhhe.com/shukla/eem1

Acknowledgements

Work on this book started three years back. We received many suggestions from students, faculty members and colleagues, which we have tried to include here.

Dr Shukla would like to take this opportunity to express his gratitude to his guide, Prof. Ashok Kumar, Head of the Department of Physics, Harcourt Butler Technological Institute, Kanpur, who inspired him to write this book. His profound regards are due to the late Prof. A K Vashistha, Ex-Director of this institute, who always encouraged him to carry out such academic works. Dr Shukla would also like to express extreme gratitude to his wife, Mrs Sangeeta Shukla; Mr Alok Shukla (electrical engineer), Mr Aishwary Shukla and little Kartik. Their sacrifices and hidden support will always be special to him.

Both of us have intense regard for our parents (Late Sri K S Shukla and Smt. Krishna Kumar Shukla, and Dr T K Pal and Mrs Heeramati) for their motivation and constant encouragement.

We would also like to thank the numerous reviewers who sent us useful comments and suggestions after going through the manuscript. Their names are given below:

Bharti Dwivedi	*Institute of Engineering and Technology, Lucknow, Uttar Pradesh*
Jyoti Bajpai	*Krishna Institute of Technology, Kanpur, Uttar Pradesh*
Mahesh Gupta	*Vidya Niketan Samiti Institute of Technology (VNSIT), Bhopal, Madhya Pradesh*
Arkendu Mitra	*Narula Institute of Technology, Kolkata*
Abhimanyu Mohapatra	*College of Engineering and Technology (CET), Bhubaneswar, Odisha*
S B L Seksena	*National Institute of Technology (NIT), Jamshedpur, Jharkhand*
Tushar Kanti Gangopadhyay	*J C Ghosh Polytechnic, Kolkata*
Pradeep Ranjan Tripathy	*Kalinga Institute of Industrial Technology (KIIT) Polytechnic, Bhubaneswar, Odisha*
A P Kinge	*Pune Institute of Computer Technology, Pune*
Sachin V Rajani	*Vyavasayi Vidya Pratishthan (VVP) Engineering College, Rajkot, Gujarat*
R Pandian	*Velammal College of Engineering and Technology, Madurai, Tamil Nadu*
S Senthil Rani	*Velammal College of Engineering and Technology, Madurai, Tamil Nadu*
S Suresh Kumar	*Kalasalingam University, Krishnankoil, Tamil Nadu*
S R Manohara	*Siddaganga Institute of Technology, Tumkur, Karnataka*
Y S Reddy	*Chaitanya Bharathi Institute of Technology, Hyderabad*

Finally, we feel a deep sense of gratitude for the expert team at Tata McGraw-Hill, without whose cooperation this work could not be translated into reality. Indeed, the TMH team bears a high standard of excellence, and the fulfilling discussions that we had with the members of TMH, helped us a lot in improving the contents. This will definitely reflect in the book. We do not hesitate to say that the TMH team has taught us the art of writing an excellent book!

Feedback

Every effort has been made to make the book error-free, yet some errors might still be there for which we expect a kind excuse from the readers. All the suggestions and advice from readers will be gratefully acknowledged.

R K Shukla
Archana Singh

Publisher's Note

Do you have any feedback? We look forward to receive your views and suggestions for improvement. The same can be sent to *tmh.elefeedback@gmail.com*, mentioning the title and author's name in the subject line.

1

Crystal Structure, Bonding and Defects in Solids

Chapter Objectives

- To learn various kinds of lattices and parameters related with crystals—like co ordination number, atomic radii, atomic packing factor, lattice planes, Miller Indices, interplanar separation, etc.
- To perform simple calculations about crystals.
- To know the internal structure of materials using the X-ray diffraction technique.
- To know various kinds of defects and imperfections, and their influence upon electrical and mechanical properties of the materials.

1.1 INTRODUCTION

Solids are classified into three categories, namely crystalline (single crystal), **amorphous** and **polycrystalline**. Each type is characterised by the size of an ordered region within a material. An ordered region is a spatial volume in which atoms or molecules have a regular geometric arrangement or periodicity. **Amorphous materials have order only up to few atomic dimensions, while polycrystalline materials have a high degree of order over many atomic dimensions (long-range order).** These ordered or single-crystal regions vary in size and orientation with respect to one another. The single-crystal regions are called grains and are separated from one another by grain boundaries. Single-crystal materials have a high degree of order or geometric periodicity throughout the volume of the material. The advantage of a single-crystal material is that, in general, its electrical properties are superior as grain boundaries tend to degrade the electrical properties. Figure. 1.1 shows the three kinds of solids.

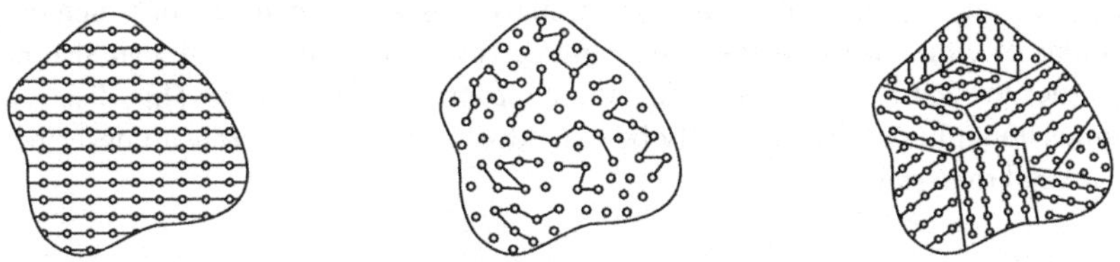

(a) Crystalline (b) Amorphous (c) Polycrystalline

Fig. 1.1 *Representation of the three types of crystals*

1.2 SPACE LATTICE

We will first discuss the single crystal with its regular geometric periodicity of atoms. A representative unit or group of atoms is repeated at regular intervals in each of the three dimensions to form the single crystal. The regular arrangement of atoms in a crystal appears '**net-like**' and is known as a **lattice**.

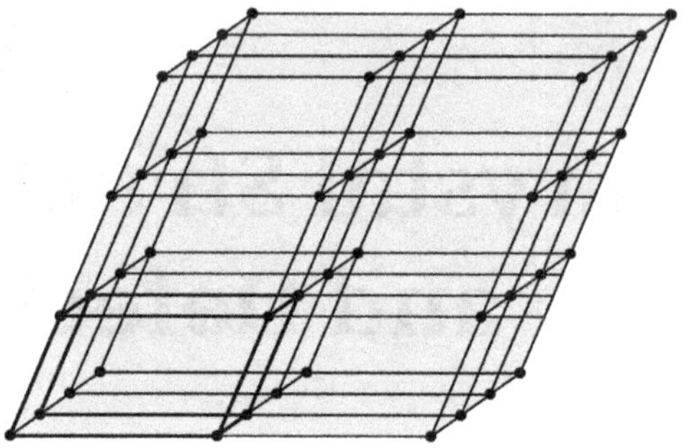

Fig. 1.2 *A space lattice*

We can represent a particular atomic array by a dot that is called a lattice point. Figure 1.2 shows an infinite two-dimensional array of lattice points. The simplest means of repeating an atomic array is by translation. Each lattice point in Fig. 1.3 can be translated a distance a_1 in one direction and a distance b_1 in a second non-collinear direction to generate the two-dimensional lattice. A third non-collinear translation will produce the three-dimensional lattice. The translation directions need not be perpendicular.

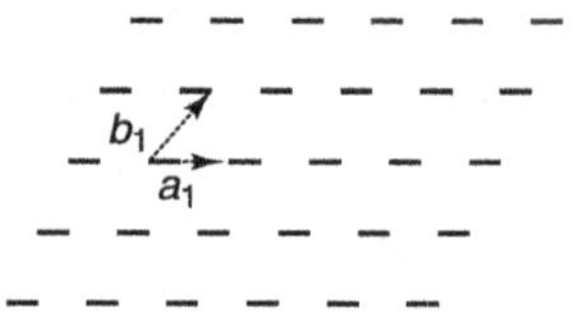

Fig. 1.3 *Two-dimensional representation of a single-crystal lattice*

Since the three-dimensional lattice is a periodic repetition of a group of atoms, we do not need to consider the entire lattice, but only a fundamental unit that is being repeated. A unit cell is a small volume of the crystal that can be used to reproduce the entire crystal. A unit cell is not a unique entity. Figure 1.4 shows several possible unit cells in a two-dimensional lattice.

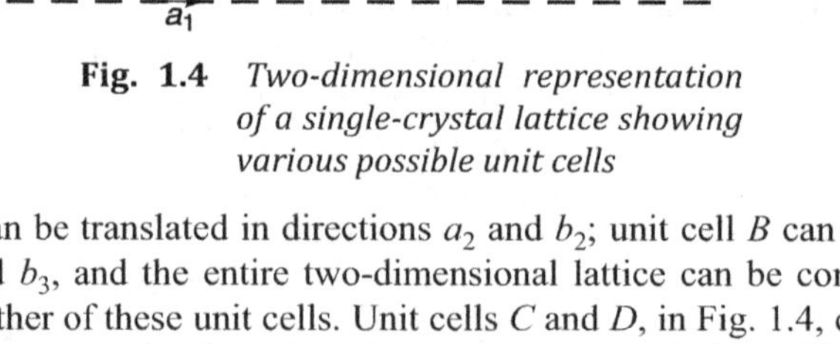

Fig. 1.4 *Two-dimensional representation of a single-crystal lattice showing various possible unit cells*

Unit cell A can be translated in directions a_2 and b_2; unit cell B can be translated in directions a_3 and b_3, and the entire two-dimensional lattice can be constructed by the translations of either of these unit cells. Unit cells C and D, in Fig. 1.4, can also be used to construct the entire lattice by using the appropriate translations. This discussion of two-dimensional unit cells can easily be extended to three dimensions to describe a real single-crystal material.

1.3 PRIMITIVE CELL

A primitive cell is the smallest unit cell that can be repeated to form the lattice. In many cases, it is more convenient to use a unit cell that is not a primitive one. Unit cells with orthogonal sides may be chosen, for example; however, the sides of a primitive cell may be non-orthogonal.

A generalised three-dimensional unit cell is shown in Fig. 1.5. The relationship between this cell and the lattice is characterised by three vectors $\vec{a}$, $\vec{b}$ and $\vec{c}$, which need not be perpendicular and which may or may not be equal in length. Every equivalent lattice point in the three-dimensional crystal can be found using the vector

$$\bar{r} = p\bar{a} + q\bar{b} + s\bar{c}$$

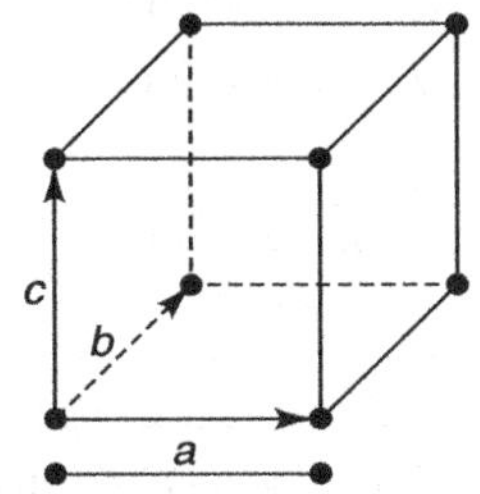

Fig. 1.5 *A generalised primitive unit cell*

where p, q and s are integers. Since the location of the origin is arbitrary, we will let p, q and s be positive integers for simplicity.

1.4 BASIC CRYSTAL STRUCTURES AND THEIR CHARACTERISTICS

Let us consider three-crystal structures that determine some of the basic characteristics of these crystals. Figure 1.6 shows the simple cubic, body-centred cubic and face-centred cubic structures. For these simple structures, we may choose unit cells such that the general vectors $\vec{a}$, $\vec{b}$ and $\vec{c}$ are perpendicular to each other and the lengths are equal. The simple cubic (SC) structure has an atom located at each corner; the body-centred cubic (BCC) structure has an additional atom at the centre of the cube; and the face-centred cubic (FCC) structure has additional atoms on each face plane.

By knowing the crystal structure and lattice dimensions of a material we can determine several characteristics of the crystal. For example, we can determine the volume density of atoms.

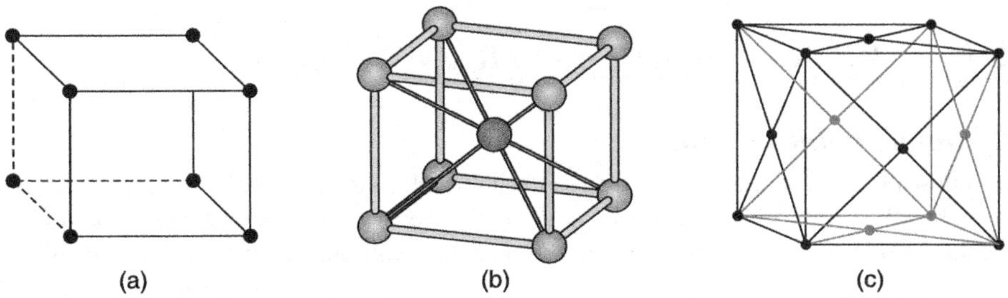

Fig. 1.6 *Three lattice types: (a) Simple cubic (b) Body-centred cubic (c) Face-centred cubic*

1.5 BRAVAIS SPACE LATTICE

Auguste Bravais, in 1948, showed that there are 14 types of lattices under seven crystal systems (Table 1.1).

Table 1.1 *Various kinds of lattices*

Name of system	*Relation between primitive*	*angles*	*Number of possible lattices*	*Examples*
1. Cubic	$a = b = c$	$\alpha = \beta = \gamma = 90°C$	3	NaCl
2. Monoclinic	$a \neq b \neq c$	$\alpha = \beta = 90°C\ \gamma = 90°C$	2	Na_2SO_4 or $FeSO_4$
3. Triclinic	$a \neq b \neq c$	$\alpha \neq \beta \neq \gamma \neq 90°C$	1	$CuSO_4$
4. Tetragonal	$a = b \neq c$	$\alpha = \beta = \gamma = 90°C$	2	$NiSO_4$
5. Orthorhombic	$a \neq b \neq c$	$\alpha = \beta = \gamma = 90°C$	4	KNO_3
6. Rhombohedral	$a = b = c$	$\alpha = \beta = \gamma \neq 90°C$	1	$CaSO_4$
7. Hexagonal	$a = b \neq c$	$\alpha = \beta = 90°C\ \gamma = 120°C$	1	Quartz
			Total: 14	

In simple crystal, i.e. in elements there is a single atom or ion at each lattice point. In others (compounds), there is a group of two or more atoms or ions at each lattice point. This group of atoms in known as basis. It is an assembly of one or more atoms which, when attached identically to each lattice point, give rise to crystal structure. Every basis is identical in composition, arrangement and orientation. In crystal-like aluminium and barium, the basis is a single atom, while in NaCl and KCl the basis is diatomic. Basis is repeated in space to form crystal structure. When basis is added to each lattice point in space, crystal structure is obtained. One can write the relation as—

Lattice + Basis = Crystal structure

Formation of crystal structure by the addition of basis is shown in Fig. 1.7.

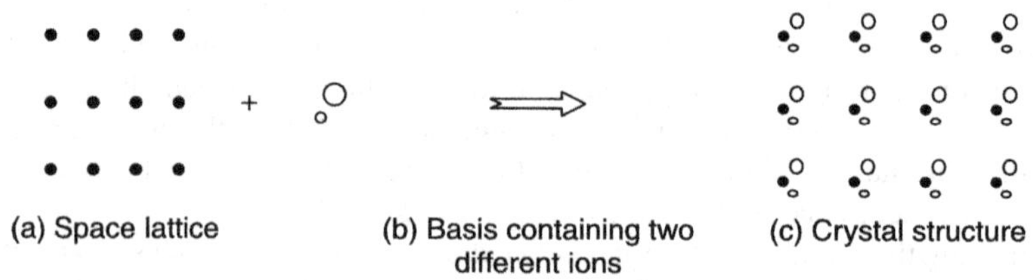

Fig. 1.7 *Formation of crystal structure by the addition of basis*

1.6 NUMBERS OF ATOMS PER UNIT CELL

(i) Simple Cubic Crystal

By knowing the arrangement of atoms in a unit cell, one can get the number of atoms in each cell. In case of SC lattice, there are eight atoms located at each corner, and each corner atom is shared by eight unit cells. Hence, the share of each unit cell is equal to 1/8 to an atom. In this way, the total number of atoms in one unit cell = $8 \times \frac{1}{8}$ = one atom.

(ii) Body-centred Cubic Crystal

In this kind of lattice, there are eight atoms located at each corner exactly as in the simple-cubic crystal along with one atom located at the body centre. Therefore, the total number of atoms in one unit cell = $8 \times \frac{1}{8} + 1 = 2$ atoms.

(iii) Face-centred Cubic Crystal

In face centred cubic crystal, there are eight atoms located at each corner which are shared by eight unit cells, contributing, therefore, $8 \times \frac{1}{8} = 1$ atom to each cell; besides each face-centred atom is shared by two unit cells i.e., contribution = $6 \times \frac{1}{2} = 3$ atoms to each cell. In this way, the total numbers of atoms per unit cell = 1 + 3 = 4 atoms.

1.7 CO-ORDINATION NUMBER

Each atom in a crystalline structure is surrounded by other atoms. **The number of equidistant nearest neighbours that an atom has in a given structure is named as co-ordination number.** Higher the co-ordination number, more closely packed will be the structure and *vice versa*. Following are the co-ordination numbers of three main kinds of lattices.

(i) Simple Cubic Lattice

In simple cubic lattice, there is one atom located at each corner of the cube. Any corner atom has four nearest neighbours in the same plane and two nearest neighbours (one exactly above and the other exactly below) in a vertical plane.

Hence, co-ordination number = 4 + 2 = 6.

If a is a side of the cube, then the distance between the nearest neighbours will be equal to a.

(ii) Body-centred Cubic Lattice

In this kind, one atom is located at each corner of the cube and one atom at the centre of each face of the cube. For any corner atom of the unit cell, the nearest atoms are the atoms which are at the centre of unit cell having eight body-centred atoms; hence, co-ordination number is eight. Similarly, by considering the centred atom of each unit cell, we can say that the co-ordination number is eight because every centred atom is surrounded by eight equidistant neighbours.

Hence, co-ordination number for BCC lattice = 8.

The nearest distance between two atoms = $\frac{\sqrt{3}a}{2}$.

(iii) Face-centred Cubic Lattice

In FCC lattice, there is one atom at each corner of the cube and one atom at the centre of each face of the cube. For any corner atom of the unit cell, the nearest are the face-centred atoms. For any corner atom, there will be four face-centred atoms of the surrounding unit cell in its own plane, four face-centred atoms below this plane and another four face-centred atoms above this plane.

Hence, co-ordination number of FCC lattice = 4 + 4 + 4 = 12.

The distance between two nearest neighbours = $\frac{a}{\sqrt{2}}$.

1.8 ATOMIC RADII

If it is assumed that atoms are spherical in shape and are in contact, then one can readily calculate their radii from the knowledge of crystal structure of the solid and the lattice parameter. Figure 1.8 shows a face-centred cubic unit cell in which lattice constant is a side (length of each edge) of the cube. If the radius of each atom is r, the length of diagonal on any face is $4r$ and

$$(4r)^2 = a^2 + a^2$$

or $$r = \frac{\sqrt{2}a}{4} \quad \text{.......................... FCC (structure)}$$

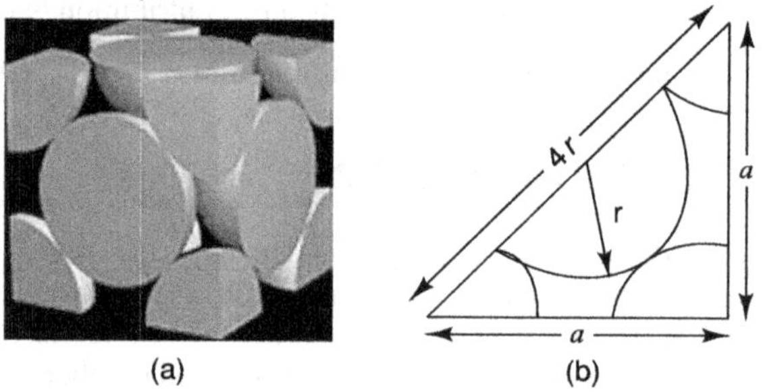

Fig. 1.8 *(a) Cluster of atoms (b) Section of a cube*

Similarly, for other cubic systems the atomic radii in terms of a are given below:

$$r = \frac{a}{2} \quad \text{———— SC Structure}$$

$$r = \frac{\sqrt{3}a}{4} \quad \text{———— BCC Structure}$$

$$r = \frac{\sqrt{3}a}{8} \quad \text{———— Diamond Structure}$$

For copper, which has FCC structure and lattice constant $a = 3.61$ Å, the atomic radius is

$$r = \frac{\sqrt{2} \times 3.61 \text{ A}}{4} = 1.28 \text{ Å}$$

The atomic radii, which is discussed above, in real sense is **'half inter-nuclear distance of closest approach'**, since it is hardly correct to consider an atom a hard sphere.

1.9 IONIC RADII

Ionic radii can be determined from the data on ionic crystal in a similar way. Table 1.2 shows the ionic radii of some positive and negative ions. These radii differ from those of the corresponding neutral atoms, often quite considerably. This is because the removal of one or more electrons to form a positive ion leads to reduction in size as the remaining electrons are more tightly bound; and if all the electrons are removed from outermost shell, the size shrinks to one smaller than that of the inert gas atom of the preceding period.

Table 1.2 *Ionic radii of some positive and negative ions*

Positive ions	*Radius, A*	*Negative ions*	*Radius, A*
Li^+	0.60	H^-	2.08
Na^+	0.95	F^-	1.36
K^+	1.33	Cl^-	1.81
Rb^+	1.48	Br^-	1.95
Cs^+	1.69	I^-	2.16
Be^{++}	0.31	O^-	1.40
Mg^{++}	0.65	S^-	1.84
Ca^{++}	0.99	N^{3-}	1.71
Sr^{++}	1.13	P^{3-}	2.12
Ba^{++}	1.35	C^{4-}	2.60
Al^{3+}	0.50	Si^{4-}	2.71

For example, the radius of Na atom is 1.86 Å, while that of Na^+ ion is only 0.95 A. The radius is less by 1.1 Å (radius of Na atom having structure similar to that of Na^+).

On the other hand, a negative ion containing excess of electrons is found to have a larger size comparable to the parent atom. For example, radius of S atom is 1.27 Å, while of S^- is 1.84Å.

1.10 CALCULATION OF LATTICE CONSTANT

Let us consider the case of a cubic lattice having side a (lattice constant). If ρ be the density of crystal under consideration then—

$$\text{Volume of unit cell} = a^3 \tag{1.1}$$

$$\text{Mass of each unit cell} = \rho a^3 \tag{1.2}$$

Assuming that there are n lattice points (molecules) per unit cell, M be the molecular weight and N the Avogadro number then—

$$\text{Mass of each molecule} = \frac{M}{N}$$

$$\text{Mass in each unit cell} = n \times \frac{M}{N} \tag{1.3}$$

From Eqs (1.2) and (1.3),

$$\rho a^3 = \frac{n.M.}{N}$$

or

$$a^3 = \frac{n.M}{\rho N}$$

or

$$a = \left(\frac{n.M}{\rho N}\right)^{1/3} \tag{1.4}$$

From the above formula one can get lattice constant a by substituting all other parameters.

Example 1.1

Calculate the lattice constant '*a*' of a substance having FCC lattice, molecular weight 60.2 and density 6250 kg/m³. Given: $N = 6.02 \times 10^{26}$/kg mole

Solution Given molecular weight $M = 60.2$, density $\rho = 6250$ kg/m³, and $N = 6.02 \times 10^{26}$/kg mole.

For FCC lattice $n = 4$

$$\text{Lattice constant } a = \left(\frac{nM}{N\rho}\right)^{1/3}$$

$$\therefore \quad a = \left[\frac{4 \times 60.2}{6250 \times 6.02 \times 10^{26}}\right]^{1/3}$$

$$= \left[\frac{240.8}{37265 \times 10^{26}}\right]^{1/3}$$

$$= 4 \text{ Å}$$

Ans.

Example 1.2

A monochromatic beam of X-rays of wavelength 2.0 Å is reflected by cubic crystal of KCl. Determine the inter-planar distances for (1 0 0), (1 1 0) and (1 1 1) planes. Given: density of KCl = 1980 kg / m³ and molecular weight $M = 74.5$ and Avogadro's number, $N = 6.023 \times 10^{26}$/kg mole.

Solution Given $M = 74.5$, $\rho = 1980$ kg/m³ and $N = 6.023 \times 10^{26}$/kg mole.

Formulae used are

$$a^3 = \frac{nN}{N\rho}$$

$$\text{and } d_{hkl} = \frac{a}{\sqrt{h^2 + k^2 + l^2}} \text{ for cubic crystal.}$$

$$\therefore \quad a^3 = \frac{nM}{N\rho} = \frac{4 \times 74.5}{6.023 \times 10^{26} \times 1.98 \times 10^3} = 24.99 \times 10^{-29}$$

$$a = [249.9 \times 10^{-30}]^{1/3} = 6.30 \times 10^{-10} = 6.3 \text{ Å}$$

$$d_{100} = \frac{a}{\sqrt{1^2 + 0^2 + 0}} = \frac{a}{1} = \frac{6.3 \text{ A°}}{1} = 6.3 \text{ Å}$$

$$d_{110} = \frac{a}{\sqrt{1^2 + 1^2 + 0}} = \frac{a}{\sqrt{2}} = 4.38 \times 10^{-10} \text{ m} = 4.38 \text{ Å}$$

$$d_{111} = \frac{a}{\sqrt{1^2 + 1^2 + 1^2}} = \frac{a}{\sqrt{3}} = \frac{6.3 \times 10^{-10}}{\sqrt{3}} = 3.64 \times 10^{-10} \text{ m} = 3.64 \text{ Å}$$

Ans.

1.11 LATTICE PLANES AND MILLER INDICES

A **crystal lattice can be considered to be made up of a large number of parallel equidistant planes named as 'lattice planes' which pass through lattice points**. These lattice planes can be chosen in a number of ways, as shown in Fig. 1.9.

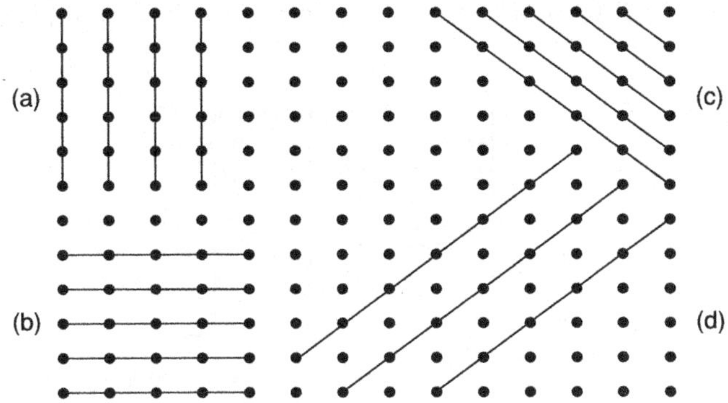

Fig. 1.9 *Various kinds of lattice planes*

Each family of these planes has a particular orientation with reference to the axes of unit cell. **The orientation of these planes is specified by the so-called Miller indices (*hkl*)** whose values for any plane can be obtained by using the following procedure:

- Find the intercepts of the plane on the three-crystal axes (OX, OY, OZ). Let these be pa, qb and rc; where a, b, c are corresponding primitives and p, q, r be integer values.
- Write the reciprocal of the number p, q, r as $\frac{1}{p}, \frac{1}{q}, \frac{1}{r}$
- Find the smallest possible integers h, k, l such that

$$h : k : l = \frac{1}{p} : \frac{1}{q} : \frac{1}{r}$$

It is done by multiplying the reciprocals by the LCM of their denominators.

These numbers (h, k, l) are known as Miller indices and the corresponding lattice plane is specified as $[hkl]$.

It should be noted that these indices refer to a set of parallel planes and not a particular plane.

Example 1.3

Find the Miller indices of plane which cuts the intercepts on axes as 4, 1, 2.

Solution Reciprocals are $\frac{1}{4}, 1, \frac{1}{2}$

LCM of denominator is 4.

$$\text{Miller indices} = 4\frac{1}{4}, 4\times 1, 4\times\frac{1}{2} = [1, 4, 2]$$ **Ans.**

In case an intercept is at infinity, the corresponding indices will be zero. One or more indices may be negative when the corresponding intercepts are negative. They are represented with a bar as— $[\bar{h}, kl]$ or $[h, \bar{k}, l]$, etc.

In order to make the idea more clear, we consider the Miller indices of some planes in a cubic crystal shown in Fig. 1.10. It should be noted that when a face is parallel to an axis, its intercept on the axis is infinite. Consider plane *CGBF*; it cuts the X-axis at a point *F* and is parallel to *Y* and *Z* axes. Assuming the side of cube to be unity, the intercepts made by this plane on the three axes are $1 : \infty : \infty$. Their reciprocals are $1 : 0 : 0$. Hence, the Miller indices of this face are (100). The face *HAF* cuts the three axes by equal length. Hence, the Miller indices are (111). Similarly, the Miller indices of plane *DHGC* are (011).

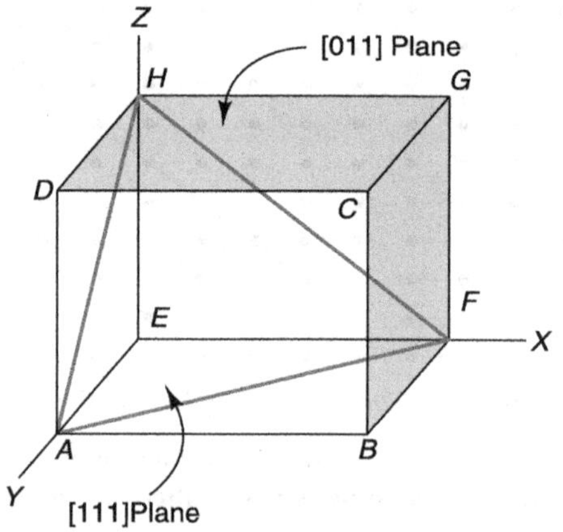

Fig. 1.10 *Miller indices of some planes in a cubic crystal*

Example 1.4

A plane cuts intercepts 3*a*, 4*b* and 2*c* along the crystallographic axes in a crystal. Determine the Miller indices of the plane.

Solution Intercept are $3a$, $4b$ and $2c$.

From the law of rational indices, we have

$$3a : 4b : 2c = \frac{a}{h} : \frac{b}{k} : \frac{c}{l}$$

or
$$\frac{1}{h} : \frac{1}{k} : \frac{1}{l} = 3 : 4 : 2$$

or
$$h : k : l = \frac{1}{3} : \frac{1}{4} : \frac{1}{2} = 4 : 3 : 6$$

Therefore, the Miller indices of the plane are (4, 3, 6). **Ans.**

Example 1.5

In a triclinic crystal, a lattice plane makes intercepts at a length *a*, 2*b* and –3*c*/2. Find the Miller indices of the plane.

Solution Intercepts are a, $2b$ and $-3c/2$.

$$\therefore \quad a : 2b : \frac{-3c}{2} = \frac{a}{h} : \frac{b}{k} : \frac{c}{l}$$

or $$\frac{1}{h}:\frac{1}{k}:\frac{1}{l} = 1:2:\frac{-3}{2}$$

or $$h:k:l = 1:\frac{1}{2}:\frac{-2}{3} = 6:3:-4$$

Therefore, the Miller indices of the given plane are $(6\ 3\ \bar{4})$. **Ans.**

Example 1.6

Deduce the Miller indices for planes in each of the following sets which intercept $\vec{a}$, $\vec{b}$ and $\vec{c}$ at (i) 4*a*, 4*b*, 2*c* (ii) *a*, 2*b*, ∞ (iii) *a*, *b*/2, *c*.

Solution

(i) Intercepts are 3*a*, 3*b*, 2*c*.

Then, $$4a:4b:2c = \frac{a}{h}:\frac{b}{k}:\frac{c}{l}$$

or $$4:4:2 = \frac{1}{h}:\frac{1}{k}:\frac{1}{l}$$

or $$h:k:l = \frac{1}{4}:\frac{1}{4}:\frac{1}{2}$$

or $$h:k:l = 2:2:4$$

Therefore, the Miller indices are (2 2 4). **Ans.**

(ii) Intercepts are 2*a*, *b*, ∞.

Then,

$$2a:b:\infty = \frac{a}{h}:\frac{b}{k}:\frac{c}{l}$$

or $$\frac{1}{h}:\frac{1}{k}:\frac{1}{l} = 2:1:\infty$$

or $$h:k:l = \frac{1}{2}:\frac{1}{1}:\frac{1}{\infty} = 1:2:0$$

Therefore, the Millers indices are (2 1 0). **Ans.**

(iii) Intercepts are *a*, *b*/2, *c*.

Then,

or $$h:k:l = 2:1:2$$

Therefore, the Miller indices are (2 1 2). **Ans.**

1.12 ATOMIC PACKING FACTOR: PACKING EFFICIENCY

The atoms in a lattice are bonded strongly to their nearest neighbours and they tend to come close to each other as far as possible. The close packing of atoms in a lattice has limitations depending upon the structure of the lattice and the relative size of the bonding atoms. The co-ordination between the nearest neighbouring atoms is different in different structures.

Packing factor is a measure of how closely the atoms are packed in a crystal and **is defined as the ratio of the volume occupied by the atoms in a unit cell to the volume of the unit cell.**

$$\text{Atomic packing factor} = \frac{\text{Volume occupied by atoms in unit cell}}{\text{Volume of unit cell}}$$

(i) Atomic Packing Factor of Simple Cubic Structure

Figure 1.11 shows the unit cell of the **simple cubic** lattice.

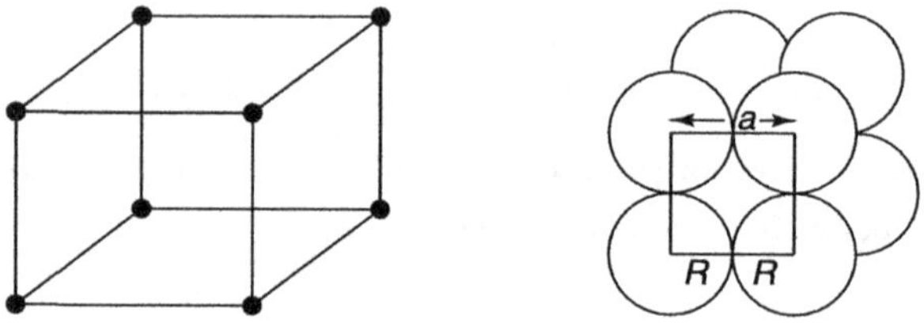

Fig. 1.11 *Unit cell of simple cubic lattice*

In the simple cubic mono-atomic lattice, each atom is bonded to its adjacent atom so that the number of nearest neighbours (co-ordination number) for each atom is six. The dimension of unit cell is twice that of the atomic radius—

$$a = 2R$$

There are eight corner atoms; so, the total number of atoms per unit cell is

$$\frac{1}{8} \times 8 = 1$$

$$\text{Volume occupied by atoms} = 1 \times \frac{4}{3}\pi R^3$$

$$\text{Volume of unit cell} = a^3 = (2R)^3$$

$$\text{Therefore, APF} = \frac{\frac{4}{3}\pi R^3}{(2R)^3} = 0.52$$

Hence one can conclude that packing of atoms in SC lattice is not so tight. Only 52% (which is also called packing efficiency) of the space is occupied. This is the reason why cubic structure is not so common among mono-atomic materials (metals). Among the elements, only polonium is simple cubic.

(ii) Atomic Packing Factor of FCC Lattice

Figure 1.12 shows a FCC lattice in which each atom is bonded closely to the neighbouring atom along the face diagonal of unit cell. Each atom has 12 nearest neighbours. The dimensions of unit cell is related to atomic radius as—

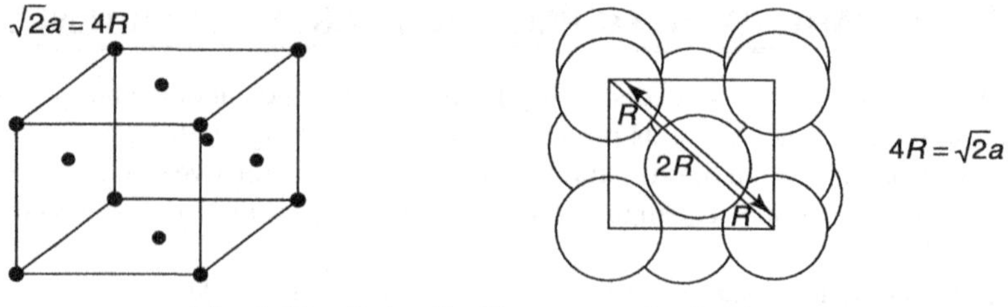

Fig. 1.12 *Unit cell of face-centred cubic lattice*

There are eight corner atoms and six atoms at the centre of faces. The number of atoms per unit cell

$$= \left(\frac{1}{8}\times 8\right)+\left(\frac{1}{2}\times 6\right)=4$$

Volume occupied by four atoms $= 4\times\frac{4}{3}\pi R^3$

Volume of unit cell $a^3 = \left(\frac{4R}{\sqrt{2}}\right)^3$

Therefore Packing factor for FCC lattice $= \dfrac{\frac{16}{3}\pi R^3}{\left(\frac{4R}{\sqrt{2}}\right)^3} = 0.74$

Hence, **packing factor of FCC lattice is quite high. In fact, it is one of the most closely-packed structures as its packing efficiency in 74%. Most metals (e.g. Al, Cu, Ag, Ni) have FCC structure.**

(iii) Packing Factor of BCC Lattice

Similarly, one can find the packing factor for BCC structure, which is 0.68. It is more compared to SC structure. Examples of BCC structures are Fe, Cr, W, etc. Figure 1.13 shows a BCC lattice.

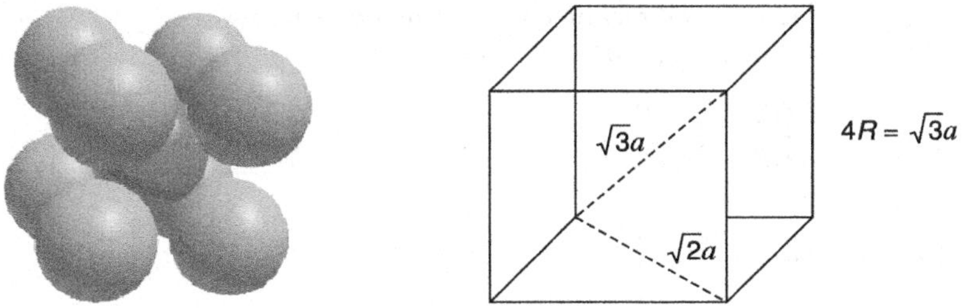

Fig. 1.13 *Unit cell of body-centred cubic lattice*

The dimensions of unit cell is related with atomic radius as

$$\sqrt{3}a = 4R$$

There are eight corner atom and one atom at the body centre.

The number of atoms per unit cell $= \left[\frac{1}{8}\times 8\right]+1=2$

Volume occupied by atoms $= 2\times\frac{4}{3}\pi R^3$

Volume of unit cell $= \left(\frac{4R}{\sqrt{3}}\right)^3$

Therefore, packing factor for BCC lattice $= \dfrac{\frac{8}{3}\pi R^3}{\left(\frac{4R}{\sqrt{3}}\right)^3} = 0.68$

1.13 INTERPLANAR SPACING (d_{hkl})

The separation between two successive lattice planes of cubic, tetragonal, and orthorhombic crystals, for which $\alpha = \beta = \gamma = 90°$, can be calculated as follows:

Let *OX*, *OY*, *OZ* be the three axes parallel to the crystal axes (Fig. 1.14). Let the plane *ABC* have intercepts $OA = \frac{a}{h}$, $OB = \frac{b}{k}$ and $OC = \frac{c}{l}$. Let us suppose that the origin *O* lies in a plane adjacent to *ABC*. Then *ON*, the length of the normal from the origin to the plane *ABC* is equal to interplanar distance *d*. Sometimes it is also represented as d_{hkl}.

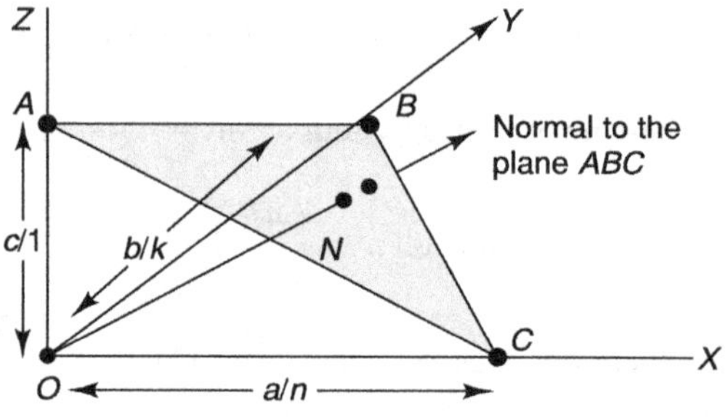

Fig. 1.14 *Interplanar spacing*

Let θ_a, θ_b, θ_c be the angels which *ON* makes with three crystallographic axes, respectively. Then the direction cosines of *ON* are:

$$\cos\theta_a = \frac{ON}{OA}, \qquad \cos\theta_b = \frac{ON}{OB}, \qquad \cos\theta_c = \frac{ON}{OC}$$

We know that the sum of direction cosines is unity; therefore

$$\cos^2\theta_a + \cos^2\theta_b + \cos^2\theta_c = 1$$

or

$$\left(\frac{ON}{OA}\right)^2 + \left(\frac{ON}{OB}\right)^2 + \left(\frac{ON}{OC}\right) = 1$$

$$\left[\frac{d}{a/h}\right]^2 + \left[\frac{d}{b/k}\right]^2 + \left[\frac{d}{c/l}\right]^2 = 1$$

$$\Rightarrow \qquad d = \frac{1}{\sqrt{\frac{h^2}{a^2} + \frac{k^2}{b^2} + \frac{l^2}{c^2}}}$$

For a cubic crystal $a = b = c$; hence

$$d = \frac{a}{\sqrt{b^2 + k^2 + l^2}}$$

For (100) plane of simple cubic crystal,

$$d_{100} = a$$

For (110) plane of simple cubic crystal,

$$d_{110} = \frac{a}{\sqrt{2}}$$

Similarly, for (111) plane in simple cubic crystal,

$$d_{111} = \frac{a}{\sqrt{3}}$$

Therefore, the ratio of interplanar spacing in the above planes for a simple cubic crystal—

$$d_{100} : d_{110} : d_{111} = 1 : \frac{1}{\sqrt{2}} : \frac{1}{\sqrt{3}}$$

1.14 SPACING BETWEEN LATTICE PLANES

(i) Spacing between Lattice Planes in BCC Crystal

Figure 1.15 shows BCC lattices. Comparing with SC lattice, it is found that in BCC lattice, there exist additional planes halfway between the (100) planes and also between the (111) planes.

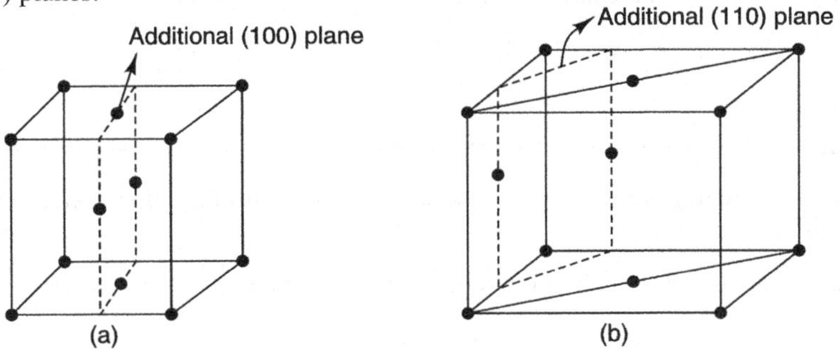

Fig. 1.15 *(a) Additional 100 plane in BCC lattice (b) Additional 110 plane in BCC lattice*

Comparing with simple cubic system,

$$d_{100} = \frac{1}{2}[d_{100}]_{SC} = \frac{a}{2}$$

$$d_{110} = \frac{1}{2}[d_{110}]_{SC} = \frac{a}{\sqrt{2}}$$

$$d_{111} = \frac{1}{2}[d_{111}]_{SC} = \frac{1}{2}\cdot\frac{a}{\sqrt{3}}$$

Therefore, the ratio between used plane in BCC crystal,

$$d_{100} : d_{110} : d_{111} = \frac{a}{2} : \frac{a}{\sqrt{2}} : \frac{a}{2\sqrt{3}} = \frac{1}{2} : \frac{1}{\sqrt{2}} : \frac{1}{2\sqrt{3}} = 1 : \sqrt{2} : \frac{1}{\sqrt{3}}$$

(ii) Spacing between Lattice Planes in FCC Crystal

Figure 1.16 shows the FCC lattice. On comparing with SC system, we find that in FCC system there exist additional planes halfway between the (100) planes and also between the (110) planes.

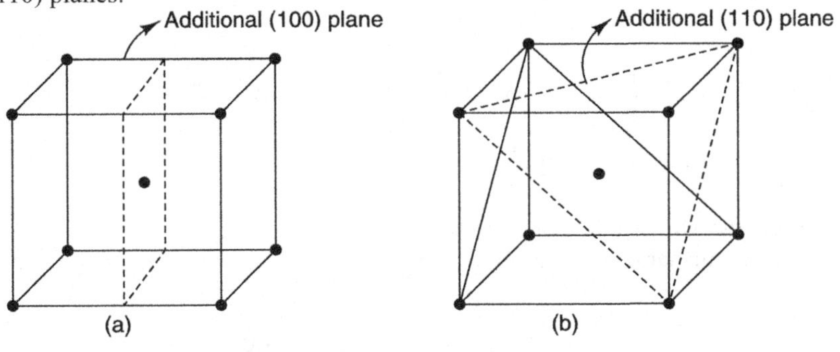

Fig. 1.16 *(a) Additional (100) plane in BCC crystal (b) Additional (110) plane in BCC crystal*

$$d_{100} = \frac{a}{2}$$

$$d_{110} = \frac{a}{2\sqrt{2}}$$

$$d_{111} = \frac{a}{\sqrt{3}}$$

Hence, ratio between the above planes

$$d_{100} : d_{110} : d_{111} = 1 : \frac{1}{\sqrt{2}} : \frac{2}{\sqrt{3}}$$

Example 1.7

Calculate the spacing between (1 0 0) and (1 1 1) planes of a cubic system of lattice parameter *a*.

Solution Spacing between the planes of a cubic system of lattice parameter a.

$$d_{hkl} = \frac{a}{\sqrt{h^2 + k^2 + l^2}}$$

For plane (1 0 0), $$d_{100} = \frac{a}{\sqrt{1^2 + 0^2 + 0^2}} = a$$

and for plane (1 1 1), $$d_{111} = \frac{a}{\sqrt{1^2 + 1^2 + 1^2}} = \frac{a}{\sqrt{3}}$$ **Ans.**

Example 1.8

Deduce the Miller indices of a set of parallel planes which make intercepts in the ratio of $a : 2b$ on the *X* and *Y* axis and are parallel to *Z* axis, $\vec{a}$, $\vec{b}$, $\vec{c}$ being primitive vectors of lattice. Also, calculate the interplanar distance *d* of the plane taking the lattice to be cubic with $a = b = c = 10$ Å.

Solution The parallel planes are parallel to Z axis. It means that their intercepts on the Z axis are infinite. Thus, the intercepts are a, $2b$ and ∞. And, lattice constant are $a = b = c = 10$ Å. Then,

$$a : 2b : c = \frac{a}{h} : \frac{b}{k} : \frac{c}{l}$$

or $$\frac{1}{h} : \frac{1}{k} : \frac{1}{l} = 1 : 2 : \infty$$

or $$h : k : l = 1 : \frac{1}{2} : \frac{1}{\infty} = 2 : 1 : 0$$

Therefore, the Miller indices are (210).
And, the interplanar distance is

$$d = \frac{a}{\sqrt{h^2 + k^2 + l^2}} = \frac{5 \times 10^{-10}}{\sqrt{2^2 + 1^2 + 0^2}}$$

$$= \frac{10 \times 10^{-10}}{\sqrt{5}} = 2\sqrt{5} \text{ Å}$$ **Ans.**

Example 1.9

Calculate the interplanar spacing for (2 3 1) plane of a FCC structure whose atomic radius is 0.175 nm.

Solution Given plane = (2 3 1) and atomic radius (r) = 0.125 nm.

Atomic radius (r) of FCC structure

$$= \frac{\sqrt{2}}{4}a$$

and interplanar spacing $$d = \frac{a}{\sqrt{h^2+k^2+l^2}}$$

$$\because \quad r = \frac{\sqrt{2}a}{4} \Rightarrow a = \frac{4r}{\sqrt{2}} = \frac{4\times 0.125\times 10^{-9}}{\sqrt{2}}$$

$$\therefore \quad d_{231} = \frac{a}{\sqrt{h^2+k^2+l^2}} = \frac{4\times 0.125\times 10^{-9}}{\sqrt{2}\times\sqrt{2^2+3^2+1}}$$

$$= \frac{4\times 0.125\times 10^{-9}}{\sqrt{2}\times\sqrt{14}} = 0.354\times 10^{-9}\ \text{m}$$

$$d_{231} = 0.354\ \text{nm}$$ **Ans.**

Example 1.10

Calculate the distance between two atoms of basis of the diamond structure if the lattice constant of the structure is 4 Å.

Solution Given lattice constant a = 4 Å.

The distance between two atoms is equivalent to the nearest neighbour distance.

For diamond structure, nearest neighbour distance $= \frac{\sqrt{3}}{4}a$

$\therefore$ distance between two atoms $= \frac{\sqrt{3}}{4}\times 4\ \text{Å} = \frac{1.732\times 4}{4} = 1.732\ \text{Å}$ **Ans.**

1.15 X-RAY DIFFRACTION AND BRAGG'S LAW

This is a very popular technique which is used to determine the internal structure of the crystal or to test whether a given material is amorphous or crystalline. The basic principle of this technique is diffraction of X-rays. Interplanar separation (lattice parameter) can be measured using X-ray diffraction analysis. Max VonLaue thought to utilise a 'single crystal' to replace the grating, assuming the arrays of atoms identical to rulings in a grating and ultimately succeeded in demonstrating, experimentally, the 'diffraction' phenomenon with these radiations. Since then this technique is being widely used to determine the internal structure of the materials. The details of his experiment are described in the forthcoming section.

(i) Laue Pattern/Spots

Laue passed a narrow beam of X-rays through ZnS crystal and allowed the emerging beam of X-rays to fall on a photographic plate. The developed photographic plate contained a central spot surrounded by a group of other symmetrically arranged spots having weak intensity. Whatever appeared on the photographic plate was a 'diffraction pattern' also known as Laue pattern/ spots. By analysing the pattern so obtained, the internal structure of the crystal could be well understood. Figure 1.17 shows the Laue pattern.

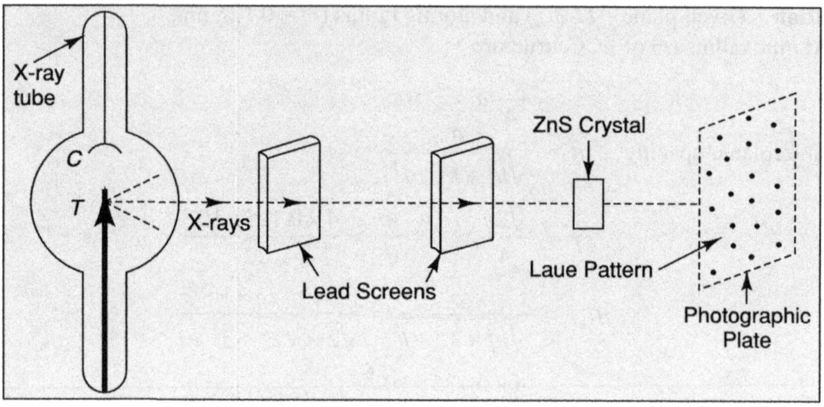

Fig. 1.17 *Laue pattern*

(ii) Bragg's Study of the Pattern

A detailed study of Laue pattern was conducted by W L Bragg. He noticed that when X-rays fall upon the atoms in a crystal lattice, each atom acts as a source of scattered radiation of the same wavelength; therefore, it is possible to construct the reflected wave-front in accordance with the Huygens' principle. As the atoms in a crystal are regularly arranged, a beam of X-rays is reflected from the plane of atoms following the ordinary laws of reflection. The portion of the X-rays falling on inter-atomic space is transmitted into the crystal and is reflected by the successive inner planes. These reflected beams can combine in a constructive or destructive manner depending upon their relative path differences. Each diffraction spot (in Laue pattern) is produced by interference occurring between these beams coming from different parallel planes.

(iii) Bragg's Law

As discussed earlier, a single crystal can play the role of a grating for X-rays. This idea was used by Bragg to demonstrate X-rays diffraction (XRD). Consider a set of parallel planes of a crystal having interplanar separating d. Let a beam of X-rays incident upon a crystal lattice at a glancing angle θ. These waves are reflected by different parallel planes.

In Figure 1.18, such two rays AB and CD are shown, which are meeting the planes at an angle θ. The reflected rays in the figure are named as BF and DH. θ is known as glancing angle.

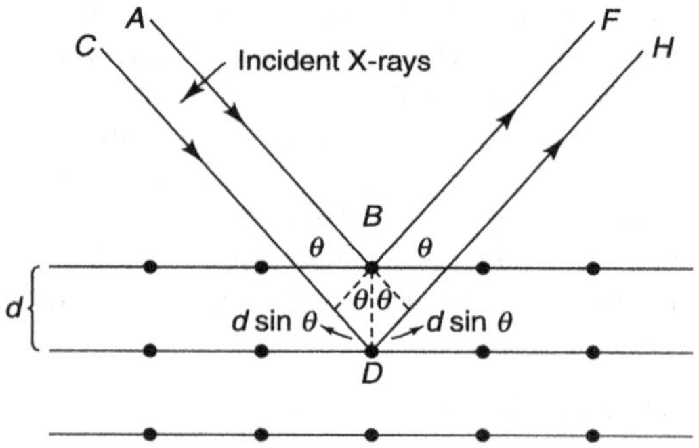

Fig. 1.18 *Bragg's reflections*

Path difference between reflected rays BF and $DH = 2d \sin \theta$. If this path difference is integral multiple of λ, then a constructive interference will occur, giving the maximum intensity of the reflected beam.

Therefore,

$$2d \sin \theta = n\lambda \qquad \text{where } n = 0, 1, 2, 3, \ldots$$

The above equation is known as Bragg's equation and the law is named as Bragg's law. Bragg's law is the backbone of crystallography and XRD is a commercial technique which is used to understand insight of a material under investigation.

1.16 BRAGG'S X-RAY SPECTROMETER

The Bragg's spectrometer works on the principle of Bragg's law. Its commercial name is X-ray diffractometer (XRD). Figure 1.19 shows the schematic diagram of Bragg's spectrometer.

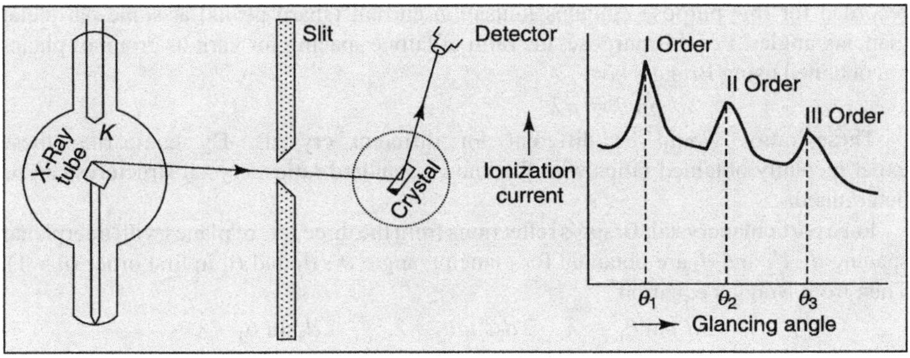

Fig. 1.19 *Experimental set-up of Bragg's spectrometer*

(i) Main Components

Following are the main components of Bragg's X-ray spectrometer:

- X-ray source
- Graduated circular table
- Ionisation chamber (detector)

(ii) X-ray Source

The purpose of X-ray tube is to produce and to collimate X-rays into a fine beam which is then allowed to incident upon the crystal. The glancing angle can be set by rotating the crystal and the reading can be seen on the graduated arc.

(iii) Graduated Circular Table

Glancing angle can be recorded from the scale present over the circular table.

(iv) Detector

The reflected beam of X-rays then enters the ionisation chamber which ionises the gas causing a current in the associated circuit. The rate of ionisation depends upon the intensity of X-rays. The resulting ionisation current therefore measures the reflected intensity.

For taking the observation, the glancing angle is varied in small steps and the ionisation current is recorded, which is then plotted with the glancing angle. Figure 1.19 shows a graph for NaCl, which indicates an abrupt increase of ionisation current for some fixed values of n. These peaks are obtained for those values of n that satisfy Bragg's equation:

$$2d \sin \theta = n\lambda$$

The different peaks obtained correspond to different orders ($n = 1, 2, 3, \ldots$)

$$2d \sin \theta_1 = \lambda, 2d \sin \theta_2 = 2\lambda, 2d \sin \theta_3 = 3\lambda,$$

$$\Rightarrow \quad \sin\theta_1 : \sin\theta_2 : \sin\theta_3 = 1 : 2 : 3$$

1.17 DETERMINING THE CRYSTAL STRUCTURE USING BRAGG'S LAW

Bragg's spectrometer is most widely used **to determine the crystal structure and to test whether the material is crystalline or amorphous.** This is done by measuring the interplanar separation, which is also known as lattice parameter. The pattern which is recorded for this purpose contains ionisation current (sharp peaks) at some particular glancing angles. For this purpose, the ratio of lattice spacing for various group of planes are obtained using Bragg's law.

$$2d \sin \theta = n\lambda$$

These ratios would be different for different crystals. By comparing these experimentally obtained ratios with the known standard ratios, crystal structures can be determined.

For a particular crystal, Bragg's reflections from the three sets of planes with interplanar spacing d_1, d_2 and d_3 are obtained for glancing angle θ_1, θ_2 and θ_3 in first order ($n = 1$). Then from Bragg's equation,

$$2d_1 \sin \theta_1 = \lambda, \quad 2d_2 \sin \theta_2 = \lambda, \qquad 2d_3 \sin \theta_3 = \lambda$$

$$d_1 : d_2 : d_3 = \frac{1}{\sin\theta_1} : \frac{1}{\sin\theta_2} : \frac{1}{\sin\theta_3}$$

For KCl crystal, Bragg obtained strong reflection of X-rays at glancing angles of 5.23°, 7.37° and 9.25° using three different reflecting planes: (100), (110) and (111).

Therefore, for KCl

$$d_{100} : d_{110} : d_{111} = \frac{1}{\sin 5.23^\circ} : \frac{1}{\sin 7.37^\circ} : \frac{1}{\sin 9.25^\circ}$$

or

$$\boxed{d_{100} : d_{110} : d_{111} = 1 : \frac{1}{\sqrt{2}} : \frac{1}{\sqrt{3}}}$$

Example 1.11

In an NaCl crystal, the spacing between the successive (1 0 0) plane is 1.41 Å, X-ray incident on the surface of the crystal is found to give rise to first order Bragg reflection at glancing angle 8.8°. Calculate the wavelength of X-ray.

Solution Given $d = 1.41$ Å, $\theta = 8.8^\circ$ and $n = 1$.

Formula used is $2d \sin \theta = n\lambda$

$$2 \times 1.41 \times 10^{-10} \times \sin 8.8 = \lambda$$

$$\Rightarrow \quad \lambda = 2.82 \times 10^{-10} \times 0.153$$

$$= 0.431 \text{ Å}$$

Ans.

Example 1.12

The first-order diffraction is found to occur at a glancing angle of 9°. Calculate the wavelength of X-ray and the glancing angle for second order diffraction, if the spacing between the adjacent plane is 2.5 Å.

Solution Given $n = 1$, $\theta = 9°$ and $d = 2.5$ Å = 2.5×10^{-10} m

Formula used, $2d \sin\theta = n\lambda$

Therefore, $\lambda = 2 \times (2.5 \times 10^{-10}) \times \sin 9° = 0.7853$ Å

For $n = 2$

$$\theta = \sin^{-1}\left(\frac{2\lambda}{2d}\right) = \sin^{-1}\left(\frac{\lambda}{d}\right)$$

$$= \sin^{-1}\left(\frac{0.7853 \times 10^{-10}}{2.5 \times 10^{-10}}\right)$$

$$= 18.1°$$ **Ans.**

Example 1.13

X-rays of wavelength 2.0 Å make a glancing angle 60° in the first order, when diffracted from NaCl crystal. Find the lattice constant of NaCl.

Solution Given $\lambda = 2.0$ A°, $\theta = 60°$ and $n = 1$.

Formula used is $2d \sin\theta = n\lambda$

$$\therefore \quad d = \left(\frac{n\lambda}{2\sin\theta}\right) = \frac{1 \times 2.0 \times 10^{-10}}{2 \sin 60°} = \frac{2.0 \times 10^{-10} \times 2}{2 \times \sqrt{3}}$$

$$= 1.51 \text{ Å}$$ **Ans.**

Example 1.14

X-ray of wavelength 1.4 Å is found to be Bragg reflected from the (1 1 1) plane of an FCC structure. If the lattice parameter of the crystal is 2 Å, find the angle at which the X-ray is incident on the (1 1 1) plane of the crystal.

Solution Given $\lambda = 1.4$ Å, lattice parameter of FCC structure (a) = 5 Å and the plane of the FCC structure = (111)

Interplanar spacing $$d_{hkl} = \frac{a}{\sqrt{h^2 + k^2 + l^2}}$$

and $$2d \sin\theta = n\lambda$$

$$\therefore \quad d_{111} = \frac{a}{\sqrt{3}} = \frac{2 \times 10^{-10}}{\sqrt{3}} = 1.15 \times 10^{-10} \text{ m}$$

and so, $$\theta = \sin^{-1}\left(\frac{n\lambda}{2d_{111}}\right) = \sin^{-1}\left[\frac{1 \times (1.4 \times 10^{-10})}{2 \times 1.15 \times 10^{-10}}\right]$$

$$= \sin^{-1} 0.60$$ **Ans.**

Example 1.15

From the following data, calculate the wavelength of neutron beam and its speed. Spacing between successive planes is 3.84 Å, glancing angle 45° and the order of Bragg reflection = 1.

Solution Given $d = 3.84$ Å $= 3.84 \times 10^{-10}$ m, $\theta = 45°$ and $n = 1$

Formula used are $$2d \sin \theta = n\lambda \text{ and } \lambda = \frac{h}{mv}$$

Thus, $$2 \times 3.84\ 10^{-10} \times \sin 45° = \lambda$$

or $$\lambda = 2 \times 3.84 \times 10^{-10} \times \frac{1}{\sqrt{2}} = 5.41 \text{ Å}$$

$\because$ $$\lambda = \frac{h}{mv} \quad \text{[according to de Broglie relation]}$$

$\therefore$ $$\lambda = \frac{h}{mv} = \frac{6.62 \times 10^{-34}}{1.67 \times 10^{-27} \times 5.41 \times 10^{-10}}$$

$$\lambda = 0.732 \times 10^{-3} \text{ m}$$ **Ans.**

Example 1.16

Electrons are accelerated from the rest by 200 V are reflected from an FCC crystal. The reflection maximum is observed at 22°. Determine the lattice parameter, if the Bragg reflection occurs from the (1 1 1) plane.

Solution Given $V = 200$ volts, $\theta = 22°$ and $n = 1$.

Formula used are $$\lambda = \frac{h}{(2meV)^{1/2}} \qquad \left[\because \frac{1}{2}mv^2 = eV\right]$$

and $$2d \sin \theta = n\lambda$$

$\therefore$ $$\lambda = \frac{h}{\sqrt{2meV}}$$

$$= \frac{6.6 \times 10^{-34}}{\sqrt{2 \times 9.1 \times 10^{-31} \times 1.6 \times 10^{-19} \times 200}} = 0.865 \times 10^{-10} \text{ m}$$

so, $$d_{111} = \frac{n\lambda}{2 \sin \theta} = \frac{1 \times 0.865}{2 \times \sin 22°} = 1.15 \text{ Å}$$

and $$d_{111} = \frac{a}{\sqrt{3}}$$

$\therefore$ $$a = d_{111} \times \sqrt{3} = \sqrt{3} \times 1.15 \text{ Å} = 1.99 \text{ Å}$$ **Ans.**

1.18 BONDING IN SOLIDS

(i) Introduction

The bonding between atoms takes place to acquire minimum energy configuration (stable configuration). Except the noble gas atoms, no atom exists in isolation. Bonding occurs due to attractive electrostatic interaction between negative charges of the electrons and positive charges of the nuclei. When atoms are sufficiently apart, the force between them is mainly attractive. As the atoms come closer to form a bond, a repulsive force acts at very close distances. This force increases rapidly as the atoms come in closer proximity.

(ii) Bond Length

At a particular distance between the atoms, the attractive and repulsive forces balance each other and net force become zero. **This equilibrium distance between the two bonding atoms is known as 'bond length'**.

The formation of bond between the atoms to acquire stable configuration can take place in many ways. For example, by transfer of electrons from one atom to the other, by equal sharing of electrons or by some other means. Therefore, the strength of bonds will be different in different situations. The chemical bonds, therefore, are classified in two categories (Fig. 1.20).

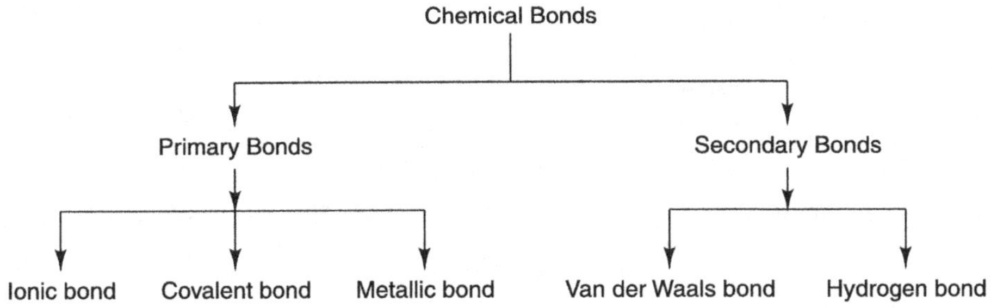

Fig. 1.20 *Types of chemical bonds*

1.19 PRIMARY BONDS

Primary bonds are strong bonds. Let us discuss them in detail.

(i) Ionic Bond

The Ionic bond is developed between those two atoms where one atom has low ionisation energy and the other has high electron affinity. The atoms exchange electrons to become oppositely charged ions and consequently develop the strong attractive electrostatic force. **Therefore, ionic bond is an electrostatic force of attraction between two oppositely charged ions formed by the process of electron transfer between the related atoms.**

In other words, the ionic bond is formed between two atoms where one atom has one excess electron from the closed-shell configuration and the other is one electron short. For example,

- Li— $1s^2\ 2s^1$ $\longrightarrow$ One excess electron
- Na— $1s^2\ 2s^2\ 2p^6\ 3s^1$ $\longrightarrow$ One excess electron
- F— $1s^2\ 2s^2\ 2p^5$ $\longrightarrow$ One less electron
- Cl— $1s^2\ 2s^2\ 2p^6\ 3s^2\ 3p^5$ $\longrightarrow$ One less electron

The electrostatic force between the atoms to acquire stable configuration is also named as 'Coulomb interaction'. The bonding energy in case of NaCl can be calculated very easily:

For NaCl, distance between closest oppositely charged ions is 2.81×10^{-10} m.

The bonding energy $U_{\text{cohesive}} = \dfrac{-e^2}{4\pi\varepsilon_0 r} = \dfrac{-(1.6\times10^{-19})^2}{4\pi\times8.85\times10^{-12}\times2.81\times10^{-10}}$

$$= -5.14 \text{ eV}$$

- **Cohesive Energy of NaCl** Cohesive energy of an ionic crystal is defined as the energy that would be liberated in the formation of the crystal from individual neutral atoms, or the energy needed to break the crystal into individual atoms.

The Coulomb energy of interaction between Na^+ and Cl^- ions separated by a distance r is

$$U_{\text{coulomb}} = \frac{\alpha e^2}{4\pi\varepsilon_0 r}$$

Where α is known as Madelung constant and its value depends upon crystal structure. For NaCl, $\alpha = 1.7475$

The repulsive energy term $= \dfrac{B}{r^m}$

B is constant and $m = 9$

Net potential energy $U = \dfrac{\alpha e^2}{4\pi\varepsilon_0 r} + \dfrac{B}{r^m}$

At the equilibrium separation r_0, U must be minimum. This suggests that

$$\text{at } r = r_0, \quad \frac{dU}{dr} = 0$$

Hence, $\left[\dfrac{dU}{dr}\right]_{r=r_0} = \dfrac{\alpha e^2}{4\pi\varepsilon_0 r^2} - \dfrac{mB}{r_0^{m+1}} = 0$

$\Rightarrow$ $B = \dfrac{\alpha e^2}{4\pi\varepsilon_0 m} r_0^{m-1}$

The potential energy at $r = r_0$ is thus

$$U = -\frac{\alpha e^2}{4\pi\varepsilon_0 r_0} + \frac{\alpha e^2}{4\pi\varepsilon_0 m r_0}$$

or $$\boxed{U = -\frac{\alpha e^2}{4\pi\varepsilon_0 r_0}\left[1 - \frac{1}{m}\right]} \qquad (1.5)$$

This is the total potential energy per ion pair in the crystal. In the NaCl crystal, the equilibrium distance between Na^+ and Cl^- is $r_0 = 2.81$ A° and $\alpha = 1.7475$. Taking $m = 9$, the potential energy per ion-pair becomes (on substituting these parameters in Eq. 1.5).

$$U = -7.96 \text{ eV}$$

Hence, the contribution per ion of the cohesive energy of the crystal is

$$U_{\text{cohesive}} = \frac{U}{2} = -\frac{7.96 \text{ eV}}{2} = -3.98 \text{ eV}$$

This is known as lattice energy per ion in NaCl crystal. But some energy is needed to transfer an electron from Na-atom to Cl-atom to form $Na^+ - Cl^-$ pair.

$$= +5.14 - 3.61 \text{ eV} = +1.53 \text{ eV}$$

Hence, each atom contributes $\dfrac{+1.53}{2}$ eV $= 0.77$ eV to the cohesive energy.

∴ the total cohesive energy per atom in NaCl crystal is

$$U_{\text{cohesive}} = -3.98 + 0.77$$
$$= -3.21 \text{ eV/atom.}$$

❖ Characteristics of Ionic Solids

- These compounds are crystalline in nature.
- They usually crystallise in relatively close-packed structures.
- Ionic crystal are rigid, brittle and exhibit high melting and boiling points.
- They are highly soluble in solvents like water and liquid ammonia, and insoluble in non-polar solvents.
- They are generally non-conductors of electricity, but their melts and solutions conduct electricity.
- They are transparent for all frequencies up to fundamental absorption frequencies and are opaque to higher ones.
- Some ionic crystals strongly absorb electromagnetic radiations in far infrared radiation regions.

(ii) Covalent Bond

When two similar atoms share equal electrons, the bond between them in known as a covalent bond. As these electrons circulate between the atoms, they spend more time between the atoms than elsewhere, which produces an electric force.

Covalent bonds are strong bonds and directional in nature in covalent crystals, cohesive energies lie between 6 to 12 electron volt per atom.

They are of three types:

- Sigma bond ⟶ Formed as a result of end-to-end overlapping of orbits.
- Pi bond ⟶ Formed due to lateral overlapping of orbits.
- Hybrid bonds ⟶ Formed when interaction between S and P orbital occurs (e.g., sp^3, sp^2, or sp hybridisation).

❖ Characteristics of Covalent Solids

- They are extremely hard and difficult to deform.
- They are poor conductors of heat and electricity as they carry no free electrons.
- Many covalent solids are transparent.
- They exhibit rigid electronic structure.

(iii) The Metallic Bond

In a metal, free electrons are present in a very large number and they behave like electron gas. Many phenomena associated with metals confirm the above fact. The interaction between electron gas and the positive metal ions lead to a strong cohesive force and give rise to metallic bond. The presence of such free electrons in a metal account nicely for the high electrical and thermal conductivities, heat capacity, surface lustre and other unique properties. Also these electrons never move inside a metal or in any solid with total freedom. All of them are influenced, to some extent, by the other particles (+ ions) present inside and when their presence is taken into consideration—one finds a very good resemblance with experimental results. If we look deeply, we will find that a metallic bond is a kind of unsaturated covalent bond.

❖ Characteristics of Metallic Solids

- Metallic bonds are generally non-directional, i.e. the bond strength is equal in all directions.

- Such bonds are generally found in alkali metals.
- Bond energy and melting point of a solid having this bond (alkali metal) is always low.
- As the number of bonding electrons increases, the bond energies and melting point also increase.
- The electrons in a solid (metal) forget to which of the two ions it really belongs.
- Material (metals) involving metallic bond are very good conductors of heat and electricity.
- The strength of these materials varies in a wide range. Stronger the bond greater will be the strength of material.
- Metals exhibit unique lustre, commonly known as metallic lustre.

1.20 SECONDARY BONDS

As compared to primary bonds, secondary bonds are relatively weak. The driving force for such bonding arises due to the attraction of electric dipoles contained in atoms or molecules. These dipoles, can be either fluctuating or permanent. In both of these, secondary bonds are there between the atom and molecule.

(i) Van der Waals Bonds

Atoms, molecules or ions of all substances exert attractive force on each other when they are brought close. The bond so developed is known as Van der Waals bond. The forces that act are electrostatic in nature and are much weaker than those found in ionic and covalent bonds. Van der waals force between the molecules or atom (including rare gases) is proportional to r^{-7}, where r is the separation between the atoms or molecules. These forces were proposed by Van der Waals, a Dutch scientist, to explain the departure of real gases from ideal behaviour. If these forces would not be there then the gases (rare gases) would not condense into liquids or solids.

❖ Characteristics of Van der Waals Bonds

- The crystals exhibit low melting and boiling points.
- The mechanical strength of crystals (having this bond) is low.
- Cohesive energy is also low.
- Many features like friction, surface tension, viscosity, adhesion, cohesion arise due to Van der Waals forces.
- Joule-Thomson effect is also an outcome of Van der Waals forces.
- Crystals formed due to Van der Waals bond are soft. Dry CO_2 has Van der Waals bonding.

(ii) Hydrogen Bond

Certain compounds, notably water, have much higher boiling and melting points than would be expected. For example, the non-metallic hydrides H_2Te, H_2Se and H_2S have the respective boiling points of $-2°$, $-42°$ and $-60°$; while H_2O, with a still smaller molecular weight, has a boiling point of $100°$. Similar anomalies are found in NH_3 and HF. The intermolecular forces in H_2O, NH_3 and HF are due to hydrogen bond, which is stronger than Van der Waals bond, but weaker than ionic or covalent bonds.

❖ Characteristics of Hydrogen Bond

- Due to intermolecular hydrogen bonding, two or more molecules of substance may get associated resulting in the formation of clusters of molecules (e.g., HF, water and ice).

- The compounds containing H-bond exhibit abnormal higher melting and boiling points.
- Intermolecular hydrogen bonding may also influence the physical state of a substance. For example, O and S belong to same group of periodic table, but their hydrides H_2O and H_2S exhibit entirely different state; H_2O is water and H_2S is gas.
- Compounds whose molecules are able to form hydrogen bonding with water are soluble in it. For example, alcohols—in spite of being covalent—are soluble in water. Many other covalent compounds are also soluble in water.

1.21 MIXED BONDING

Atoms or molecules can involve in more than one type of bonding. It could be a combination of ionic-covalent, metallic-covalent, metallic-ionic and ionic-covalent-metallic.

For example, ionic-covalent mixed bonding results due to the difference in the electro-negativities of the elements forming the mixed bonding. Examples of compounds having mixed iono-covalent bonding are GaAs, zinc sulphide, zinc telluride, and many other chalcogenide glasses. We have established this fact using X-ray K-absorption shift studies in chalcogenide glasses.

1.22 DEFECTS IN CRYSTALS

Crystals exhibit perfect regularity (order) at 0 K. Such arrangement corresponds to lowest-energy state of the system. As the temperature goes up, a deviation (departure) from the perfect-ordered state begins, thereby producing some disorder. This disorder increases further, if the crystal contains impurities. A departure from perfect-ordered state in a crystal is known as imperfection (defect) in the crystal. A crystal is rarely found perfect in nature; sometimes the defects play a positive role and improve the physical, electrical or mechanical properties and vice versa. Some times defects or imperfections are deliberately created to improve the properties of crystal or solid. Presence of defects gives rise to new properties in the crystal. Imperfections are classified mainly in two categories:

- Electronic imperfections
- Atomic imperfections

Drawbacks of Imperfections

Imperfections lead to several disadvantages in solids. Certain undesirable effects which are found due to presence of imperfections in the solids are

- The mechanical strength of the material decreases
- The material does not attain its theoretical strength, as well

1.23 ELECTRONIC IMPERFECTIONS

In the lowest energy state (i.e. at 0 K), the energy levels in a covalent or ionic crystal are completely filled. At a higher temperature, some electrons may acquire high-energy levels. Some electrons may be released from covalent bonds of the crystal (e.g. pure Si crystal); these are the electrons which freely move in the crystal causing a current to conduct. The electrical conductivity shown by silicon is attributed to this phenomenon.

The release of such electrons is also simultaneously accompanied by a vacancy called 'hole'. Holes are also responsible for conduction of current. Such free electrons and holes in a solid are known as electronic imperfections. In pure (intrinsic) materials the number of holes and electrons are always equal. A small amount of impurity added to semiconductor materials leads to a drastic change in conductivity, making these materials extremely promising and suitable for fabrication of solid state devices.

1.24 ATOMIC IMPERFECTIONS

The defects which arise due to an irregularity in the arrangement of atoms or ions in a crystal are known as atomic imperfections. The atomic imperfections are further classified into four types:

- Point imperfections
- Line imperfections
- Planer imperfections
- Volume imperfections

The classification of atomic imperfections in summarised in Fig. 1.21.

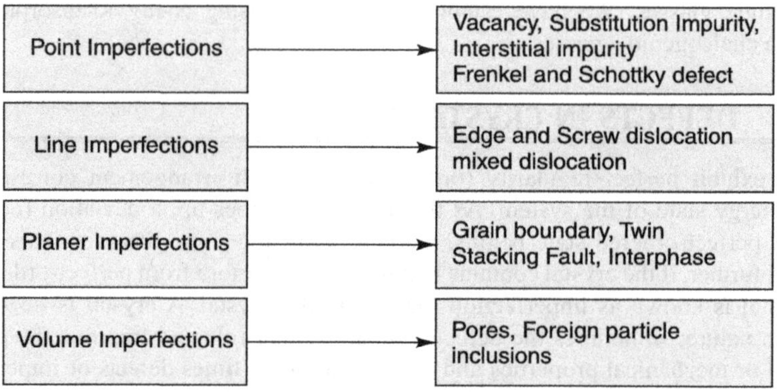

Fig. 1.21 *Classification of atomic imperfections*

1.25 RECENT ADVANCEMENT IN MATERIALS

Materials in the form of wire, fibre, etc., are termed as **whiskers**. A piece of mild steel or copper, a piece of semiconductor or a block of reinforced cement concrete (RCC) are available in bulk form. If bulk is changed to the form of a wire, fibre or whisker, strength of the material improves considerably. However, imperfections and dislocations are major hindrances; due to them theoretical strength is not achieved in the materials. To achieve the goal of $\sigma_{ult} \approx E$, material in the form of whisker is a must.

Whiskers are available in metallic and non-metallic, and organic and inorganic materials. History of whiskers is as old as 1970, when synthetic whiskers were produced. Whiskers, in the natural form, were known to the mankind since historical days. Spider's web, bamboo flakes, human-bone flakes, some kinds of grass, etc. Are all known natural whiskers.

1.26 DIFFERENCE BETWEEN BULK, FIBRE AND WHISKER FORMS OF MATERIALS

Whiskers are obtained by elongating a single crystal into fibrous form. Due to this action, defects such as grain boundaries are fully eliminated. Figure 1.22 shows a bulk polycrystalline material. It may be elongated to the form of a fibre. A single crystal A is taken out of it and is elongated to produce a whisker. Thus, the differences among bulk, fibre and whisker form of a material lies in its manufacturing, which will further affect its strength.

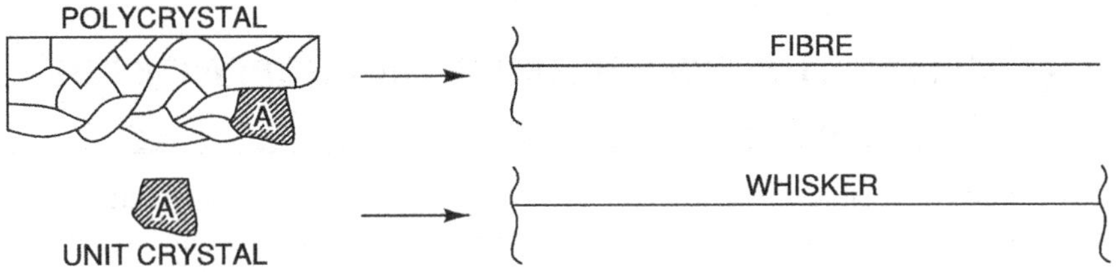

Fig. 1.22 *(a) Bulk polycrystal elongated to fibre form (b) Unit crystal elongated to whisker form*

1.27 EFFECTS OF SIZE OF WHISKERS ON MECHANICAL PROPERTIES OF MATERIALS

Diameter of whiskers varies between 2 to 20 microns. Greatest strengths are usually observed in whiskers of smallest diameters as shown in Fig. 1.23. Strongest whiskers are almost free from dislocations.

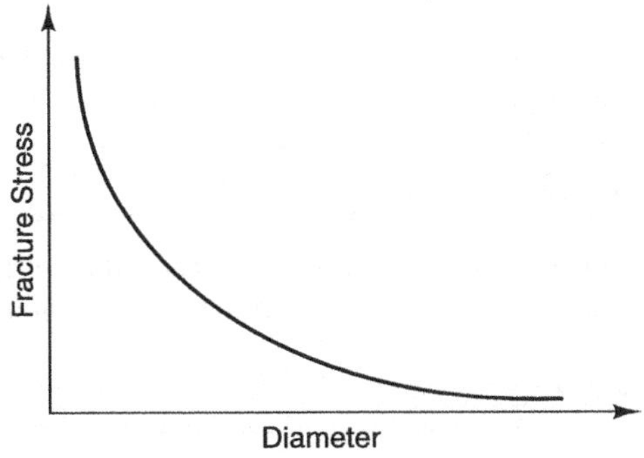

Fig. 1.23 *Effect of whisker diameter on fracture strength*

Strengths nearing 2.7 GPa and 13.2 GPa have been achieved in copper and iron whiskers, respectively. A comparison between whisker and bulk form for strength σ, Young's modulus E and elastic deformation are shown in Table 1.3.

Table 1.3 *Strength and Young's modulus of some whiskers and their bulk forms*

Material	*Strength (GPa)*		*Young's modulus (GPa)*		*Elastic (deformation)*	
	Bulk	*Whisker*	*Bulk*	*Whisker*	*Bulk*	*Whisker*
Tungsten	1.5	14.0	400	2400	0.01%	0.3%
Mild steel	0.48	12.8	200	1000	0.2 %	5%
Alumina	5.0	14.0	390	2250	0	3.5%
Graphite	2.6	19.7	27	703	—	—
Silicon carbide	10.0	20.7	450	840	0	—

1.28 EFFECTS OF TEMPERATURE ON PROPERTIES OF WHISKERS

The temperature dependence of strength and modulus of the whisker and bulk forms are shown in Fig. 1.24. It shows a much higher value for whisker as compared to the bulk form materials.

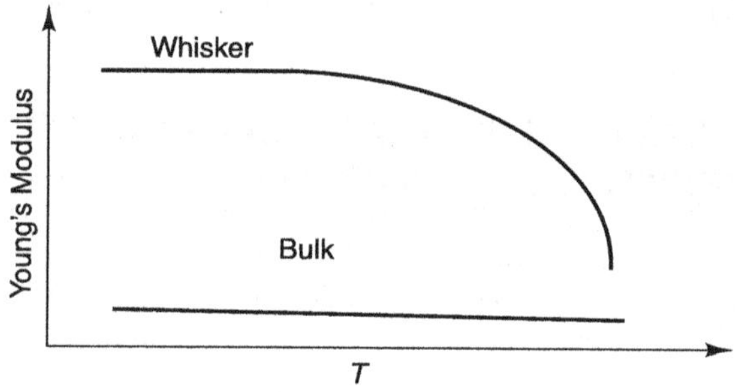

Fig. 1.24 *Temperature dependence of whisker and bulk form for their strength and Young's modulus*

1.29 VOLUME DEFECTS

Apart from the above-mentioned imperfections, certain three-dimensional defects are also there, for example, cracks, inclusion, perceptible particles and large voids pores.

1.30 COLOUR CENTRES

A colour centre is a lattice defect that absorbs a visible light. It is possible to colour an alkali halide crystal just by adding a suitable impurity. These may be transition element ions or the ions whose salts are normally coloured. When a crystal is coloured, it is known to contain a colour centre.

1.31 METHODS TO PRODUCE COLOUR CENTRES

The following are certain ways to produce colour centres:

- By doping a chemical impurity
- By the bombardment of neutron, electron X-ray or γ-ray

- By introducing excess of metal ions, e.g., Na or K in NaCl or KCl, etc.
- By electrolysis

1.32 KINDS OF COLOUR CENTRES

There are two kinds of colour centres:

- F-centres
- V-centres

(i) F-centres

F-centre is the simplest kind of colour centre that can be produced by heating the crystal in excess of alkali vapour. This is an electron bound at the negative ion vacancy. The optical absorption of F-centre arises from an electric dipole transition to a bound excited state of the centre.

(ii) V-centres

It is also possible to colour the alkali halide crystals by treating them in excess halogen ions. In such crystals, the coloured centres are the positive ion vacancies trapping holes as mentioned above in F-centres. In order to distinguish these colour centres, they are called V-centres. They are not as stable at room temperature as F-centres.

Summary

- Material exists mainly in three forms: crystalline, polycrystalline and amorphous.
- A crystal is described by a lattice in which each lattice point is occupied by one atom or a group of atoms. The group of atoms occupying a lattice point is called the basis of the structure.
- A unit cell is a fundamental block which, when repeated in three directions, generates the entire lattice.
- Direction indices and Miller indices are used to specify crystal directions and crystal planes, respectively. Crystal planes are given by three indices enclosed in brackets as (hkl).
- Packing factor is defined as the ratio of the volume occupied by the atoms in a unit cell to the volume of the unit cell. FCC and HCP structures have the highest packing factor of 0.74.
- X-rays are used to determine the structure and lattice parameter of crystals.
- Bragg's law is the key relation used in the determination of crystal structure by X-ray diffraction.
- Laue method is used to find crystal structure of materials using single crystals.
- Bragg's diffractometers are used in the determination of crystal structure.
- Chemical bonds in solids may be classified in five categories; Van der Waals, covalent, ionic, metallic and hydrogen bond.
- Van der Waals bond occurs between two closed configuration atoms or between molecules in organic solids;it is relatively a weak bond. The bond is due to dipole-dipole interaction. The dipoles may be permanent dipoles, as in organic molecules or induced dipoles, as in inert gas atoms.

- Covalent bond is a very strong bond formed by sharing of electrons between the atoms.
- The electrons in the molecule occupy the molecular orbital obeying Pauli's exclusion principle just as in atoms.
- Molecules in which the outermost valence electrons occupy bonding-molecular orbital are stable.
- Ionic bond is formed between two atoms by transfer of electrons from one atom to the other making one positively charged and the other negatively charged. The Coulomb force of attraction is responsible for the bond.
- When two atoms come close to form a bond, the charge clouds of the two atoms overlap and this results in a repulsive interaction between the two atoms.
- The bonding energy can be easily evaluated for an ionic crystal by taking the sum of interaction of all the ions in the crystal with all their neighbouring ions. Both the repulsive and the attractive energies are considered. For the repulsive energy, only the nearest neighbours are counted; whereas, for the attractive energy all the ions in the crystal are counted.
- Most crystals have bonds that are partially covalent and partially ionic. This is because in the covalent bond, in which the electrons are shared by two atoms, one of the atoms may have stronger affinity for electrons. As a result, the electrons tend to be closer to one of the atoms, making the bond partially ionic.
- Metallic bond found in metals is due to the interaction between free conduction electrons and the ionic cores in the lattice. Electrons are not localised in the metallic bond. Metals are characterised by high electrical and thermal conductivity because of the free electrons.
- Hydrogen bond is the bond between two atoms formed by a hydrogen atom between the two. The bond is very strong and largely ionic in character. The electron of the hydrogen atom will lie closer to the atom which has stronger affinity for the electron. A typical example of hydrogen bonding is between two fluorine atoms in HF and the bonding between water molecules in ice.
- The type of chemical bonding and crystal structure are two factors that determine the properties of materials. Impurities and defects in crystals also critically affect certain properties.
- The presence of defect affects the properties of the crystal. Some defects are desirable and are sometimes introduced deliberately to control the properties of crystals.
- The four types of point defects are: lattice vacancy, Schottky defect, Frenkel defect and Impurity atoms. Schottky defects are lattice vacancies. Frenkel defect are atoms in interstitial positions in the lattice. These types of defects are normally found in ionic and metallic crystals.

Exercise

1. What is crystalline state of matter? Give seven systems of crystals and their Bravais lattices.
2. What is space lattice? Describe the seven systems of crystals. Explain with examples the types of lattices in cubic system.

3. Explain the concept of unit cell. Describe with suitable diagrams the characteristic of unit cell in all three types of crystal lattices.
4. Explain the following terms: Primitive cell, Basis, Miller Indices, Angles α, β and γ, and Bravais space lattices.
5. What are different systems of lattice? Deduce an expression for the lattice constant of a face-centred cubic lattice.
6. Calculate the co-ordination number of lattice points per unit cell in FCC type of cubic lattice.
7. Enumerate the seven crystal systems pointing out their characteristic features. Draw diagrams to show simple, body-centred and face-centred cubic lattices. Give suitable examples of the substances which crystalline with these structures.
8. In relation to crystal lattices, define the following terms:
   (i) Crystal structure
   (ii) Translation vector
9. In NaCl structure the lattice is FCC, but the co-ordination number is that for simple cubic lattice. Explain.
10. Show that in a simple cubic lattice the spacing between the successive lattice planes (100), (110) as (111) are in the ratio of 1 : 0.71 : 0.58.
11. Show that for a cubic lattice, the lattice constant 'a' is given by

$$a = \left(\frac{nM}{N\rho}\right)^{1/3}$$

12. For a body-centred cubic lattice, find the number of nearest neighbours, nearest neighbour distance and the spacing of (110) planes.
13. What do you mean by unit cell, atomic packing factor, co-ordination number lattice constant and Miller Indices? Explain their role in crystal structure analysis.
14. What are Miller indices and how are they obtained for a given plane in a crystal? Draw the (112) and (111) planes in simple cubic crystal.
15. Derive expression for interplanar spacing between two successive lattice planes in cubic lattices.
16. Derive formula for determining the separation between successive lattice planes for a cubic crystal.
17. Calculate the number of lattice points per unit cell, and the co-ordination number in a simple, body-centred and face-centred cubic lattice.
18. Writes short notes on:
    (i) Lattice constant and Miller Indices
    (ii) Lattice constant and co-ordination number
    (iii) Bravais lattices
    (iv) Miller Indices
19. Why doesn't electrical neutrality change in cases of Frenkel's and Schottky's defects?
20. From where do the point imperfections originate within the crystalline solid? What are their effects on properties of materials?
21. What is meant by imperfections in a crystal? Explain different types of imperfections in solids.

22. (a) Write short notes on:
    (i) Miller Indices
    (ii) X-ray crystallography
    (b) Calculate C/a ratio for an ideally close-packed HCP crystal.
23. Why is the name 'screw dislocation'? Show Burgers circuit and Burgers vector on a crystal which is having a screw dislocation.
24. What are various sources of dislocations? How do they affect the properties of materials? Suggest ways and means to remove or minimise them.
25. Differentiate between the following:
    (a) Climb-up and climb-down,
    (b) Cross-slip and jog,
    (c) Point and surface imperfections, and
    (d) High and low angle boundary defects.
26. Write notes on the following:
    (a) Twist boundary,
    (b) Interphase boundary,
    (c) Stacking fault, and
    (d) Volume imperfections.
27. What are whiskers? Why do they possess much higher strength as compared to their conventional form? Name some natural whiskers?
28. Compare fibre, whisker and the bulk form of a material. How do the cross-section and the temperature affect the properties of whiskers?
29. What is meant by crystal growth? How is crystallisation influenced by rate of cooling? Differentiate between cold and hot working.
30. Discuss 'recovery', 'recrystallisation', and 'grain growth'. What is the advantage of crystal growth in preferred direction?
31. What is atomic packing factor? Calculate its values for simple and body-centred cubes.
32. Draw the (112) and (111) planes in a simple cubic cell.

Numerical Problems

1. X-ray of wavelength 2.0 Å are incident on a crystal having 1.5 Å as inter-atomic spacing. Calculate the maximum number of orders. **[Ans. 2]**
2. The Bragg's angle corresponding to first-order reflection from (1 1 1) plane in a crystal is 30°, when X-rays of wavelength 2.5 Å are used. Calculate the inter-atomic spacing. **[Ans. 4.32 Å]**
3. Find the Miller indices of a plane which makes intercepts at $3a, \frac{b}{2}, 3c$, in a simple cubic crystal. **[Ans. 161]**
4. Calculate the value of d_{200} and d_{111} in lead. It has a FCC structure and $a = 5.0$ Å. **[Ans. 2.94 Å]**
5. In a tetragonal lattice $a = b = 3.5$ Å and $c = 1.8$ Å. Calculate the lattice spacing between (1 1 1) planes. **[Ans. 1.49 Å]**
6. The electrons are accelerated to 728 volts and are reflected from a crystal. The first reflection maximum occurs at glancing angle of 10°. Determine the interplanar spacing of the crystal. **[Ans. 1.32 Å]**
7. Calculate the cohesive energy of copper, assuming the bond energy between two copper atoms to be 55.5 kJ/mol. **[Ans. 333 KJ/mole]**

8. Find the de Broglie wavelength of electrons accelerated to a potential difference of 100 volts. **[Ans. 1.22 Å]**
9. X-rays of wavelength 1.0 Å are diffracted by a Bragg's crystal spectrograph at a glancing of 4.8°C. Find the interplanar separation of atomic planes in the crystal. **[Ans. 5.98Å]**
10. Calculate the lattice constant of potassium bromide. Density of potassium is 2.7×10^3 kg/m^3 and belongs to a FCC lattice. (Molecular weight of potassium bromide is 119.0 and Avogadro number is 6.02×10^{26} per kg.) **[Ans. 6.64×10^{-10} m = 6.64 Å]**
11. For a BCC lattice of identical atoms having lattice constant 4 Å C, calculate the maximum atomic packing fraction the radius of atoms treated as hard spheres with the nearest neighbours touching. **[67.5%]**
12. Determine the number of atoms per unit cell of lead which has a FCC structure. Atomic weight of Pb = 207.2, density of Pb = 11.36×10^3 kg m^{-3}, a = 3.2 Å and Avogadro's number = 6.023×10^{26}/kg mole. **[Ans. 1.082 ≈ 1]**
13. Calculate the lattice constant of the rock salt crystal from the following data:
    Molecular weight of NaCl = 58.45
    Density of NaCl = 2.17×10^3 kg/m^3
    Avogadro's number = 6×10^{26} **[Ans. 5.64 A]**
14. Determine the Miller indices of plane parallel to the z axis and cut intercepts of 2 and 2/3 along x and y axes, respectively. **[Ans. (1 3 0)]**
15. An alkali halide of molecular weight 74.6 having NaCl structure has an inter-atomic distance of 0.30 nm. Calculate its density if it contains: (i) 0.1% Schottky defect, and (ii) 0.1% Frenkel defect. **[Ans. (i) 2.28×10^3, (ii) No change]**

Self-quiz Questions

1. What is meant by unit cell and lattice constant? Find the number of atoms per unit cell in FCC lattice.
2. "Periodicity of structure is an essential feature of a crystal." Explain.
3. What are the differences between amorphous and crystalline solids?
4. What do you understand by the unit cell of a space lattice? Deduce the number of lattice points per unit cell in various types of cubic unit cells.
5. What do you mean by the term 'co-ordination number' of a crystal lattice? Calculate the co-ordination number and the distance between the nearest neighbouring atoms for simple cubic, BCC and FCC lattices.
6. Describe NaCl crystal structure and deduce the co-ordination number.
7. What are lattice planes? How are they represented in terms of Miller indices?
8. What do you understand by Miller indices of a crystal plane?
9. What is the difference between a primitive cell and a unit cell? What is the utility of these concepts?
10. What is the difference between a lattice and a crystal?
11. What do you mean by mixed bonding?
12. What is the meaning of lattice parameter?

Objective-type Questions

1. Which of the following materials normally exist(s) in amorphous forms?
   (a) copper
   (b) window glass
   (c) quartz
   (d) sugar
2. A crystal system, the one with the least symmetry is
   (a) tetragonal
   (b) monoclinic
   (c) triclinic
   (d) rhombohedral
3. The direction along the face diagonal of a unit cell of a cubic crystal is denoted by
   (a) [111] (b) [100]
   (c) [110] (d) [112]
4. In a BCC lattice which one of the following directions has the maximum linear density?
   (a) [110] (b) [100]
   (c) [211] (d) [111]
5. A crystal plane intercepts the crystal axes at $\frac{a}{2}$, b and is parallel to the c axis. The Miller indices of the plane are
   (a) (120) (b) (210)
   (c) (012) (d) (102)
6. The plane with the highest planar density in a FCC structure is
   (a) FCC (b) simple cubic
   (c) BCC (d) HCP
7. Number of atoms per unit cell in diamond cubic structure are
   (a) four (b) six
   (c) eight (d) twelve
8. The ionic radius of potassium is 1.33 and that of chlorine is 1.81. The stable configuration of KCl structure is
   (a) triangular (b) tetrahedral
   (c) octahedral (d) cubic
9. The number of nearest neighbour in octahedral configuration is
   (a) four (b) six
   (c) eight (d) twelve
10. In a BCC lattice, in which of the following planes there is no Bragg reflection?
    (a) (110) (b) (211)
    (c) (220) (d) (221)
11. Germanium exhibit
    (a) rock-salt structure
    (b) diamond-cubic structure
    (c) caesium-chloride structure
    (d) zinc-sulphide structure
12. Among the following bonds which one is the weakest?
    (a) Covalent bond
    (b) Ionic bond
    (c) Metallic bond
    (d) Van der Waals bond
13. What type of bonding exists between water molecules in ice?
    (a) Covalent bond
    (b) Ionic bond
    (c) Hydrogen bond
    (d) Metallic bond
14. Van der Waals bond is formed due to
    (a) electrons and ion cores interaction
    (b) dipole-dipole interaction
    (c) Coulomb interaction
    (d) sharing of electrons
15. The bonding in silicate crystals is
    (a) ionic
    (b) covalent.
    (c) partially ionic and partially covalent
    (d) Van der Waals.
16. In hexagonal crystal, the angles between the axes are
    (a) $\alpha = \beta = \gamma = 90°$
    (b) $\alpha \neq \beta \neq \gamma \neq 90°$
    (c) $\alpha = \beta = 90°, \gamma = 120°$
    (d) $\alpha = \gamma = 90° \neq \beta$

17. The co-ordination number of BCC lattice is
    (a) six (b) eight
    (c) twelve (d) three
18. The number of atoms per unit cell in the FCC lattice is
    (a) one (b) two
    (c) three (d) four
19. The packing fraction of diamond crystal structure is
    (a) 74% (b) 68%
    (c) 52% (d) 34%
20. X-rays can be deflected by
    (a) magnetic fields
    (b) electric fields.
    (c) large size obstacles
    (d) none of these
21. The Miller indices of a set of parallel planes which make equal intercepts on the three axes are
    (a) (121) (b) (111)
    (c) (100) (d) (101)
22. Hexagonal closed pack structure is
    (a) simple cubic structure
    (b) body-centred cubic structure
    (c) face-centred cubic structure
    (d) base-centred cubic structure
23. Which one of the following crystals is an example of monoclinic?
    (a) NaCl (b) Mg
    (c) $FeSO_4$ (d) $CuSO_4, 5H_2O$
24. The atomic radius for FCC lattice is
    (a) $\frac{a}{2}$ (b) $\frac{a}{2\sqrt{2}}$
    (c) $\frac{\sqrt{3}}{4}a$ (d) $\frac{a}{\sqrt{2}}$
25. The packing fraction has maximum value for
    (a) FCC structure
    (b) SC structure
    (c) HCP structure
    (d) BCC structure
26. NaCl crystal has
    (a) FCC structure
    (b) HCP structure
    (c) SC structure
    (d) BCC structure
27. Electrons are transferred from one type of atom to the other type in
    (a) covalent crystals
    (b) ionic crystals
    (c) metallic crystals
    (d) molecular crystals
28. Which of the following bonds has high electrical conductivity?
    (a) Ionic bond
    (b) Covalent bond
    (c) Metallic bond
    (d) H-bond
29. The energy required to separate the atoms of a molecule is called
    (a) ionisation energy
    (b) Coulomb energy
    (c) cohesive energy
    (d) dissociation energy
30. IF ***n*** be the number of atoms in unit cell of the cubic system, N and M be the Avogadro's number and atomic weight, respectively, and ρ be the density of the element, then the lattice constant is given by
    (a) $\left[\frac{nM}{N\rho}\right]^{1/3}$ (b) $\left[\frac{nM}{N\rho}\right]^{-1/3}$
    (c) $\left[\frac{M\rho}{nN}\right]^{1/3}$ (d) $\frac{nM}{N\rho}$
31. Presence of carbon atom in FCC iron (γ–iron) at 1100° is an example of
    (a) interstitial impurity
    (b) substitutional impurity
    (c) vacancy
    (d) phonons
32. In Frenkel's defect, the displaced atom is called
    (a) anion (b) cation
    (c) hole (d) real atom
33. Imperfections in alkali halides are known as
    (a) substitutional
    (b) Frenkel's defect
    (c) Schottky's defect
    (d) both (b) and (c)

34. Point imperfections during interaction with each other
    (a) lower their total energy
    (b) are not affected at all
    (c) are thermodynamically stable
    (d) both (a) and (c)
35. Number of vacancies at absolute zero (temperature) will be equal to
    (a) $+\infty$
    (b) any finite value
    (c) zero
    (d) none of these
36. Directions of edge dislocation and Burgers vector are
    (a) perpendicular to each other
    (b) parallel to each other
    (c) inclined at 45° to each other
    (d) concurrent
37. On application of thermal energy, a dislocation vanishes
    (a) at the crystal surface
    (b) at the nodal point
    (c) as a loop
    (d) both (a) and (b)
38. Motion of an edge dislocation on a plane perpendicular to the glide plane is known as
    (a) cross-slip
    (b) vacancy diffusion
    (c) climb motion
    (d) creep
39. The grain boundary area decreases when a polycrystalline material is heated
    (a) below 0.5 T_m
    (b) below 0 Kelvin
    (c) above room temperature
    (d) above 0.5 T_m
40. Ideal crystals have
    (a) no imperfections
    (b) only screw dislocations
    (c) real crystals
    (d) Frenkel's defect
41. Influence of dislocations on mechanical strength of a crystal is to
    (a) increase it
    (b) keep it uniform
    (c) decrease it
    (d) none of these
42. The imperfections may be minimised by
    (a) thermal energy
    (b) making metal in fibre form
    (c) surface treatment
    (d) all of these
43. Volume imperfections can be of the type of
    (a) Schottky's defect
    (b) twin
    (c) stacking fault
    (d) non-crystalline regions
44. The overall electrical neutrality is maintained in
    (a) Frenkel's defect
    (b) screw dislocation
    (c) low angle boundary defect
    (d) none of these

Answers

1. (b)	2. (c)	3. (c)	4. (d)	5. (b)
6. (a)	7. (c)	8. (d)	9. (b)	10. (d)
11. (b)	12. (d)	13. (c)	14. (b)	15. (c)
16. (c)	17. (b)	18. (d)	19. (d)	20. (d)
21. (b)	22. (c)	23. (c)	24. (d)	25. (a)
26. (c)	27. (b)	28. (c)	29. (d)	30. (d)
31. (a)	32. (b)	33. (c)	34. (d)	35. (c)
36. (a)	37. (d)	38. (c)	39. (d)	40. (a)
41. (c)	42. (a)	43. (d)	44. (a)	

Fill in the Blanks

1. There are .... type of lattices under seven crystal system.
2. The co-ordination number of BCC lattice is ....
3. Atomic radii of diamond .... structure is $\frac{\sqrt{3}}{8}a$.
4. Packing factor of a FCC structure is .... and is equal to 0.74.
5. XRD analysis provides the .... of crystals.
6. The equilibrium distance between the two bonding atoms is known as ....
7. Materials involving .... bond are very good conductor of heat and electricity.
8. The hydrogen bond and Van der Waals bond are .... bonds and both are weak.
9. Among the atomic bonds, the one which is highly directional is ....
10. The non-directional bond is also known as ....
11. The unit of cohesive energy is ....
12. The type of bonding that occurs between the molecules in a molecular solid is ....
13. The type of bonding in silicate crystals is ....
14. Example of a crystal with perfect covalent bonding is ....
15. Long-range order is found in .... materials.
16. The structure of NaCl crystal is ....
17. The packing in diamond is relatively very ....
18. The packing factor of the HCP lattice is same as that of the .... lattice.
19. Germanium, silicon crystallise in .... structure.
20. Glass exhibits a .... order which is limited to very few atomic dimensions.
21. Presence of .... vacancies is energetically more favourable.
22. The number of vacancies in a crystal .... with decrease in temperature.
23. The slip planes in crystals are normally planes of .... atomic packing.
24. The movement of dislocation is easier in .... crystals.

Answers

1. 14	2. eight	3. diamond	4. highest
5. internal structure	6. bond length	7. metallic	8. secondary
9. covalent bond	10. metallic bond	11. KJ/mole	12. Van der Waals bond
13. partially ionic and partially covalent	14. diamond	15. crystalline	
16. face-centred cubic	17. loose	18. FCC lattice	19. diamond cubic
20. short range	21. pair	22. decrease	23. highest
24. FCC			

Reading List

1. Streetman Ben G, Banerjee, Sanjay Kumar *Solid state Electronic Devices.* 6th ed. New Delhi: PHI Learning Pvt Ltd.
2. Vijaya, M S; Rangarajan G *Material Science.* New Delhi: McGraw-Hill Education Pvt Ltd.

3. Beiser, Arthur *Perspectives of Modern Physics*. New Delhi: McGraw-Hill Education Pvt Ltd.
4. Shukla, R K *Fundamentals of Engineering Physics*. Kanpur: Ira Books.
5. Rajput, R K *Electrical Engineering Materials*. New Delhi: Laxmi Publications Pvt Ltd.
6. Neamen, Donald A (1992) *Semiconductor Physics and Devices: Basic Principles*. Boston: Homewood IL Irwin.
7. Higgins R A (1998) *Properties of Engineering Materials*. 2nd ed. New Delhi: Viva Books Pvt Ltd.
8. Dekker A J (1998) *Solid State Physics*. New Delhi: Macmillan India Ltd.
9. Gupta, Navneet (2000) *Engineering Practical Physics*. 2nd ed. New Delhi: Dhanpat Rai & Co.
10. Dekker A J (2000) *Electrical Engineering Materials*. New Delhi: PHI Learning Pvt Ltd.
11. Kittel, Charles (1996) *Introduction to Solid State Physics*. 7th ed. New York: John Willey & Sons.
12. Gupta, Navneet; Shahdev S K (2001) *A Text book of Basic Electronics*. Ist ed. Dhanpat Rai & Co.

2

Band Theory of Solids

Chapter Objectives

- To give an overview of free-electron theory and its limitations.
- To understand the formation of bands in materials and the classification of materials on the basis of band picture.
- To provide a knowledge of Brillouin zones and a discussion regarding the formation of allowed and forbidden energy bands.
- To understand the difference in electrical properties of different materials on the basis of band theory.
- To explain the insulating and metallic behaviour of materials on the basis of band theory.
- To discuss the concept of hole—a new contribution of the band theory.

2.1 INTRODUCTION TO FREE-ELECTRON THEORY

The free-electron theory was put forward by Lorentz and Drude to explain the electronic structure of metals. It assumes that the valence electrons of the constituent atoms move about freely throughout the metal specimen, as if they were electron gas. The mutual repulsion between the negative electrons is not taken into account and the forces between the conduction electrons and ion cores are neglected in the free-electron model. The conduction electrons are assumed to be completely free to roam everywhere within the specimen. The total energy is all kinetic; potential energy is neglected.

Although the free-electron model of metals provides a good insight into many properties, like low specific heat of metal thermal conductivity, electrical conductivity and paramagnetic susceptibility, yet it is silent about why some materials are metals, while some are insulators or semiconductors. Experimentally, the conductivity of materials (semiconductors and metals) is strongly dependent upon temperature, but free-electron model has no answer to this.

To understand the basic difference between insulators and conductors we must extend the free-electron model, considering the periodicity of the lattice. There may be a possibility of bands as a result of solid formation rather than sharp energy levels as we find in isolated atom. We will see that the crystal behaves as an insulator, if the allowed energy bands are either filled or empty, for then no electron can move on an applied electric field. If one or more bands are partially filled, the crystal behaves as a metal. The same electron, which was confined to a certain fixed energy level in the atom, becomes free to roam inside the whole band in a solid material. The response of this free-electron inside the band to the external electric field is also quite interesting. Inside a crystal lattice, the electron behaves wave-like and undergoes Bragg's reflection for some values of k that give rise to a forbidden band. We shall discuss these things in detail in the forthcoming sections.

Free-electron model considers a constant potential for the electron inside the lattice. The permitted energies for the electron in such potential (in terms of k) are given by

$$E_k = \frac{\hbar^2 k^2}{2m} \qquad (2.1)$$

The above relation clearly shows that there is no upper limit for the electron's energy, i.e. it can have any value beginning from zero to infinity, as shown in Fig. 2.1. But, the reality is somewhat different. Moving electron within the lattice exhibit only certain permitted values of energy.

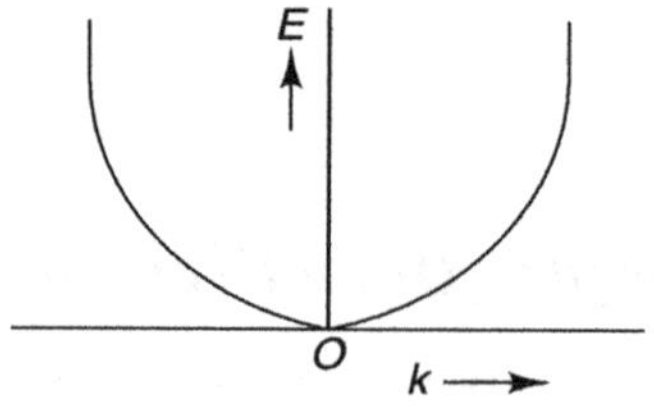

Fig. 2.1 *k vs. E for free electron moving in constant potential*

Example 2.1

An electron is bound by a potential which closely approaches an infinite square well of 2.5×10^{-10} m width. Calculate the lowest three permissible quantum energies the electron can have.

Solution In case of infinite potential well

$$E_n = \frac{n^2 k^2}{8ma^2}$$

Here, $h = 6.63 \times 10^{-34}$ Js, $m = 9.1 \times 10^{-31}$ kg, $a = 2.5 \times 10^{-10}$ m.

Now,
$$E_1 = \frac{h^2}{8ma^2} = \frac{(6.63\times10^{-34})^2}{8\times91\times10^{-31}\times(2.5\times10^{-10})^2}$$
$$= 9.63 \times 10^{-19}\ \text{J}$$
$$= 6.0\ \text{eV}$$
$$E_2 = n^2E_1 = 4 \times 6.0 = 24\ \text{eV}$$
$$E_3 = n^2E_1 = 9 \times 6.0 = 54\ \text{eV}$$ **Ans.**

2.2 KRONIG-PENNEY MODEL

An attempt to explain the above was made by Kronig and Penney—in 1930—by considering the interaction of electrons with the positive ions of the lattice, which was totally ignored in the free-electron model. Kronig and Penney were first to consider the motion of electrons in a periodic potential of the positive ion cores of the crystal lattice, taking into consideration the quantum mechanical ideas. Considering a one-dimensional case for simplicity, the potential energy of an electron within the lattice is shown in Fig. 2.2. The highest potential is halfway between the ions and tends to infinity as the position of ion is approached.

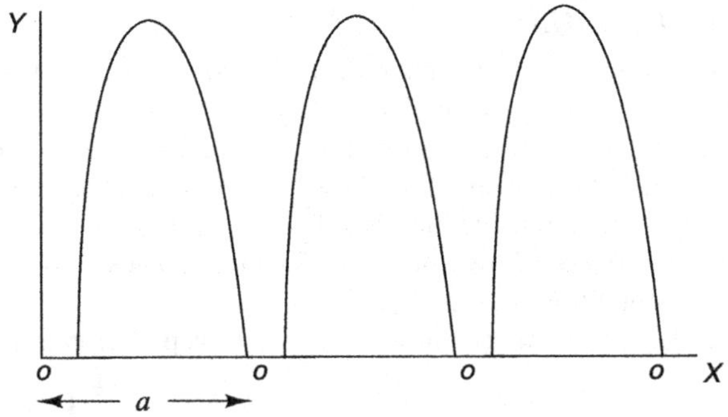

Fig. 2.2 *The variation of electrons' potential energy in a one-dimensional lattice*

The situation can be easily described by considering a potential as shown in Fig. 2.3, which is simpler and still displays the essential features of the above function.

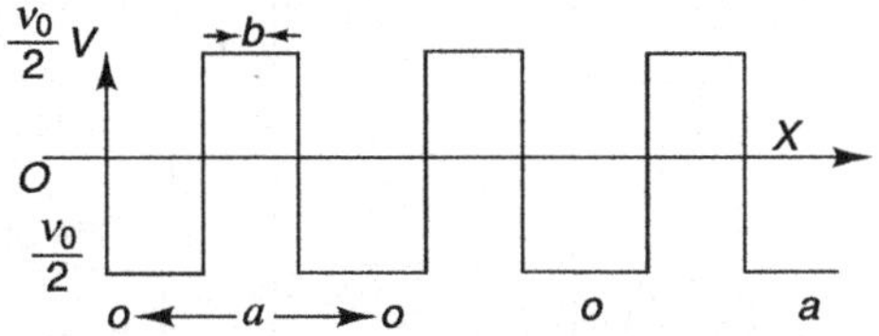

Fig. 2.3 *A simpler variation of potential energy for mathematical analysis*

The Schrödinger's equation for the above potential function can be written as

$$\frac{d^2\psi}{dx^2} + \frac{2m}{\hbar^2}(E - V)\psi_{(x)} = 0 \quad (2.2)$$

One can solve the above equation very easily for $V_{(x)} = \frac{V_0}{2}$ and $V_{(x)} = \frac{V_0}{2}$ regions separately, taking into account the boundary condition and also bearing in mind that the solution is periodic. The solution to the above equation is assumed to be of the form

$$\psi_k(x) = u_k(x)e^{ikx} \quad (2.3)$$

On using it and simplifying, we find that a solution to the above equation exists if k is related to energy E by the following equation:

$$\cos ka = \frac{P \sin \alpha a}{\alpha a} + \cos \alpha a \quad (2.4)$$

where $$P = \frac{ma}{\hbar^2}V_0.b \quad \text{and} \quad \alpha = \sqrt{\frac{2mE}{\hbar^2}} \quad (2.5)$$

The plot of Eq. (2.4) is given in Fig. 2.4.

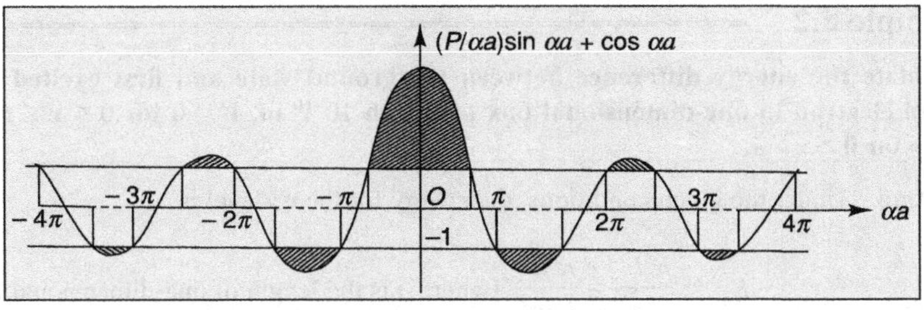

Fig. 2.4 *E vs. K vector for an electron moving in a periodic potential in a mono-atomic linear lattice*

Interpretation of Solution

The slope of variation decreases at higher values of αa. Since the Eq. (2.4) has to be satisfied, those portions of the curve which lie beyond the limits of + 1 and – 1 are not permitted. In the graph, such portions are shown as shaded regions. The values of αa that match with the shaded region are, therefore, not allowed. These are the same values for which the moving electron inside the crystal lattice undergoes Bragg's reflection and gives rise to discontinuities as shown in Figs 2.5 and 2.6. Those which are allowed are marked by dark horizontal lines in Fig. 2.5.

Using the above Eq. (2.4), one can now plot E vs. k, as per Kronig–Penney model.

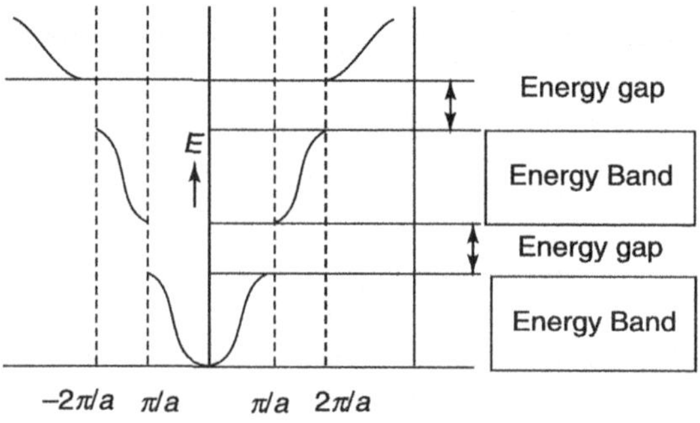

Fig. 2.5 *Discontinuity curve*

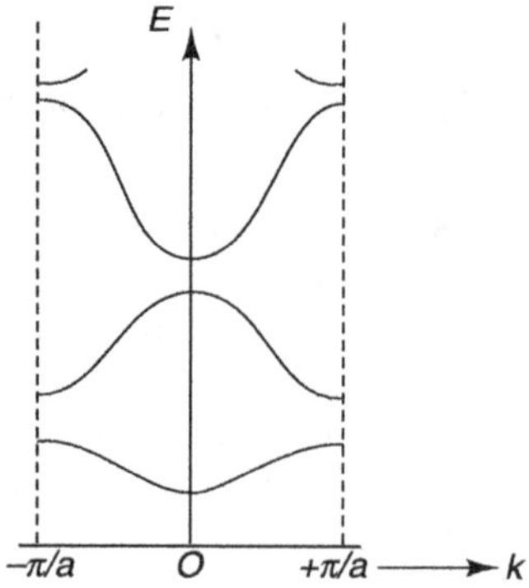

Fig. 2.6 *E vs. k curve in reduce zone scheme*

The following conclusions can be drawn from the above plot:

- There exist allowed energy bands rather than sharp levels separated by a gap known as forbidden band.
- The function E_k is periodic in k.
- Energy gap (forbidden bands) occur for $k = \pm\frac{n\pi}{a}$, where n has integral values, a being the lattice parameter (inter-atomic separation).
- The separation between energy levels is almost negligible within the allowed band.
- The region $-\frac{\pi}{a}$ to $+\frac{\pi}{a}$ in k-space is known as first Brillouin zone of the lattice. Similarly, the second Brillouin zone occurs between $-\frac{2\pi}{a}$ to $+\frac{2\pi}{a}$.

Example 2.2

Calculate the energy difference between the ground state and first excited state for an electron in one-dimensional box of length 10^{-10} m, $V = 0$ for $0 \le x \le a$ and $V = \infty$ for $0 > x > a$.

Solution Under the given conditions, the energy for the nth level is

$$E_n = \frac{\hbar^2 k^2}{2m} = \frac{n^2 h^2}{8ma^2} \text{ [where } a \text{ is the length of one-dimensional box]}$$

so, $$E_1 = \frac{h^2}{8ma^2}(n=1)\left[\text{and } k = \frac{n\pi}{a}\right]$$

and $$E_2 = \frac{4h^2}{8ma^2}(n=2)$$

The energy difference between first excited and ground state is

$$\Delta E = E_2 - E_1 = \frac{3h^2}{8ma^2} = \frac{3\times(6.62\times10^{-34})^2}{8\times9.1\times10^{-31}\times(10^{-10})^2}$$

$$= 1.806 \times 10^{-17} \text{ J}$$

$$= 112.87 \text{ eV}$$

Ans.

Example 2.3

Find the first three energy levels of an electron in an infinite potential well. The width of the potential well is 5 Å. Comment on the result.

Solution The nth energy level of the electron is given by

$$E_n = \frac{\pi^2 n^2 \hbar^2}{2ma^2}$$

$$= \frac{n^2(1.054\times10^{-34})^2\pi^2}{2(9.11\times10^{-31})(5\times10^{-10})^2}$$

$$= n^2 (2.41 \times 10^{-19}) \text{ J}$$

or $$E_n = \frac{n^2(2.41\times10^{-19})}{1.6\times10^{-19}} = n^2(1.51)\text{eV}$$

Then,

$$E_1 = 1.51 \text{ eV}, \quad E_2 = 6.04 \text{ eV}, \quad E_3 = 13.59 \text{ eV}$$

The above calculation shows that the energy of a bound electron cannot be continuous.

Ans.

Example 2.4

Determine the lowest allowed energy bandwidth assuming that the coefficient $P = 10$ and width of potential $a = 5$ Å.

Solution To find the lowest allowed energy bandwidth, we have to find the difference in αa values, as ka changes from 0 to π. For $ka = 0$ Eq. (2.4) becomes

$$1 = 10 = \frac{\sin \alpha a}{\alpha a} + \cos \alpha a$$

This gives $\alpha a = 2.628$ rad.

We see that for $ka = \pi, \alpha a = \pi$

For $\alpha a = \pi$, we have

$$\sqrt{\frac{2mE_2}{\hbar^2}}.a = \pi$$

Therefore,

$$E_2 = \frac{\pi^2\hbar^2}{2ma^2} = \frac{\pi^2(1.054\times10^{-34})^2}{2(9.1\times10^{-31})(5\times10^{-10})^2}$$
$$= 2.407 \times 10^{-19} \text{ J}$$
$$= 1.50 \text{ eV}$$

For $\alpha a = 2.628$, we find that $E_1 = 1.68 \times 10^{-19}$ J $= 1.053$ eV. The allowed energy bandwidth is then

$$\Delta E = E_2 - E_1 = 1.50 - 1.053 = 0.447 \text{ eV}$$ **Ans.**

2.3 EXPLANATION FOR DISCONTINUITIES IN *E* VS. *k* CURVE (CONCEPT OF BRILLOUIN ZONES)

The Brillouin zone is a representation of permissive values of k of the electrons in one, two or three dimensions. The origin of allowed and forbidden energy bands can be explained by the concept of Brillouin zones. The electron in solids moves under a periodically varying potential rather than a constant one and it behaves wave-like within the crystal lattice. The de Broglie wavelength, associated with free electron inside the crystal, can be expressed by

$$\lambda = \frac{h}{\sqrt{2\,mE}} \tag{2.6}$$

From the above relation, it is evident that only those electrons that have low values of k (low energy electrons) will not undergo diffraction, as the wavelength associated with these will be quite large compared to lattice parameter a of the crystal lattice. Such electrons can roam freely throughout the lattice.

On the other hand, the electrons with sufficient high energy (of order of Fermi energy) will definitely exhibit a wavelength comparable with the inter-atomic separation of the crystal lattice and undergo Bragg's diffraction. Such electrons will not move freely within the crystal. As a result of Bragg's reflection, only certain ranges of momenta, therefore, are permitted, giving rise to allowed energy bands. Those values of momenta which are not permitted give rise to forbidden energy bands. Therefore, the discontinuities in E vs. k curve are a consequence of Bragg's reflection (diffraction) of electron waves inside the crystal lattice. Figure 2.7 givens the one-dimensional lattice (for simplicity) having inter-atomic separation a.

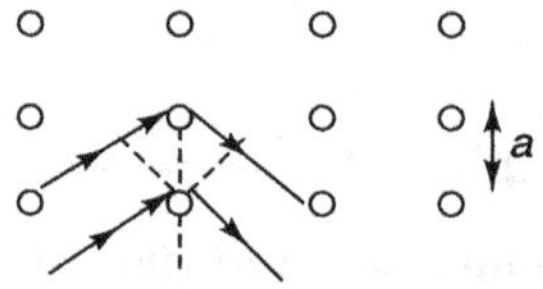

Fig. 2.7 *Bragg's reflection in one-dimensional lattice*

The variation between energy E vs. k is quite significant, as it describes the behaviour of the electron in the solid, when it gets accelerated. The variation of k signifies the variation in momentum ($P = \hbar k$) and hence, acceleration of the electron. To understand the transport mechanism of an electron in a crystal, it is important to know how its energy changes, as the electron gets accelerated. The energy values which correspond to continuous k-values form the allowed energy band and those between discontinuities form the forbidden band.

Applying Bragg's law,

$$2d \sin\theta = \pm n\lambda \tag{2.7}$$

For present case, taking $d = a$, the equation becomes

$$2a \sin \theta = \pm n\lambda$$

where θ is the glancing angle, which can be taken 90° for an electron wave travelling along the crystal lattice and therefore,

$$\lambda = \frac{\pm 2a}{n}$$

This means that as the electron gets accelerated, its momentum increases and so λ changes. Whenever momentum acquires values that satisfy the above condition $\left(\lambda \approx \pm\frac{2a}{n}\right)$, the electrons get Bragg reflected or diffracted. These are those values of k or λ at which the discontinuities (gaps) in E vs. k curve occur. The values of k not satisfying the above condition travel almost freely. Therefore,

$$2a = \pm n\lambda$$

$$\frac{1}{2a} = \pm\frac{1}{n\lambda}$$

or
$$\frac{2\pi}{2a} = \pm\frac{2\pi}{n\lambda} = \frac{k}{n}$$

i.e.
$$k = \pm\frac{n\pi}{a} \qquad \text{where } n = 1, 2, 3\ldots$$

The region in k-space that the electrons of low energy can occupy without being diffracted is known as first Brillouin zone. The boundary of first zone occurs at $k = \frac{\pm\pi}{a}$, where the electrons are Bragg reflected. The second Bragg's reflection occurs for $k = \frac{\pm 2\pi}{a}$ moving in the $\pm X$ and $\pm Y$ direction. The region with k-values between $-\frac{2\pi}{a}$ and $-\frac{\pi}{a}$ and between $+\frac{\pi}{a}$ and $+\frac{2\pi}{a}$ is called second Brillouin zone and so on. The discontinuity in energy values occurs at the boundary of Brillouin zones and, therefore, we get forbidden gaps at the boundaries, where the electrons suffer Bragg's reflection. The first and second Brillouin zones of a two-dimensional square lattice are shown in Fig. 2.8.

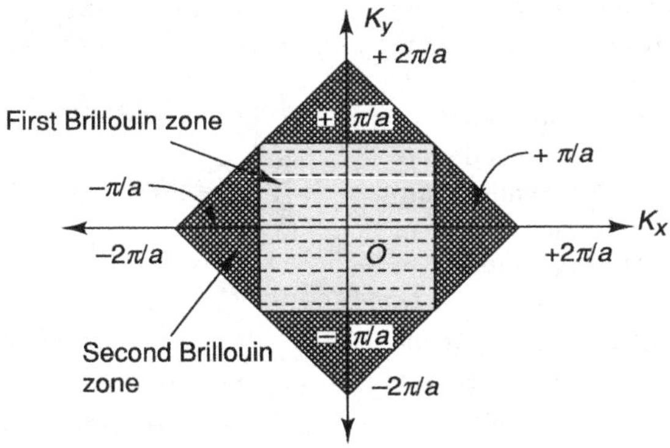

Fig. 2.8 *Brillouin zones in a two-dimensional square lattice*

Example 2.5

Consider a two-dimensional square lattice of side 3.0 Å. At what electron momentum values do the sides of first Brillouin zone appear? What is the energy of free electron with this momentum?

Solution Given $a = 3.0 \times 10^{-10}$ m
The momentum of electron is given by

$$p = \hbar k$$

For first Brillouin zone $k = \pm\dfrac{\pi}{a}$, then

$$p = \frac{h}{2\pi}\frac{\pi}{a} = \frac{h}{2a} = \frac{6.62 \times 10^{-34}}{2 \times 3 \times 10^{-10}}$$

Energy

$$E = \frac{p^2}{2m} = \frac{(1.1 \times 10^{-10})^2}{2 \times 9.1 \times 10^{-31}}$$

$$= 6.688 \times 10^{-19}\ \text{J}$$

$$= \frac{6.688 \times 10^{-19}}{1.6 \times 10^{-19}}$$

$$= 4.2\ \text{eV}$$

Ans.

2.4 FORMATION OF SOLID MATERIAL

A solid material may be assumed to come to existence when the assembly of atoms is brought together in such a way that the distance between any two atoms becomes of the order of inter-atomic separation. During this process, the Pauli exclusion principle plays a vital role. When two atoms are far apart, there is no interaction of electron-wave function between them; they can have identical electronic structure. Therefore, isolated atoms exhibit sharp energy levels. When a large number of atoms come in close proximity, it has been observed that the interaction takes place between the outermost atomic orbitals; the innermost orbital remains almost unaffected to the said interaction. As a result a splitting and overlapping between the energy levels takes place, giving rise to formation of bands. Therefore, the bands are characteristic of a solid, while sharp energy levels are characteristics of isolated atoms. The formation of bands in various materials (i.e. metals, semiconductors and insulator) will be discussed in forthcoming sections. The splitting in various energy levels is shown in Fig. 2.9.

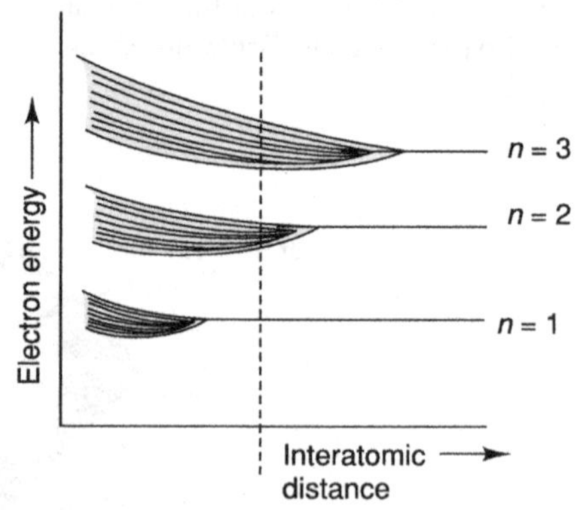

Fig. 2.9 *Splitting in various energy levels during solid formation*

2.5 FORMATION OF BAND IN METALS

The electrical behaviour of crystalline solid is determined by its energy bands' structure, as well as how these bands are normally occupied by the electrons. In the following discussion, two cases of metals, namely sodium and magnesium, are discussed from the point of view of the energy band structure.

It is to be kept in mind that overlapping of conduction and valence band is not a characteristic feature of all the metals. Sodium is exceptional; there is no overlapping and yet it is a metal. To clarify this, let us inspect this material on the basis of band picture. The electronic configuration of isolated atom of Na is given by

$$\mathbf{Na_{11} \rightarrow 1s^2\ 2s^2\ 2p^6\ 3s^1\ 3p^0}$$

The nucleus of sodium atom has charge +11e. The potential energy of an electron in the field of nucleus is therefore

$$V(r) = -\frac{1}{4\pi\varepsilon_0}\cdot\frac{Ze^2}{r}$$

The corresponding potential experienced by the electron of sodium atom is given in Figs 2.10 and 2.11.

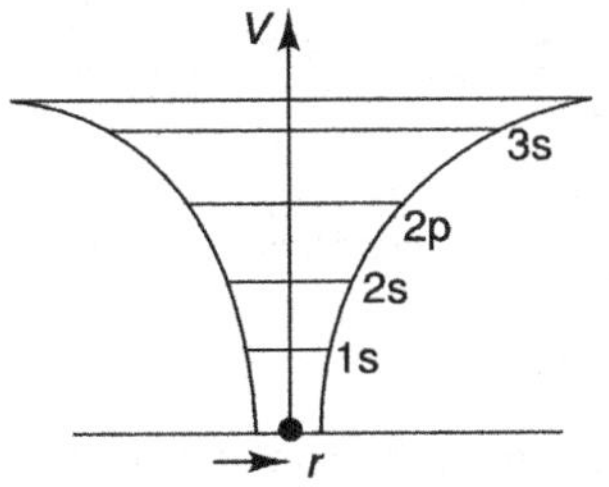

Fig. 2.10 *Potential energy as a function of r in single isolated atom of sodium*

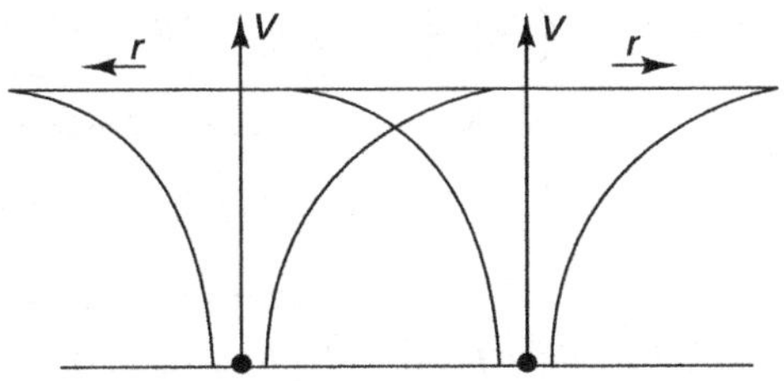

Fig. 2.11 *Two close atoms of sodium*

Figures 2.10 and 2.11 show potential energy as a function of r in single isolated atom of sodium, and two close atoms of sodium, respectively. The figures indicate that potential energy is highly negative for the distances close to the nucleus and approaches zero for larger ones. The potential energies of various levels are indicated in the figures for single atom and two close atoms of sodium. Now, let us see what happens to the energy levels during solid formation. A solid may be assumed to come to existence when numerous such isolated atoms come in close proximity to each other. When this happens, the outermost energy levels 2p and 3s spread into energy. This spreading takes place only in outer energy levels, while the inner ones remain unaffected. The corresponding potential energy distribution for a solid, therefore, appears as shown in Fig. 2.12.

Suppose N-atom participates in solid formation in the above case of sodium, then there are N-3s electrons in all. The occupying capacity of S-orbital is two, (2N-states are available in solid sodium). Therefore, only half states are filled and half remain empty in the same band. Since the width of a given band depends on inter-atomic spacing, closer the atoms, wider will be the band.

In such a situation, even a very little energy can compel the electron to go to the nearby state in the same band causing a conduction. All other shells (bands) 3p, 3d, etc., are completely empty. So in this case, the 3s band, which is topmost partially filled is of interest, as it contains the valence electrons. The situation of the solid sodium can be shown by a simplified band picture in Fig. 2.13.

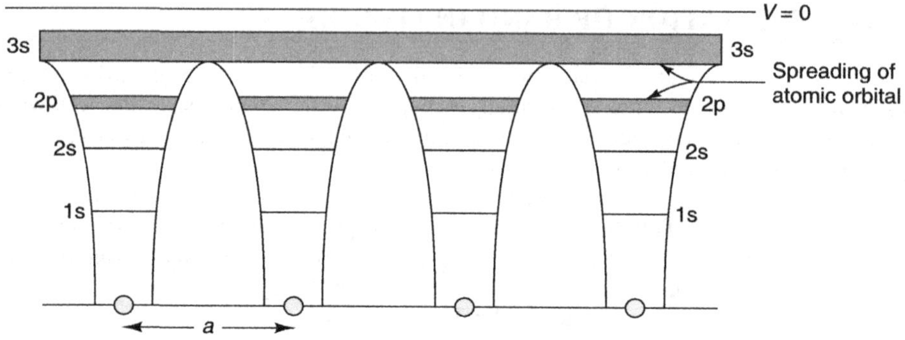

Fig. 2.12 *Potential energy of an electron in a 1-D lattice*

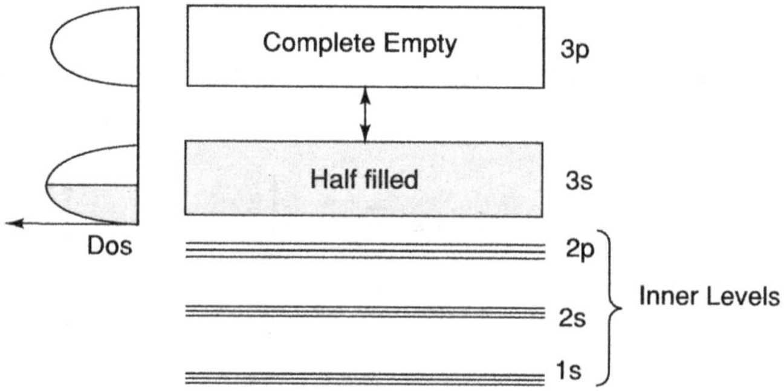

Fig. 2.13 *Band picture of solid sodium (no overlapping between 3s and 3p bands)*

This is the reason why, even in the absence of overlapping of bands, sodium is a metal. On the other hand, the case of magnesium is different. The electronic configuration of magnesium is

$$\mathbf{Mg_{12} \longrightarrow 1s^2\ 2s^2\ 2p^6\ 3s^2\ 3p^0}$$

Exactly on the basis of the above discussion, the band picture of magnesium is shown in Fig. 2.14.

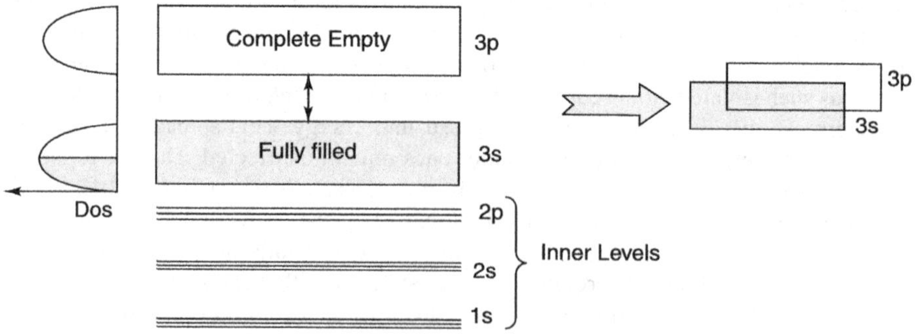

Fig. 2.14 *Overlapping of 3s and 3p bands in magnesium*

Hence, to understand the properties of a solid, prime consideration is given to the topmost partially-filled band and the immediate next upper band. Therefore, properties of solid are mainly decided by the following factors:

- Occupation state of the topmost band
- Separation between topmost occupied band and the immediate empty band
- The density of states
- The presence of impurities which are mainly responsible for additional states of forbidden gap

2.6 FORMATION OF BANDS IN SEMICONDUCTORS AND INSULATING MATERIALS

In the last section, we discussed how the bands originate during the solid formation. In a nutshell, this happens as a result of interaction, splitting, overlapping and redistribution of energy levels (atomic wave function) when atoms come in close proximity during the solid formation. Let us now investigate how this happens in the case of semiconductors and insulating materials.

Most of the metals and semiconductors exhibit a crystalline structure. A crystal exhibits perfect regularity of atoms in three dimensions. During solid formation, the innermost energy levels remain unaffected, while the outer ones are affected utmost due to the interaction between atoms forming a solid. As a consequence, overlapping takes place between outermost shells which leads to formation of bands.

Consider a semiconducting material (Si, Ge, etc.) consisting of N-atoms. When the atoms are sufficiently apart, the interaction between them is almost negligible and the energy levels are quite similar to that of an isolated atom. The atom of a semiconducting material has two s-electrons and two p-electrons. For a crystal containing N-atoms, therefore, these will be 2N-s and 2N-p electrons, respectively. The maximum capacity of s-orbital is two (i.e. it can accommodate two electrons), while the occupying capacity of p-orbital is six. So in absence of overlapping, s-band will be fully-occupied and p-band will be only one third. As the atoms come closer (≈ atomic separation), there takes place overlapping of spreaded s and p atomic orbitals. This means $(2N + 6N)$ states become available to ($2N$-s electrons + $2N$-p electrons). That is, overall $8N$ states become available to $4N$ electrons. These $8N$ states are then redistributed equally into two bands containing $4N$ closely states each. The lower energy $4N$ states are then occupied by the available $4N$ electrons that constitute a valence band. The remaining $4N$ states, which contain no electron, constitute a conduction band. These two bands are separated by a gap nearly ≈1 eV, which is known as a forbidden band. The insulators also undergo the same mechanism of overlapping except that the order of magnitude of forbidden gap is nearly 5 eV or above.

The splitting, overlapping and redistribution of energy levels is shown in Figs. 2.15 and 2.16.

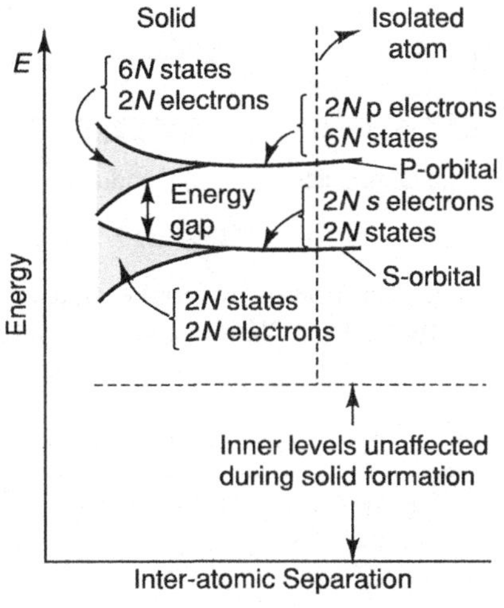

Fig. 2.15 *Splitting of energy levels during solid formation*

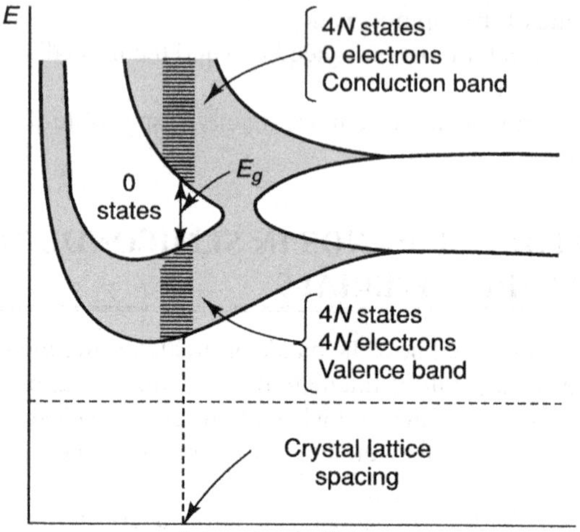

Fig. 2.16 *Formation of bands in materials*

2.7 CLASSIFICATION OF MATERIALS ON THE BASIS OF BAND STRUCTURE

The materials can be placed into four categories depending upon the band structure characteristics as shown in Fig. 2.17.

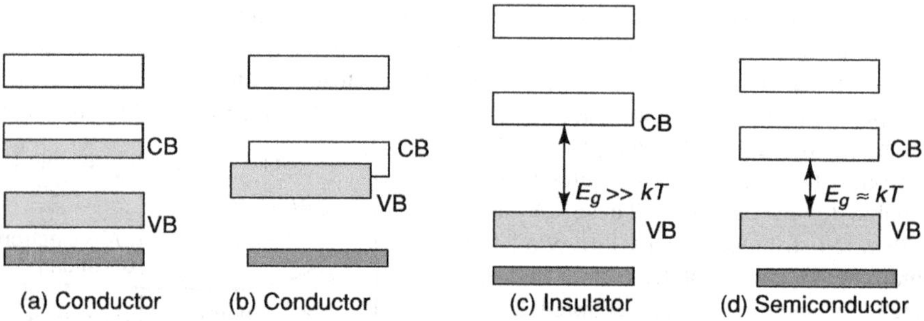

Fig. 2.17 *Band structure of same materials*

The first kind of material is that in which the highest occupied energy band is only partially filled at 0K (Na-like material). The quantum states of Na vapour are given in Table 2.1.

Table 2.1 *States available and occupied in sodium*

Energy level	*No. of electrons*	*Total states available*	*Total states occupied*
1S	2	2 N	2N
2S	2	2N	2N
2P	6	6N	6N
3S	1	2N	N
3P	0	6N	0

In sodium like these materials, the electronic configuration is such that the outermost sub-shell contains an odd number of electrons. Therefore, an electric field can provide sufficient energy to the electrons of this partially filled band, so that these can move to available vacant states in the same band. The materials of this kind are, therefore, good conductor of electricity.

The second kind are those which have the highest occupied level completely filled at 0K and the next higher level is completely empty when the atoms are well separated. But, when the atoms come closer during the solid formation, these energy levels split into the bands and the bands overlap each other as shown in Fig. 2.17(b). Zinc and magnesium are example of this kind. The quantum states of magnesium are given in Table 2.2

Table 2.2 *States available and occupied in magnesium*

Energy level	No. of electrons	Total states available	Total states occupied
1S	2	2N	2N
2S	2	2N	2N
2P	6	6N	6N
3S	2	2N }	2N }
3P	0	0	6N

In this case as well, there are empty states at energies close to occupied states and, therefore, these materials are also good conductors.

The third class of materials is that in which the highest occupied energy band is completely filled and the next higher band (which is empty) is well above it, as shown in Fig. 2.17(c). The gap between these two bands is quite large. When the electric field is applied in such materials, the electrons from valence band cannot reach the conduction band where they get empty energy states to show conduction. Hence, such materials behave like insulators. Diamond is an example of this kind in which the separation between valence and conduction band is nearly 6 eV.

The fourth class of materials is the one which has the highest occupied band (which is completely filled) at 0 K, but the next higher band (which is empty) is only slightly above the filled band. The band structure is very similar to that of an insulator, but the gap is very small—nearly 1.0 eV (in silicon). Also, these materials cannot show conduction on applying electric field at 0 K. However, at a temperature more than 0 K, thermal collisions may push some of the electrons from the highest occupied band to the next empty band. These few electrons, in the otherwise empty band, can respond to even a weak electric field because they receive a large number of empty states just above their existing energy. As electrons from filled (valence band) band are pushed up in energy to land into a higher energy band, some empty states are created in this band, as well. These empty states allow some movement of electrons in the same band and thus promote conduction. As the total number of electrons that receive energy from electric field is quite small, the conductivity of these materials is small, as well. Such materials are known as semiconductors. Table 2.3 shows the conductivity of some materials.

Table 2.3 *Conductivity of materials at room temperature (in Ohm^{-1} m^{-1})*

Metals		*Insulator*		*Semiconductors*	
Material	*Conductivity*	*Material*	*Conductivity*	*Material*	*Conductivity*
Copper	6.0×10^7	Glass	10^{-11}	Silicon	5×10^{-4}
Gold	5.0×10^7	Mica	10^{-15}	Germanium	2.0×10^{-4}
Silver	6.0×10^7			Gallium Arsenide	1.0×10^{-6}

2.8 EXPLANATION FOR DIFFERENCE IN THE ELECTRICAL PROPERTIES OF DIFFERENT MATERIALS

Why do different materials exhibit different electrical properties? Free-electron model has no reply to this. If the band picture of different material is examined critically, we could definitely observe that each material exhibits its own structure and this is the reason why different materials exhibit different electrical properties. We have illustrated this by taking the example of silicon, which is a good insulator in the diamond lattice. The simplified band structure of silicon is shown in Fig. 2.18.

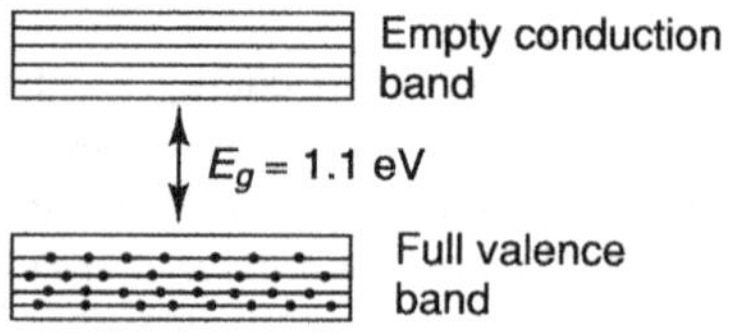

Fig. 2.18 *Band picture of silicon at 0 K*

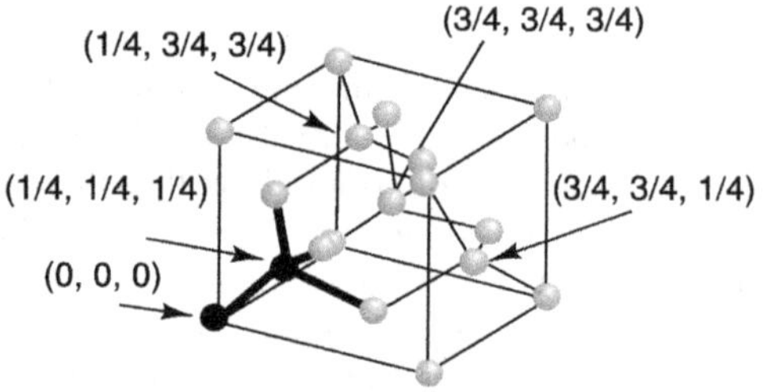

Fig. 2.19 *Diamond structure*

The conduction process can be understood with the help of Fig. 2.18. Silicon exhibits a valence band which is completely full with electrons with no empty state available at 0 K. Also, it exhibits a conduction band that is totally empty having a large number of vacant states (allowed energy states).

It is clear that there can be no charge transport within the valence band as there are no vacant states available. On the other hand, the conduction band has no charge carriers; so no transport of charge can be thought there, as well. This is the reason why silicon exhibits a high resistivity. Now, if an external electric field is applied, it compels some electrons to occupy vacant state of conduction band, leaving behind some vacancies in the parent valence band. These vacancies act like a positive charge carrier known as a hole. The 'hole' is not a separate entity, but its existence in linked with the missing electron of valence band. One can understand a hole as a moving air bubble in a glass filled with water. An empty glass does not mean, therefore, that it is filled with the holes only.

Semiconductors at 0 K exhibit basically a similar structure as that of insulators—a filled valence band separated by an empty conduction band that has no allowed energy states. The difference lies in the magnitude of forbidden band gap, which is smaller in semiconductors—typically nearly 1.1 eV compared with 5 eV for diamond. The relatively small band gaps of semiconductors allow for excitation of electrons from valence band to conduction band by reasonable amount of thermal or optical energy. For example, at room temperature, a semiconducting material with 1 eV band gap exhibits enough number of electrons excited thermally across the energy gap into the conduction bands, while in insulator having more then 5 eV gap, this number is extremely small (negligible). This is a remarkable different between a semiconductor and insulator—the number of charge carriers can be greatly increased in a semiconductor, while it is not possible in an insulator.

A band picture insulator is shown in Fig. 2.20.

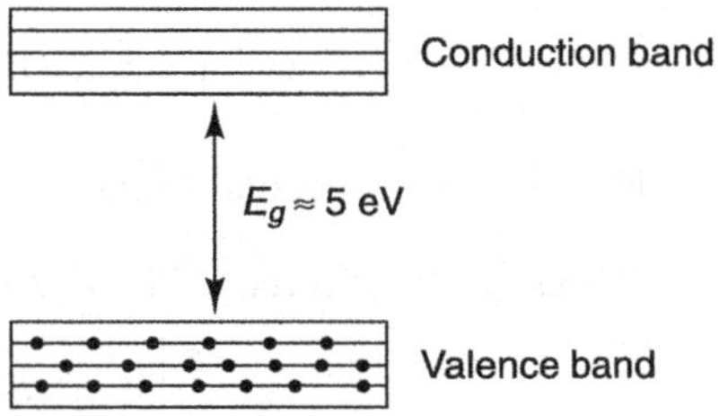

Fig. 2.20 *Band picture of an insulator*

2.9 IMPORTANT CHARACTERISTIC OF A BAND ELECTRON

A band electron can be described by a block function which can travel through the crystal without any attenuation of wave function. Thus, as long as the things remain completely periodic, the band electron has an infinite mean free path. Therefore, quantum mechanics can explain the total absence of elastic scattering by the array of positive ion cores of a lattice. The mean free path of an electron ends where it faces an interruption from the perfect periodicity. Any such interruption renders a block function unsuitable and encourages scattering. Interruptions which result in a finite mean free path are caused by crystal surfaces, static point defects and lattice vibrations.

2.10 NUMBER OF ENERGY STATES PER BAND

Consider a crystal lattice with lattice constant a and apply the periodic boundary condition to the wave functions over the crystal length. The crystal described above is made up of primitive cells having lattice constant a.

$$\psi(x+L) = \psi(x)$$

Taking block functions $\psi_k = U_k(x)e^{ik\cdot x}$ this requires

$$e^{ik(x+L)}U_k(x+L) = e^{ik\cdot x}.U_k(x)$$

As U_k is periodic, we have

$$U_k(x+L) = U_k(x)$$

Therefore, the boundary condition requires

$$e^{ik(x+L)} = e^{ik\cdot x}$$

or $$e^{ikL} = 1$$

or $$kL = 2n\pi$$

$$k = \frac{2n\pi}{L}$$

where $n = 0, \pm 1, \pm 2, \pm 3, \ldots$

In the first Brillouin zone, k ranges from $-\frac{\pi}{a}$ to $+\frac{\pi}{a}$

$$\therefore \quad \frac{2n\pi}{L} = \frac{\pi}{a}$$

$$n = \frac{L}{a} = \frac{\text{Number of primitive cells}}{2} = \frac{N}{2}$$

where $N = \frac{L}{a}$

The value $\frac{2n\pi}{L} = \frac{\pi}{a}$ is not counted because it is just the same as $\frac{2n\pi}{L} = +\frac{\pi}{a}$, as two are connected by $\frac{2\pi}{a}$ [Kronig–Penney model $K' = k + \frac{2n\pi}{a}$].

Therefore, the allowed values of k are

$$k = \frac{2\pi}{L}\left[0, \pm 1, \pm 2, \pm 3 \ldots + \frac{N}{2}\right]$$

or
$$k = 0, \pm\frac{2\pi}{L}, \pm\frac{4\pi}{L}, \pm\frac{6\pi}{L}, \ldots \pm (N-2)\frac{\pi}{L}, +\frac{N\pi}{L}$$

The total number of values is exactly N, the number of primitive cells.

For example, if $N = 10$, then the allowed values of k are

$$k = 0, \pm\frac{2\pi}{L}, \pm\frac{4\pi}{L}, \pm\frac{6\pi}{L}, \pm\frac{8\pi}{L}, \pm\frac{10\pi}{L}$$

Therefore, one can conclude that

"Each primitive cell contributes exactly one independent value of k to each energy band."

The above result can also be extended to three dimensions, taking into consideration the two allowed values of spin of electron $m_s = \pm\frac{1}{2}$, there are 2N independent states of 2N orbitals in each energy band, i.e. number of states in each band = twice the number of primitive cells.

2.11 EXPLANATION FOR INSULATING AND METALLIC BEHAVIOUR OF MATERIALS

The question for which free-electron model could not provide a reply can be explained now in the following way. For an atom having valency 1 (like Na) in each primitive cell, the energy band will be half filled.

To this electron many other states in the same band are available, which it can occupy when the material is subjected to electric field. Hence, this will be a good conductor.

On the other hand, if each atom contributes two electrons (as in Mg) to the band, it will be completely filled. There will be no conductivity and the material will behave like an insulator. If there are two atoms of valency 1, again the energy band will be exactly filled. This is true many times. Such conclusions are only valid provided there is no overlapping of atomic orbitals (wave functions). Metals sometimes exhibit overlapping of bands. This depends upon the structure (electronic configuration a material) of the atoms constituting crystal lattice. Mg is one such material which shows a metallic behaviour, although it is not a good metal and exhibit, overlapping of bands.

An electron from a filled valence band can move to an unfilled conduction band by absorbing energy either by thermal means or by incidence of light. Such an electron will contribute to a current in the conduction band under the action of electric field, leaving behind a vacant side in the valence band. The vacancy so produced also behaves as a positive charge carrier known as a 'hole'. This hole can move about and contribute to a current as positive charge carrier.

The bands which are formed by the overlapping of atomic orbitals containing even number of electrons are always full and contain no allowed energy state to which these electrons can move under the application of electric field. On the other hand, the bands which originate from atoms containing odd number of valence electrons always exhibit some allowed vacant energy sites to which the electrons can move under application of external electric field. In such materials there is no overlapping of atomic orbitals, while in the earlier discussed case overlapping is a must.

One can therefore conclude that a solid may be an insulator provided it contains even number of electron per primitive cell and the uppermost band must be separated from the band above it by an energy gap $Eg >> kT$ at the temperature of interest. The distinction between an insulator and a semiconductor is given as under:

$E_g >> kT$ .... Insulator

$E_g \approx kT$ .... Semiconductor

2.12 CONCEPT OF HOLE: A REMARKABLE CONTRIBUTION OF BAND THEORY

The existence of holes is one of the most significant and interesting contribution of the band theory. Vacant orbitals in a band are known as holes or hole states. The existence of hole alone is meaningless. It is a byproduct of a missing electron in the valence band.

At normal temperatures ($T > 0$ K), some electrons in the valence band gain enough thermal energy and reach the conduction band. This creates some mobile electrons in conduction band and some vacant states in valence band as shown in Fig. 2.21.

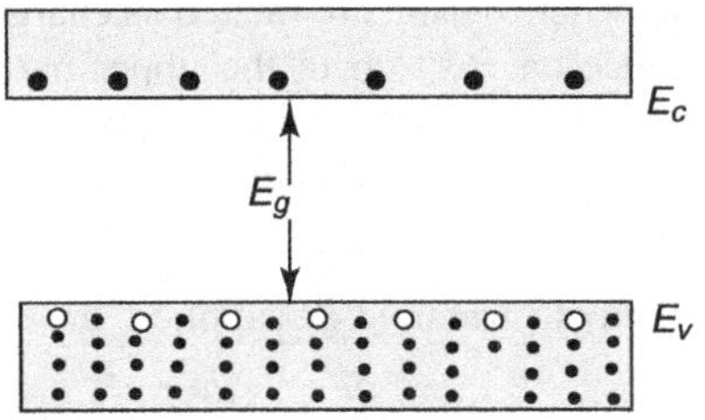

Fig. 2.21 *Electron-hole pairs in intrinsic semiconductor*

Such empty state in the valence band is referred as hole. If the conduction band electron and the hole are created by the excitation of a valence band electron to the conduction band, they are called an Electron-Hole Pair (EHP).

After excitation to the conduction band, an electron is surrounded by a large number of vacant energy states. For example, the equilibrium number of electron-hole pairs in pure Si at room temperature is only about 10^{10} EHP/cm^3 as compared to the Si atom density of 5×10^{22} atoms/cm^3. Thus, the few electrons in the conduction band are free to move about via the many available empty states.

The charge transport in the valence band is somewhat complicated. However, it is possible to show that the effects of current in a valence band containing holes can be accounted for by simply keeping track of the holes themselves.

In a filled band, all available energy states are occupied. For every electron moving with a given velocity, there is an equal and opposite electron motion elsewhere in the band. If we apply an electric field, the net current is zero because for every electron j moving with velocity v_j, there is a corresponding electron j' with velocity $-v_j$. Figure 2.22 illustrates this effect in terms of the electron energy vs. wave vector plot for the valence band. Since k is proportional to electron momentum, it is clear that the two electrons have oppositely directed velocities. With N electrons/cm^3 in the band, we express the current density using an overall sum of the electron velocities and including the charge q on each electron. In a unit volume,

$$J = (-q)\sum_{i}^{N} v_i = 0 \text{ (filled band)} \qquad (2.8)$$

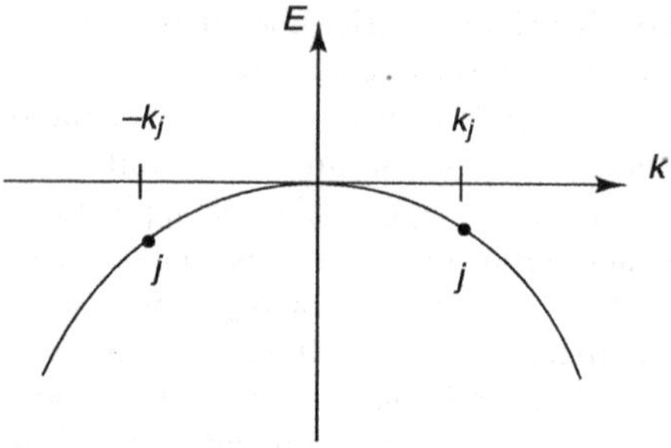

Fig. 2.22 *A valence band with completely filled states*

If we create a hole by removing the jth electron, the net current density in the valence band involves the sum over all velocities minus the contribution of the electron we have removed.

$$J = \underbrace{(-q)\sum^{N} v_i - (-q)v_j}_{qv_j} \text{ (jth electron missing)} \qquad (2.9)$$

Since the first term is zero, from Eq. (2.8), one can conclude that the current contribution of the hole equivalent to that of a positively charged particle with velocity v_j, that of the missing electron. Of course, the charge transport is actually due to the motion of the new uncompensated electron (j'). Its current contribution $(-q)(-v_j)$ is equivalent to that of a positively charged particle with velocity $+ v_j$. For convenience, vacant states in the valence band are treated as charge carriers with positive mass.

Since an electron in the upper part of the band has a negative effective mass m^*, removing an electron from the upper part of the band corresponds to adding mass. A hole, thus, has a positive effective mass.

$$m_h^* = -m_e^*$$

If an electric field ξ is applied, the rate of change of current I' due to the field is,

$$\frac{dI'}{dt} = q\frac{dv}{dt} = a \text{ (acceleration)}$$

$$= q\left(\frac{-e\xi}{m_j^*}\right) \qquad (2.10)$$

$$\left(\because \text{Acceleration} = \frac{\text{force}}{\text{mass}} = \frac{e\xi}{m_j^*}\right)$$

where m_j^* is the mass of an electron occupying the state j.

Now, since it is the tendency of hole to remain in the upper part of a nearly filled band, where the mass of electron m_j^* is negative, the RHS becomes + ve. It could also be said that the current is carried by the hole, which behaves as a positive charge of effective mass $m_h^* = |m_e^*|$ because in that case the force $e\xi$ in the above equation is + ve and with m_h^* + ve the RHS of the equation is again positive. Thus, we can say that the motion of the hole is the same as that of a particle of + ve charge q and a positive effective mass m_h^*.

So, we can say that vacant states near the top of an otherwise filled band behave in every way as particle of + ve charge e. Treating the hole in this manner results in a great simplification in that the motion of all electrons in the band has been reduced to that of a single particle.

Example

The following simple example is enough to understand the behaviour of a hole. If we have two bottles, one completely filled with water and one completely empty, we can ask ourselves "Will there be any net transport of water when we tilt the bottles?" The answer is "no". In the case of the empty bottle, the answer is obvious. In the case of the completely full, there cannot be any net motion of water because there is no empty space for water to move into. Similarly, an empty conduction band, completely devoid of electrons or a valence band completely full of electrons, cannot give rise to a net motion of electrons, and thus can not contribute to current conduction.

Next, we imagine transferring some water droplets from the full bottle into the empty bottle, leaving behind some air bubbles and ask ourselves the same question. When we tilt the bottles, there will be net transport of water. The water droplets will roll downhill in one bottle and the air bubbles will move uphill in the other. Similarly, a few electrons in an otherwise empty conduction band move opposite to an electric field, while holes in an otherwise filled valence band move in the direction of the field. The bubble analogy is imperfect, but it may provide a physical feel for why the charge and mass of a hole have opposite signs from those of an electron.

In all the discussions, we shall focus on the electrons in the conduction band and on the holes in the valence band. We can provide explanation for the current flow in a semiconductor by the motion of both types of charge carriers. Refer the sketch of valence band as given in Fig. 2.22.

It should be kept in mind that in the valence band hole energy increases oppositely to electrons energy because the two carriers have opposite charge. Thus, hole energy increase downward as shown in figure and holes seeking the lowest energy state available are generally found at the top of valence band. In contrast, an electron of valence band acquires the bottom of conduction band.

Summary

- The isolated atoms of a material are characterised by well-defined sharp and discrete energy levels, while energy bonds correspond to a solid material.
- Materials are broadly classified as metals and non-metals.
- Lattice and atomic structure of constituent elements decide whether the material is metal, insulator or semiconductor.
- The widths of the energy bands and the energy gap depend on the type of atomic bond and the inter-atomic separation, i.e. the crystal structure.
- For an electron moving in a periodic potential, the E vs. k curve has discontinuities at values of k given by $k = \pm n\pi/a$, where as the lattice constant and n takes values 1, 2, 3, ..., etc.
- The discontinuities in the E vs. k curve correspond to the forbidden energy bands. The range of k values for which electrons have allowed energy values are called Brillouin zones.
- Those k-values associated with electrons which do not follow Bragg's condition constitute the allowed energy band.
- The discontinuity in the energy at k values equal to $\pm n\pi/a$ can be entitled to the Bragg reflection of electron waves at the Brillouin zone boundaries.
- The concept of effective mass observes that electron inside a material behave in an entirely different way.

- If in a completely filled valence band some of the electrons get thermally excited to the conduction band, the vacant sites left in the valence band act like positive charges with negative mass. These are called holes.
- The effective mass relates the motion of charge particle in a lattice to an externally applied force, and takes into account the effect of crystal lattice to the motion of charge particle.
- An electron is a negatively charged particle with a positive effective mass existing at the bottom of conduction band. A hole is a positively charged particle with a positive effective mass existing at the top of valence band.
- Energies within an allowed energy band are actually discrete and the band contains a finite number of quantum states.

Exercise

1. Explain the electron energy bands formation in a solid?
2. How will you determine whether a given substance is an insulator, a conductor or a semiconductor? Explain with necessary theory.
3. How are the conductors, semiconductors and insulators distinguished on the basis of energy band theory?
4. Show how the theory explains the temperature variation of electrical conductivity of a semiconductor.
5. Explain briefly the formation of allowed and forbidden energy bands in crystals. How this helps in classifying materials into conductors and insulators.
6. Give an outline of the basis of Kronig–Penney model. Explain the extreme conditions when energy levels are either discrete or continuous.
7. Explain why the semiconductors are considered opaque to wavelength below a certain value, whereas the conductors are opaque to all wavelengths.
8. What is the basic difference between the free-electron theory and the band theory regarding the energy of electrons in solids? Explain why the elements behave as good conductors, semiconductors or insulators, according to band theory. How is the valency of element related to the filling of these bands?
9. Using one-dimensional Kronig–Penney potential, show that the energy spectrum of the electrons in a solid consists of a number of allowed energy bands separated by forbidden regions.
10. Show that the maximum number of electrons that can be accommodated in an energy band is equal to twice the number of primitive cells. How is this conclusion used in distinguishing between metals, insulators and semiconductors?
11. Derive an expression for the effective mass charge carriers and discuss its significant in the band structure of solids.
12. Discuss Kronig–Penney model. Using the model, show the energy spectrum of electron consisting a number of allowed energy bands separated by forbidden bands.
13. What is the effect of periodic potential on the energy of electrons in a metal? Explain it on the basis of Kronig–Penney model including the formation of energy bands.
14. Discuss how the concept of bands was originated in solids. Give necessary theory. What are *E-K* diagrams and what do you conclude from them?
15. What are Brillouin zones? Explain using *E-K* diagrams. Discuss the formation of Brillouin zones for (i) linear lattice, and (ii) two-dimensional lattice.

16. Discuss the motion of electrons in a periodic field of a crystal and show that the effective mass of an electron in a crystal is inversely proportional to the derivative of *E-K* curve. Under what conditions, the effective mass of an electron can become + ve, – ve and infinity?
17. Based on band theory of solids, distinguish between conductors, semiconductors and insulators.

Numerical Problems

1. Find the energy of an electron moving in one-dimension in an infinitely high potential box of width 1 Å. The mass of the electron is 9.11×10^{-31} kg and $h = 6.63 \times 10^{-34}$ Js. **[Ans. 6.02×10^{-18} J]**
2. An electron is bound in a one-dimensional potential box which has a width 2.5×10^{-10} m. Assuming the height of the box to be infinite, calculate the lowest two permitted energy values of the electron. **[Ans. 6.0 eV, 24.0 eV]**
3. Compute the lowest energy of a neutron confined to the nucleus, where nucleus is considered as a box with a size of 10^{-14} m. **[Ans. 3.28×10^{-13} J]**
4. Calculate the energy difference between the ground state and the first excited state for an electron in a one-dimensional rigid box of length 10^{-8} cm. **[Ans. 112.8 eV]**

Self-quiz Questions

1. Explain how the atomic energy levels split into bands when a number of atoms are brought close together to form a crystal?
2. What are Brillouin zones? Explain using *E-k* diagrams.
3. Discuss the formation of Brillouin zones for (i) linear lattice, and (ii) two-dimensional lattice.
4. Explain origin of bands in solids.
5. Discuss the band theory of solids and explain the formation of bands and concept of holes.
6. Discuss the origin of energy band gap due to periodic potential experienced by electrons in solids.
7. What is the effect of periodic potential on the energy of electron in a metal? Explain it on the basis of Kronig–Penney model along with the formation of energy bands.
8. How do you differentiate between semiconductors and insulators?
9. Explain the concept of hole in a semiconductor.
10. Explain the significance of holes in the band theory of solids.
11. Explain the term 'forbidden gap' in energy bands.

Objective-type Questions

1. In the following, which are the correct statements of Ohm's law? (J is current density, E is electric field, σ is electrical conductivity, ρ is electrical resistivity, I is current, R is resistance, L is length of the conductor and A is the area of cross section of the conductor)
   (a) $J = \sigma E$ (b) $\rho = J/E$
   (c) $I = EL/R$ (d) $I = VA/\rho L$
2. Which of the following statements are true? The electrical conductivity of a metal is
   (a) directly proportional to the probability of collision of the electrons with the lattice

(b) inversely proportional to the relaxation time
(c) directly proportional to the density of free electrons
(d) directly proportional to the mean free path

3. Which of the following processes increase the resistivity of a metal?
(a) The metal is heated
(b) The metal is purified
(c) The metal is cooled
(d) Imperfections are introduced into the metal

4. In an one-dimensional lattice of inter-atomic separation a, the maximum wavelength of the electron that gets Bragg reflected is
(a) $a/4$ (b) $a/2$
(c) a (d) $2a$

5. Which of the following statements are true?
(a) A solid is an insulator if it has $E_g \cong 1$ eV.
(b) A solid is a semiconductor if it has $E_g > 3$ eV.
(c) A solid is a metal if its valence band is partially filled.
(d) A solid is a metal if its conduction and valence bands overlap.

6. The range of the Brillouin zones corresponds to the ranges of
(a) forbidden energy bands in a solid.
(b) allowed energy bands in a solid.
(c) discontinuous regions in the E vs. k curve in solids.
(d) continuous regions in the E vs. k curve in solids.

7. The energy value in a free-electron model is given by
(a) $E = \dfrac{h^2}{8\pi^2 k^2 m}$ (b) $E = \dfrac{h^2 k^2}{2m}$
(c) $E = \dfrac{h^2 k^2}{2m}$ (d) $E = 2mh^2 k^2$

8. The first Brillouin zone is defined between the region
(a) $k = 0$ to $\dfrac{\pi}{a}$
(b) $-\dfrac{2\pi}{a}$ to $\dfrac{\pi}{a}$
(c) $k = -\dfrac{\pi}{a}$ to $\dfrac{2\pi}{a}$
(d) $-\dfrac{\pi}{a}$ to $\dfrac{\pi}{a}$

9. The second Brillouin zone is defined between the region
(a) $k = \dfrac{-\pi}{a}$ to $\dfrac{+\pi}{a}$
(b) $k = \dfrac{-2\pi}{a}$ to $\dfrac{-\pi}{a}$ and $\dfrac{\pi}{a}$ to $\dfrac{2\pi}{a}$
(c) $k = \dfrac{-\pi}{a}$ to $\dfrac{2\pi}{a}$
(d) $k = \dfrac{-2\pi}{a}$ to $\dfrac{2\pi}{a}$

10. Bragg's condition for diffraction is given by the relation
(a) $2d \sin \theta = n\lambda$ (b) $2 \sin \theta = \lambda$
(c) $2 \sin \theta = n\lambda$ (d) $a \sin \theta = n\lambda$

11. The group velocity of the electron in an one-dimensional lattice is defined as
(a) $v_g = \dfrac{dk}{d\omega}$ (b) $v_g = \dfrac{d\omega}{dk}$
(c) $v_g = \dfrac{\omega}{k}$ (d) $v_g = \omega k$

12. The energy gap between the valence and conduction bands in a semiconductor is of the order of
(a) 26 eV (b) 1.0 eV
(c) 7.0 eV (d) 0.001 eV

Answers

1. (a)	2. (c)	3. (a),(d)	4. (d)
5. (c),(d)	6. (a),(c)	7. (b)	8. (d)
9. (d)	10. (a)	11. (b)	12. (b)

Fill in the Blanks

1. According to free-electron theory of metals, an electron moves under .... potential within the crystal.
2. Splitting of atomic orbitals take place when the atoms come .... to each other.
3. The metals are characterised by .... valence and conduction band.
4. In .... metal no overlapping is there in between valence and conduction band.
5. Kronig–Penney model utilises .... to describe the motion of electron within the crystal.
6. The separation between valence and conduction band is nearly .... for semiconductors.
7. Silicon exhibits .... structure.
8. The .... Brillouin zone extends from $-\frac{\Pi}{a}$ to $+\frac{\Pi}{a}$
9. Forbidden bands exhibit those values of wavelengths for which electron undergoes ....
10. The material in which 3*s* and 3*p* bands overlap is ....

Answers

1. constant	2. closer	3. overlapped	4. sodium
5. step potential	6. one electro volt	7. diamond	8. first
9. Bragg's reflection	10. magnesium		

Reading List

1. Vijaya, M S; Rangarajan, G *Material Science*. New Delhi: McGraw-Hill Education Pvt Ltd.
2. Dekker, A J (1998) *Solid State Physics*. New Delhi: Macmillan India Ltd.
3. Dekker, A J (2000) *Electrical Engineering Materials*. New Delhi: PHI Learning Pvt Ltd.
4. Kittel, Charles (1996) *Introduction to Solid State Physics*. 7th ed. New York: John Willey & Sons.
5. Nemen, Donald A (1992) *Semiconductor Physics and Devices: Basic Principles*. Boston: Homewood IL Irwin.
6. Streetman, Ben G; Banerjee, Sanjay Kumar *Solid State Electronic Devices*. 6th ed. New Delhi: PHI Learning Pvt Ltd.
7. Baiser, Arthur *Perspectives of Modern Physics*. New Delhi: McGraw-Hill Education Pvt Ltd.

3

Magnetic Properties of Materials

Chapter Objectives

- To know the basic terms related to magnetism.
- To understand the classification of magnetic materials on the basis of their behaviour in an external magnetising field.
- To know the relation among various magnetic parameters.
- To know the properties and applications of all kind of magnetic materials.
- To understand the theories associated with dia, para and ferromagnetic materials.
- To study the hysteresis curve and its significance in the development of magnetic materials for various applications in the industry.
- To understand the phenomenon of magnetostriction and to apply it in the production of ultrasonic waves.
- To know about hard and soft magnetic materials and their application.

3.1 INTRODUCTION

In 1982, Oersted pioneered in showing that an electric current is surrounded by a magnetic field when it flows through a wire. The strength magnetic field so produced is also dependent upon with magnetic material.

The magnetic materials and their properties play a key role while fabricating the magnets for various purposes. Magnetic effect of current is a well-known phenomenon and will prove very useful if applied to technical instruments. In some situations, we need a magnetic field that persists for a long time and is free from mechanical mishandlings. The magnets needed for this purpose are called permanent magnets. The magnetic field produced by such magnets is not too high, but is long lasting. However, many other situations may demand a very high magnetic field. This can be achieved by using the above principle—magnetic effect of current. Such high magnetic fields can only be produced by electromagnets. The field so produced is not permanent because when current is switched off the magnetic field ceases. Electromagnets, therefore, are termed as 'temporary magnets'. Strength of the magnetic field, as well as various parameters depend upon a range of magnetic properties of the material, like initial permeability, susceptibility, coercivity, hysteresis, loss, etc., which we shall discuss later. Permanent magnets are used in various appliances like galvanometer, voltmeter, ammeters, speakers, etc., while electromagnets are used in appliances where a high magnetic field is required. Magnetic properties are not exhibited only by iron and its alloys, but every material exhibits—little or more—magnetic properties. Iron and its alloys are most common as they are used for the construction of many types of electrical equipment. The prime need is to discuss the other materials too which exhibit magnetic properties.

3.2 ORIGIN OF MAGNETISM—MICROSCOPIC POINT OF VIEW

Revolving electron of an atom behaves like a 'tiny current loop' and therefore, exhibits a magnetic moment. Hence, each revolving electron has a definite magnetic moment. These magnetic moments are randomly oriented in materials commonly known as elementary magnets or dipoles and therefore, the net value of magnetic moment is found to be zero. The magnetic moment, due to orbital motion, can be calculated as given below.

In the case of simplest atom, hydrogen, that contains a proton and a single electron current produced due to an electron $i = \frac{e}{T}$

Time period $T = \frac{2\pi r}{v}$, where v is speed of electron and r is radius of orbit.

Therefore, current $$i = \frac{ev}{2\pi r} \tag{3.1}$$

Magnetic moment due to orbital motion, $\mu = iA = \frac{ev}{2\pi r} \times \pi r^2$

or $$\mu = \frac{evr}{2} \tag{3.2}$$

From Bohr's theory of H-atom—'the angular momentum of revolving electron is quantised' and can have some definite values only.

i.e. $$L = m_e v. r = \frac{nh}{2\pi}; n = 1, 2, 3, \ldots$$

where $m_e \rightarrow$ mass of electron

Therefore, $$v.r = \frac{nh}{2\pi m_e}$$

Substituting this in Eq. (3.2),

Magnetic moment $$\mu = \frac{e}{2} \cdot \frac{nh}{2\pi m_e}$$

or $$\mu = n\left[\frac{eh}{4\pi m_e}\right] \quad \text{where } n = 1, 2, 3, \ldots$$

Therefore, smallest magnetic moment will be $\frac{eh}{4\pi m_e}$, commonly known as 'Bohr magneton' and is represented by μ_B. Its value is 9.28×10^{-24} Amp m^2.

Besides the above, spinning of electrons also contributes to the magnetic moment, which is—many times—greater than the orbital or nuclear motion. Hence, the net magnetic moment of an atom is mainly due to the spinning of electrons.

Example 3.1

In a hydrogen atom, an electron revolves around a nucleus in an orbit of 0.53 Å radius. If the frequency of revolution of an electron is 6.6×10^{15} Hz, find the magnetic moment of the orbiting electron and calculate the value of Bohr magneton.

Solution Given $r = 0.53 \times 10^{-10}$ m and $n = 6.6 \times 10^{15}$ Hz
Magnetic moment $\mu = i\,A$

$$i = \frac{e}{T} = en = 1.6 \times 10^{-19} \times 6.6 \times 10^{15} \text{ Å}$$

Area $= \pi r^2 = 3.14 \times (0.53 \times 10^{-10})^2$

$\therefore$ $\mu = iA = 1.6 \times 10^{-19} \times 6.6 \times 10^{15} \times 3.14 \times (0.53 \times 10^{-10})^2$

$= 9.314 \times 10^{-24}\ \text{Am}^2$

Bohr magneton is the smallest value of the orbital magnetic moment of the electron. For $n = 1$, Bohr magneton

$$\mu_B = \frac{eh}{4\pi m_e}$$

$$= \frac{1.6\times10^{-19}\times6.6\times10^{-34}}{4\times3.14\times9.1\times10^{-31}}$$

$$= 9.239 \times 10^{-24}$$

$$= 9.24 \times 10^{-24}\ \text{J/T}$$ **Ans.**

Magnetic materials can be classified on the basis of presence of permanent dipoles, as given in Table 3.1.

Table 3.1 *Classification on the basis of permanent dipoles*

Class of Material	*Status of permanent dipoles*	*Interaction between neighbouring dipoles*
Diamagnetic	No	...NIL...
Paramagnetic	Yes	Negligible
Ferromagnetic	Yes	Parallel orientation
Antiferromagnetic	Yes	Antiparallel orientation of equal moments
Ferrimagnetic	Yes	Antiparallel orientation of unequal moments

3.3 BASIC TERMS IN MAGNETISM

Given below are some important magnetic parameters that will be used frequently in the ongoing discussion:

- Magnetic flux ϕ_m
- Magnetisation vector M
- Flux density B
- Magnetic permeability μ_m
- Magnetic susceptibility χ_m

(i) Magnetic Flux ϕ_m

The magnetic flux across any surface, placed in a magnetic field, is given by

$$\phi_m = \oint_s \overline{B}.\overline{ds} \tag{3.3}$$

It is also defined as magnetic field times an area and is expressed in webers.

Since the isolated magnetic pole never exists in nature; if we consider a closed surface in a magnetic field, the inward flux will be equal to outward flux. It is because the same number of lines of force that enter the surface must come out.

i.e. 'The net magnetic flux across a closed surface is always zero.'
Mathematically, the above statement can be expressed as

$$\oint_S \overline{B}.\overline{ds} = 0 \tag{3.4}$$

It is also called the Gauss's law of magnetism.

(ii) Magnetisation Vector $\overline{M}$

Revolving electrons of an atom can be treated as small magnetic or atomic dipoles, having a definite magnetic moment. These magnetic dipoles are present in material in a 'random manner'. On subjecting the material to external magnetising field, these dipoles tend to align partially, giving rise to a net magnetic moment in the direction of magnetising field in a small volume of the material. This magnetisation (or magnetic moment) per unit volume is named as 'magnetisation vector'. Sometimes it is also known as intensity of magnetisation. The degree of alignment of the dipoles is also dependent on external applied magnetic field.

$$\text{Magnetisation vector} = \frac{\text{Magnetic moment}}{\text{Volume of specimen}} \tag{3.5}$$

For a bar magnet having pole strength m and separation $2l$ (between north and south pole) along with cross section A, the magnetisation vector M is given by

$$M = \frac{2m.l}{A.2l} = \frac{m}{A} \tag{3.6}$$

i.e. pole-strength per unit area.

Example 3.2

An iron rod of 10^{-3} metre^3 volume and relative permeability of 1150 is placed inside a long solenoid wound with 5 turns per cm. If a current of 0.5 amp is passed through the solenoid, find the magnetic moment of the rod.

Solution

$$B = \mu_0(H + M) \quad \text{or} \quad M = \frac{B - \mu_0 H}{\mu_0}$$

If μ is the permeability of the material, then

$$B = \mu H \quad \therefore M = \frac{(\mu - \mu_0)H}{\mu_0} = \left(\frac{\mu}{\mu_0} - 1\right)H$$

But, $\mu/\mu_0 = \mu_r \quad \therefore M = (\mu_r - 1)H$

If i current is flowing through a solenoid having n turns per unit length, then magnetic intensity H is given by

$$H = ni \quad \therefore M = (\mu_r - 1)ni$$

Here, $\mu_r = 1150$, $n = 5$ turns/cm = 5000 turns/m and $i = 0.5$ amp

$$M = (1150 - 1)\ 500 \times 0.5 = 2.87 \times 10^5 \text{ A/m}$$

We know that the magnetisation M is the magnetic moment per unit volume

$\therefore$ magnetic moment, $m = M \times V$, where V is the volume

$m = 2.87 \times 10^5 \times 10^{-3} = 2.87 \times 10^{-2}$ A – m^2 **Ans.**

Example 3.3

An 2.0 m long, 10 mm in diameter iron rod of permeability 90 is placed inside a long solenoid wound with 300 turns/metre. If a current of 0.5 ampere is passed through the solenoid, find the magnetic moment of the rod.

Solution We know that the intensity of magnetisation M is

$$M = (\mu_r - 1)ni$$

Hence, $\mu_r = 100$, $n = 300$ turns/metre and $i = 0.5$ amp

$$\therefore \quad M = (90 - 1) \times 300 \times 0.5 = 1.3 \times 10^4 \text{ A/m}$$

Hence, magnetic moment, $m = M \times V$, where V is the volume of the rod

$$V = \pi r^2 l = 3.14 \times (5 \times 10^{-3})^2 \times 0.2 = 15.7 \times 10^{-6} \text{ m}$$

$$\left(\because r = \frac{D}{2} = \frac{10 \text{ mm}}{2} = 5 \times 10^{-3} \text{m}\right)$$

$$\therefore \quad m = 1.3 \times 10^4 \times 15.7 \times 10^{-6} = 20.4 \times 10^{-2} \text{ A} - \text{m}^2$$ **Ans.**

Example 3.4

Calculate the change in magnetic moment of a revolving electron in an applied magnetic field of 2.0 T acting perpendicular to the plane of the orbit. Assume $r = 5.29 \times 10^{-11}$ m for the normal state of H-atom.

Solution The magnetic moment of H-atom is given by

$$\mu = \frac{1}{2}erv = \frac{1}{2}er^2\omega$$

Therefore, change in magnetic moment $\Delta\mu = \frac{1}{2}er^2\Delta\omega$

$$= \frac{1}{2}er^2 . \left(\pm\frac{eB_0}{2m}\right)$$

$$= \pm\frac{e^2 B_0 r^2}{4m}$$

$$= \pm\frac{(1.6 \times 10^{-192}) \times 2.0 (5.29 \times 10^{-11})^2}{4 \times 9.1 \times 10^{-31}}$$

$$= \pm 3.9 \times 10^{-29} \text{ J/T}$$ **Ans.**

(iii) Flux Density B: Magnetic Induction

When a magnetic material is kept in an external magnetising field, the material gets magnetised. The net magnetic field inside the material is the sum of applied magnetic field and the magnetic field produced by alignment of atomic dipoles. If the material is ferromagnetic, the net magnetic field inside the material increases, and if the material is diamagnetic the net magnetic field inside the material decreases. That is, the density of magnetic flux lines inside the material either goes up or down, depending on whether it is ferromagnetic or diamagnetic. This magnetic field inside the material is expressed as flux-density $\overline{B}$. The flux density inside ferromagnetic and diamagnetic material is shown in Figs 3.1 and 3.2.

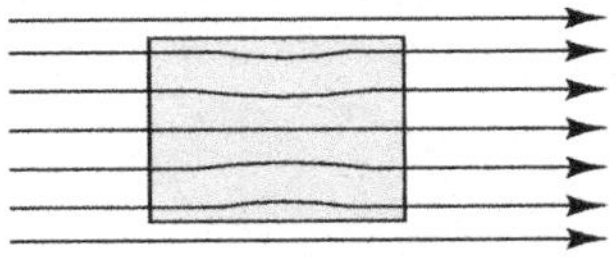

Fig. 3.1 *Flux density inside ferromagnetic material*

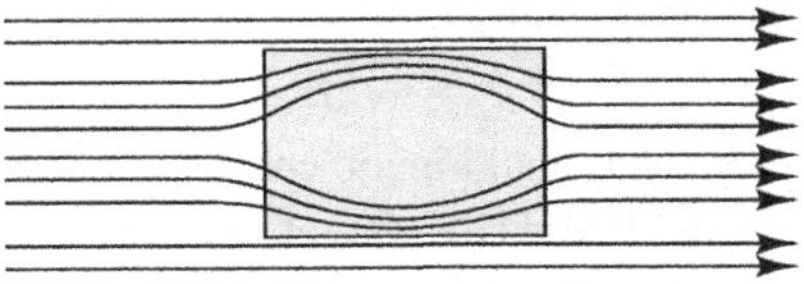

Fig. 3.2 *Flux density inside diamagnetic material*

(iv) Magnetic Permeability μ_m

Magnetic permeability of a material gives an indication of the number of magnetic flux lines that penetrate into the material on subjecting it to external magnetising field (H). It is defined as the ratio of magnetic flux density to the external magnetising field H.

i.e.
$$\mu = \frac{\text{Magnetic flux density}}{\text{Magnetising field}} = \frac{B}{H} \tag{3.7}$$

The unit of magnetic permeability is Henry/m.

Another important term of magnetism is known as 'relative permeability' denoted by μ_r, which is defined as

$$\mu_r = \frac{\mu}{\mu_0} \tag{3.8}$$

It is a dimensionless quantity being the ratio of two similar quantities.

(v) Magnetic Susceptibility χ_m

The magnetic susceptibility χ_m of a material expresses the response of a medium to an external magnetic field like electrical susceptibility χ_e. It is related to the properties of atoms and molecules of the medium. It is defined as the ratio of magnetisation vector M to the external magnetising field H.

$$\chi_m = \frac{\text{Magnetisation vector } (M)}{\text{Magnetising field } (H)} \tag{3.9}$$

It is again a pure number and dimensionless quantity.

Example 3.5

Prove that a superconductor ($B = 0$) is a perfect diamagnet.

Solution By definition, superconductor excludes magnetic flux completely from its bulk; $B = 0$

We know that, $B = \mu_0(H + M) = 0$, $\therefore H + M = 0$ or $M = -H$

or susceptibility, $\chi = M/H = -1$

It is true for diamagnet. Hence, a superconductor is a perfect diamagnet.

Example 3.6

A paramagnetic substance is composed of atoms with a magnetic dipole moment of 3.3. It is placed in a magnetic field of 52 T. To what temperature must the substance cooled, so that the magnetic energy of each atom would be as large as mean translational kinetic energy per atom? Assume $\mu = 924 \times 10^{-24}$.

Solution The magnetic energy of a dipole in an external magnetic field is given by

$$U = -\bar{\mu}.\bar{B} \qquad ...(i)$$

Mean translational kinetic energy per atom $= \frac{3}{2}kT$ ...(ii)

From Eqs (i) and (ii), we get

$$\frac{3}{2}kT = -\mu.B$$

$$\Rightarrow \quad T = \frac{\mu B}{\frac{3}{2}k} = \frac{\mu B}{1.5k} = \frac{3.3\times 9.24\times 10^{-24}\times 5.2}{1.5\times 1.38\times 10^{-23}} = 7.6 \text{ K}$$ Ans.

3.4 RELATION BETWEEN MAGNETIC PERMEABILITY AND SUSCEPTIBILITY

When a magnetic material is subjected to external magnetising field H, the magnetic flux density produced is given by

$$B = \mu_0\mu_r H \qquad (3.10)$$

We know that B, H and M are related as

$$B = \mu_0(H + M) \qquad (3.11)$$

$$\mu_0\mu_r H = \mu_0(H + M)$$

$$\mu_r = 1 + \frac{M}{H} = 1 + \chi_m$$

Therefore, $$\boxed{\mu_r = 1 + \chi_m} \qquad (3.12)$$

Example 3.7

Determine the magnetisation vector and flux density in silicon, if its magnetic susceptibility is -4.2×10^{-6} and the magnetic field in it is 1.15×10^5 Am^{-1}. What would be the value of relative permeability of the material?

Solution Given $\chi_m = -4.2 \times 10^{-6}$ and $H = 1.15 \times 10^5$ Am^{-1}.

Magnetisation $M = \chi_m H = -4.2 \times 10^{-6} \times 1.15 \times 10^5$ Am^{-1}
$= -0.48$ Am^{-1}

Flux density $B = \mu_0(H + M)$
$= 4\pi \times 10^{-7} \times (1.19 \times 10^5 - 0.48)$
$= 0.150$ T

Relative permeability

$$B = \mu H = \mu_0 H\left(1 + \frac{M}{H}\right)$$

or

$$\mu_r = \frac{\mu}{\mu_0} = 1 + \frac{M}{H}$$

$$= 1 + \frac{-0.48}{1.15 \times 0^5}$$

$$= 1 - 0.41 \times 10^{-5}$$

$$= 0.999$$ **Ans.**

Example 3.8

Find the percentage increase in magnetic induction when the space within a current-carrying toroid is filled with magnesium. Given χ_m for magnesium as 1.4×10^{-5}.

Solution Given $\chi_m = 1.4 \times 10^{-5}$

Magnetic flux density

$$B = \mu_0 H \tag{i}$$

When the free space is magnesium filled, then

$$B' = \mu_r \mu_0 H \tag{ii}$$

and

$$\mu_r = 1 + \chi_m \tag{iii}$$

From eqns (ii) and (iii)

$$B' = (1 + \chi_m)B \tag{iv}$$

Hence, the percentage increase in magnetic induction

$$= \frac{B' - B}{B} \times 100$$

By using Eqs (i) and (iv), the above eq becomes

$$= \frac{(1+\chi_m)B - B}{B} \times 100 = \chi_m \times 100 \tag{v}$$

$$= 1.4 \times 10^{-5} \times 100$$

$$= 1.4 \times 10^{-3}\,\%$$

$$= 0.0014\%$$

Ans.

Example 3.9

Determine the magnetisation vector and flux density of the diamagnetic material if its magnetic susceptibility is -0.2×10^{-5} and the magnetic field in it is 10^4 A m^{-1}.

Solution Given $\chi_m = -0.4 \times 10^{-5}$ and $H = 10^4$ A m^{-1}.

Magnetisation vector

$$M = \chi_m H$$

$$= -0.2 \times 10^{-5} \times 10^4$$

$$= 0.02 \text{ Am}^{-1}$$

Magnetic flux density

$$B = \mu_0(H + M)$$

$$= 4\pi \times 10^{-7} \times [10^4 - 0.02]$$

$$= 0.0116 \text{ T}$$

Ans.

Example 3.10

The magnetic susceptibility of aluminium is 2.1×10^{-5}. Find its permeability and relative permeability.

Solution Given $\chi_m = 2.1 \times 10^{-5}$

Permeability

$$\mu = \mu_0 \mu_r \text{ and}$$

Relative permeability

$$\mu_r = 1 + \chi_m$$

$$\therefore \quad \mu_r = 1 + 2.1 \times 10^{-5}$$

$$\mu_r = 1.000021$$

$$\mu = \mu_0 \mu_r = 4\pi \times 10^{-7} \times 1.000021$$

$$= 12.56 \times 10^{-7} \text{ N/A}^2$$

Ans.

Example 3.11

The magnetic susceptibility of a medium is 940 × 10^{-4}. Calculate its absolute and relative permeability.

Solution Given $\chi_m = 0.084$.

Absolute permeability $\mu = \mu_0 \mu_r$

Relative permeability $\mu_r = 1 + \chi_m$

so, $\mu_r = 1 + 0.084 = 1.084$

$$\mu = \mu_0 \mu_r = 4\pi \times 10^{-7} \times 1.084$$

$$= 13.61 \times 10^{-7} \text{ N/A}^2$$

Ans.

Example 3.12

The maximum value of the permeability of a material is 0.126 N/A². What is its relative permeability and magnetic susceptibility?

Solution Given $\mu = 0.126$ N/A²

Relative permeability $\mu_r = \dfrac{\mu}{\mu_0} = \dfrac{0.126}{4\pi \times 10^{-7}}$

$$= 10^5$$

$$\chi = \mu_r - 1 = 10^5 - 1$$

$$= 99999$$

Ans.

The magnetic parameters discussed above are analogues to the dielectric parameters. A comparison between electrical and magnetic parameters is shown below in Table 3.2, which plays similar role.

Table 3.2 *Comparisons of analogous electrical and magnetic parameters*

S.No.	*Electrical Parameters*	*Magnetic Parameters*
1.	Electric Field E	Magnetising field H
2.	Electric dipole moment P	Magnetic dipole moment μ
3.	Electric displacement vector D	Magnetic induction B
4.	Permittivity ε	Constant Permeability μ
5.	Relativity permittivity or dielectric constant	Relative permeability μ_r
6.	Electric susceptibility χ_e	Magnetic susceptibility χ_m
7.	Polarisation vector P	Magnetisation vector M
8.	Non-polar dielectric	Diamagnetic
9.	Polar dielectric	Paramagnetic
10.	Ferroelectric hysteresis	Ferromagnetic hysteresis
11.	Spontaneous polarisation	Spontaneous magnetisation
12.	Saturation polarisation	Saturation magnetisation
13.	Remanant polarisation	Remanant magnetisation (retentivity)
14.	Coercive electric field	Magnetic coercivity
15.	Ferroelectric domains	Ferromagnetic domains
16.	Ferroelectric Curie temperature Tc	Ferromagnetic Curie temperature Tc
17.	Converse piezo electricity	Magnetostriction

3.5 CLASSIFICATION OF MAGNETIC MATERIALS

The magnetic materials are classified into three major categories and five in all. These are

- Diamagnetic materials
- Paramagnetic materials
- Ferromagnetic materials
- Antiferromagnetic materials
- Ferrimagnetic materials

The above classification is based on the behaviour of magnetic materials when they are subjected to an external magnetising field.

Diamagnetic material gets feebly magnetised in a direction opposite to the external magnetising field and is characterised by small negative susceptibility. Paramagnetic substances acquire feeble magnetisation in the direction of applied magnetic field and carry a small, but positive susceptibility. Therefore, if a paramagnetic material specimen is freely suspended in a magnetic field, it comes to rest and aligns itself parallel to the magnetic field. This behaviour is not seen in a diamagnetic specimen which sets itself perpendicular to the magnetic field. Ferromagnetic materials are characterised by a high value of susceptibility and get strongly magnetised in the direction of magnetic field. Besides these, Table 3.3 distinguishes the materials on the basis of other magnetic parameters.

Table 3.3 *A comparison of various kinds of magnetic materials*

Property /Parameter	*Diamagnetic*	*Paramagnetic*	*Ferromagnetic*
Response to external magnetising field	Gets feebly magnetised opposite to magnetic field (feebly repelled)	Gets feebly magnetised in the same direction of magnetic field (feebly attracted)	Gets strongly magnetised in the field direction
Status of elementary dipoles in absence of external magnetic field	No permanent dipoles	Permanent dipoles randomly oriented	Permanent dipoles aligned within a domain
Relative permeability μ_r	Slightly less than 1	Slightly greater than 1	Very high
Magnetic susceptibility χ_m	Small negative of order of 10^{-8}	Small positive of order of $\sim 10^8$	Positive and very high
Temperature dependence of magnetic susceptibilty	Independent of temperature	Inversely proportional to temperature $\chi_m = \frac{C}{T}$	Above critical temperature Tc; $\chi_m = \frac{C}{T - Tc}$ Above Tc Ferromagnetic becomes paramagnetic

3.6 CHARACTERISTICS OF DIAMAGNETIC MATERIALS

1. Most substances in nature are diamagnetic.
2. The atoms of a diamagnetic material acquire feeble magnetisation in a direction opposite to external magnetising field.

3. They are analogues to the induced electric fields in electrostatics.
4. In diamagnetic materials, atoms having no permanent magnetic dipole moment acquire induced dipole moments when they are placed in an external magnetic field.
5. In these materials the vector sum of orbital and spin magnetic moment is zero, causing net atomic dipole-moment equal to zero.
6. Such materials exhibit a small and negative magnetic susceptibility which is independent of temperature.
7. Permeability is also small and less than one. Because of this induced field, the material opposes the increase of flux.
8. A diamagnetic specimen sets itself perpendicular to the applied magnetic field.
9. In diamagnetic materials, the electrons in an atom are completely paired, i.e. for each electron spinning in one direction, there is also present an electron spinning in opposite direction. Similar situation is also found in orbital motion of electron.

Table 3.4 gives the relative permeability of some diamagnetic materials.

Table 3.4 *Relative permeability of some diamagnetic materials*

Material	*μ_r(At Room Temperature)*
Mercury	-3.2×10^{-5}
Silver	-2.6×10^{-5}
Bismuth	-1.7×10^{-5}
Ethyl Alcohol	-1.3×10^{-5}
Copper	-9.7×10^{-6}

3.7 CHARACTERISTICS OF PARAMAGNETIC MATERIALS

1. These substances acquire feeble magnetisation in the direction of applied magnetic field.
2. Paramagnetism occurs in those materials whose atoms have permanent magnetic dipole moment.
3. These materials are characterised by small positive value of permeability.
4. Susceptibility is small and positive and varies inversely as temperature.
5. In a paramagnetic specimen, the net atomic dipole moment is zero because the dipoles are randomly oriented in space within the material.
6. A paramagnetic sample, when subjected to external field, aligns itself in the direction of magnetic field.
7. In large applied fields, the magnetisation approaches its maximum value, which occurs when all the dipoles are aligned parallel to the magnetic field.
8. On removing the magnetic field, the atomic dipoles of paramagnetic materials again becomes random. The magnetic force between the atoms is so weak that it is not able to prevent randomisation. This effect can be used to achieve cooling in a process known as '**adiabatic demagnetisation**.'

Table 3.5 gives the relative permeability of some paramagnetic materials.

Table 3.5 *Relative permeability of some paramagnetic materials*

Material	μ_r *(At Room Temperature)*
Chromium	3.3×10^{-4}
Tungsten	6.8×10^{-5}
Aluminium	2.2×10^{-5}
Magnesium	1.2×10^{-5}
Oxygen (1 atm)	1.9×10^{-6}
$CuCl_2$	3.6×10^{-7}

3.8 CHARACTERISTICS OF FERROMAGNETIC MATERIALS

1. These materials are distinguished by a large value of permeability, and therefore are capable of attaining a high degree of magnetisation.
2. These substances are capable of developing north and south poles, when kept in a magnetic field.
3. In these materials, induced magnetisation is not proportional to the magnetising force.
4. Susceptibility decreases steadily with the rise in temperature up to a critical temperature. Above this temperature, ferromagnetic character disappears and the substance becomes paramagnetic.
5. The variation of permeability with temperature for a ferromagnetic material is shown in Fig. 3.3, which shows that near the Curie point loss of magnetic properties is very sudden as μ shows a drastic change.
6. These materials experience a magnificent characteristic known as hysteresis, which is of extreme interest in the industry for deciding the magnetic materials used for various purposes.

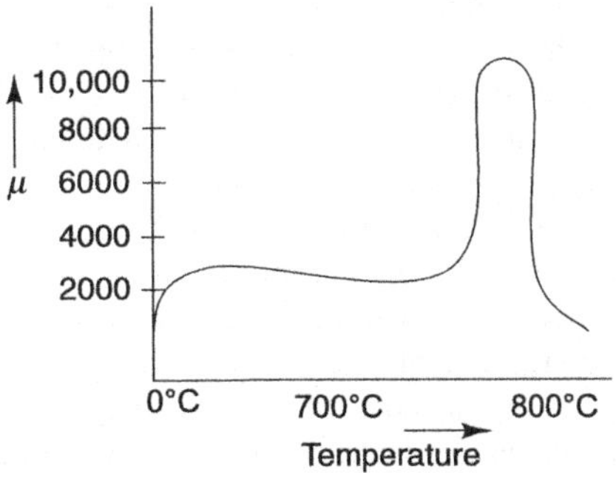

Fig. 3.3 *Variation of permeability with temperature for a ferromagnetic material*

3.9 CHARACTERISTICS OF ANTIFERROMAGNETIC MATERIALS

1. Magnetic moments are aligned antiparallel to each other resulting in a net zero magnetic dipole.
2. Antiferromagnetic property is observed only below a certain temperature known as **Néel temperature**, which is due to ordered antiparallel alignment of spin magnetic moments.
3. Above Néel temperature, antiferromagnetic materials become paramagnetic.

4. This behaviour is exhibited only in very few elements like, Mn, Cr and ferrous and nickel oxides.

3.10 CHARACTERISTICS OF FERRIMAGNETIC MATERIALS

1. These are a special class of ferromagnetic materials.
2. They exhibit all properties of ferromagnetic materials, like high permeability, high saturation magnetisation, hysteresis, etc.
3. They also exhibit spin magnetic moments, which are aligned in the crystal.
4. These are suitable for high frequency applications in special magnetic devices.
5. In ferrimagnetic specimen, all the magnetic moments are not aligned in the same direction. Some of them are oriented in opposite direction, giving rise to the net finite magnetic moment.

3.11 LANGEVIN'S THEORY OF DIAMAGNETISM

This theory was proposed by Langevin in 1905. Before going into further details, let us first talk about an atom. It contains a positively charged nucleus and an electron (charged particle) that revolves around it. In a nutshell, an atom behaves like a 'tiny current loop' which exhibits some magnetic moment that can be calculated. In a material, there are numerous such tiny magnetic moments that are oriented in a random manner. Therefore, the vector sum of magnetic moments of all the atoms cancel each other in the absence of external magnetic field.

As soon as the material is subjected to an external magnetising field, the electronic motion of all the atoms gets affected and the orbit of electron begins a precessional motion (motion of tilted spinning top) without changing the shape of the orbit. This means the motion of the system is the same except that a uniform rotation around an axis parallel to the field is superimposed. Therefore, perfect balance of magnetic moment with the atom gets disturbed. The electronic orbits shift each other a little giving rise to a negative magnetic moment of atom as a whole, and the material is seen negatively magnetised in the presence of magnetic field.

In the absence of external magnetic field, a single electron moving in an atom of diamagnetic material with velocity v_0 under the action of centripetal force F_e

$$F_e = \frac{m_e v_0^2}{r} = m_e \omega_0^2 r \tag{3.13}$$

where ω is angular velocity in the absence of magnetic field.

On applying a uniform magnetic field B perpendicular to the plane of the orbit, an additional magnetic force acting on electron is given by

$$F_m = e\,(v \times B)$$

or

$$F_m = e\,(\omega r B) \tag{3.14}$$

where ω is angular velocity in the presence of magnetic field.

Due to the influence of F_m, the angular velocity would change. Thus, ω and ω_0 are different.

When an external magnetic field is applied perpendicular to the plane of the orbit, then it will induce an emf ε in the loop, given by

$$\varepsilon = \frac{d\phi}{dt} \tag{3.15}$$

This emf results in an electric field E which acts tangentially to the direction of electronic motion. Induced emf may also be expressed as

$$\varepsilon = \oint \overline{E}.\overline{dl} = 2\pi r E$$

Therefore, electric field $E = \dfrac{\varepsilon}{2\pi r}$ (3.16)

The electrons in the atom are accelerated by this field and experience a force e, and therefore velocity of electron changes.

Hence, $$m_e \frac{dv}{dt} = eE = e.\frac{\varepsilon}{2\pi r} = \frac{e}{2\pi r}.\frac{d\phi}{dt} \tag{3.17}$$

Change in the velocity of electron, when flux changes from $0 - \phi$ is given by

$$m_e \int_{v_0}^{v_0+\Delta v} \frac{dv}{dt} dt = \frac{e}{2\pi r} \int_0^{\phi} \frac{d\phi}{dt} dt$$

On integrating and simplifying, we get

$$\Delta v = \frac{e}{2\pi m_e r}.\phi \tag{3.18}$$

Therefore, change in angular velocity of the electron due to magnetic field

$$\Delta\omega_L = \frac{\Delta v}{r} = \frac{e}{2\pi m_e r^2}\phi = \frac{eB.A}{2m_e A}$$

$$\Rightarrow \qquad \boxed{\Delta\omega_L = \frac{e.B}{2m_e}} \tag{3.19}$$

$\Delta\omega_L$ is called *Larmor angular frequency.*

The above change in the angular velocity causes a precessional motion of the orbit without change in its form.

Earlier in the absence of a magnetic field, the angular frequency of electron was $w_{0.}$ It becomes $\omega_0 + \Delta\omega_L$ in the presence of a magnetic field, which is applied perpendicular to the plane of orbit.

If still electron revolves in the same orbit of radius r, the centripetal force must increase. The new centripetal force is given by

$$F' = m_e r(\omega_0 + \Delta\omega_L)^2 = m_e r\omega_0^2 + m_e r\Delta\omega_L^2 + 2m_e r\omega_0.\Delta\omega_L$$

Since $\Delta\omega_L$ is very small ($\Delta\omega_L << \omega_0$), using this approximation

$$F' = m_e r\omega_0^2 + 2m_e r\omega_0\Delta\omega_L \tag{3.20}$$

When magnetic field is applied, an additional force acts on the electron. The magnitude of this force is given by

$$F_B = evB = er\omega.B = e.r.(\omega_0 + \Delta\omega_L)B \tag{3.21}$$

Substituting the value of B from Eq. (3.19) in Eq. (3.21)

$$F_B = e.r.(\omega_0 + \Delta\omega_L).\frac{2m_e}{e}.\Delta\omega_L$$

$$= 2m_e.r\omega_0\Delta\omega_L + 2m_e r\Delta\omega_L^2$$

Neglecting $\Delta\omega_L^2$ as ($\Delta\omega_L << \omega_0$), we get

$$F_B = 2m_e r\omega_0\,\Delta\omega_L \tag{3.22}$$

From equation (3.20)

$$F' - m_e r \omega_0^2 = 2m_e r \omega_0 \, \Delta\omega_L \tag{3.23}$$

On comparing (3.22) and (3.23), we can conclude that "the increase in the centripetal force is completely balanced by additional inward magnetic force." Hence, radius of orbit remains the same.

The magnetic moment of a revolving electron is given by

$$m = n \; i \; A$$

where $n \rightarrow$ number of revolutions per second.

The change in magnetic moment due to change in angular velocity, when the material is subjected to an external magnetising field

$$dm = e.A.\ \Delta n = e.A.\frac{\Delta\omega_L}{2\pi}$$

$$= \frac{e.A.}{2\pi}.\frac{eB}{2m_e} = \frac{e^2.\pi r^2 B}{4\pi m_e}$$

or $$dm = \frac{e^2 r^2 B}{4m_e}$$

Since this is an induced effect, this magnetic moment will always be opposite to the magnetic field, therefore

$$d\overline{m} = -\frac{e^2 r^2 \overline{B}}{4m_e} \tag{3.24}$$

When all atoms are taken into consideration (assuming N molecules per unit volume), the total magnetic moment per unit volume

$$\overline{m} = -\frac{e^2 N.B}{6m_e}\Sigma\overline{r}^2$$

where $r \rightarrow$ projection of r in a plane perpendicular to field.

For spherical symmetry $$\overline{x}^2 = \overline{y}^2 = \overline{z}^2 = \frac{\overline{r}^2}{3}$$

Magnetic susceptibility $$\chi_m = \frac{\text{Magnetic / moment / per unit volume}}{\text{Magentising/field}}$$

$$\boxed{\chi_m = -\frac{e^2 B}{6m_e}\Sigma\overline{r}^2} \tag{3.25}$$

From the above relation, the following conclusions could be drawn:

1. Diamagnetic susceptibility is negative.
2. Diamagnetic susceptibility is independent of temperature.
3. It is an induced effect which disappears as and when the field is removed.
4. It is the intrinsic property of all materials.

3.12 EXPLANATION OF DIA, PARA AND FERROMAGNETISM

(i) Diamagnetism

Those materials whose atoms contain even number of electrons with a pair of opposite spins, so that the magnetic moment of one electron gets cancelled by the other are called diamagnetic materials. As a result, the net magnetic moment of an atom of diamagnetic material is zero.

On subjecting a diamagnetic material to the external magnetising field, magnetic force begins to act upon the electron and its motion gets modified. One electron of each pair is slowed down, while the other gets accelerated. Therefore, the magnetic moments of the electron spinning in the direction is not completely neutralised by the magnetic moment of the other electron having opposite spin. As such the material acquires a small induced magnetic moment in a direction opposite to applied field.

(ii) Paramagnetism

This is found in those materials whose atoms have excess electrons spinning in same direction, and therefore atoms behave like tiny magnets. In the absence of external magnetic field, these tiny magnets or magnetic dipoles are oriented in a random manner in a specimen giving rise to a net zero magnetic dipole moment. Hence, a paramagnetic substance is magnetically neutral. Figures 3.4 and 3.5 show a paramagnetic specimen in the absence and presence of a magnetic field, respectively.

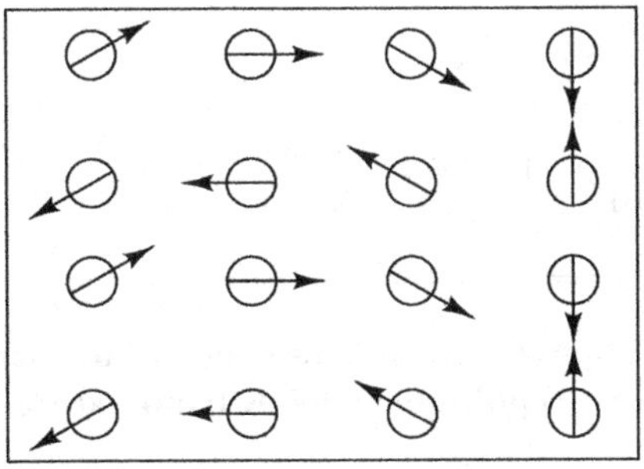

Fig. 3.4 *Paramagnetic specimen in absence of magnetic field*

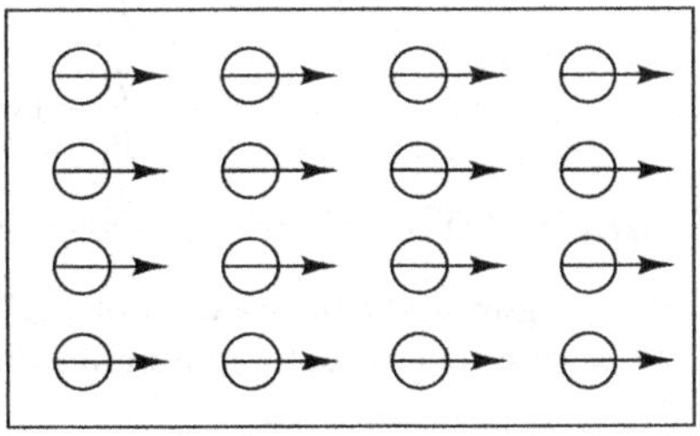

Fig. 3.5 *Paramagnetic specimen in presence of magnetic field*

When a paramagnetic substance is placed in an external magnetic field, each atomic dipole experiences a couple, which tends to orient the magnet in the direction of the magnetic field.

As a result, a paramagnetic substance is magnetised in the direction of the magnetising field, but due to thermal agitation the magnetism of a paramagnetic substance is very weak.

Example 3.13

Find the diamagnetic susceptibility of He, assuming that the two electrons are contributing to its diamagnetism. Consider the mean radius of the atom as 0.6 Å and $N = 28 \times 10^{26}$ per m^3.

Solution Given $N = 28 \times 10^{26}$ per m^3 and $R = 0.6 \times 10^{-10}$ m

Susceptibility of diamagnetic material

$$\chi_{\text{dia}} = -\frac{\mu_0 Ze^2 NR^2}{6m}$$

$$= -\frac{(4\pi \times 10^{-7}) \times 2(1.6 \times 10^{-19}) \times 28 \times 10^{26} \times (0.6 \times 10^{-10})^2}{6 \times 9.1 \times 10^{-31}}$$

$$= -11.872 \times 10^{-8}$$ **Ans.**

Example 3.14

A magnetising field of 2000 A/m produces a magnetic flux of 2×10^{-5} Wb in a bar of iron of 0.2 cm^2 cross-sections. Calculate permeability and susceptibility of the bar.

Solution Given $H = 10^3$ A/m, $\phi = 2 \times 10^{-5}$ Wb and $A = 0.2 \times 10^{-4}$ m^2.

Magnetic flux density $$B = \frac{\phi_m}{A}$$

Permeability $$\mu = \frac{B}{H}$$

Susceptibility $$\chi_m = \mu_r - 1 = \frac{\mu}{\mu_0} - 1$$

$\therefore$ $$B = \frac{\phi}{A} = \frac{2\times10^{-5}}{2\times10^{-5}} = 1.0 \text{ Wb/m}^2$$

$$\mu = \frac{B}{H} = \frac{1}{2\times10^{3}} = 0.5\times10^{3} \text{ N/A}^2$$

and $$\chi_m = \frac{\mu}{\mu_0} - 1 = \frac{0.5\times10^{-3}}{4\pi\times10^{-7}} - 1 = 3924$$ **Ans.**

Example 3.15

A paramagnetic substance contains 6.5×10^{25} atoms per m^3 and the magnetic moment of each atom is one Bohr magneton. Find the susceptibility at 300 K temperature.

Solution Given $N = 6.5 \times 10^{25}$ atoms/m^3.
T corresponding to room temperature = 27 + 273 = 300 K

Susceptibility $$\chi_m = \frac{\mu_0 N M^2}{3kT} \quad (3.26)$$

The magnetic moment of each atom

$$M = n\left(\frac{eh}{4\pi m}\right)$$

$$= 1\left(\frac{1.6\times10^{-19}\times6.6\times10^{-34}}{4\times3.14\times9.1\times10^{-31}}\right)$$

$$= 9.24 \times 10^{-24} \text{ Am}^2$$

From Eq. 3.26, we get

$$\chi_m = \frac{(4\pi\times10^{7})\times6.5\times10^{25}\times(9.24\times10^{-24})^2}{3\times1.38\times10^{-23}\times300}$$

$$= 5.612099 \times 10^{-7}$$
$$= 5.612 \times 10^{-7}$$ **Ans.**

(iii) Ferromagnetism—Domain Theory

The magnetism attained by a ferromagnetic substance in an external magnetic field is intense. Like paramagnetic substances, atoms of ferromagnetic substances are magnetic dipoles, having a permanent magnetic moment. But, in a ferromagnetic material the atoms experience strong mutual interactions, due to which even in the absence of an external field, the magnetic moments of nearby atoms gets aligned in the same direction.

As a result, in a ferromagnetic substance innumerable small regions are formed in which the magnetic moments of magnetic dipoles are aligned in the same direction. These small regions are called domains. Each domain consists of 10^{27} to 10^{21} atoms and has maximum linear dimension of 10^{-5} m. Each domain is itself in a stage of magnetic saturation. In the absence of any external magnetic field, different domains are randomly oriented in such a way that the net magnetic moment of the substance is zero. That is why a piece of ferromagnetic substance is magnetically neutral or unmagetised. Figure 3.6 represents the probable directions of magnetic moments of four domains of a ferromagnetic substance (iron).

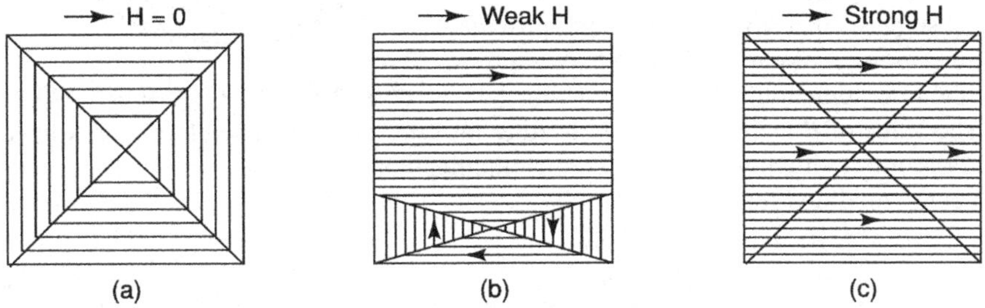

Fig. 3.6 *Probable directions of magnetic moments of four domains of iron*

When the substance is placed in an external magnetic field, its net magnetic moment increases. There could be two possible reasons

- ❖ **By Displacement of Boundaries of Domains** The domains having magnetic moments in the direction of magnetic field increase in size, while those having magnetic moments oppositely directed decreased in size. This occurs when magnetic field is weak.

- ❖ **By Rotation of Domains** All domains rotate in such a way that their magnetic moments get aligned in the field direction (see Fig. 3.6). This occurs when the magnetic field is strong.

3.13 DEMAGNETISATION

If we beat a magnet with a hammer or pass an alternating current or heat it above Curie temperature (*Tc*), then it loses its magnetism. It is because in any of the above-mentioned processes, the random motion of a atoms or molecules of a substance increases. Therefore, the domain having magnetic moment in the direction of external magnetic field is totally disturbed and the net magnetic moment of the magnet becomes zero, i.e. the substance is demagnetised.

3.14 AMPÈRE'S LAW IN THE PRESENCE OF MAGNETIC MATERIAL: ($\vec{H}$, $\vec{B}$ AND $\vec{M}$ VECTOR)

So far, we have discussed that if a magnetic material is placed in a magnetising field, it modifies the original field. This happens because each dipole present inside the material tries to align itself along the direction of applied field. As a result it modifies the original field.

Let us now apply the Ampère's law to a toroid filled with soft iron, which contains no turns, the current being I_0, radius r_0. To find the field at any point on the dotted circular path, we have to apply Ampère's law

$$\oint_c \bar{B}d\bar{l} = \mu_0 I \quad (3.27)$$

$I \rightarrow$ current enclosed by the Amperian path

i.e. $B.\ 2\pi r_0 = \mu_0 N_0 I_0$

$d\bar{l} \rightarrow$ small elements of differential paths which suggests that B will be constant, provided that the strength of current is not changing. But, the reality is something different.

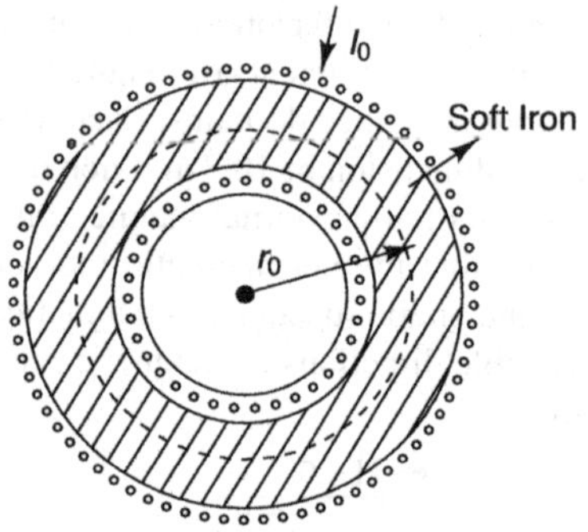

Fig. 3.7 *Toroid filled with soft iron*

The value of B increases due to inserting the core material. Therefore, our Ampère's law fails at this situation and needs some modification.

Let the modified form is

$$\oint \bar{B}.d\bar{l} = \mu_0(I + I_m)N_0 \quad (3.28)$$

$$= \mu_0 N_0 I + \mu_0 N_0 I_m \quad (3.29)$$

The term I_m naturally seems to be related to magnetisation. It is equal to that amount of current which has a magnetic dipole moment equal to magnetic moment produced due to alignment of the dipoles.

Consider a small element dl.

The magnetic moment of this element $d\bar{m} = \bar{M}.(Adl)$ (3.30)

But magnetic moment = NIA

i.e. $$NIA = \frac{N_0}{2\pi r_0} dl.I_m.A \quad (3.31)$$

$$\bar{M}.A.d\bar{l} = \frac{N_0}{2\pi r_0} dl.I_m.A \quad (3.32)$$

$$2\pi r_0 \bar{M} = N_0\, I_m$$

Substituting this value in relation (3.27)

$$\oint \bar{B}d\bar{l} = \mu_0 N_0 I_0 + \mu_0 N_0\, I_m$$

$$= \mu_0 N_0 I_0 + \mu_0\, 2\pi r_0\, M$$

$$2\pi r_0.\bar{B} - \mu_0\, 2\pi r_0 M = \mu_0 N_0 I_0$$

$$2\pi r_0.(\bar{B} - \mu_0.\bar{M}) = \mu_0 N_0 I_0$$

$$\frac{(\bar{B} - \mu_0 \bar{M})}{\mu_0} 2\pi r_0 = I_0 N_0$$

$$\oint \left(\frac{B - \mu_0 \bar{M}}{\mu_0}\right) dl = N_0 I_0 \qquad \boxed{\oint \bar{H}.d\bar{l} = N_0 I_0}$$

Restatement of Ampère's Law

"Line integral of magnetising field vector over a closed path is equal to the actual current enclosed by the paths."

Where $$H = \frac{\bar{B} - \mu_0 \bar{M}}{\mu_0}$$

$$\mu_0 \bar{H} = \bar{B} - \mu_0 \bar{M}$$

$$\boxed{\overline{B} = \mu_0(\overline{H} + \overline{M})} \tag{3.33}$$

Example 3.16

An iron rod of 2.0 m length and 4 sq. cm cross-section is in the form of a closed ring. If the permeability of iron is 50 × 10^{-4} Hm^{-1}, show that the number of ampere turns required to produce a magnetic flux of 4 × 10^{-4} Wb through the closed ring is 200.

Solution Given $L = 2.0$ m, $A = 4 \times 10^{-4}$ m^2, $\mu = 50 \times 10^{-4}$ H/m and $\phi = 4 \times 10^{-4}$ Wb

Magnetic flux density $B = \dfrac{\phi}{A}$

$$= \frac{4\times10^{-4}}{4\times10^{-4}} = 1.0 \text{ Wb/m}^2$$

Also, $B = \mu NI$

∴ ampere turn $NI = \dfrac{B}{\mu}$

$$= \frac{1.0}{50\times10^{-4}} = 200 \text{ A/m}$$ **Ans.**

Example 3.17

A bar magnet has a coercivity of 5 × 10^3 A/m. It is desired to demagnetise it by inserting it inside a 10 cm long solenoid having 500 turns. What current should be sent through the solenoid?

Solution

Here, coercivity $H = 5 \times 10^3$ A/m, $l = 10$ cm and total turns = 50.

Turns per metre $N = 500 / 10 = 5000$ turns / m

Now, $H = Ni$

$5 \times 10^3 = 5000 \times i$

or $i = 1.0$ A **Ans.**

Example 3.18

An iron rod of 50 cm length and 6 sq. cm cross-section area is in the form of a circular ring. If the permeability of iron is 65 × 10^{-4} H/m, compute the number of turns required to produce a flux of 4 × 10^{-5} Wb.

Solution Here, $A = 6$ cm^2 $= 6 \times 10^{-4}$ m^2, $l = 50$ cm $= 0.5$ m,

$\mu = 65 \times 10^{-4}$ H/m and $\phi = 4 \times 10^{-5}$ Wb

As $H = \dfrac{N_i}{l}$ or $N_i = Hl$

$$B = \frac{\phi}{A} = \frac{4\times10^{-5}}{6\times10^{-4}} = 0.06 \text{ Wb/m}^2$$

∵ $H = \dfrac{B}{\mu} = \dfrac{0.06}{65\times10^{-4}} = 9.23$ A turns/m

Number of turns = $N = Hl = 9.23 \times 0.5 = 4.6$ **Ans.**

Example 3.19

A magnetising field of 500 Am^{-1} produces a magnetic flux of 2.4 × 10^{-5} Weber in an iron bar of 0.2 cm^2 cross-sectional area. Compute the permeability and susceptibility of the bar.

Solution: Given $\phi = 2.4 \times 10^{-5}$ Wb and $A = 0.2$ cm^2 = 0.2×10^{-4} m^2.
The magnetic flux density is given by

$$B = \frac{\phi}{A} = \frac{2.4\times10^{-5}}{0.2\times10^{-4}} = 1.2 \text{ NA}^{-1}\text{m}^{-1}$$

The permeability is given by

$$\mu = \frac{B}{H}$$

or

$$\mu = \frac{1.2}{500} = 2.4\times10^{-3} \text{ N/m}$$

The susceptibility is given by

$$f_m = \frac{\mu}{\mu_0} - 1$$

$$= \frac{2.4\times10^{-3}}{4\times3.14\times10^{-7}} - 1 = 1910 - 1$$

$$= 1909$$

Ans.

3.15 HYSTERESIS AND HYSTERESIS LOSS

Ferromagnetic materials exhibit a remarkable characteristic which is very interesting and useful in the industry during the selection of magnetic materials for various purposes. It is observed that when a magnetic material is carried out under a process of magnetisation, the flux density B, as well as magnetisation vector M, lag behind magnetising field H. This is known as hysteresis.

The energy spent in the process of aligning the elementary dipoles inside the material is not completely recoverable and appears in the form of heat in the specimen. The amount of heat appeared is found to be equal to area enclosed by B-H curve and is called hysteresis loss.

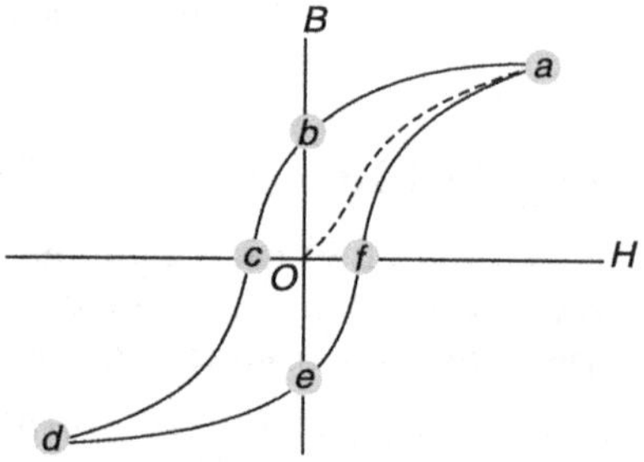

Fig. 3.8 *B-H curve*

Hysteresis loss = $\oint H.dB$ = Area of B-H curve

(i) Calculation of Hysteresis Loss

Consider a magnetic material in the toridal shape. The number of turns in the toroid being N and I be the current flowing through it.

Using Ampère's law,

$$\oint H.dB = NI$$

or

$$H.l = NI \tag{3.34}$$

Let us calculate the amount of work done in establishing a current I in the toroidal coil that produces magnetising field H and gives rise to flux density B.

Flux linked with the coil $\phi = NAB$ (3.35)

When current increases, the flux linked with the coil also increases, and therefore gives rise to induced emf.

Induced emf $$e = -\frac{d\phi}{dt} = -\frac{d}{dt}(NBA) = NA\frac{dB}{dt} \quad (3.36)$$

Rate of doing work $$\frac{dW}{dT} = e.I = NAI\frac{dB}{dt} = HlA\frac{dB}{dt} \quad (3.37)$$

Rate of doing work per unit volume of the specimen, where volume = lA

$$\frac{dW}{dt} = H.\frac{dB}{dt}$$

$$\Rightarrow \quad dW = H.dB \quad (3.38)$$

Therefore, energy transferred per unit volume per cycle

$$= \oint H.dB = \text{Area of } B\text{-}H \text{ loop}$$

$$= \frac{1}{2}B.H. = \frac{1}{2}\mu H^2$$

A similar feature is also observed if the magnetisation vector $\overline{M}$ is plotted vs. $\overline{H}$.

We know that

$$B = \mu_0 H + \mu_0 M$$

On differentiating,

$$dB = \mu_0 H + \mu_0 dM$$

and $$HdB = \mu_0.H.dH + \mu_0\, dM$$

On integrating over a complete cycle

$$\oint H.dB = \mu_0 \oint H.dM$$

Area of B-H curve = μ (Area of M-H curve) (3.39)

(ii) Core Loss

The loss due to eddy currents along with hysteresis loss is known as core loss.

3.16 LANGEVIN'S THEORY OF PARAMAGNETISM

In paramagnetic substances each atom exhibits a permanent magnetic dipole moment. But, because of random distribution, the net dipole moment as a whole in a material is found to be zero. On subjecting the specimen to external magnetising field, this random distribution of magnetic dipoles tend to convert into a partially aligned one and is maximum at a low temperature for a given field. The magnetic susceptibility of many paramagnetic materials is found inversely proportional to the temperature for ordinary fields.

i.e. $$\chi_m = \frac{C}{T} \text{ .... Curie's law}$$

which is not valid for higher fields.

(i) Explanation

Maxwell–Boltzmann statistics was used by Langevin to provide explanation of paramagnetism. He assumed that each atomic dipole experiences a force only due to external field B. If a given atomic dipole is making an angle θ with B, its magnetic potential energy is given by

$$W = -m.B = -mB\cos\theta \tag{3.40}$$

when m is the magnetic moment of elementary atomic dipole.

The number of atoms / molecules per unit volume lying in the energy range W and $W + dw$ is

$$dn = Ce^{-W/kT}\,dW \tag{3.41}$$

where C is a constant to be determined. From Eq. (3.40),

$$dW = mB\sin\theta d\theta$$

Therefore, Eq. (3.41) becomes

$$dn = Ce^{-W/kT}.mB\sin\theta d\theta \tag{3.42}$$

On integrating (3.42) for all possible orientations $(0 \rightarrow \pi)$,

total number of atoms or molecules per unit volume of the substance

$$n = \int dn = \int_0^\pi Ce^{mB\cos\theta/kT}.mB\sin\theta d\theta$$

and, therefore

$$C = \frac{n}{\int_0^\pi e^{mB\cos\theta/kT}.mB\sin\theta d\theta} \tag{3.43}$$

The component of each dipole along magnetic field direction is $m\cos\theta$, and therefore the resultant magnetic moment due to all dn atoms or molecule along the field will be $dn.m\cos\theta$

Therefore, intensity of magnetisation (magnetic moment per unit volume),

$$M = \int_0^\pi m\cos\theta.dn$$

where $dn \rightarrow$ number of atoms present in the elemental area under consideration.

$$= C\int_0^\pi m^2 Be^{mB\cos\theta/kT}.\sin\theta\cos\theta d\theta$$

On substituting value of C in the above integral,

$$M = nm\frac{\int_0^\pi e^{mB\cos\theta/kT}.\sin\theta\cos\theta d\theta}{\int_0^\pi e^{mB\cos\theta/kT}.\sin\theta d\theta}$$

Putting $\frac{mB}{kT} = x$ and $\cos\theta = y$, we get

$$M = nm\frac{\int_{-1}^{+1} e^{xy}ydy}{\int_{-1}^{+1} e^{xy}dy}$$

On simplifying,

$$M = M_s\left[\coth x - \frac{1}{x}\right] \tag{3.44}$$

where $M_s = n.m$

Ms is the saturation value, when all the atoms are aligned in the field direction.

i.e. $M = M_s.L(x)$

$L(x)$ is named as Langevin's function. The plot of $\frac{M}{M_s}$ vs. $L(x)$ is shown in Fig. 3.9 which shows that its value is unity (all the dipoles are aligned) and for ordinary magnetic fields it is $\frac{x}{3}$

i.e. $$M = M_s.\frac{x}{3} = \frac{n.m^2 B}{3kT}$$

$$\left\{\begin{array}{l}\sin ce\ M_s = mn \\ \text{and } x = \dfrac{mB}{kT}\end{array}\right.$$

Therefore, $$\overline{M} = \frac{n.m^2}{3kT}.\overline{B} = \frac{nm^2\mu_0 H}{3kT}$$

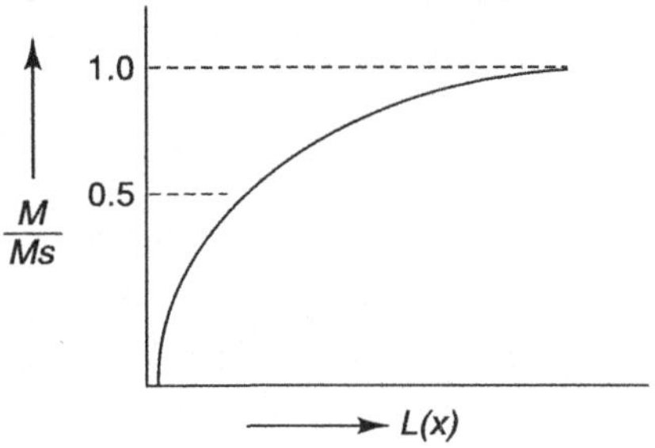

Fig. 3.9 *M/Ms vs. L(x)*

The volume susceptibility $\chi_m = \frac{M}{H} = \frac{n^2m^2\mu_0}{3nkT} = \frac{Ms^2\mu_0}{3nkT}$ (3.45)

$$\chi_m = \frac{C}{T}$$

where $C = \frac{\mu_0 M_s^2}{3nk}$; known as Curie constant.

Therefore, paramagnetic susceptibility varies inversely as absolute temperature.

(ii) Limitations of Langevin's Theory

The following are some limitations of the above theory

1. It fails to explain a relationship between paramagnetism and ferromagnetism.
2. It is unable to explain why several paramagnets, like highly compressed and cool gases, and highly concentrated salt solution do not obey the above-said Curie law.
3. It applies strictly to gases where molecules are far apart with almost negligible interaction between their molecules.

3.17 MODIFICATION IN LANGEVIN'S THEORY (CURIE–WEISS LAW)

The modification in the Langevin's theory was proposed by Weiss in 1907 to overcome the above controversies. He suggested a new concept of internal molecular field.

According to Weiss, there must exist a molecular field (internal) at the site of atomic dipoles by all the neighbouring molecules which is proportional to the magnetisation vector M, i.e.

Molecular field $= \alpha M$

Therefore, the actual field acting upon the dipole is the sum of the applied field and the ones arising from all the neighbouring dipoles, i.e.

$$H_i = H + \alpha M \quad (3.46)$$

where α is the molecular field coefficient.

From Langevin's theory

The magnetisation vector $M = \frac{nm^2\mu_0 H_i}{3KT}$ (3.47)

H is replaced by H_i.

$$M = \frac{\mu_0 nm^2(H + \alpha M)}{3kT} = \frac{\mu_0 Ms^2(H + \alpha m)}{3nkT}$$

Therefore, volume susceptibility

$$\chi_m = \frac{M}{H} = \frac{\mu_0 nm^2}{3kT} + \frac{\mu_0 nm^2 \alpha}{3kT} \chi_m$$

$$\chi_m \left[1 - \frac{\mu_0 nm^2 \alpha}{3kT}\right] = \frac{\mu_0 nm^2}{3kT}$$

$$\chi_m = \frac{\dfrac{\mu_0 nm^2}{3kT}}{1 - \dfrac{\mu_0 nm^2}{3kT}.\alpha} = \frac{\dfrac{\mu_0 nm^2}{3k}}{T - \dfrac{\alpha \mu_0 nm^2}{3k}} = \frac{C}{T - \textcircled{H}}$$

i.e.
$$\chi_m = \frac{C}{T - \textcircled{H}} \tag{3.48}$$

Equation (3.48) represents Curie–Weiss law, where $\textcircled{H}$ is known as ***Curie temperature***. The above equation is valid only for $T > \textcircled{H}$. For most paramagnetic materials, the Curie temperature $\textcircled{H}$ is so low that even the Curie law $\chi_m = \frac{C}{T}$ is valid at room temperature.

3.18 ANTI-FERROMAGNETISM AND NÉEL TEMPERATURE

In some materials, when the distance between the interacting atoms is small, the exchange forces produce a tendency for antiparallel alignment of electron spins of neighbouring atoms. Such kind of interaction is also found in ferromagnetic materials. It is quite interesting to note that the prediction of antiferromagnetic materials were done well in advance as a result of theoretical study, and some years later these materials were discovered experimentally. In this series, a large number of properties were predicted by Néel in 1930s and for the first time antiferromagnetism was discovered in 1938, by Bizette, Squire and Tsai. These are the materials which are associated with sharp maxima in the susceptibility vs. temperature plot as given in Fig. 3.10.

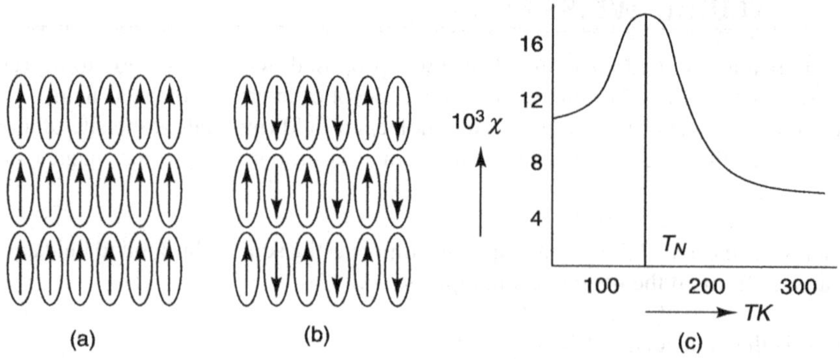

Fig. 3.10 *(a) Alignments of spins in ferrromagnetic material (b) Alignment of spins in antiferromagnetic materials (c) Susceptibility vs. temperature in MnF_2*

The temperature at which this maxima occurs is known as Néel temperature. Above T_N material becomes paramagnetic. The Néel temperature plays the same role in antiferromagnetic materials as Curie temperature in ferromagnetic materials. Therefore, above Néel temperature, the susceptibility follows the relation

$$\chi = \frac{C}{T + T_N}$$

The reciprocal of susceptibility vs. temperature is discussed in Fig. 3.11, showing a comparison between para, ferro and antiferromagnetic materials.

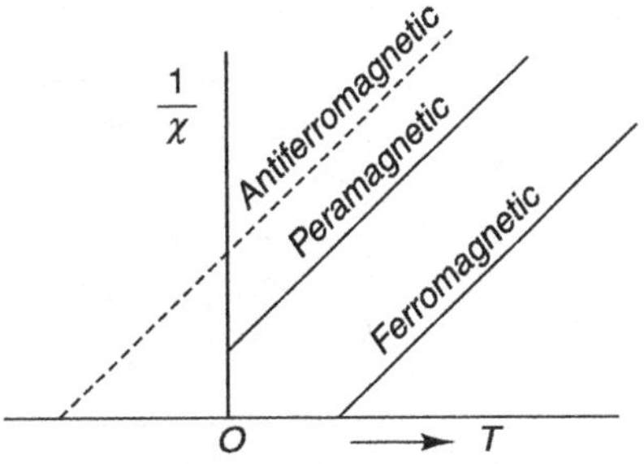

Fig. 3.11 *Reciprocal of susceptibility vs. temperature*

3.19 FERRIMAGNETIC MATERIALS

The term ferrimagnetism was originally used to describe ferromagnetic parallel spins, combined with some antiparallel spins, in such a way that the net magnetic moment is not cancelled totally as in antiferromagnetic materials, as shown in Fig. 3.12.

Néel described this type of spin arrangement for an important class of magnetic oxides known as ferrites. These materials are of great interest to an electrical engineer. They exhibit all properties of ferromagnetic materials along with some additional marvellous characteristics, which make them highly intriguing. Their *d-c* resistivity is quite high as compared to ferromagnetic materials. Also, they behave like a semiconductor from conductivity point of view.

The reciprocal of susceptibility vs. temperature is given in Fig. 3.13 for a ferromagnetic material.

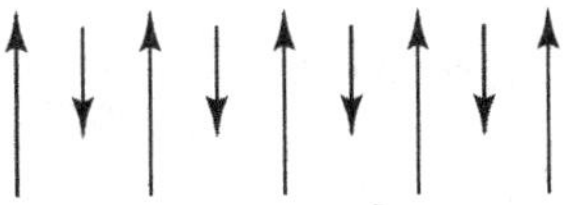

Fig. 3.12 *Alignment of dipoles in ferrimagnetic specimen*

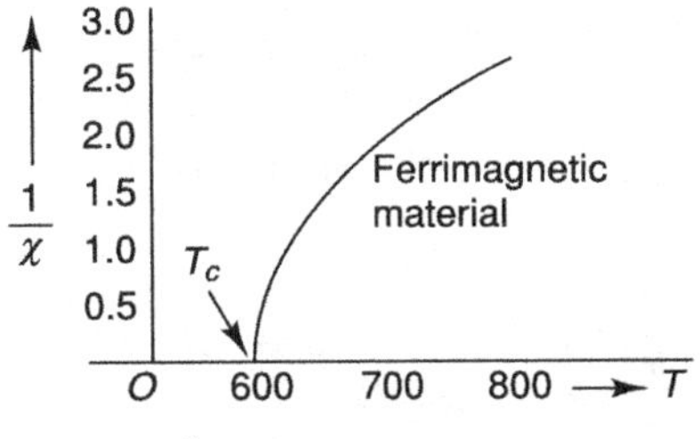

Fig. 3.13 $\frac{1}{\chi}$ *vs. T plot for a ferrimagnetic materials*

On the basis of certain excellent properties, these are used in various devices as shown in Fig. 3.14.

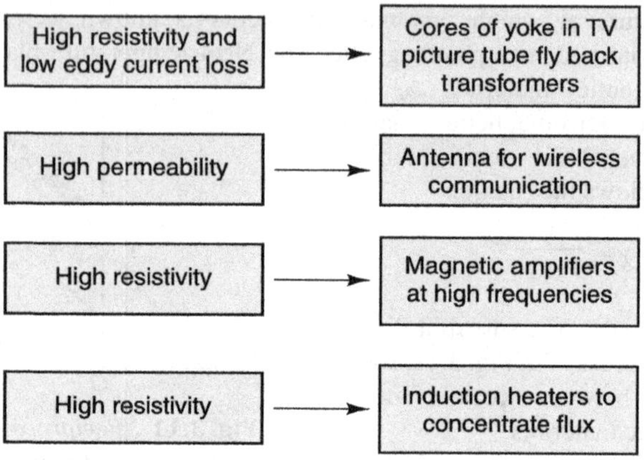

Fig. 3.14 *Use of ferrimagnetic materials*

Other applications of ferromagnetic materials are given in the preceding section.

3.20 PROPERTIES OF SOME IMPORTANT MAGNETIC MATERIALS

This following equation is valid in a ferromagnetic region.

When T < (H), the magnetisation (M) of ferromagnetic materials is stress-dependent. For example, a compressive stress increases the magnetisation M for nickel, while a tensile stress reduces M, and therefore magnetic moment μ also increases. In nickel–iron alloys, a tensile stress increases the M or μ.

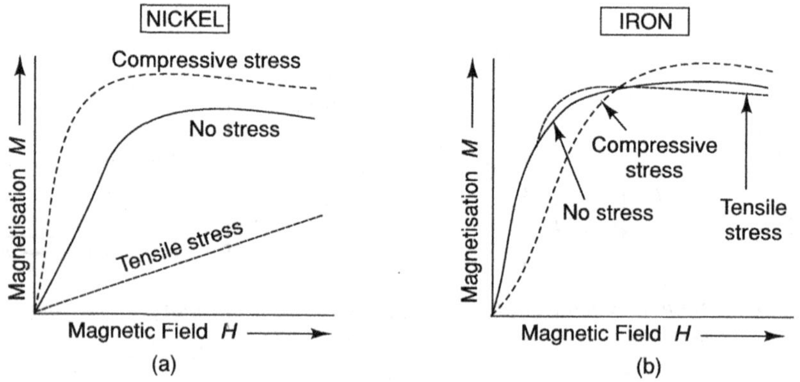

Fig. 3.15 *Schematic representation of the stress dependence on (a) nickel, and (b) iron*

The most important ferromagnetic materials are iron (Fe), cobalt (Co), nickel (Ni) and gadolinium (Gd). The ferromagnetic properties of the transition elements Fe, Co and Ni are due to the spins of the inner unpaired electrons. In Fe, Co and Ni the unpaired inner 3d electrons are responsible for the ferromagnetism. The spins of the 3d electrons of adjacent atoms align in a parallel direction by a phenomenon called spontaneous magnetisation, Ms. Table 3.6 shows the electronic configuration, M_s and (H) of ferromagnetic materials.

Table. 3.6 *Electronic configuration of ferromagnetic materials*

Element	*Electronic configuration*	M_s *at 0 K Amp. turn/m*	*(H) K*
Fe	$3d^6 4s^2$	1.69	1043
Co	$3d^7 4s^2$	1.36	1404
Ni	$3d^8 4s^2$	0.47	631
Gd	$4f^7 5d^1 6s^2$	5.66	289

The parallel alignment of atomic magnetic dipoles occurring in microscopic regions are called the magnetic domain. The parallel alignment of the magnetic dipoles of atoms of Fe, Co and Ni is due to the creation of energy known as exchange interaction.

Exchange interaction energy is the net gain in energy $E_{\text{unmagnetised}} - E_{\text{magnetised}}$ and is a function of the ratio of the atomic diameter to the 3*d* orbital diameter (D/d). For parallel alignment to occur, this ratio must be in the range between 1.4 is 2.7, as shown in Fig. 3.16.

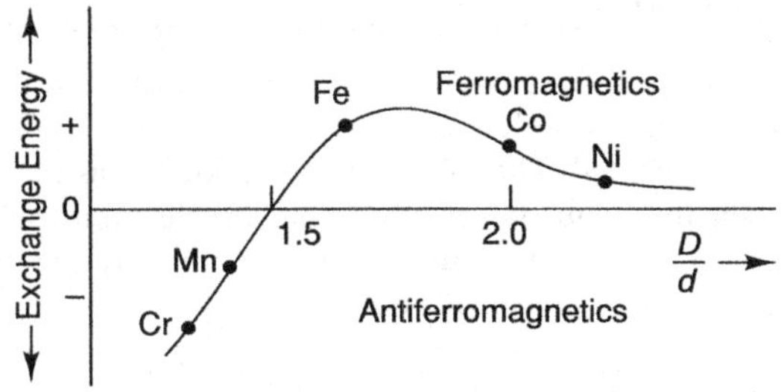

Fig. 3.16 *Parallel alignment*

The exchange energy is positive among the common metals, only, Fe, Co and Ni have positive exchange energy. *Mn* is antiferromagnetic, but on alloying, the interatomic distances are changed and the exchange energy becomes positive and the spins of neighbouring atoms can be aligned. If *Mn – Mn*, distance is increased such that D/d falls in the range 1.4 to 2.7, for example, in Heusler alloys, Cu_2MnAl and Cu_3MnSn, which are ferromagnetic.

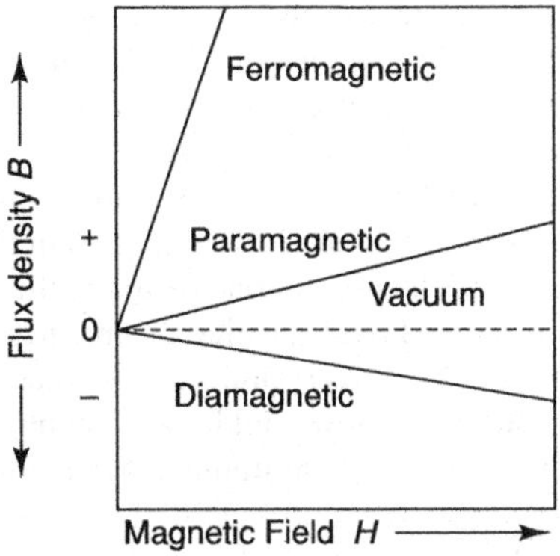

Fig. 3.17 *B vs. H plot comparison between dia, para and ferromagnetic materials*

The difference between dia, para and ferromagnetic material is shown in Fig. 3.17. Ferromagnetic materials can be sub-classed as

- Antiferromagnetic materials
- Ferrimagnetic materials

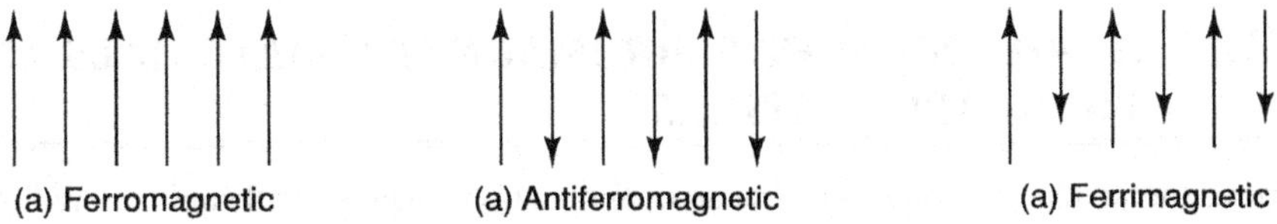

Fig. 3.18 *Sub-classes of ferromagnetic materials*

Coupling of electron spins in atoms is denoted by arrows. The length of an arrow is a measure of the magnetic moment of atom.

When single crystals of ferromagnetic materials are magnetised by applying external magnetic field, the crystal gets easily magnetised in certain preferred crystallographic directions. For example, iron crystal, which is BCC, is easily magnetised in the <100> direction. Easily magnetised means that it attains saturation magnetisation at a much lower applied magnetic field. Such a direction in which a crystal can be easily magnetised is called the easy direction. Other directions in which it is more difficult to magnetise a crystal are called hard directions.

For nickel (which is FCC), the easy direction is <111>, and for cobalt (which is HCP) the easy direction is <001>.

3.21 MAGNETOSTRICTION AND MAGNETOSTRICTIVE MATERIALS

Similar to dielectrics, when a ferromagnetic crystal is magnetised in the easy direction of magnetisation, the crystal gets elongated in that direction. If it is magnetised in the hard direction, it gets contracted. This phenomenon is known as magnetostriction. This happens because when the magnetic electron spin dipole moments of the atom in a solid are rotated into alignment, the length of the bonds between the atoms changes. The fields of the dipole themselves affect the atomic spacing as they may attract or repel each other. Therefore, the shape and volume of a ferromagnetic solid changes. Although the change in dimensions is quite small, yet it plays an important role in domain geometry and in practical use of transformer material. For example, iron gets elongated when magnetic field is applied in the <100> direction and gets contracted when the field is applied in the <111> direction. The longitudinal strain in the direction of the applied field is called the magnetostriction constant. In polycrystalline iron there will be elongation at low fields and contraction at high fields. The longitudinal strain for various materials is shown in Fig. 3.19.

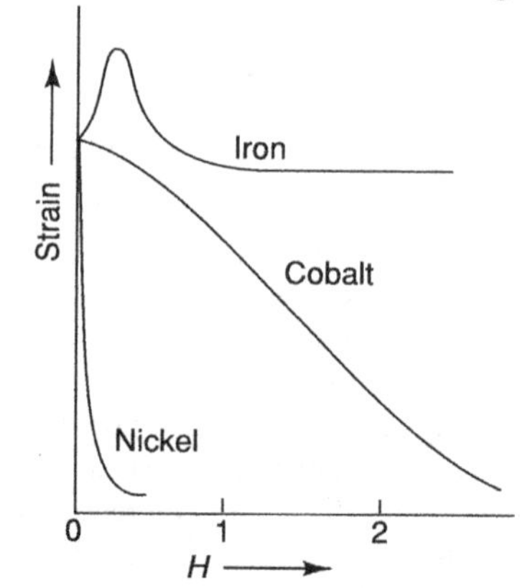

Fig. 3.19 *Longitudinal strain vs. magnetic field*

When a long rod of ferromagnetic material, with axis along the easy direction, is subjected to alternating field, the rod vibrates at twice the frequency of the applied ac field. This is because the rod gets magnetised and demagnetised twice in one cycle of the applied ac field. When the rod vibrates, it produces acoustic waves at the vibrating frequency. This principle is made use of in the generation of ultrasonic waves.

3.22 HARD AND SOFT FERROMAGNETIC MATERIALS AND THEIR APPLICATIONS

Ferromagnetic materials have wide range of applications in almost all engineering fields. Depending on their characteristic parameter values, these materials are classified as soft and hard ferromagnetic materials.

(i) Soft Ferromagnetic Materials

Soft ferromagnetic materials are those that can be easily magnetised and demagnetised. This implies that a soft ferromagnetic material has low coercive field. In many applications, low coercive field and high saturation magnetisation are desirable. Soft magnetic materials satisfy this condition. The hysteresis loop for a soft magnetic material will be thin and long, as shown in Fig. 3.20(a). Consequently, the energy loss (which is equal to the area of the hysteresis loop) is very low for a soft material. Examples of soft magnetic materials are iron-silicon alloys (3–4% or Si), iron-cobalt-manganese alloy, etc. These materials are used in transformer cores, motors, generators, sensors, etc.

(ii) Hard Magnetic Materials

Hard magnetic materials are those that are difficult to magnetise and demagnetise. They are used in applications where permanent magnets with high magnetic fields are required. A hard magnetic material has high coercive field and hence, a wide hysteresis loop as shown in Fig. 20(b). Examples of hard magnetic materials are alnico, rare earth metal alloys with Mn, Fe, Co, Ni, etc.

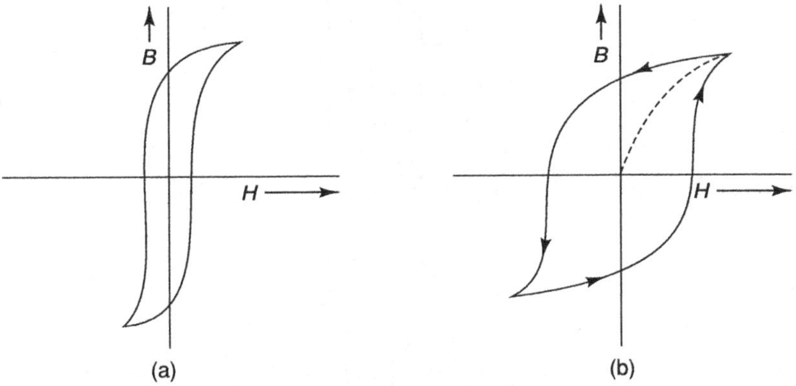

Fig. 3.20 *Hysteresis loop for a (a) soft ferromagnetic material, and (b) hard ferromagnetic material*

3.23 FERRIMAGNETIC MATERIALS OR FERRITES

Ferrimagnetic materials are a special class of ferromagnetic materials. They exhibit all the properties of ferromagnetic materials like, high permeability, saturation magnetisation, hysteresis, etc. They are different from ferromagnetic materials in the way the spin magnetic moments are aligned in the crystal. They are of great practical interest because they have special characteristics which make them highly suitable for high frequency applications in special magnetic devices. The general electric and magnetic characteristics of ferrites are given below:

1. They exhibit a high value of resistivity, normally more than 10^5 ohm cm.
2. The dielectric constant of ferrites is of order of 10^{-12}.
3. They exhibit extremely low dielectric loss.
4. Permeability in quite large.
5. They exhibit appreciable saturation magnetisation, but smaller than ferromagnetic materials.
6. They exhibit low coercivity.

7. Curie temperature varies from 100°C to several hundred 0°C.
8. They are mechanically hard, brittle and difficult to machine.

The special feature of ferrites is that in the magnetised state all the spin magnetic moments are not oriented in the same direction. Some of them are oriented in the opposite direction. But, since the spin magnetic moments are of two types with different values, the net magnetic moment will be finite.

Ferrites have the molecular formula $Me^{2+}O.Fe_2^{3+}O_3$, where Me stands for a divalent metal such as Fe, Mn, Co, Ni, Cu, Mg, Zn or Cd. The crystal structure of ferrite is the inverse spinel structure.

The net spin *S* due to the divalent ions may be calculated by the electronic configuration and using Hund's rule, which is given in Table 3.7.

Table 3.7 *Calculation of net spin*

Ion	*Inner core electronic configuration*	*Spin orientation*	*S*	*Magnetic moment*
Mn^{++}	$3d^5$	↑ | ↑ | ↑ | ↑ | ↑	5/2	$5\mu_B$
Fe^{++}	$3d^6$	↑↓ | ↑ | ↑ | ↑ | ↑	2	$4\mu_B$
Co^{++}	$3d^7$	↑↓ | ↑↓ | ↑ | ↑ | ↑	3/2	$3\mu_B$
Ni^{++}	$3d^8$	↑↓ | ↑↓ | ↑↓ | ↑ | ↑	1	$2\mu_B$
Cu^{++}	$3d^9$	↑↓ | ↑↓ | ↑↓ | ↑↓ | ↑	1/2	$1\mu_B$

(i) Magnetic Materials for Memory Storage Devices

The features of hysteresis curve of ferrite are very unique. Itis square in shape as shown in Fig. 3.21.

The demand for low-priced technology for information storage is continuously increasing. Traditionally, high permeability soft Mn–Zn ferrites and alloys like permalloy have been used for heads. The square shape of the hysteresis loop implies that when subjected to a magnetic field, ferrite crystal can exist in two states, either $+M_S$ or $-M_S$. The reversal of magnetisation occurs over a very narrow range of applied field strengths. This characteristic of the ferrite makes it a very suitable digital storage device, i.e. digital memory devices in computer. The two states of magnetisation of ferrites correspond to the two binary digits 0 and 1.

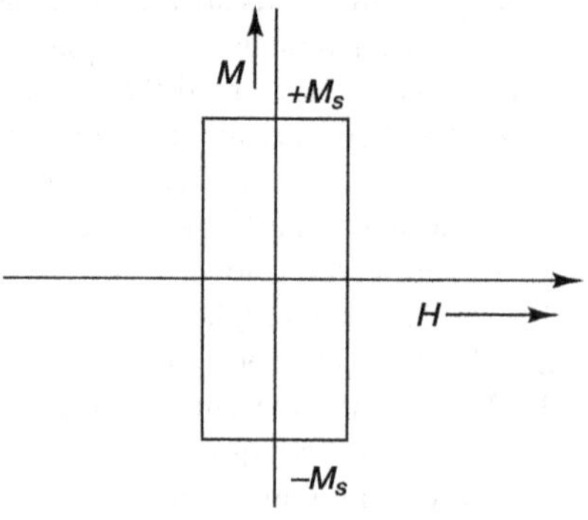

Fig. 3.21 *Hysteresis curve of a ferrite*

(ii) Magnetic Material for Computer Hard Disks, Floppy Disks, Credit Cards

Many significant compositions have been developed [like (Mn Zn) Fe_2O_4, (Mn Cu) Fe_2O_4, etc.] which exhibit quite low core loss (nearly 400 times less). Ferrites are very versatile magnetic materials with some special characteristics suitable for many device applications. The reduction in size has also become possible by using ferrites. There are two types of ferrites—soft ferrites and hard ferrites.

The ferrites of the type MeO. Fe_2O_3, described above, are soft. They exhibit many applications, such as in high frequency transformer cores and computer memories. Some of the very common soft ferrite devices are computer hard disks, floppy disks, credit cards, audio cassettes, video cassettes, recorder-heads and transformer cores.

Ferrites used in magnetic tapes and disks are made of very small needle-like particles of Fe_2O_3 or CrO_2, which are coated on a thin polymeric film or polymeric disk. Each particle is a single domain of size 10–100 nm. The two states of magnetisation are one along the axis of the particle and the other in the opposite direction.

Hard ferrites, which are used in permanent magnets, have the general formula MeO. Fe_2O_3. The metal ion can be barium or strontium. These hard ferrites find wide range applications in generators, relays, motors, loud speakers, telephones, etc.

(iii) Magnetic Materials for Non-volatile Memories

In Non-volatile memories, magnetic bubbles are magnetic domains formed in a thin film of magnetic material. The easy direction of magnetisation is normal to the plane of the film. When a critical magnetic field is applied normal to the plane of the thin magnetic film, small cylindrical domains are formed with the axis of the cylindrical domains normal to the surface of the film. Below the critical field, the bubbles are unstable and as they collapse the film becomes unmagnetised. The magnetic films are made of ferrites or garnets. They can be used to record data. A sequence of bubbles can be shifted by electromagnetic signals. They can be used as non-volatile memories.

3.24 APPLICATIONS OF HYSTERESIS CURVE

Hysteresis curve plays an important role in deciding the magnetic material for various applications in the industry. The selection of a magnetic material is done on the basis of the following magnetic parameters:

- Magnetic susceptibility
- Magnetic permeability
- Retentivity
- Coercivity
- Hysteresis loss

We need permanent magnets in some appliances, while in other situations we need temporary magnets or electromagnets. Electromagnets provide us a very high magnetisation which only persists as long as the current flows through the coils. Magnetic field ceases when the current through the coils is switched off. Permanent magnets carry small magnetic field—though it is long-lasting and least affected by mechanical mishandling. In Fig. 3.22 the *B-H* curve for two materials is given, from which a decision on a suitable material could be easily made.

The overall properties of the material desired for various purposes are given below:

(i) Permanent Magnets

The magnetic materials required for fabrication of permanent magnets must have

- a high retentivity, so that a strong magnetic field is obtained.
- a high coercivity, so that the magnetic property is retained for a long time, unaffected by mechanical mishandling.

From the hysteresis curve, it is clear that steel will be suitable for the purpose. Practically, cobalt-steel and Alnico are more suitable than soft iron. The hysteresis loss is immaterial, as the material in this case is never put to cyclic changes of magnetisation and demagnetisation.

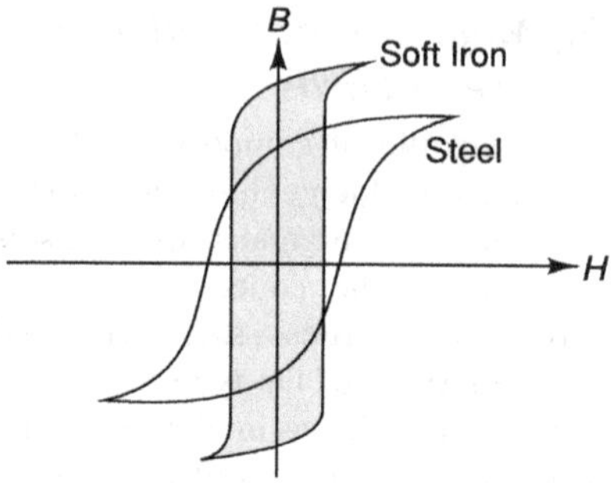

Fig. 3.22 *A comparison of hysteresis curve for steel and iron*

(ii) Temporary Magnets (Electromagnets)

Temporary magnets are used to produce a high magnetic field. The magnetic field persists only when the current flows through the coil. It ceases when the current is switched off. The soft magnetic materials exhibiting the following features are, therefore, used for this purpose:

- High initial permeability
- High retentivity
- Low coercivity
- Low hysteresis loss

The above requirements are fulfilled by soft iron, and therefore, it is used to fabricate electromagnets.

3.25 MAGNETO MATERIALS FOR ELECTRICAL MACHINES

The chief disadvantage with ferromagnetic materials, which are used in electrical machines, is that they exhibit low resistivity. The laminations used in electrical machines exhibit resistivity around 1.4×10^{-6} ohm cm, while the maximum value obtainable of a ferromagnetic alloy is only 10^{-4} ohm cm. This drawback of ferromagnetic materials limit their application into high frequency operations, leading to high eddy current losses and poor magnetic utilisation, even at low frequencies. This disadvantage can be resolved up to certain extent by using ferrites which exhibit favourable properties (like high resistivity), and therefore can be used for frequencies up to microwave in transformer cores.

(i) Transformer Cores, Generators and Telephone Diaphragms

The magnetic material suitable for this purpose must exhibit the following characteristic:

- High initial permeability
- Low hysteresis loss, so that the material could magnetise and demagnetise easily
- Core-loss should be low

Soft iron, permalloy and 4% Si-Fe alloy are suitable to fabricate above.

(ii) Magnetic Shielding

The properties which are looked upon for this purpose are:

- High saturation magnetisation
- Very low coercivity

Permalloy (22% Fe + 78% Ni) can serve this purpose.

3.26 LOSSES IN TRANSFORMERS: IRON LOSS AND CORE LOSS

Iron loss or core loss is the energy which is dissipated as heat in the core material, when it is subjected to alternating electric fields.

The eddy current loss and hysteresis loss are two traditional components of iron loss or core loss. The eddy current loss occurs in the magnetic material when it is subjected to changing magnetic field. Like hysteresis loss, this also increases the temperature of the material which is proportional to i^2R, where i is the eddy current and R is the resistance of eddy current path. The power loss due to eddy currents, at low frequencies is given by

$$P_e = \frac{\pi^2 B_m^2 t^2 f^2}{\rho\beta} = K_e f^2 B_m^2$$

where $$K_e = \frac{\pi^2 t^2}{6\rho}$$

In the above equation, K_e is eddy current coefficient, t is thickness, ρ is the resistivity and β is the coefficient which depends upon geometry [equal to 6 for this thin laminations]. At 50 Hz, the use of iron is divided into thin lamination (nearly between 0.35 – 0.65 mm), which is insulated from each other to minimise core loss as shown in Fig. 3.23.

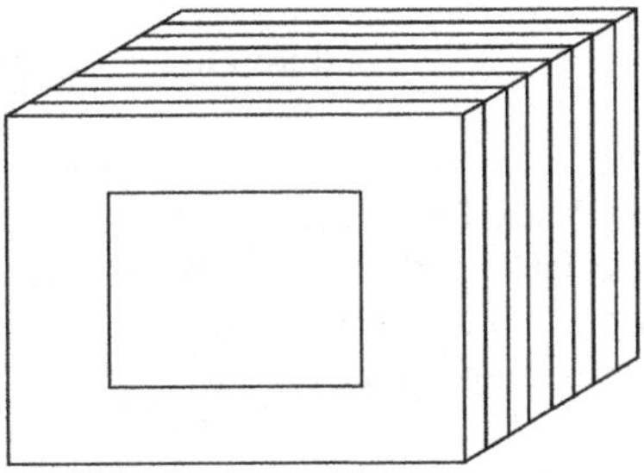

Fig. 3.23 *A laminated core of transformer*

Example 3.20

Calculate the loss of energy per hour in a ferromagnetic specimen when it is subjected to 50 Hz magnetisation. The specimen weighs 50 kg and area of hysteresis loop is 200 joules per m³. Density of iron = 7500 kg/m³.

Solution Volume of 1 kg iron $= \frac{1}{\text{density}} = \frac{1}{7500}$

Loss of energy per kg $= 200 \times \frac{1}{7500}$ Joules/cycle

Number of reversals/s = 50

Weight of specimen = 50 kg

Hysteresis loss/s $= 50 \times 200 \times \frac{1}{7500} = \frac{200}{3}$ Joules/second

Loss per hour $= \frac{200}{3} \times 60 \times 60 = 2.5 \times 10^5$ J **Ans.**

Example 3.21

Determine the loss per kg in a material for a maximum flux density of 1.1 Wb/m² and frequency of 50 Hz using 0.5 mm thick sheets. The resistivity of the material is 30×10^{-8} ohms m. The density is 7800 kg/m³. Hysteresis loss per cycle is 380 W-S/m².

Solution

$$\text{Eddy current loss } P_e = \frac{\pi^2 f^2 B_m^2 t^2}{6\rho} \text{ watt per m}^3$$

$$= \frac{\pi^2 f^2 B_m^2 t^2}{6\rho \times 7800} \text{ watt/kg}$$

$$= \frac{\pi^2 \times 50 \times 50 \times (1.1)^2 \times (0.0005)^2}{6 \times 30 \times 10^{-8} \times 7800}$$

$$= 0.5316 \text{ W/kg}$$

$$\text{Hysteresis loss} = \frac{380 \times 50}{7800} = 2.43 \text{ watt / kg}$$

$$\text{Total iron loss} = \text{Eddy current loss + hysteresis loss}$$

$$= 0.5316 + 2.43$$

$$= 2.9616 \text{ watt / kg}$$

Ans.

When frequency is high enough (nearly 400 Hz), thinner laminations made from special alloys or from grain-oriented electrical steels are used. The following below are given some important soft magnetic materials useful for transformers and rotating electrical machines.

3.27 SOFT IRON SILICON-IRON ALLOY—CORE MATERIAL FOR TRANSFORMER AND ROTATING MACHINES

The materials which are used in electrical machines require high value of permeability, high magnetic saturation and low core losses. Practically, it is not possible to achieve all of these characteristics in a single material. It is observed that there is a conflict to get low loss and high permeability. Presence of carbon increases coercively and hysteresis loss, which can be removed by annealing the material in presence of hydrogen.

The magnetic characteristics of material can be improved and optimised by small addition of silicon (i.e. up to 3%). However, it leads to a marginal decrease in saturation magnetisation. Addition of more silicon makes the material brittle and extremely difficult to work with. The two basic parameters which are taken into consideration are

- Technical requirement
- Economy

The core material of transformer must exhibit low eddy current loss, which is achieved by laminating the thin material sheets to get insulation from one another. Eddy current loss can also be minimised by increasing the electrical resistivity of the core material. This is the main reason of using on alloy of iron and silicon (up to 4%).

Cold rolled (0.35 mm thick) iron-silicon alloy provides a high value of flux density, typically 1.8 Tesla as compared to 1.5 Tesla in case of hot rolled alloys. The cold rolled oriented steel uses phosphate base coating. For high rating transformers, kaolin or varnish is also applied on the laminations.

3.28 MATERIAL FOR AIRCRAFT GENERATORS

Iron-cobalt alloys (permendur) exhibit highest saturation magnetisation—nearly ~ 2.35 tesla. A material with 24% Co, the permendur alloy can be casted as it allows getting complex shapes such as, magnetic pole pieces with 49% cobalt and 2% vanadium hot and cold rolling of the 49 permendur becomes possible with improved mechanical properties. Such an alloy, although costly, finds important applications in aircraft generators and some 400 Hz motors, where high saturation magnetisation is desirable.

3.29 GRAIN ORIENTATION

Grain-oriented magnetic materials provide good magnetic properties. Favourable grain orientation also leads to an improvement in magnetic properties, and therefore saturation magnetisation can be achieved at lower field strength in easy direction.

The metal sheets obtained by rolling and heating provide a texture of grains. It is found that in iron and iron alloys, the easy direction is parallel to rolling direction, and therefore the magnetic properties of grain oriented steels are best in this direction.

3.30 NON-ORIENTED AND GRAIN ORIENTED STEELS

Electrical steels are classified in two types:

- Non-oriented electrical steels
- Grain-oriented electrical steels

Non-oriented electrical steels with 0–3% Si are isotropic. They are obtained in strips, having thickness between 0.35 to 0.8 mm. These are laminated to provide insulation.

Grain-oriented silicon iron exhibits quite low core loss (nearly 50% as compared to non-oriented). The value of saturation flux density is about two times to that of non-oriented materials.

3.31 HARDEND STEEL

The following steels are normally accepted as hardened steel:

- Carbon steel
- Tungsten steel
- Chrome steel
- Cobalt steel

The high physical strength, good workability and low cost are the characteristics of these steels. They can be used in the form of hot rolled bars, rods or strips.

Carbon steel is quite popular as it bears low cost. It exhibits low coercivity and high residual induction, and is therefore used where high flux density is not required.

Tungsten steel, containing 5% tungsten and 0.7% carbon, is better option compared to carbon steel, as it bears improved magnetic properties (having saturation flux density 1.2 Wb/m^2 and coercivity 8000 AT/m) required to be used in dc meter magnets and in other devices where comparatively large size is permissible. Its high cost is the biggest hurdle these days. Besides, it exhibits high magnetic reluctance, excellent water and shock resistance, and toughness.

Chrome steel, containing 3.5% chromium, is the best material to be used in permanent magnets from the point of view of its low cost per unit magnetic flux. It exhibits saturation magnetisation 1.5 wb/m^2 and coercivity 7500 AT / m.

Cobalt steels are the materials exhibiting high coercivity. Material containing 3–38% cobalt along with 2–8%. Chromium and tungsten come in this group. Alnico has replaced high cobalt percentage steel, as it exhibits high tensile strength and wide range of magnetic properties. It has a saturation flux density equal to 2.4 wb/m^2 and coercivity 10^{-4} AT/m. It is adequately hard and strong. Due to a high content of cobalt, it stands expensive and is used in motors, and heavy-duty instruments of commercial importance.

Summary

- Orbital and spin motion of an electron gives rise to orbital-magnetic moment and spin-magnetic moment, and both of these contribute to total magnetic moment of an atom.
- Only partially filled shells and unpaired electrons contribute to the atomic magnetic moment.
- Magnetic materials are classified on the basis of their behaviour in the external applied field. They are diamagnetic, paramagnetic and ferromagnetic.
- Diamagnetic character is exhibited by all the materials and is a fundamental one.
- Diamagnetic susceptibility is small, negative and is independent of temperature.
- Paramagnetic susceptibility is positive, small and is inversely proportional to temperature.
- Ferromagnetic materials obey Curie-Weiss law. Such materials become paramagnetic. above curie temperature
- The internal field in ferromagnetic materials arises as a result of strong interaction between adjacent spins called the exchange interaction.
- Ferromagnetic materials exhibit hysteresis. Their flux density *B* lags behind *H,* and therefore *B* vs. *H* is non-linear.
- Hard magnetic materials have high retentivity and high value of coercivity. Their hysteresis loop is wide. Such materials are useful for making permanent magnets.
- Soft magnetic materials have high retentivity, but small coercivity. Their hysteresis loop is narrow. These materials are suitable for electromagnets and high-frequency magnetic applications.
- Ferrites are a special type of ferromagnetic materials. In them, some spins are oriented parallel and others are oriented antiparallel. They are ceramics and have molecular formula $MeO.Fe_2O_3$, where *Me* is a divalent metal ion. These have a wide range of applications in many magnetic devices.

Exercise

1. What is meant by magnetic induction and magnetisation vector?
2. Explain the terms magnetic susceptibility and magnetic permeability. What is the range of values for these in the case of para and ferromagnetic substances?
3. Explain the cause of magnetism in the substances.
4. Distinguish between para, ferro and diamagnetic substances on the basis of to their characteristics.
5. Define permeability and magnetic susceptibility, and prove:

$$\mu = \mu_0(1 + \chi_m)$$

6. Define $\vec{B}, \vec{M}$ and $\overline{H}$ and prove that $\vec{B} = \mu_0\vec{H} + \mu_0\vec{M}$.

7. Prove that the magnetic moment due to orbital motion of an electron must be an integral multiple of $\frac{eh}{4\pi m}$.
8. What is a diamagnetic substance? Give Langevin's theory of diamagnetism.
9. What is a magnetic curve? With the help of this curve, explain the concepts of residual magnetism (retentivity), coercive force and magnetic hysteresis.
10. What is meant by hysteresis and cycle of magnetisation? Prove that the energy dissipated per cc of the magnetisation is proportional area of *M-H* loop.
11. Discuss the importance of the hysteresis curve. How is it useful in the selection of the materials for permanent magnets, electromagnets, and cores of transformer and motors?
12. How does Ohm's law apply to a magnetic circuit? Write the necessary formulae for a magnetic circuit with reference to this law.
13. Give Langevin's theory of paramagnetism. How does it account the main characteristic of paramagnetic materials?
14. What are Curie law and Curie–Weiss law?
15. Discuss domain theory of ferromagnetism.
16. Distinguish between ferro, antiferro and ferrimagnetic substances.
17. What is meant by hysteresis? Explain hysteresis loss. How would you use the hysteresis curves to select material for the construction of permanent magnets?
18. What is hysteresis? Describe the ballistic galvanometer method for tracing the hysterics loop of a sample of available in the form of an anchor ring.
19. Explain the difference between ferromagnetism and antiferromagnetism.
20. Discuss the importance of hysteresis curve. How would you use the hysteresis curve for selecting the material for the construction of a permanent magnet?
21. Diamagnetically distinguish between paramagnetic and ferromagnetic materials. Giving one example of each; describe their uses.
22. What is hysteresis curve? Explain residual magnetism and coercive force. Prove that the energy dissipated per cycle per cc of magnetisation is μ_0 times the area of *M-H* curve.
23. What are diamagnetic, paramagnetic and ferromagnetic substances? Define magnetic permeability and magnetic susceptibility; establish the relation $B = \mu_0(H + I)$.
24. Discuss Langevin's theory of diamagnetism. Show that the diamagnetic susceptibility is negative and independent of temperature and field strength.
25. Explain diamagnetism, paramagnetism and ferromagnetism on the basis of magnetic dipoles of the atoms.

Numerical Problems

1. Determine the energy lost per hour in a material of iron subjected to magnetisation at 50 cycles per second. The material weighs 50 kg and the area of hysterics loop is 200 joules/m^3. Density of iron = 7500 kg/m^3. **[Ans. 2.5×10^5 J]**
2. Magnetic susceptibility of aluminium at room temperature is 0.82×10^{-8}. Calculate the magnetisation when a field of 0.20 T is applied. **[Ans. 0.0013 A/m]**
3. A paramagnetic soft contains 10^{28} ions/m^3 with magnetic moment 0.1 μ_B. Calculate the magnetisation in a field of 1 T at 300 K. **[Ans. 0.00020 A/m]**
4. The magnetic susceptibility of a medium is 50×10^{-11}. Determine the permeability and relative permeability. **[Ans. $1 + 950 \times 10^{-11}$]**

5. A magnetic material has a magnetisation of 2200 A/m and flux density of 0.0033 Wb/m^2. Calculate the magnetising force and the relative permeability of the material. **[Ans. 203 A/m, 5.14]**
6. The area of hysteresis curve obtained with a certain specimen of iron was 12.5 cm^2. The coordinates were such that 1 cm = 1,000 AT/m and 1 cm = 0.2 T. Calculate (i) the hysteresis loss per cubic metre per cycle, and (ii) the hysteresis lost per cubic metre at a frequency of 50 Hz. **[Ans. 2500 J/m^3/ cycle, 125 Kw]**
7. The hysteresis loop of a sample of sheet steel subjected to a maximum flux density of 1.3 Wb/m^3 has an area of 88 cm^2, the scales being 1 cm = 0.1 Wb/m^2 and 1 cm = 50 AT/m. Calculate the hysteresis loss in watts when 1,500 cm^3 of the same material is subjected to an alternating flux of 1.3 Wb/m^3, peak value of a frequency of 65 Hz. **[Ans. 42.9 W]**
8. When operating at a frequency of 50 Hz, the eddy current loss in transformer is 100 W. Determine the eddy current loss, if it operates at a frequency of (a) 60 Hz, and (b) 10 Hz. **[Ans. 121 W, 484 W]**
9. A coil of wire 250 mm long and having 400 turns carries a current 15 A. Calculate the magnetic field strength H and flux density B, if the coil is in vacuum. Permeability in vacuum is 1.257×10^{-6} H/m. **[Ans. 0.03 Wb m^3]**
10. Relative permeability of supermalloy is 200000. It has a magnetisation of 6000 A/m. Determine the strength of the magnet thus produced. Take $\mu_0 = 4\pi \times 10^{-7}$ henry/metre. **[Ans. 0.001 T]**
11. Prove that the hysteresis loss per unit volume per cycle is given μ_0 times are area of *H-M* loop. **[Ans. –]**
12. Calculate the hysteresis loss of energy E per hour in the iron core of a transformer, if the area of the *B-H* loop is 250 J/m^3 and the frequency of ac is 60 Hz. Mass of the core is 9.0 kg and the density of iron is 7500 kg/m^3. **[Ans. 18 Joules/s]**

Self-quiz Questions

1. What is magnetic hysteresis?
2. Explain the terms 'antiferromagnetism' and 'ferrimagnetism' with reference to magnetic dipoles of atoms.
3. Differentiate between ferromagnetism and ferrimagnetism.
4. What is magnetostriction?
5. What are eddy currents?
6. Give the values of magnetic susceptibility for ferromagnetic, paramagnetic, diamagnetic and non-magnetic materials.
7. What is intensity of magnetisation?
8. Give the relationship between magnetic field intensity H and magnetic flux density.
9. Define magnetic flux density.
10. What are ferrites?
11. What is Curie temperature?

Objective-type Questions

1. The area of *B-H* hysteresis is *d* notes
   (a) the retentivity of the material
   (b) the energy dissipated per cycle
   (c) the permeability of the medium
   (d) the susceptibility of the substance

2. The susceptibility of a magnetic material is found to depend on temperature and strength of the magnetising field. The material is
   (a) a superconductor
   (b) paramagnetic
   (c) ferromagnetic
   (d) diamagnetic
3. Magnetic lines of force
   (a) cannot intersect at all
   (b) intersect at infinity
   (c) intersect within the magnet
   (d) intersect at the neutral point
4. Which of the following substance in not a magnetic material?
   (a) Brass (b) Cobalt
   (c) Nickel (d) Iron
5. Relative permeability of iron is 5500. Its magnetic susceptibility is
   (a) 5500×10^7 (b) 5499
   (c) 5501 (d) 5500×10^{-7}
6. Gauss is the unit of
   (a) magnetic flux density
   (b) magnetic flux
   (c) magnetic induction
   (d) none of these
7. Domain formation is a necessary formation of
   (a) ferromagnetic
   (b) diamagnetics
   (c) paramagnetics
   (d) all of these
8. Soft iron is used for making electromagnets because its
   (a) magnetic saturation limit is high, and retentivity and coercivity are small
   (b) retentively is high
   (c) coercivity is high
   (d) area of hysteresis loop is large
9. Which of the following does not have permanent magnetic dipoles?
   (a) Paramagnetic materials
   (b) Diamagnetic materials
   (c) Ferromagnetic materials
   (d) Antiferromagnetic materials
10. The absence of a magnetic monopole is explained by
   (a) Coulomb's law
   (b) Lorentz law
   (c) Gauss's law
   (d) Fleming's law
11. The susceptibility of a ferromagnetic material is *K* at 27°C. At what temperature will its ability be 0.5 K?
   (a) 600°C (b) 237°C
   (c) 54°C (d) 372°C
12. The dimensional formula of magnetic susceptibility is
   (a) $[M^0 L^0 T^0]$ (b) $[LTA^{-1}]$
   (c) [AT] (d) $[A^3 T^{-1}]$
13. Pick up the correct choice.
   (a) Diamagnetism occurs in all materials
   (b) Diamagnetism results from the partial alignment of permanent magnetic moment
   (c) The magnetising field intensity *H* is always zero in free space
   (d) none of these
14. Which of the following materials is used for making permanent magnet?
   (a) Platinum-cobalt
   (b) Alnico V
   (c) Carbon-steel
   (d) All the these
15. The energy loss per unit volume of the material per cycle is equal to μ_0 times the area of
   (a) *B-H* loop (b) *M-H* loop
   (c) *I-H* loop (d) λ-*H* loop
16. The core of an electromagnet is made of soft iron because soft iron has
   (a) small susceptibility and small retentivity
   (b) large susceptibility and small retentivity
   (c) large density and large retentivity
   (d) small density and large retentivity
17. In the case of a bar magnet, lines of magnetic induction

(a) start from the north pole and end at the south pole
(b) run continuously through the bar and outside
(c) emerge in circular paths from the middle of the bar
(d) are produced only at north pole like rays of light from a bulb

18. Which of the following instruments is used to measure magnetic field?
(a) A thermometer
(b) A pyrometer
(c) A flux meter
(d) A hygrometer

19. Small piece of an ummagnetised substance gets repelled when it is brought near a powerful magnet. Then the substance can be
(a) paramagnetic
(b) ferromagnetic
(c) non-magnetic
(d) diamagnetic

20. The hysteresis loop for the material of a permanent magnet is
(a) short and wide
(b) tall and narrow
(c) tall and wide
(d) short and narrow

21. In a ferromagnetic material, susceptibility is
(a) very small and positive
(b) very small and negative
(c) very large and positive
(d) very large and negative

22. A magnetic needle is kept in a non-uniform magnetic field. It experiences
(a) a force and a torque
(b) a force, but not torque
(c) a torque, but not force
(d) neither a force, nor torque

23. In a normal iron piece
(a) there are no magnetic dipoles
(b) magnetic dipoles are arranged in specific direction
(c) magnetic dipoles are arranged in random direction
(d) none of these

24. The magnetic moment of atomic neon is
(a) zero (b) μ_B
(c) $\frac{1}{2}\mu_B$ (d) $2\mu_B$

25. The material for making permanent magnets should have
(a) high retentivity and high coercivity
(b) high retentivity and low coercivity
(c) low retentivity and high coercivity
(d) low retentivity and low coercivity

26. A steel wire of length l has a magnetic moment M. It is then bent into a semicircular arc. The new magnetic moment is
(a) M (b) M/l
(c) $2M/\pi$ (d) $m \times l$

27. A hydrogen molecule is diamagnetic. A hydrogen atom is a
(a) paramagnetic
(b) ferrimagnetic
(c) diamagnetic
(d) ferrimagnetic

28. Which of the following is ferromagnetic?
(a) Hydrogen
(b) Platinum
(c) Water
(d) Gadolinium

29. Which of the following is paramagnetic?
(a) Bismuth (b) Antimony
(c) Chromium (d) Water

30. Which of the following is suitable for the core of the electromagnets?
(a) Air (b) Soft iron
(c) Steel (d) Cu-Ni alloy

31. Ratio of the magnetic induction to the intensity of the magnetising field is called
(a) susceptibility
(b) retentivity
(c) relative permeability
(d) absolute permeability

32. When a ferromagnetic material is placed in an external magnetic field, the magnetic domains
    (a) increase in size
    (b) decrease
    (c) may increase or decrease
    (d) have no relation with the field
33. In cassette tapes, the sound is recorded in the form of
    (a) variable electrical resistance of the tape
    (b) sound wave on the tape
    (c) electrical energy
    (d) magnetic field on the tape
34. Property possessed by ferromagnetic substance is
    (a) directional property.
    (b) attracting magnetic substances.
    (c) hysteresis
    (d) susceptibility independent of temperature
35. The SI unit of intensity of magnetisation is
    (a) Am
    (b) Am^{-1}
    (c) Wbm^{-1}
    (d) none of these
36. Magnetic susceptibility x is defined as
    (a) M/H (b) H/M
    (c) $M \times H$ (d) none of these
37. An example of diamagnetic substance is
    (a) nickel (b) aluminium
    (c) copper (d) iron
38. Metals which are magnetised along the direction of external magnetising field by orientation of their atomic magnets are called
    (a) paramagnetics
    (b) diamagnetics
    (c) antiferromagnetics
    (d) ferromagnetics
39. The hysteresis is
    (a) lagging of the magnetising force H with respect to the flux density B
    (b) lagging of the flux density B with respect to the magnetising force H
    (c) the ratio of B and H
    (d) none of these
40. Diamagnetic materials possess
    (a) permanent magnetic dipoles
    (b) no permanent magnetic dipoles
    (c) induced dipole moment
    (d) none of these
41. The hysteresis curve of a magnetic material is generally studied for
    (a) diamagnetic materials
    (b) ferromagnetic materials
    (c) paramagnetic materials
    (d) dielectric materials
42. If the magnetic flux is expressed in weber, then magnetic induction can be expressed as
    (a) weber/m^2 (b) weber/m
    (c) weber-m (d) weber–m^2

Answers

1. (b)	2. (c)	3. (a)	4. (a)	5. (b)
6. (a)	7. (a)	8. (a)	9. (b)	10. (c)
11. (d)	12. (a)	13. (a)	14. (d)	15. (b)
16. (b)	17. (b)	18. (c)	19. (d)	20. (c)
21. (c)	22. (a)	23. (c)	24. (a)	25. (a)
26. (c)	27. (a)	28. (d)	29. (c)	30. (b)
31. (d)	32. (c)	33. (d)	34. (c)	35. (b)
36. (a)	37. (c)	38. (a)	39. (b)	40. (b)
41. (b)	42. (a)			

Fill in the Blanks

1. The memory of a computer can be increased by .... the diameter of magnetic bubble.
2. The permeability of iron can be increased by .... it.
3. The residual magnetic flux is more in case of ....
4. The poles of alternates are normally made of ....
5. A material suitable for very high frequency application is.................
6. Air exhibits .... behaviour.
7. The structure of garnet is ....
8. High initial permeability .... area of *B-H* loop.
9. The recent developed hard magnetic material is ....
10. Ferrites exhibiting .... loop from soft magnets.
11. Ferrites exhibit .... and low power at high frequencies.
12. Magnetic flux lines form a .... path.
13. .... materials have strong magnetic reluctance.
14. Iron laser should be kept as .... as is economically possible.
15. All organic substances are .... in character.
16. The magnetic property of ferromagnetic materials .... with cold working.
17. The magnetic flux density left in the ferromagnetic materials on removal of external magnetic field is known as ....
18. The smallest magnetic moment is known as .... magnetron.
19. The magnetically soft materials exhibit .... permeability.
20. Paramagnetic susceptibility varies .... as temperature of material.

Answers

1. decreasing	2. purifying	3. metallic magnets	4. alnico
5. silicon steel	6. paramagnetic	7. cubic	8. reduces
9. hycomax	10. narrow hysteresis	11. high electrical resistivity	
12. closed	13. magnetically hard	14. small	15. diamagnetic
16. deteriorate	17. retentivity	18. Bohr	19. high
20. inversely			

Reading List

1. Marcelo Alonso; Edward J. FIN *Fundamental University Physics*. Addison–Wesley Publishing Co. Inc.
2. Vijaya, M S; Rangarajan G *Material Science.* New Delhi: McGraw-Hill Education Pvt Ltd.
3. Rajput, R K *Electrical Engineering Materials.* New Delhi: Laxmi Publications Pvt Ltd.
4. Dekker A J (2000) *Electrical Engineering Materials.* New Delhi: PHI Learning Pvt Ltd.
5. Indulkar, C S *Electrical Engineering Materials.* New Delhi: S. Chand Publication Pvt Ltd.
6. Shukla, R K *Principles of Engineering Physics*. Kanpur: Ira Books.
7. Seth, S P *Electrical Engineering Materials.* New Delhi: Dhanpat Rai & Co.

8. Kittel, Charles (1996) *Introduction to Solid State Physics*. 7th ed. New York: John Willey & Sons.
9. Lal, K C; Ahmad S I *A Textbook of Modern Physics*. New Delhi:S.Chand & Company Pvt Ltd.
10. Malik, Hitendra K; Singh, A K *Engineering Physics*. New Delhi: McGraw-Hill Education Pvt Ltd.

4

Behaviour of Dielectric Materials in ac and dc Fields

Chapter Objectives

- To discuss various kinds of dielectric materials and the polarisation mechanism.
- To understand the behaviour of dielectric material in steady and time varying fields.
- To understand static and complex dielectric constant.
- To get a knowledge of dielectric loss and its significance.
- To discuss the temperature dependence of dielectric constant.
- To understand the various kinds of materials—piezoelectric, pyroelectric and ferroelectric—and their application.
- To understand the breakdown mechanism in dielectric materials and its consequences.

4.1 INTRODUCTION

The materials which are capable of retarding the flow of electricity or heat through them are known as dielectrics or insulators. Heat and electricity are two major sources of power or energy which are widely utilised in a large number of industrial operations. The safe handling of these sources is almost impossible without the use of an insulator. The material when used to prevent the loss of electrical energy and provide a safety in its operation is named as electrical insulating material. The properties which are taken into consideration for an insulator are the operating temperature and breakdown voltage. However, when it is used to store the electrical charge, it is known as a dielectric material.

The electrical conductivity of these materials is quite low (nearly 10^{-6} mho per cm or less) and the band gap energy is more than 3 eV. This is the reason why the current cannot flow through them. The concept of dielectric material was given for the first time by Michael Faraday. He discovered that the capacity of a capacitor can be increased by filling it with a dielectric material. Although these materials appear less relevant, they play a crucial role in electrical engineering. These materials exhibit magnificent properties and their behaviour is quite interesting in dc and time-varying (alternating) fields. Some properties of the dielectric materials are listed in the next page.

Characteristics of Dielectric Materials

1. Dielectric materials exhibit a high value of insulation resistance.
2. They exhibit polarisation through a progressive orientation of permanent or induced dipoles. Since the charges are not free to move, they do not flow, but the interior of the material becomes non-equipotential.
3. They can store electrostatic energy (which is dependent upon dielectric constant). Dielectric constant is principally of importance in communication circuit as a good electrical insulating material should have a low dielectric constant. However, a high value of dielectric constant is desirable when the material is used in capacitors.
4. When a dielectric material is subjected to an alternating field, the absorption of electrical energy takes place in the material, which appears in the form of heat in the material.
5. They exhibit mechanical strength which is measured as tensile and compressive strength of the material.
6. The electrical proprieties of insulating materials are adversely affected by porosity of the material.
7. Insulating materials should exhibit least thermal expansion and contraction.

4.2 CLASSIFICATION OF DIELECTRIC MATERIALS AT MICROSCOPIC LEVEL

The dielectric materials are classified into two categories on the basis of constituent atoms or molecules present in them. These are

- Polar dielectric materials
- Non-polar dielectric materials

These materials can further be classified on the basis of their behaviour when subjected to an electric field which is shown in Fig. 4.1.

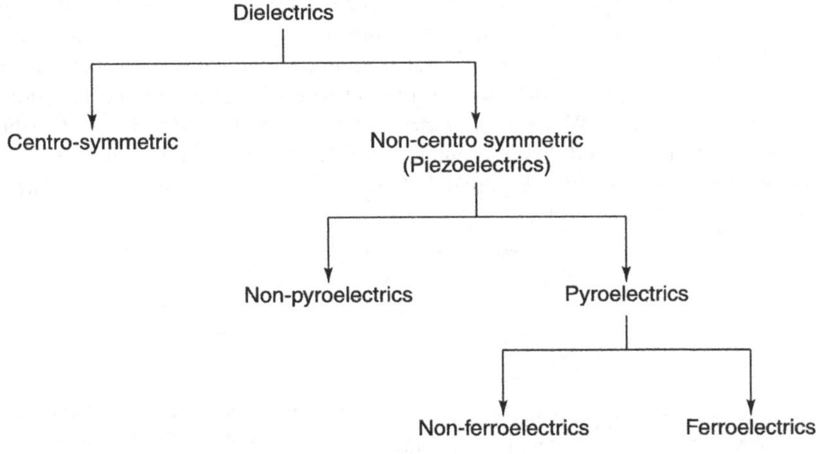

Fig. 4.1 *Classification of dielectric materials on the basis of their behaviour*

A summary of the behaviour of the above materials is given in the last section.

4.3 POLAR DIELECTRIC MATERIALS

In this category, each constituent molecule exhibits a permanent electric dipole moment. In other words, each molecule itself is a dipole, but these are oriented in a random manner inside the material, and therefore the net dipole moment in the material is zero in the absence of an electric field. Also, the centre of gravity of positive charge does not coincide with the centre of gravity of negative charge. Examples of this category are H_2O, N_2O, HCl, CO and NH_3 molecule. As soon as the material is put in an external electric field, the dipoles tend to align in direction of the field. Also, this alignment increases with lowering the temperature. Figure 4.2 shows the orientation of dipoles in the absence and presence of an electric field.

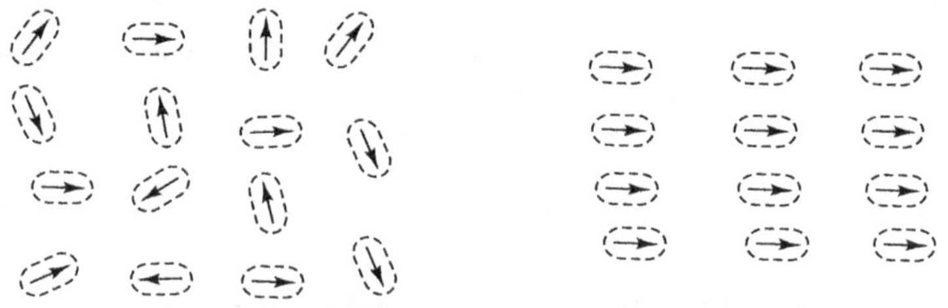

(a) Random orientation of dipoles in the absence of an electric field

(b) Alignment of dipoles in the presence of an electric field

Fig. 4.2 *The orientation of dipoles in the absence and presence of an electric field*

4.4 NON-POLAR DIELECTRIC MATERIALS

Normally, diatomic molecules come under the category of non-polar dielectric materials. In such molecules, the centre of gravity of positive charge coincides with the centre of gravity of negative charge. Therefore, initial electric dipole moment of each molecule is zero. This feature is usually observed in the molecules of symmetrical structure, e.g. N_2, O_2, H_2, CO_2, CH_4, etc. On putting these in an external electric field, negative charge electron cloud gets displaced with respect to positive charge and the molecule acquires a dipole moment. When the electric field is switched off, the charge distribution is again restored the induced dipole moment, so created, varies with electric field and is independent of the temperature. Figure 4.3 shows the charge distribution in the presence and absence of an electric field.

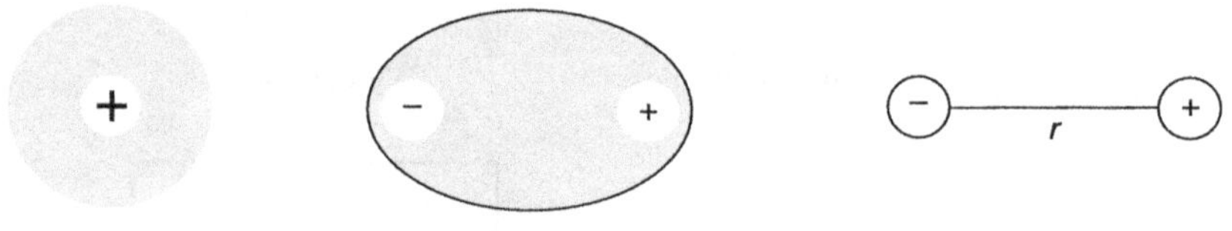

(a) In the absence of an electric field

(b) In the presence of an electric field (induced dipole moment is not zero)

(c) Centres of positive and negative (induced dipole moment is zero) charges displaced by distance r in electric field

Fig. 4.3 *The charge distribution in the presence and absence of an electric field*

Therefore, the dielectrics which are polarised only when they are placed in an electric field are known as non-polar dielectrics.

Induced dipole moment $P = \alpha E$

where α is a constant known as polarisability of atom or molecule and $P = -(ze).r.$

4.5 KINDS OF POLARISATIONS

There are four kinds of polarisations. They are described here:

(i) Electronic Polarisation P_e

This kind of polarisation occurs in the materials containing non-polar molecules and monoatomic gases. When the material is subjected to external electric field, a displacement in the negatively charged electron cloud is observed. The molecule thus acquires induced dipole moment. On the removal of electric field, this induced polarisation is lost. The polarisation of this kind is independent of temperature.

(ii) Ionic Polarisation P_i

This is achieved when electric field is applied to an ionic material. Cations and anions present in the materials are displaced in opposite directions. This is also independent of temperature. Its value is 0.1 α_e where α_e is the electronic polarisability. Alkali halides come in this category.

(iii) Orientational Polarisation P_0

The polarisation of this kind is found in the materials containing molecules having permanent electric dipole moment or polar molecules. On applying external electric field, each molecule of the material (elementary dipole) tends to align itself completely in the field direction, thus acquiring some net electric dipole moment. The alignment of the elementary dipoles is partial only at ordinary temperature, which becomes complete at a low temperature. Examples of this category are polymer, water, ceramics, etc.

(iv) Space Charge Polarisation P_s

When an electric field is applied to a dielectric material at a high temperature, the electric charge gets accumulated at the interface. Due to a sudden change in conductivity, there is a tendency of redistribution of charges in the dielectric medium in the presence of an electric field. The space charge polarisation is not significant in most of the materials.

(v) Total Polarisation or Polarisability

Total polarisability is the sum of electronic, ionic and orientational polarisation, i.e.

$$\alpha = \alpha_e + \alpha_i + \alpha_0$$

or

$$\alpha = \alpha_d + \alpha_0$$

where α_d is known as deformation polarisability, is equal to $(\alpha_e + \alpha_i)$ and is caused by electric field. Non-polar dielectrics exhibit only deformation polarisability, while polar dielectrics have deformation, as well as, orientation polarisabilities. According to Langevin,

$$a_0 = \frac{P_0^2}{3kT}$$

Therefore,

$$\alpha = \alpha_e + \alpha_i + \frac{P_0^2}{3kT}$$

The total polarisation P can be expressed as

$$P = P_e + P_i + P_0$$

or

$$P = N\left(\alpha_e + \alpha_i + \frac{P_0^2}{3kT}\right)E$$

The total polarisation P of a multiphase material is equal to the sum of all kinds of polarisation, including space charge polarisation is

$$P = P_e + P_i + P_0 + P_s$$

For a single phase dielectric, $Ps = 0$

4.6 BEHAVIOUR OF DIELECTRIC MATERIALS

The behaviour of dielectric material is very interesting in static (dc) electric field and time varying (alternating) electric fields. We will discuss now the behaviour of dielectric material in both cases.

(i) Behaviour of Dielectric Materials (non-polar) in dc Field (Electronic Polarisation)

When a dielectric materials is kept in dc electric field, the electron cloud surrounding the atom is shifted, causing a separation between positive and negative charge, and therefore the molecule acquires induced polarisation. This is known as electronic polarisation. This separation between the charges leads to a Coulombian force of attraction, which opposes the displacement between the charges and hence equilibrium is established. The order of magnitude of displacement is $\sim 10^{-17}$ m. A dielectric in which there is no ionic bonding or permanent dipoles, the polarisation is only due to electronic. Examples are inert gases, diamond and phosphorous.

Volume density of charge $\rho = \dfrac{-Ze}{\frac{4}{3}\pi R^3}$

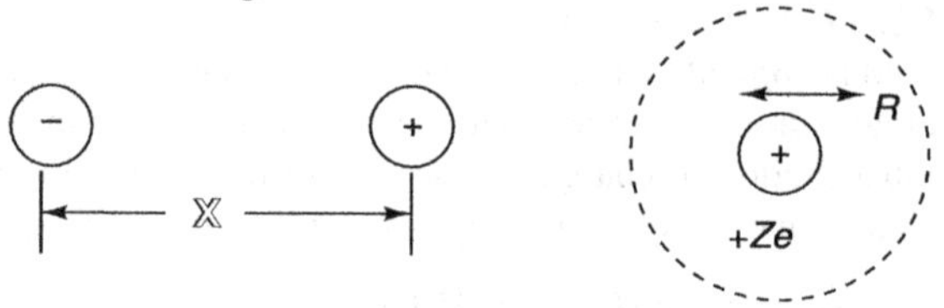

Fig. 4.4 *Separation of charges and coulombian force of attraction*

Assuming the separation between charges to be x,

the force on electron cloud in electric field $= -Ze.\,E$ (4.1)

Coulmb force between the charges

$$= \frac{1}{4\pi\varepsilon_0}\cdot\frac{(Ze)(\text{Charge enclosed by sphere of radius } x)}{x^2}$$

$$= \frac{1}{4\pi\varepsilon_0}\frac{(Ze).\frac{4}{3}\pi x^3.\bar{\rho}}{x^2} = \frac{1}{4\pi\varepsilon_0}\frac{(Ze).\frac{4}{3}\pi x^3.\frac{-Ze}{\frac{4}{3}\pi R^3}}{x^2}$$

$$= \frac{1}{4\pi\varepsilon_0} \cdot \frac{Z^2 e^2 x}{R^3} \qquad (4.2)$$

At equilibrium,

$$-ZeE = -\frac{1}{4\pi\varepsilon_0} \frac{Z^2 e^2 x}{R^3}$$

$$x = \frac{4\pi\varepsilon_0 R^3 E}{Ze}$$

Induced dipole moment, $P = Ze.\,x = 4\pi\varepsilon_0 R^3$

Since $P = \alpha E \Rightarrow \alpha_e = 4\pi\varepsilon_0 R^3$

$$\overline{P}_e = n\overline{p}e = n\alpha E$$

$$P_e = \epsilon_0\, E(\epsilon_r - 1)$$

$$\epsilon_0 = E(\epsilon_r - 1) = n\alpha E$$

$$\epsilon_r - 1 = \frac{n\alpha_0}{\epsilon_0}$$

(ii) Theory of Orientational Polarisation (Langevin–Debye Theory)

When a polar dielectric material is subjected to a static electric field, two things normally happen:

- The material acquires an induced electric dipole moment.
- In addition to (a), the dipoles experience a torque and tend to align in the field direction and this alignment gains a saturation value, if the field is strong enough.

The alignment of the dipoles is also temperature dependent and increases if the temperature of specimen is lowered. Thermal energy of the molecules is responsible for the random orientation of the dipoles. The above situation may be tackled using the statistical mechanics for the calculation of net electric dipole moment at a temperature T. In this derivation, it is assumed that molecules of liquid or gas exert no force of attraction on each other and are quite far from each other. The figure below shows the orientation of electric dipoles in the absence and presence of the electric field.

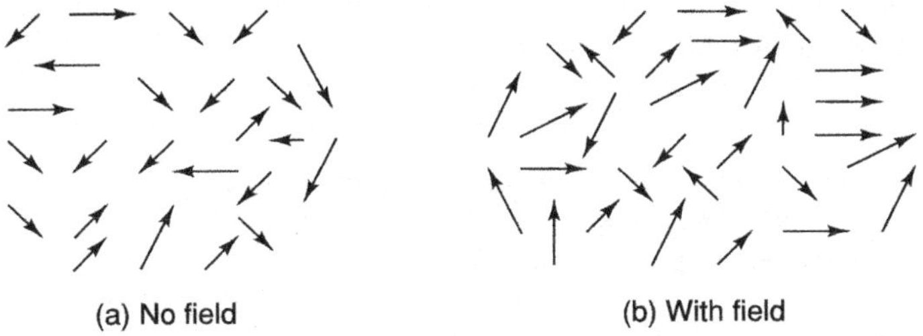

Fig. 4.5 *Random and partial alignment of dipoles*

Let us calculate the net electric dipole moment due to alignment of dipoles in the electric field. Assuming the number of molecules per unit volume N and θ the angle made by the permanent dipole moment P_0 with the field E. The potential energy of the molecules in this field is

$$U = \overline{P}_0.\overline{E} = -P_0 E\cos\theta \qquad (4.3)$$

The number of molecules dN (dipoles), having energy U and oriented at angles between θ and $\theta + d\theta$ with respect to the field direction may be given using Maxwell-Boltzmann statistics.

$$dN = ce^{-U/kT}d\omega \tag{4.4}$$

where $d\omega$ is the solid angle included between θ and $\theta + d\theta$ with field E and C as a constant. From Fig. 4.7, one can calculate $d\omega$, as well.

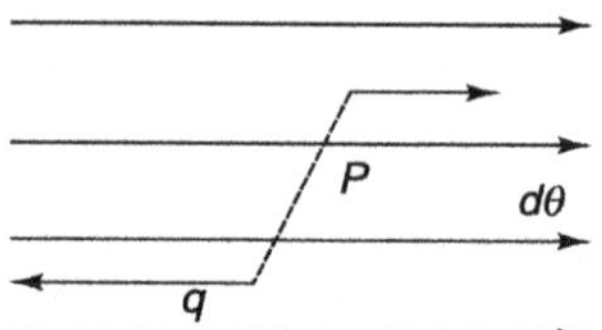

Fig. 4.6 *Dipole in an electric field*

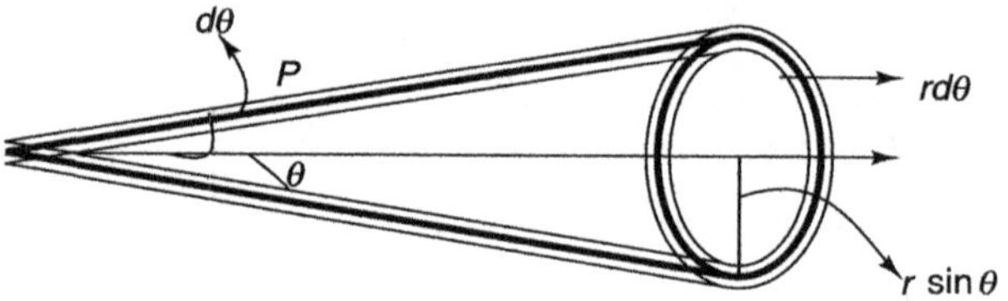

Fig. 4.7 *Calculation of dω*

(iii) Calculation of dω

$$d\omega = \frac{\text{Annular area between } \theta \text{ and } \theta + \text{d}\theta}{r^2}$$

$$d\omega = \frac{2\pi(r\sin\theta).rd\theta}{r^2} \tag{4.5}$$

or $\quad d\omega = 2\pi \sin\theta d\theta$

On substitution this equation in (4.4)

$$dN = ce^{+\frac{P_0 E \cos\theta}{kT}}.2\pi\sin\theta d\theta \tag{4.6}$$

On putting $2\pi c = A$ and $\frac{P_0 E}{kT} = x$ in above, we get

$$dN = Ae^{x\cos\theta}\sin\theta\, d\theta \tag{4.7}$$

Therefore, the total number of molecular dipoles per unit volume

$$N = A\int_0^{\pi} e^{x\cos\theta}.\sin\theta d\theta$$

On evaluating the above integral, we get

$$N = \frac{2A}{x}\sinh x \tag{4.8}$$

The dipole moment of dN molecules along the electric field will be

$$dP = P_0 \cos\theta dN$$

Total dipole moment along the field direction

$$P = P_0 A\int_0^{\pi} e^{x\cos\theta}\sin\theta\cos\theta d\theta \tag{4.9}$$

On evaluating,

$$P = \frac{2P_0 A}{x}\left[\cosh x - \frac{\sinh x}{x}\right] = NP_0\left[\coth x - \frac{1}{x}\right]$$

From Eq. (4.8),

$$\frac{2A}{x} = \frac{N}{\sinh x}$$

Therefore,

$$P = \frac{NP_0}{\sinh x}\left[\cosh x - \frac{\sinh x}{x}\right]$$

$$= NP_0\left[\coth x - \frac{1}{x}\right]$$

$\left[\coth x - \frac{1}{x}\right]$ is known as Langevin function and is denoted by $L(x)$. Therefore,

$$L(x) = \left[\coth x - \frac{1}{x}\right] \quad \text{where} \quad x = \frac{P_0 E}{kT}$$

Hence, $$P = NP_0 L(x) \tag{4.10}$$

Figure 4.8 shows how Langevin function is dependent upon x or (external electric field). From the graph one cannot see that $L(x)$ becomes 1 at high field and low temperature. Under the situation, the net dipole moment attain a saturation value.

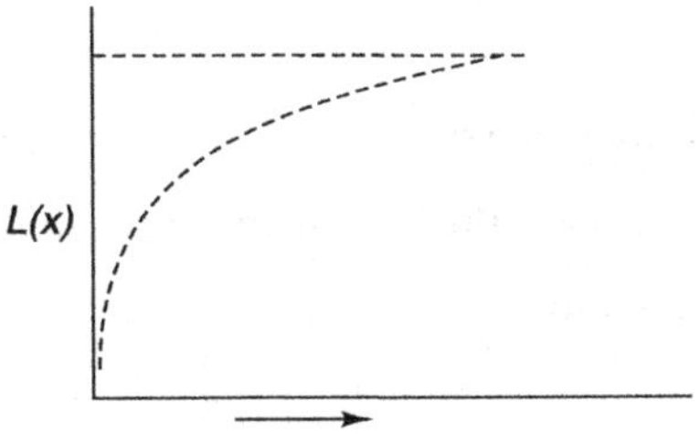

Fig. 4.8 *Langevin function*

Therefore,

$$L(x) = \coth x - \frac{1}{x} = 1$$

$$P = NP_0 \tag{4.11}$$

At an ordinary temperature and at fields near to dielectric breakdown value, x is nearly 10^{-3} for many dielectrics. Therefore, under these situations,

$$L(x) = \frac{x}{3}$$

So, $$P = NP_0\frac{x}{3} = \frac{NP^2 E}{3kT} = N\alpha_0 E \tag{4.12}$$

where α_0 is orientational polarisability and is given as follows:

$$\alpha_0 = \frac{P}{NE} = \frac{P_0^2}{3kT}$$

Hence, orientational polarisability is directly proportional to the external electric field.

Example 4.1

The polarisability of argon is 1.8 × 10^{-40} Fm^2. Calculate the dielectric constant of argon at NTP.

Solution

$$P = \varepsilon_0 E(\varepsilon_r - 1) = Np$$

where p is the dipole moment of the argon atom and N is the number of argon atoms per unit volume at NTP.

$$\varepsilon_r - 1 = Np/\varepsilon_0 E = (N/\varepsilon_0)\alpha$$

Number of argon atoms per unit volume at NTP

$$N = 6.02 \times 10^{23}/22.4 \times 10^3 = 2.689 \times 10^{19}/\text{cm}^3$$

$$= 2.689 \times 10^{25}/\text{m}^3$$

$$\varepsilon_r = 1 + (2.689 \times 10^{25}/8.85 \times 10^{-12})1.8 \times 10^{-40}$$
$$= 1.0005469$$ **Ans.**

Example 4.2

Estimate the shift of the electron cloud with respect to the nucleus in argon atom when a field of 2×10^5 V/m is applied. The polarisability of argon is 1.8×10^{-40} Fm^2.

Solution

$$p = \alpha E = 1.8 \times 10^{-40} \times 2 \times 10^5 = 3.6 \times 10^{-35} \text{ Coul.m}$$

For argon, $z = 18$

So $P = 18\, e.x$

where x is the displacement

$$x = p/18e = 3.6 \times 10^{-35}/18 \times 1.6 \times 10^{-19}$$
$$= 12.5 \times 10^{-18} \text{ m}$$ **Ans.**

Example 4.3

Compare the local field acting on a given molecule of benzene and water.

Solution

$$E_{10C} = E_1 + \frac{P}{3\varepsilon_0}$$
$$= \left(E_0 - \frac{P}{\varepsilon_0}\right) + \frac{P}{3\varepsilon_0}$$
$$= E_0 + \frac{-2P}{3\varepsilon_0}$$

Benzene, $$E_{10C} = 300 \times 10^2 - \frac{2 \times 3.398 \times 10^{-7}}{3 \times 8.85 \times 10^{-12}} = 4.4 \times 10^3 \text{ V/m}$$

Water, $$E_{10C} = 300 \times 10^2 - \frac{2 \times 2.124 \times 10^{-5}}{3 \times 8.85 \times 10^{-12}} = -1570 \times 10^3 \text{ V/m}$$ **Ans.**

Example 4.4

Find the polarisabilities of benzene and water.

Solution

$$p = \alpha\varepsilon_{10C}$$
$$\alpha = \frac{p}{\varepsilon_{10C}}$$

For benzene, $$\alpha = \frac{5.12 \times 10^{-34}}{4.4 \times 10^3} = 1.16 \times 10^{-38} \text{ Fm}^2$$

Water, $$\alpha = \frac{6.34 \times 10^{-34}}{1570 \times 10^3} = 4.04 \times 10^{-40} \text{ Fm}^2$$ **Ans.**

Example 4.5

Benzene has static dielectric constant 2.28 while water has 81 at 300 K. Find the polarisation when the plates of capacitor are immersed into these liquids at 300 K and an electric field is applied having strength of 600 V/cm.

Solution

Polarisation, $P = \epsilon_0 E(\epsilon_r - 1)$

For benzene, $P_B = 8.85 \times 10^{-12} \times 600 \times 10^2 \times 1.28$

$= 6.78 \times 10^{-7}$ c/m^2

For water, $P_w = 8.85 \times 10^{-12} \times 600 \times 10^2 \times 80$

$= 4.2 \times 10^{-5}$ c/m^2 **Ans.**

Example 4.6

NaCl has a static dielectric of 5.6 and an optical index of refraction of 1.5. Account for the difference between $\varepsilon_r(0)$ and n^2. Calculate the percentage contribution of ionic polarisability.

Solution

$$\varepsilon_r(0) = 5.6$$

$$n = 1.5;\ n^2 = \varepsilon_{r\infty} = 5.6 - 2.25 = 3.35$$

The difference gives the contribution due to ionic polarisability.

Percentage contribution from ionic polarisation

$$= \frac{3.35}{5.6} \times 100 = 59.8\%$$ **Ans.**

Example 4.7

Calculate the polarisation of the He gas if placed in a field of 3×10^5 V/m. The polarisability of He is 0.18×10^{-40} Fm^2 and the concentration of the atoms is 2.6×10^{25} /m^3. Calculate the separation between positive and negative charges.

Solution

$$P = Np = N\alpha E$$

$$P = 2.6 \times 10^{25} \times 0.18 \times 10^{-40} \times 3 \times 10^5$$

$$= 1.40 \times 10^{-10} \text{ Coul/m}^2$$

$$P = Np = 1.40 \times 10^{-10} \text{ Coul/m}^2$$

$$p = \frac{1.40 \times 10^{-10}}{2.6 \times 10^{25}} = 0.53 \times 10^{-35} \text{ Coul.m}$$

Charge of He = 2 (electronic charge)

$$p = 2e \times d$$

$$d = \frac{p}{2e} = \frac{0.53 \times 10^{-35}}{2 \times 1.6 \times 10^{-19}} = 1.6 \times 10^{-17} \text{ m}$$ **Ans.**

Example 4.8

There are 10^{27} HCl molecules/m^3 in HCl vapour. Determine the orientational polarisation at room temperature if the vapour is subjected to an electric field of 10^5 V/m. The permanent dipole moment of HCl molecule is 1.04 debye.

Solution (1 debye = 3.33×10^{-30} coul.m

$$P_0 = \frac{NP^2E}{3KT} = \frac{10^{27} \times (1.04 \times 3.33 \times 10^{-31})^2 \times 10}{3 \times 1.38 \times 10^{-23} \times 300}$$

$$= 9.28 \times 10^{-8} \text{ coul/m}^2$$ **Ans.**

(iv) Clausius–Mossotti Equation (Non-polar Dielectric in dc Field)

Let us discuss the behaviour of non-polar dielectric to static (dc) field. The polarisability is an atomic (microscopic property) and dielectric constant of a material in a bulk or macroscopic property.

The relation connects the microscopic polarisability of non-polar dielectric with macroscopic dielectric constant of the material.

The electric polarisation of the dielectric containing N molecules per unit volume is given by

$$P = NP_i = N\alpha E_L \tag{4.13}$$

where $E_L \rightarrow$ Local field or internal field

According to Lorentz equation,

$$E_L = E + \frac{P}{3 \in_0}$$

Hence, $$P = N\alpha \left[E + \frac{P}{3 \in_0} \right]$$

But, $$P = (\in_r - 1) \in_0 E$$

$$(\in_r - 1) \in_0 E = N\alpha \left[E + \frac{(\in_r - 1) \in_0 E}{3 \in_0} \right] = EN\alpha \left[1 + \frac{(\in_r - 1)}{3} \right]$$

$$\in_0 [\in_r + 1] = N\alpha \left[\frac{3 + (\in_r - 1)}{3} \right] = N\alpha \left[\frac{(\in_r - 2)}{3} \right]$$

$\Rightarrow$ $$N\alpha(\in_r + 2) = 3\in_0[\in_r - 1]$$

$\Rightarrow$ $$\alpha = \frac{3 \in_0 [\in_r - 1]}{N(\in_r + 2)}$$

$\Rightarrow$ $$\alpha = \frac{3 \in_0}{N} \cdot \frac{x_e}{x_e + 3} \qquad \text{Since, } \in_r = 1 + x_e$$

α is proportional to R^3, where R is the radius of the molecule. So, if $\in_r$ and N are known for a gas at a given temperature and pressure, the approximate diameter of the atom can also be calculated using the above equation for monoatomic gases. The above equation holds good for those solids which have cubic symmetry.

Example 4.9

The atomic weight and density of sulphur are 32 and 2.08 g/cm^3, respectively. The electronic polarisability of the atom is 3.28×10^{-40} Fm^2. If solid sulphur has cubical symmetry, find relative dielectric constant.

Solution

Using Clausius–Mossotti equation,

$$\frac{\in_r - 1}{\in_r + 2} = \frac{n\alpha}{3\in_0}$$

Since $n = \frac{N.\rho}{M}$ $N \rightarrow$ Avogadro number

$M \rightarrow$ Molecular weight

$$\frac{\in_r - 1}{\in_r + 2} = \frac{N\rho\alpha}{3M\in_0}$$

$$= \frac{6.023\times 10^{26}\times 2.08\times 10^{3}\times 3.28\times 10^{-40}}{3\times 32\times 8.85\times 10^{-12}}$$

$$= 0.483$$

Therefore, $\in_r = \frac{1.966}{0.517} = 3.8$ **Ans.**

Example 4.10

The following data refers to a dielectric material; $\in_r = 4.94$ and $n^2 = 2.69$ where n is refractive index. Calculate the ratio between electronic and ionic polarisability for this material.

Solution Using Clausius–Mossotti equation,

$$\frac{\in_r - 1}{\in_r + 2} = \frac{N(\alpha_e + \alpha_i)}{3\in_0} \quad (1)$$

In optical frequency range $\in_r = n^2$ and $\alpha_i = 0$
Therefore, Eq (1) becomes

$$\frac{n^2 - 1}{n^2 + 1} = \frac{N\alpha_e}{3\in_0} \quad (2)$$

On dividing Eqs (1) and (2), we get

$$\frac{\in_r - 1}{\in_r + 2}\cdot\frac{n^2 + 2}{n^2 - i} = \frac{\alpha_e + \alpha_i}{\alpha_e}$$

or

$$1 + \frac{\alpha_e}{\alpha_i} = \frac{(4.94 - 1)}{(4.94 + 2)}\cdot\frac{(2.69 + 2)}{(2.69 - 1)}$$

On simplifying, we get

$$\frac{\alpha_e}{\alpha_i} = 1.735$$ **Ans.**

4.7 THREE ELECTRIC VECTORS

(i) Intensity of Electric Field $\overline{E}$

The intensity of electric field or strength of electric field is measured by the force experienced by a unit positive test charge and is given by

$$\overline{E} = \frac{\overline{F}}{q_t} \quad (4.14)$$

The unit of electric field strength is newton/coulomb or volt/metre. The direction of the field is same as that of the direction of force upon the positive test charge.

(ii) Dielectric Polarisation and Polarisation Vector ($\overline{P}$)

When a dielectric material is subjected to static (dc) electric field, the electric dipoles in the piece of material align either spontaneously or by the external electric field. In other words, the charge separation occurs within the material. This phenomenon is called dielectric polarisation. A non-conducting medium that can be polarised by an electric field is called a 'dielectric'. Polarisation gives rise to a net positive charge on one side of the piece of material and a net negative charge on the other side. The piece of material then becomes a large electric dipole that tends to move in a direction where the electric field increases.

The polarisation of a material is a vector quantity and is represented by *P*, which is defined as 'electric dipole moment per unit volume' of the specimen. Therefore, if p is the dipole moment induced in each atom or molecule and n is their number per unit volume, then

polarisation $\overline{p} = n\overline{p}$; where $p = q.d$

$d \rightarrow$ separation between charges

In general, the direction of dipole moment is parallel to the direction of electric field.

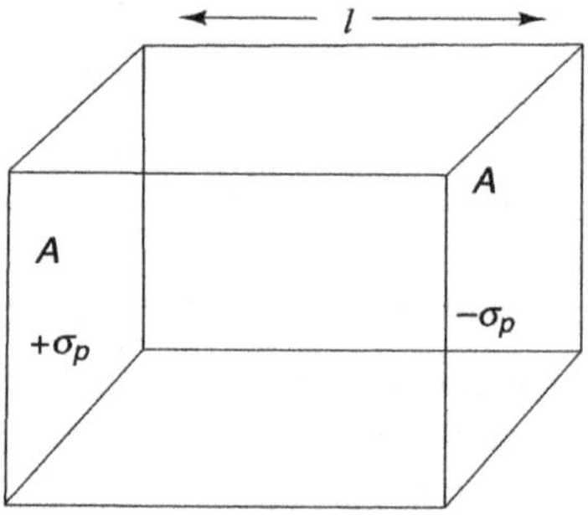

Fig. 4.9 *Charge distribution on a block of length l and A area cross-section*

Consider a rectangular block of a dielectric material having length (l) and cross-section area (A). Let surface charge densities on its faces are $+\sigma_p$ and $-\sigma_p$.

Charge on each face $\quad q = \sigma_p.A$

Induced dipole moment $\quad = ql$

$\quad = \sigma_p A.l.$

Volume of rectangular block = $A.l$

$$\text{Polarisation vector (dielectric polarisation) } \overline{P} = \frac{\text{Induced dipole movement}}{\text{Volume}}$$

$$\text{Therefore,} \qquad \overline{p} = \frac{\sigma_p Al.}{A.l.} = \sigma_p \tag{4.15}$$

Hence, the polarisation vector *P* is the surface density of charge which is produced on applying electric field *E* to a dielectric material.

(iii) Electric Displacement Vector ($\overline{D}$)

A polarised dielectric has charges on its surface. These polarisation charges are in the 'frozen State' in the sense that they are bound to specific atoms or molecules and are not free to move inside the dielectric. In other materials like ionised gas or metal, the charges are in the state of random motion and are, therefore, named as 'free charges'.

Now, when a dielectric slab is subjected to external electric field by putting it in between the plates of a capacitor, it gets polarised, i.e. the charges are separated within the material and appear on the surface of the dielectric. These are polarisation charges having a sign opposite to the nearby plate balance partially the free charges present in the plate of capacitor. If *P* is the polarisation within the slab, the surface density towards left side plate will be $\sigma_{pol} = -P$ and toward right side will be $\sigma_{pol} = +P$. Therefore, the net surface density of charge on left is

$$\sigma = \sigma_{free} + \sigma_{pol}$$

or $\sigma = \sigma_{free} + P$ with a similar opposite on the other side. These net surface charges give rise to a uniform electric field which can be given from Gauss's law as

$$E = \frac{\sigma}{\epsilon_0} = \frac{\sigma_{free} - p}{\epsilon_0}$$

or $$\sigma_{free} = \epsilon_0 E + P \tag{4.16}$$

The above relation explains the polarisation mechanism which occurs when a dielectric material is kept inside a capacitor and also correlates the free charges and induced charges (polarisation charges). The above relation suggests the introduction of a new vector field D, which is named as electric displacement vector and is defined by the relation given below

$$\overline{D} = \epsilon_0 \overline{E} + \overline{P} \tag{4.17}$$

Obviously, $\overline{D}$ is expressed in cm^{-2} and is connected with free charges.
Therefore, total free charges on the conductor

$$q_{free} = \oint_s \sigma_{free}\, ds = \oint_s \overline{D}.\overline{ds}$$

which show its validity for any closed surface. Therefore, the flux of electric displacement over a closed surface is equal to the total free charge inside the surface

$$\oint_s \sigma \overline{D} \overline{ds} = q_{free} \tag{4.18}$$

which is the Gauss's law in a dielectric material.

4.8 GAUSS'S LAW IN A DIELECTRIC

Gauss's law is applied to find the electric field under various situations. This relates the electric flux ϕ_E with the charge enclosed by the surface under consideration. The surface under consideration may be of any shape or size and is known as 'Gaussian surface'. Gaussian surface is the imaginary surface that is drawn symmetrical to the given situation so that the value of $\int \overline{E}.\overline{ds}$ could be easily obtained. The statement of the Gauss's law of electrostatics is given below:

"Flux of the electric field passing through a closed hypothetical surface (Gaussian surface) of any shape or size is equal to $\frac{1}{\epsilon_0}$ times, the charge enclosed by the surface."

i.e. $$\oint_s \overline{E}.\overline{ds} = \frac{q}{\epsilon_0} \tag{4.19}$$

and if the Gaussian surface does not enclose any charge then

$$\oint_s \overline{E}.\overline{ds} = 0 \tag{4.20}$$

This is one of the fundamental laws of electrostatics which is very useful to find the electric field under different situation like sphere (hollow and solid, conducting or non-conducting), cylinder, plane sheet of charge, etc.

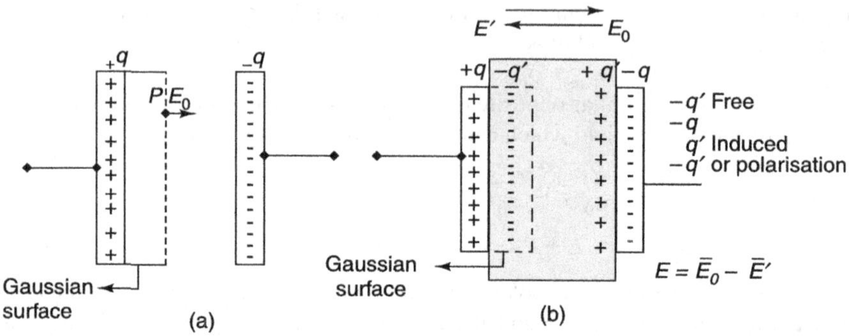

Fig. 4.10 *Parallel plate capacitor filled with air and dielectric material*

Let us apply the Gauss's law to a parallel plate capacitor field with a dielectric having dielectric constant K. Figure 4.10 shows a parallel plate capacitor in the absence and presence of dielectric:

- parallel plate capacitor in the absence of dielectric (Gaussian surface encloses only fee charges $+ q$)
- parallel plate capacitor in presence of dielectric (Gaussian surface encloses free charges and polarisation charges)

Assuming A, the surface area of plate; E_0, electric field in the absence of dielectric, using Gauss's law, we get

$$\oint_s \bar{E}.\bar{ds} = \frac{q}{\epsilon_0}$$

or $$\epsilon_0 \oint_s \bar{E}.\bar{ds} = q$$

or $$\epsilon_0 E_0.A = q$$

or $$E_0 = \frac{q}{\epsilon_0 A} \tag{4.21}$$

On filling the capacitor with dielectric, the polarisation of dielectric takes place and the opposite charges (induced charges) appear within the material (Fig. 4.10b), which are denoted by $+ q'$ and $- q'$. Hence, an induced electric field sets up within the material which reduces the original electric field. The total charge enclosed by Gaussian surface in the new situation is equal to $(q - q')$.

Again applying Gauss's law,

$$\epsilon_0 \oint_s \bar{E}.\bar{ds} = \epsilon_0 E.A = q - q'$$

or $$E = \frac{q}{\epsilon_0 A} - \frac{q'}{\epsilon_0 A} \tag{4.22}$$

The decrease in electric field is shown in the above equation.

But $$E = \frac{E_0}{K} = \frac{q}{\epsilon_0 KA}$$

Therefore, using it in above equation

$$\frac{q}{\epsilon_0 KA} = \frac{q}{\epsilon_0 A} = \frac{q'}{\epsilon_0 A} \tag{4.23}$$

or $$\frac{q}{K} = q - \frac{q'}{A}$$

or $$q' = q\left(1-\frac{1}{K}\right) \tag{4.24}$$

or $$q' < q$$

The above equation shows that the induced charge or polarisation charges will always be less than free charges. In the absence of dielectric

$$K = 1 \qquad (K \text{ and } \in_r \text{ are identical})$$

Therefore, $$q' = 0$$

Rewriting Eq. (4.22),

$$\frac{q}{K \in_0 A} = \frac{q}{\in_0 A} - \frac{q'}{\in_0 A}$$

or $$\frac{q}{\in_0 A} = \frac{q}{K \in_0 A} + \frac{q'}{\in_0 A}$$

i.e. $$\frac{q}{A} = \in_0 \left[\frac{q}{K \in_0 A}\right] + \frac{q'}{\varepsilon_0 A} \tag{4.25}$$

The quantity in the parenthesis of Eq. (4.25) is simply electric field strength $\overline{E}$ in the dielectric material. The last term in the above equation denotes the induced charges per unit area and we call it as dielectric polarisation, normally represented by $\overline{P}$.

i.e. $$\overline{P} = \frac{q'}{A} \quad \text{and} \quad \overline{E} = \frac{q}{K \in_0 A}$$

The above relation may also be written as

$$\overline{P} = \frac{q'.d}{A.d.}$$

where d is the thickness of dielectric material or separation between induced charges $+q'$ and $-q'$, which is induced dipole moment per unit volume or polarisation vector $\overline{P}$.

The Eq. (4.25) can also be written as

$$\frac{q}{A} = \in_0 . E + P$$

or $$D = \in_0 E + P \tag{4.26}$$

The above equation gives a relation between three electric vectors D, E and $\overline{P}$.

Significance of E, D and P Vectors

1. D is connected with free charges only. We can represent the vector field D exactly as we represent $\underline{E}$ vector by lines of force.
2. $\overline{E}$ is related with all kinds of charges that are actually present in the presence of dielectric material, i.e. free and induced charges.
3. $\overline{P}$ is connected with polarisation (induced) charges only vanishes outside the dielectric material.

The Gauss's law in dielectrics has also been derived in terms of surface charge densities in the preceding section.

4.9 ELECTRIC SUSCEPTIBILITY AND STATIC DIELECTRIC CONSTANT, χ_e AND $\in_r$

The electric susceptibility χ_e describes the response of a medium to the action of external electric field and is related to the properties of atoms and molecules of the medium. That is why susceptibility is different for static and time varying fields.

We know that

$$D = \epsilon_0 E + P$$

$$\epsilon_r \epsilon_0 E = \epsilon_0 E + P$$

or
$$P = \epsilon_0 E(\epsilon_r - 1) \quad (4.27)$$

Since
$$P = \chi_e \epsilon_0 E$$

Therefore,
$$\chi_e \epsilon_0 E = \epsilon_0 E(\epsilon_r - 1)$$

or
$$\boxed{\chi_e = \epsilon_r - 1} \quad (4.28)$$

This gives a relationship between electric susceptibility χ_e and dielectric constant ϵ_r. Electric susceptibility χ_e is a measure of the extent to which the material can be polarised by applying the electric field.

It can be verified that dielectric (electric) susceptibility has no unit. Therefore, the dielectric constant ϵ_r is also a measure of the extent to which the material can be polarised. The material that exhibits a high dielectric constant can be polarised easily.

Example 4.11

The permittivity of diamond is $1.46 \times 10^{-10} c^2 N^{-1} m^{-2}$. Find its dielectric constant and electrical susceptibility. (Given: $\epsilon_0 = 8.85 \times 10^{-12}\ c^2\ N^{-1} m^{-2}$)

Solution We know that

$$\epsilon_r = \frac{\epsilon}{\epsilon_0} = \frac{1.46 \times 10^{-10}}{8.85 \times 10^{-12}}$$

$$= 16.48$$

Therefore, the electrical susceptibility

$$\chi_e = \epsilon_0 (\epsilon_r - 1)$$

$$= 8.85 \times 10^{-12}\ (16.48 - 1)$$

$$= 1.37 \times 10^{-10}\ \mathrm{c^2\ N^{-1}\ M^{-2}}$$ **Ans.**

Example 4.12

If all molecular dipoles in a water drop of 0.1 cm radius are pointed in the same direction, find the polarisation. Dipole moment of water molecule is 6.0×10^{-30} cm.

Solution Number of water molecules in the water drop

$$n = \frac{N\left[\frac{4}{3}\pi r^3 \rho_{\text{water}}\right]}{M_{\text{water}}}$$

$$= \frac{6.0 \times 10^{26} \times 4.0 \times 3.14 \times (0.1 \times 10^{-2})^3 \times 10^3}{18 \times 3}$$

$$= 1.40 \times 10^{23}$$

Polarisation
$$= 1.4 \times 10^{23} \times 6.0 \times 10^{-30}$$

$$= 8.4 \times 10^{-7}\ \mathrm{cm^2}$$ **Ans.**

Example 4.13

The value of dielectric constant for the gas at 0°C is 1.000074. Calculate its dielectric susceptibility.

Solution Dielectric susceptibility

$$\chi_e = \in_r - 1$$
$$= 1.0000\,74 - 1$$
$$= 0.000074$$ **Ans.**

4.10 EFFECT OF DIELECTRIC MEDIUM UPON CAPACITANCE

We know from earlier discussion that electric field inside a capacitor in the absence of dielectric material (air capacitor) is given by

$$E_0 = \frac{q}{\in_0 A} = \frac{\sigma}{\in_0} \qquad (4.29)$$

where σ is the surface charge density.

The potential difference between the plates of a capacitor will therefore be

$$V_0 = E_0 d = \left(\frac{\sigma}{\in_0}\right) d \qquad (4.30)$$

The capacitance,
$$C_0 = \frac{q}{V_0} = \frac{\sigma A}{\left(\frac{\sigma}{\in_0} d\right)} \qquad (4.31)$$

On placing a slab of dielectric material, we notice that the charge on the plate of capacitor will be more as compared to the charge in capacitor without dielectric. Therefore, from $C = \frac{Q}{V}$, the capacity increases on filling the capacitor with a dielectric material. The above fact may be observed if we take two capacitors—one with dielectric and another without it—and if both are given the same charge, the potential difference of a capacitor with dielectric will be more than without dielectric.

i.e.
$$V = \frac{V_0}{\in_r} \qquad (4.32)$$

where $\in_r$ is known as dielectric constant.

Therefore,
$$C_0 = \frac{q}{V_0} = \frac{q}{\in_r V} = \frac{C}{\in_r}$$

or
$$C = \in_r C_0$$

Example 4.14

Two parallel conducting plates of a capacitor are separated by 2 mm and the space between them is filled with a dielectric of dielectric constant 3. The electric field intensity in the dielectric is 10^6 volts/m. Calculate

(i) the free charge per unit area on the conducting plates
(ii) the polarisation P in the dielectric
(iii) the displacement D in the dielectric

Solution

(i) The electric field in the dielectric

$$E = E_0/\varepsilon_r \text{ where } E \text{ is the applied field}$$
$$E_0 = \varepsilon_r E = 3 \times 10^6 \text{ V/m}$$

The free charge/unit area on the surface of the plates

$$\sigma = \varepsilon_0 E_0 = 8.85 \times 10^{-12} \times 3 \times 10^6$$
$$= 26.55 \times 10^{-6} \text{ Coul/m}^2$$ **Ans.**

(ii) Polarisation, P in the dielectric

$$P = \varepsilon_0(\varepsilon_r - 1)\, E_0 = 8.85 \times 10^{-12} \times 2 \times 3 \times 10^6$$
$$= 5.31 \times 10^{-5} \text{ Coul/m}$$ **Ans.**

(iii) The displacement D in the dielectric

$$D = \varepsilon_0 \varepsilon_r E_0 = 8.85 \times 10^{-12} \times 3 \times 3 \times 10^6$$
$$= 7.965 \times 10^{-5}$$ **Ans.**

Example 4.15

A parallel plate capacitor has surface area of plates 6.45×10^{-4} m² and a separation between plates 1.0×10^{-3} m. A voltage of 10 V is applied across the plate. A dielectric material having permittivity 6.0 has been placed between the plates. Find

(i) the capacitance

(ii) the charge stored on each plate

(iii) displacement vector *D*

(iv) polarisation vector *P*

Solution

(i) The capacitance of capacitor is given by

$$C = \frac{\epsilon A}{d} = \frac{\epsilon_0 \epsilon_r A}{d}$$
$$= \frac{8.85\times10^{-12}\times6.0\times6.45\times10^{-4}}{1.0\times10^{-3}}$$
$$= 3.52 \times 10^{-11} \text{ Farad}$$ **Ans.**

(ii) Charge $q = CV$

$$= 3.52 \times 10^{-11} \times 10$$
$$= 3.52 \times 10^{-10} \text{ coulomb}$$ **Ans.**

(iii) $D = \epsilon E = \epsilon_0 \epsilon_r E = \dfrac{8.85\times10^{-12}\times6.0\times10}{1.0\times10^{-3}}$

$$= 5.32 \times 10^{-7} \text{ c/m}^2$$ **Ans.**

(iv) $P = D - \epsilon_0 E$

$$= 5.32 \times 10^{-7} - \frac{8.85\times10^{-12}\times10}{2.0\times10^{-3}}$$
$$= 0.9 \times 10^{-7} \text{ c/m}^2$$ **Ans.**

Therefore, capacitance increases and becomes ϵ_r times the air capacitance.

4.11 MACROSCOPIC ELECTRIC FIELD

When dielectric is subjected to external electric field E_0, the polarisation takes place within the dielectric material and induced charges appear on the surface of the dielectric.

These induced charges produce an electric field in the opposite direction E' which is also named as 'depolarisation field'. Therefore, the net field decreases which is represented by E and can be given

$$E = E_0 - E' \tag{4.33}$$

where $$E' = \frac{q'}{\epsilon_0 A} = \frac{P}{\epsilon_0}$$

Therefore, $$E = E_0 - \frac{P}{\epsilon_0} \tag{4.34}$$

This net field is known as 'macroscopic field'. The dielectric constant ϵ_r and χ_e are related to macroscopic field, and therefore are known as macroscopic parameters.

4.12 MICROSCOPIC ELECTRIC FIELD

When a dielectric material (solid or liquid) is subjected to external electric field, each atom present in it becomes electric dipole that produces its own field which is different than the external applied field. Since the atoms in the material are closely packed, each atomic dipole is surrounded by many other atomic dipoles. The electric field at the atomic site therefore will be a sum total of applied field and the field origined from the mutual interactions between the dipoles. This resultant field is known as 'microscopic field' and is denoted by E_{micro}.

Therefore,

$$E_{micro} = E_0 - E' + E_2 + E_3 + = E + E_2 + E_3$$

where E_0 is external applied field, E' is depolarising field, E_2 and E_3 are the fields due to immediate neighbouring atoms, and E is the macroscopic field.

4.13 INTERNAL (LOCAL) FIELDS IN LIQUIDS AND SOLIDS—ONE-DIMENSIONAL CASE

When a liquid or solid dielectric material is subjected to external electric field, its atoms become electric dipoles which produce the field that is different from the applied field. This is the field which is produced at the atomic site due to alignment of dipoles. Since the atoms in a dielectric are closely packed, therefore each atom is surrounded by many other atomic dipoles. The field at the atomic site, therefore, is partly due to external field and partly due to interaction between atoms or molecules polarised under the action of applied field.

This local field can be calculated by assuming an array of atomic dipoles separated by distance a. The electric field direction is shown in Fig. 4.11. Each dipole in this array exhibits an induced dipole moment Pi.

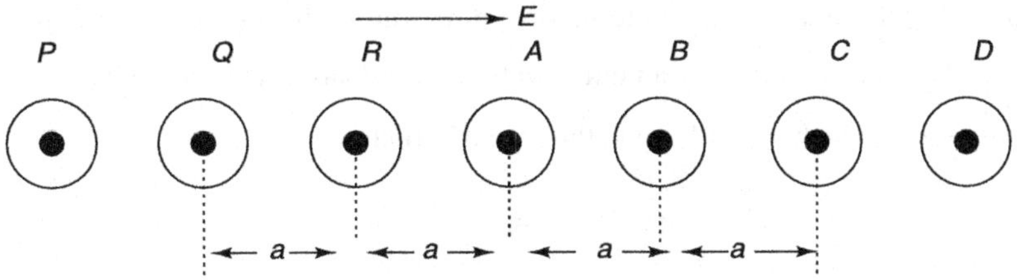

Fig. 4.11 *Array of atomic dipoles*

Let us calculate the local field as seen by the atomic dipole *A*, which has atoms on either side *B*, *C*, *D* and *P*, *Q*, *R*.

The radial and transverse components of electric field produced by dipole having moment *P* are

$$E_r = \frac{1}{4\pi\varepsilon_0}.\frac{2P\cos\theta}{r^3}\hat{n}_r$$

$$E_\theta = \frac{1}{4\pi\varepsilon_0}.\frac{P\sin\theta}{r^3}\hat{n}_\theta \tag{4.35}$$

where $\hat{n}_r$ and $\hat{n}_\theta$ are unit vectors along radial and transverse direction.

Electric field produced *A* by *R* will be ($\theta = 0, r = a$)

$$E_{AR} = \frac{1}{4\pi \in_0}.\frac{2P_i}{a^3} \quad \text{along the radial direction.}$$

Similarly $$E_{AB} = \frac{1}{4\pi \in_0}.\frac{2P_i}{a^3} \quad (\theta = \pi, r = -a)$$

Therefore, contribution of field at atom *A* due to *R* and *B* will be

$$= E_{AR} + E_{AB} = \frac{1}{4\pi\varepsilon_0}\frac{4P_i}{a^3} = \frac{P_i}{\pi\varepsilon_0 a^3} \tag{4.36}$$

Similarly, the field at *A* due to atomic dipoles *Q* and *C*

$$E_{AQ} + E_{AC} = \frac{P_i}{\pi\varepsilon_0(2a)^3} \tag{4.37}$$

and $$E_{AP} + E_{AD} = \frac{P_i}{\pi\varepsilon_0(3a)^3} \tag{4.38}$$

Therefore, total internal field at the atom *A*

$$E_L = E + \frac{P_i}{\pi\varepsilon_0(a)^3} + \frac{P_i}{\pi\varepsilon(2a)^3} + \frac{P_i}{\pi\varepsilon_0(3a)^3} + \ldots$$

$$= E + \frac{P_i}{\pi\varepsilon_0(a)^3}\left[1 + \frac{1}{2^3} + \frac{1}{3^3} + \ldots\right]$$

or $$E_L = E + \frac{P_i}{\pi\varepsilon_0(a)^3}\sum_{n=1}^{\infty}\frac{1}{n^3} \tag{4.39}$$

where $$n = 1, 2, 3,\ldots$$

$$E_L = E + \frac{1.2.P_i}{\pi\varepsilon_0(a)^3} \tag{4.40}$$

Three-dimensional Array

The internal field in a crystal would involve three similar terms. For three-dimensional case, $\frac{1}{a^3}$ should be replaced by a number *n* (number of atoms per unit volume) and $\frac{1.2}{\pi}$ by a constant γ, which depends upon the type of structure.

$$E_L = E + \frac{\gamma_n Pi}{\in_0} = E + \frac{\gamma P}{\in_0} \tag{4.41}$$

4.14 TEMPERATURE DEPENDENCE OF DIELECTRIC CONSTANT

The orientational polarisation, which is observed in polar dielectrics (like HCl, H_2O, CO, NH_3) only, is found to be temperature dependent. In polyatomic gases or liquids, the orientational polarisation decreases with increase in the temperature. Figure 4.12 shows the variation of dielectric constant with the temperature of nitrobenzene. In liquid state, the dielectric constant increases with decrease in the temperature. But, at the melting point when the liquid becomes a solid, the permanent dipoles get frozen in the oriented state or disoriented state and so the orientational polarisation ceases to contribute to the total polarisation on further reduction in temperature. Hence, ionic and electronic polarisation only contribute to the total polarisation. Therefore, dielectric constant shows a sharp fall at melting point, as shown in Fig. 4.12. Below the melting point, dielectric constant does not change with further reduction in the temperature. But, in certain polymers a little effect of temperature is observed below the melting point.

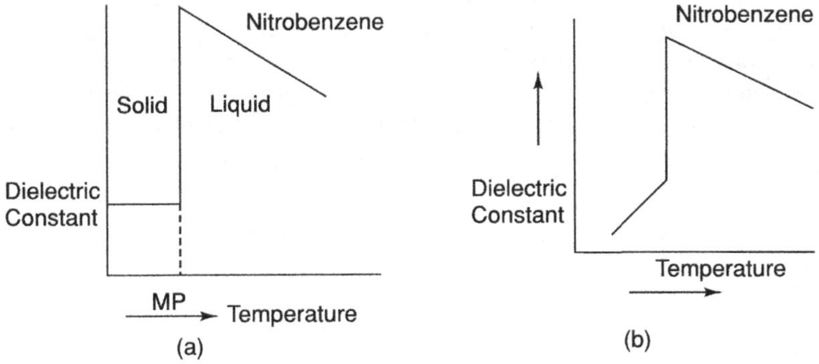

Fig. 4.12 *(a) Variation of dielectric constant in nitrobenzene with temperature*
(b) Variation of dielectric constant with temperature in polymers

4.15 RESPONSE (BEHAVIOUR) OF A DIELECTRIC TO AN ALTERNATING (TIME-VARYING) FIELD

When a dielectric material is subjected to an alternating field, the displacement of charges and orientation of diploes follow the field reversals. When the frequency is increased to a high value (i.e. when high frequency radiations incident upon a dielectric), the charges and electric dipoles in a dielectric material respond to the time-varying field of incident radiation.

From electromagnetic theory, we know that the dielectric constant is related to refractive index as given below:

$$\epsilon_e = n^2 \tag{4.42}$$

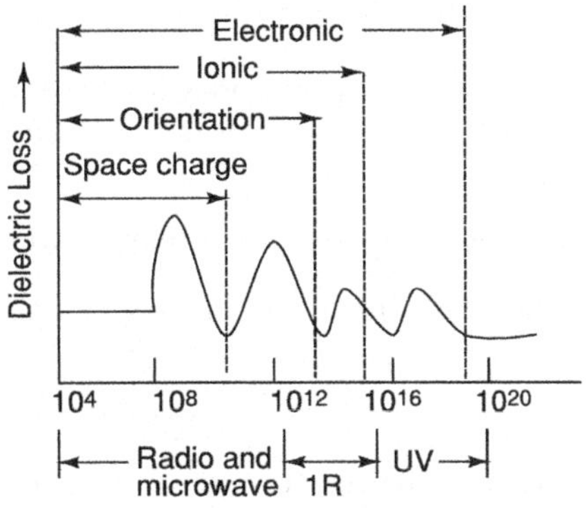

Fig. 4.13 *Dielectric loss vs. frequency*

Among all types of polarisation mechanisms, only electronic polarisation—which is a fast process—exists up to very high frequencies (even up to UV-range). The ionic and dipolar are slower processes due to inertia of ions, and molecular dipoles do not contribute to total polarisation. In other words, the contribution beyond IR is only due to electronic polarisation.

4.16 ELECTRONIC POLARISABILITY IN ALTERNATING FIELD

When time-varying field is applied, the electron cloud is set to forced vibration. Since an oscillating charge radiates em radiation, therefore electron cloud also radiates. When the frequency of time-varying field is not equal to natural frequency of the electron cloud, the forced vibrations have small amplitude and radiation is too weak. This corresponds to molecular scattering. As soon as the frequency of applied field becomes equal to natural frequency of electron cloud, a resonance takes place resulting in a much larger energy. In solid or liquid dielectric, there are strong intermolecular forces, so this energy dissipates very fast. This corresponds to true absorption. In dielectric materials, the resonance occurs at optical frequencies ($\sim 10^{15}$ Hz), therefore dielectric constant $\in_r$ is almost entirely due to electronic polarisation and occurs during each cycle of applied electric field. Let us suppose an electron cloud surrounding the nucleus of charge $+ e$, the radius of cloud being R. On subjecting this system to ac field, the cloud gets displaced by x and nucleus remains at rest. The restoring force that acts on electron cloud is given below:

$$F = -\frac{eq}{4\pi\varepsilon_0 x^2} \tag{4.43}$$

where q is total charge on electron which can be calculated as

$$q = \rho \times \frac{4}{3}\pi x^2 = \frac{e}{\frac{4}{3}\pi R^3} \cdot \frac{4}{3}\pi x^3 = \frac{ex^3}{R^3} \tag{4.44}$$

On substituting in above equation,

$$F = -\frac{e^2 x}{4\pi\varepsilon_0 R^3} = -ax \tag{4.45}$$

where a is constant.

Therefore, the above equation becomes

$$m\frac{d^2x}{dt^2} + ax = 0 \tag{4.46}$$

Solution to this equation may be written as

$$x = x_0 \sin(\omega_0 t + \delta) \tag{4.47}$$

where $\omega_0 = \sqrt{\frac{a}{m}}$ is natural frequency of the system taken under consideration and δ is the phase difference. Substituting the value, we get

$$\omega_0 = \left[\frac{e^2}{4\pi\varepsilon_0 m R^3}\right]^{1/2} \tag{4.48}$$

This ω_0 comes in UV-range, if calculated.

In the above equation, the damping effect due was not considered. The complete equation will therefore be

$$m\frac{d^2x}{dt^2} = -ax - 2b\frac{dx}{dt} \tag{4.49}$$

where the second term is damping term.

In the presence of alternating electric field, the equation becomes

$$m\frac{d^2x}{dt^2} = -ax - 2b\frac{dx}{dt} - eE_0 \cos.wt$$

or

$$m\frac{d^2x}{dt^2} = -ax - 2b\frac{dx}{dt} - e.R_e[E_0 e^{J\omega t}] \tag{4.50}$$

'Re' denotes the real part.

The solution to above equation can be written as

$$x = R_e[A^* e^{J\omega t}] \tag{4.51}$$

where A^* is the complex amplitude.

Substituting Eq. (4.51) in Eq. (4.50), we get

$$R_e\left[\left\{-\omega^2 A^* + \omega_0^2 A^* + J\frac{2b\omega}{m}A^* + \frac{e}{m}E_0\right\}e^{J\omega t}\right] = 0$$

On solving, we get

$$A^* = \frac{(e/m)E_0}{(\omega^2 - \omega_0^2) - J\left(\frac{2b\omega}{m}\right)} \tag{4.52}$$

The induced electric dipole moment, therefore

$$P_e = -e.x$$

$$= -eR_e[A^* e^{J\omega t}] \tag{4.53}$$

Replacing A^* using equation

$$P_e = +eRe\left[\frac{(e/m)E_0.e^{J\omega t}}{(\omega_0^2 - \omega^2) + J\left(\frac{2b\omega}{m}\right)}\right] \tag{4.54}$$

The complex polarisability αe^* is linked to induced dipole moment as

$$P_e = e.R_e[\alpha_e^*.E_0 e^{J\omega t}] \tag{4.55}$$

Comparing Eq. (4.53) and Eq. (4.54), we get

$$\alpha_e^* = \frac{e^2/M}{(\omega_0^2 - \omega^2) + J\left(\frac{2b\omega}{m}\right)} \tag{4.56}$$

Expressing polarisability α_e^* in real and imaginary terms,

$$\alpha_e^* = \alpha_e' - \alpha_e'' \tag{4.57}$$

On rationalising the above equation, we find

$$\alpha_e' = \frac{e^2}{m}\left[\frac{(\omega_0^2 - \omega^2)}{(\omega_0^2 - \omega^2)^2 + \left(\frac{4b^2\omega^2}{m^2}\right)}\right] \quad (4.58)$$

and
$$\alpha_e'' = \frac{e^2}{m}\left[\frac{2b\omega/m}{(\omega_0^2 - \omega^2) + \left(\frac{4b^2\omega^2}{m^2}\right)}\right] \quad (4.59)$$

The above Eqs (4.58) and (4.59) can be useful to see the frequency, and the dependence of real and imaginary polarisabilities α_e' and α_e''.

(i) Frequency Dependence of α_e'

Following points may be noted from Eq. (4.59):

- It the field is static, $\omega = 0$, and therefore

$$\alpha_e' = \frac{e^2}{m\omega_0^2}$$

- If $\omega < \omega_0$, α_e' will be positive
- If $\omega > \omega_0$, α_e' will be negative
- If $\omega = \omega_0$, α_e' will be zero
- If $\omega << \omega_0$, $\alpha_e' = \frac{e^2}{m\omega_0^2}$ and therefore, it is independent of the frequency.

When the frequency approaches ω_0, which is in UV-range, α_e' varies with frequency. When ω is very close to ω_0

then $\quad (\omega_0 - \omega) = \Delta\omega$

and $\quad \omega_0^2 - \omega^2 = (\omega_0 + \omega)(\omega_0 - \omega) = 2\omega_0\Delta\omega$

So,
$$\alpha_e' = \frac{e^2}{m}\frac{\Delta\omega/2\omega_0}{m(\Delta\omega)^2 + b^2/m^2} \quad (4.60)$$

$$\alpha_e'' = \frac{e^2 b/2m^2\omega_0}{(\Delta\omega)^2 + \frac{b^2}{m^2}} \quad (4.61)$$

(ii) Frequency Dependence of α_e''

- For statistic field, when $\omega = 0$ and $\alpha_e'' = 0$, that is, there is no imaginary dielectric constant.
- When $\omega \to \infty$, $\alpha_e'' = 0$
- When $\Delta\omega = 0$, i.e. when $\omega = \omega_0$

$$[\alpha_e'']_{\max} = \frac{e^2}{2\omega_0 b}$$

The variation of α'_e and α''_e is shown in Fig. 4.14.

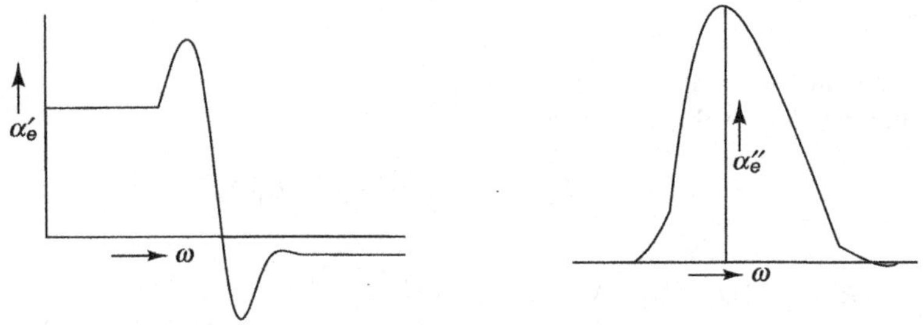

(a) Variation of α' with ω

(b) Variation of α'' with ω shows that it is directly related to power dissipation

Fig. 4.14 *Variation of α'_e and α''_e with ω*

4.17 IONIC POLARISATION IN A TIME-VARYING ELECTRIC FIELD

Ionic polarisation can be treated exactly as in dc fields. Hence, the equation of motion is for displacement of positive and negative ions with respect to each other. The restoring force in this case is proportional to displacement; so the motion is SHM. The mass involved will be the mass of ions. Since mass of ions will be much greater than that of electric, therefore w_0 will lie in the IR range. The ionic polarisability is a complex quantity and can be expressed as

$$\alpha_i = \alpha'_i - J\alpha''_i$$

Frequency dependence will be identical to that of electronic polarisation.

4.18 CLAUSIUS-MOSSOTTI EQUATION FOR AC FIELD (NON-POLAR DIELECTRIC)

In a non-polar dielectric, the polarisation is only due to electronic and ionic polarisabilities. The dielectric constant in ac field is linked with complex polarisabilities α'_e and α''_e and is also complex, denoted by $\in^*$ and is given by

$$\in_r^* = \in'_r - J \in''_r \tag{4.62}$$

Assuming that the local field is given by

$$E_L(t) = E(t) + \frac{P(t)}{3\in_0} \tag{4.63}$$

where $E(t) = E(t) = E_0 e^{Jwt}$

The polarisation $P(t)$ may be written as in static case)

$$P_{(t)} = \in_0 R_e[(\in_r^* - 1)E_0 e^{Jwt}] \tag{4.64}$$

Therefore,

$$E_L(t) = E_0 e^{Jwt} + \frac{1}{3\in_0} \in_0 R_e[(\in_r^* - 1)E_0 e^{Jwt}]$$

i.e.

$$E_L(t) = R_e \frac{[(\in_r^* + 2)(E_0 e^{Jwt})]}{3} \tag{4.65}$$

In terms of total polarisability ($\alpha_e^* + \alpha_i^*$) and local field $E_L(t)$, the polarisation may be written as

$$P(t) = N.R_e(\alpha_e^* + \alpha_i^*)E_L \tag{4.66}$$

when N is the number of atoms and ions per unit volume.

Substituting the above value of E_L, we get

$$P(t) = NR_e(\alpha_e^* + \alpha_i^*).\left\{\frac{(\epsilon_r^* + 2)}{3}.E_0 e^{Jwt}\right\} \tag{4.67}$$

or
$$\epsilon_0 R_e[(\epsilon_r^* - 1).e_0 e^{Jwt}] = NR_e(\alpha_e^* + \alpha_J^*).\left\{\frac{(\epsilon_r^* + 2)}{3}.E_0 e^{Jwt}\right\}$$

On simplifying, we get

$$\frac{\epsilon_r^* - 1}{\epsilon_r^* + 1} = \frac{1}{3\epsilon_0}(\epsilon_\theta^* + \epsilon_i^*) \tag{4.68}$$

which is Clausius–Mossotti equation for the ac field.

4.19 POLAR DIELECTRIC IN AC AND DC FIELDS

Orientational polarisation dominates in the polar dielectric. Since it is a much slower process, hence frequency dependence is quite different from that of non-polar dielectric.

(i) Rise and Decay of Polarisation in the dc Field

Let us assume a specimen of polar dielectric containing N permanent dipoles per unit volume, which is subjected to a dc electric field for a long time, so that the material acquires a saturation polarisation P_{os}. If the field is suddenly switched off at $t = 0$, the polarisation will not immediately become zero, but it will take some time to get random orientation (dipoles) again. The rate of decay of polarisation is proportional to instantaneous polarisation $P_0(t)$.

i.e.
$$\frac{dP_0(t)}{d\tau} = \frac{P_0(t)}{\tau} \tag{4.69}$$

τ, be the relaxation time which is characteristic of a given material that is defined as the time taken by the dielectric to attain a polarisation 63% of its peak value. It is different for different kinds of polarisation process.

The solution of the above equation is given by

$$P_0(t) = P_{os}e^{-t/\tau} \tag{4.70}$$

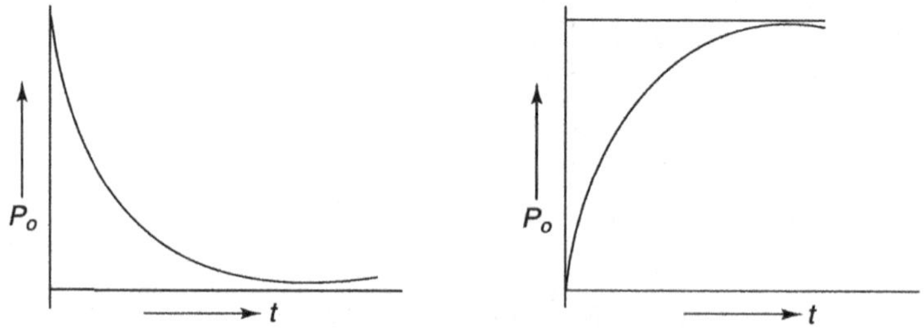

(a) Decay of polarisation (b) Rise of polarisation

Fig. 4.15 *Decay and rise of polarisation*

As $t \to \infty$ and $P_0(t) = 0$, i.e. $P_0(\infty) = 0$, so the above (4.69) equation can be written as

$$\frac{dP_0(t)}{dt} = \frac{1}{\tau}[P_0(\infty) - P_0(t)] \tag{4.71}$$

In a similar way, let the dielectric initially be in unpolarised state, i.e. $P_0(t) = 0$ and as electric field is switched on at $t = 0$, the equation for growth of polarisation can be written as

$$\frac{dP_0(t)}{dt} = \frac{P_0(t)}{\tau} \tag{4.72}$$

whose solution may be given as,

$$P_0(t) = P_{os}(1 - e^{-t/\tau}) \tag{4.73}$$

Note at $t \to \infty$, $P_0(t) \to P_{os}$

the rate equation may be written as,

$$\frac{dP_0(t)}{dt} = \frac{1}{\tau}[P_{os} - P_0(t)] \tag{4.74}$$

In this equation, since P_{os} is the polarisation when $t \to \infty$, therefore Eqs (4.91) and (4.70) are similar. Hence, the growth and decay rate of polarisation is identical.

(ii) Rise and Decay of Polarisation in the ac Field

In order to see how the rise and decay of polarisation takes place in the presence of an ac field, one has to replace ultimate polarisation $P_0(\infty)$ of Eq. (4.71) and P_{os} of Eq. (4.73) by $P_0(t)$ given below.

$$P_{o(t)} = \epsilon_0(\epsilon_r - 1)E_0 \cos \omega t \tag{4.75}$$

Therefore

$$\frac{dP_{0(t)}}{dt} = \frac{1}{\tau}[(\epsilon_r - 1)E_0 \cos \omega t - P_{0(t)}] \tag{4.76}$$

On solving the above equation, a complex dielectric constant will appear ϵ_r^* by the relation

$$P_{o(t)} = \epsilon_0 R_e[(\epsilon_r^* - 1)E_0 e^{Jwt}] \tag{4.77}$$

So that

$$\frac{dP_{o(t)}}{dt} = \epsilon R_e[(\epsilon_r^* - 1)J\omega E_0 e^{J\omega t}] \tag{4.78}$$

Upon using the Eqs (4.77) and (4.78) in Eq. (4.76) and simplifying, we get

$$\epsilon_r^* - 1 = \frac{\epsilon_r - 1}{1 + J\omega t} \tag{4.79}$$

This relates to static dielectric constant ϵ_r with complex dielectric constant ϵ_r^*. This also shows that complex dielectric constant is a function of frequency.

Equation (4.77) may be separated into real and imaginary parts

i.e.

$$\epsilon_r^* - 1 = (\epsilon_r - 1)\left[\frac{1}{1 + \omega^2\tau^2} - J\frac{\omega\tau}{1 + \omega^2\tau^2}\right] \tag{4.80}$$

The complex dielectric constant can also be written as,

$$\epsilon_r^* = \epsilon_r' - J\epsilon_r'' \tag{4.81}$$

Substituting the value of $\in_r^* - 1$ from Eq. (4.80) to Eq. (4.78),

$$P_{o(t)} = \in_0 R_e[(\in_r - 1)]\left\{\frac{1}{1+\omega^2\tau^2} - J\frac{\omega\tau}{1+\omega^2\tau^2}\right\}E_0 e^{J\omega t}$$

Replacing $E_o e^{Jwt} = E_0$ (cos ωt + J sin ωt), we get

$$P_{o(t)} = \frac{\in_0 (\in_r - 1)}{1+\omega^2\tau^2}E_0 \cos \omega t + \frac{\in_0 (\in_r - 1)\omega\tau}{1+\omega^2\tau^2}E_0 \sin \omega t \quad (4.82)$$

4.20 BEHAVIOUR OF POLAR DIELECTRIC AT HIGH FREQUENCIES

From Eq. (4.82), it is clear that first term is in phase with the applied field and the second term lags behind by $\frac{\pi}{2}$. For $w\tau >> 1$, the first term (in phase component of P) decreases to zero. This means that at very high frequencies, the dipoles cannot follow the field variation and hence polarisation vanishes. The out-phase component is a maximum for $w\tau = 1$ and approaches zero for the values on either side of 1.

4.21 DIELECTRIC LOSS

It can be shown that **imaginary part of dielectric constant $\in_r''$ is a measure of the absorption of energy by the dielectric from the alternating field.**

The current density in the material due to external alternating field is given by

$$J(t) = \frac{1}{A}\frac{dQ}{dt} = \frac{d\sigma}{dt}$$

where σ is the charge density and $D = \sigma$

$$J(t) = \frac{dD}{dt} = \frac{d}{dt}[\in_0 E_0 R_e(\in_r^* e^{J\omega t})]$$

or

$$J(t) = \in_0 E_0 \frac{d}{dt}\left[R_e\{(\in_r' - J\in_r'')e^{J\omega t}\}\right]$$

$$= \in_0 E_0 R_e\left[(\in_r' - J\in_r'')J\omega(\cos \omega t + J\sin \omega t)\right]$$

or

$$J(t) = \in_0 E_0 \omega\left[\in_r'' \cos \omega t - \in_r' \sin \omega t\right] \quad (4.83)$$

The above equation also shows that imaginary part of dielectric constant $\in_r''$ is related to the component of the current which is in phase with the applied field. This component of the current gives rise to absorption, not the other which is out-phase with the field.

The energy absorbed by the dielectric material per m^3 per second is given by

$$e(t) = \frac{1}{2\pi}\int_0^{2\pi} J(t).E(t).d(\omega t) \quad (4.84)$$

where $E(t) = E_0 \cos \omega t$

On calculating, we get

$$e(t) = \frac{\omega}{2}\in_0 \in_r'' E_0^2 \quad (4.85)$$

Thus, the energy absorbed by dielectric material is proportional to the imaginary part of dielectric constant. **This phenomenon is known as dielectric loss.**

The dielectric loss is also measured in terms of loss tangent tan δ defined by

$$\tan\delta = \frac{\epsilon_r''}{\epsilon_r'}$$

Example 4.16

For ice, the relaxation time is given equal to 18 × 10⁻⁶ s at 22°C. Calculate the frequency when the real and imaginary parts of complex dielectric constant will became equal? What will be the phase difference between current and voltage at this frequency?

Solution The real and imaginary parts of dielectric constant will be equal when $\omega\tau = 1$

or $$\omega = \frac{1}{\tau} \quad \text{or} \quad f = \frac{1}{2\pi\tau}$$

or $$f = \frac{1}{2\times 3.14\times 18\times 10^{-6}} = 8.8\text{ kHz}$$

The phase difference between the current and voltage at 8.8 kHz is given as

$$\phi = 90 - \delta$$

But $$\delta = \tan^{-1}\frac{\epsilon_r''}{\epsilon_r'} = \tan^{-1}(1) = 45° \qquad \text{Therefore, } \phi = 45°$$ **Ans.**

4.22 DIELECTRIC STRENGTH AND DIELECTRIC BREAKDOWN

Being a perfect insulator, a dielectric material is used in electrical and electronic circuits as insulator and as a medium in capacitors. When the applied electric field is increased, the potential difference across it also increases. A limit is reached when the dielectric ceases to work as an insulator and a spark occurs. **This limiting value of the voltage is known as 'breakdown voltage' which measures the 'strength of the dielectric'.**

$$\text{Therefore, dielectric strength} = \frac{\text{Breakdown voltage}}{\text{Thickness of the dielectric}}$$

In all there are three kinds of breakdown mechanisms:

- Avalanche breakdown
- Thermal breakdown
- Defect breakdown

(i) Factors Influencing the Dielectric Strength

The following are the various factor that influence dielectric strength:

- Breakdown voltage increases with the increase in the thickness of material.
- On increasing temperature dielectric strength decreases.
- The breakdown voltage decreases with increasing frequency (the rate depends on the loss factor of the material).
- On placing dielectric material in series, ac breakdown voltage decreases.
- The presence of air or gas between laminations shows a pronounced decrease in the dielectric strength.

(ii) Avalanche Breakdown

Dielectrics have a large band gap (more than 5 eV) and, therefore, when a high electric field is applied, the electrons in the valence band acquire sufficient energy to overcome the band gap and get excited to the conduction band. The mobility electrons of conduction band also acquire high acceleration and collide with other electrons of the conduction band. Thus, more and more electrons are released to the conduction band resulting in an avalanche of conduction electrons. Ultimately the material becomes highly conducting and is said to have reached a breakdown field.

(iii) Thermal Breakdown

When high frequency ac is applied to a dielectric material, there is a loss of energy known as 'dielectric loss'. This energy dissipates as heat energy. If the dissipation is not effective due to poor conductivity of the material, the material gets heated up and sometimes melts. This is known as thermal breakdown.

(iv) Defect Breakdown

This is found in those dielectric materials which contain process, cracks, etc. At high fields, the local electric field at small crack becomes so high that gas discharge will occur causing the breakdown of the dielectric.

4.23 VARIOUS KINDS OF DIELECTRIC MATERIALS

Similar to magnetic materials, dielectric materials too are classified on the basis of their behaviour in the external applied electric field. These are

- Ferroelectric materials
- Paraelectric materials
- Dielectric materials do not exist

Ferroelectric Materials

In some special types of polar dielectrics, the permanent electric dipole in the material is oriented in a specific direction, even in the absence of electric field. These dipolar dielectrics are known as ferroelectrics. Examples are rochelle salt, perovskite group (titinates and niobates such as $BaTiO_3$) and dihydrogen phosphates and arsenates (KH_2PO_4). These are a special class of pyroelectric materials which are special class of piezoelectrics.

Let us look at some characteristic properties of ferroelectrics:

- All ferroelectric materials exhibit piezoelectric effect.
- Ferroelectric dielectrics exhibit 'hysteresis', i.e. polarisation P is non-linear with E.
- They acquire spontaneous polarisation below a certain temperature.
- Their polarisation is reversible.
- These exhibit ferroelectric transition temperature.

4.24 HYSTERESIS IN FERROELECTRIC MATERIALS

When a ferroelectric material is subjected to to a time-varying electric field, the polarisation (P) lags behind the electric field. This is known as hysteresis. Given below is the nature of variation of P vs. E.

OM Remnant polarisation
OT Coercive field
Ps Saturation polarisation

As *P* vs. *E* curve is non-linear for a ferroelectric material, its dielectric constant is not a constant of *E*. Normally, the dielectric constant of a ferroelectric is measured at low fields. The value of dielectric constant of ferroelectric materials is very high (10^4 – 10^5) as compared to ordinary dielectric constants, which are only around 10.

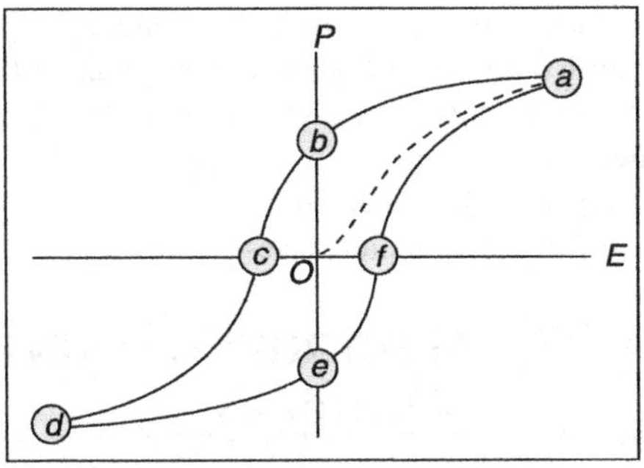

Fig. 4.16 *P vs. E curve for ferroelectric material*

4.25 APPLICATIONS OF FERROELECTRIC MATERIALS IN DEVICES

Ferroelectric materials have several device applications due to their special characteristics. Figure 4.17 summarises the various special properties of ferroelectrics and the corresponding device application.

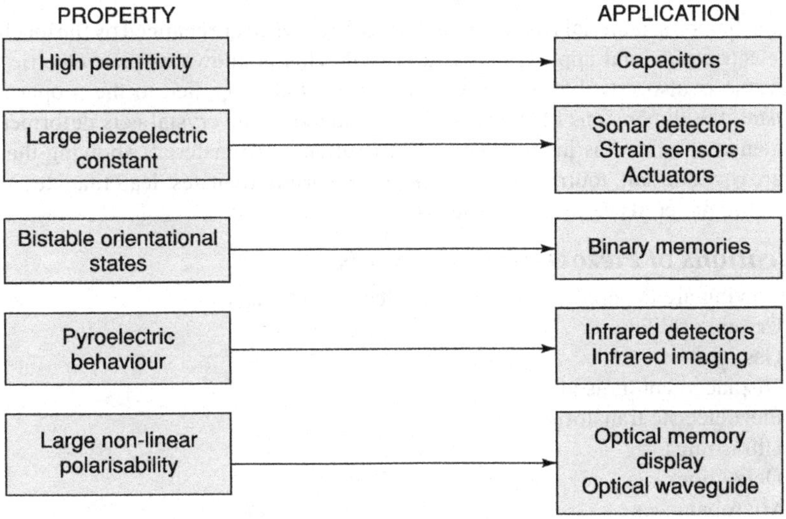

Fig. 4.17 *Special properties of ferroelectrics and their corresponding device application*

4.26 LOSS FACTOR AND ITS SIGNIFICANCE

The product of ϵ_r and tan δ is known as loss factor; tan δ is also sometimes known as loss tangent. The loss factor ϵ_r tan δ is very significant as it characterises the usefulness of a dielectric material as insulator. In every case, a low loss tangent is desirable. As a rule, dielectric losses are larger at high frequencies. This makes it more difficult to select insulating materials for high frequency applications. **In high frequency engineering the quality factor of the insulation is used (instead of tan δ), which is reciprocal of tan δ. Classification of dielectrics is also done on the basis of dielectric loss and is given below:**

(i) Low-loss dielectrics
(ii) High-loss dielectrics

4.27 ELECTROSTRICTION EFFECT AND ELECTROSTRICTIVE MATERIALS

On applying the electric field to a dielectric material, it gets polarised by shifting of electron cloud (electronic polarisation), or by shifting of positive and negative ions, or due to shifting/orientation of atomic dipoles. The dimensions of the material undergo a slight change due to the shifting of ions or orientation of dipoles. Therefore, a strain is developed in the dielectric material. This is observed in all dielectrics and is known as electrostriction effect.

4.28 DIRECT PIEZOELECTRIC EFFECT AND INVERSE PIEZOELECTRIC EFFECT

In certain dielectric materials if the dimensions of the crystal are changed by the mechanical force, electric potential appears across a crystal. This is known as piezoelectric effect. The reverse is also possible, i.e. if a varying potential is applied to the proper axis of the crystal, the dimensions of the crystal get changed or the crystal gets deformed. This phenomenon is named as inverse piezoelectric effect. The materials showing the above effect are rochelle salt, tourmaline, lead ziconate, barium titanates, lead titanates, etc. All ferroelectric materials exhibit piezoelectric effect.

Applications of Piezoelectric Materials

The following are the applications of piezoelectric materials:

- Crystal oscillators
- Gas igniters
- Displacement transducers
- Piezoelectric transformers
- Ultrasonics
- Delay lines
- Microbalances

4.29 PYROELECTRIC MATERIALS

This class is a subset of piezoelectric materials. These materials are spontaneously polarised, but they do not respond to an electric field like ferroelectrics because they require very high electric field for orienting the dipoles. The field required is so high that the material undergoes dielectric breakdown before it can get polarised.

4.30 DIFFERENCE BETWEEN FERROELECTRICITY AND PIEZOELECTRICITY

S.No	Ferroelectric Material	Piezoelectric Material
1.	Crystal gets polarised due to dipole interaction.	Crystal gets polarised by applying external stress.
2.	All ferroelectric materials are piezo electric.	All piezoelectric materials are not ferroelectric.
3.	The piezoelectric coefficient is the ratio of charge developed to the mechanical stress applied along a crystallographic axis.	The ferroelectrics have a large value of piezoelectric coefficient.
4.	These are used in accelerometres. Random access memories, generation of high voltages, electronic frequency optical electro-optical devices, high-frequency devices for medical imaging and surface acoustic devices active vibration centre, all use ferroelectric materials.	These are used in the production and detection of sound, assemblies, and several scanning probe microscopes like STM, AFM, MTA, SNOM, etc.

Summary

- Dielectrics are electrical insulators having covalent and ionic bonding.
- When a dielectric is subjected to an external electric field, positive and negative charges in the dielectric get displaced, which is known as polarisation.
- The capacitance of a capacitor with a dielectric medium between the plates C is greater than that of the capacitor without the dielectric C_0 by a factor E_r. which is called the dielectric constant of the dielectric material.
- The electric field inside the dielectric is less than the applied electric field by a factor of ε_r i.e. $E = E_0/\varepsilon_r$.
- Three vectors D, E and P are related; $D = \varepsilon_0 E + P$.
- The response of a dielectric to ac fields may be understood by representing the dielectric as a capacitor with a series or parallel resistance.
- The current through the dielectric leads the applied ac voltage by $(90 - \delta)$. This is due to the in-phase component of the ac current (loss current) through the dielectric (represented by a series or a parallel resistance). tan δ is called the loss factor.
- Due to this loss current, the dielectric constant of capacitor can be represented as a complex quantity $\varepsilon_r^* = \varepsilon_r' - j\varepsilon_r''$.
- The imaginary part of the dielectric constant ε_r'' is a measure of dissipation or power loss in the dielectric.
- Polarisation, P is a macroscopic property. Polarisation is related to the macroscopic field by the relation

$$P = \varepsilon_0(\varepsilon_r - 1)E$$

- Polarisability, α is an atomic (microscopic) property and is defined as the induced dipole moment per unit local electric field, E_{micr} at the atom. It is a characteristic of a particular atom. The induced dipole moment is given by $p_e = \alpha_e E_{micr}$.

- Clausius–Mossotti equation gives the relation between the macroscopic property ε_r and the atomic property α.
- The three types of polarisation mechanisms in a dielectric are (i) electronic, (ii) ionic, and (iii) dipolar (or orientational).
- When a dielectric is subjected to ac fields, the polarisation decreases with increasing frequency. This is because the ionic polarisation and the dipolar polarisation which are slower processes cannot follow the rapidly varying electric fields. So, at higher frequencies (visible and UV), the contribution to polarisation is only from the electronic polarisation.
- In an ac field, polarisability and the dielectric constant are represented as complex quantities. At low frequencies, the polarisability and dielectric constant are almost independent of frequency and equal to their static values. The electronic polarisability and ionic polarizability show dispersion in the ultraviolet and infrared frequency ranges, respectively.
- The maximum electric field that a dielectric can withstand without breakdown is called the dielectric strength.
- Ferroelectrics are dielectrics which exhibit spontaneous polarisation, that is, they possess polarisation even in the absence of external electric field. In a ferroelectric material, the P vs. E curve shows hysteresis. The dielectric constant of a ferroelectric is very high and its value depends on the applied electric field.
- Crystals which do not possess a centre of symmetry exhibit piezoelectricity. Piezoelectrics are those dielectrics that get polarised when subjected to mechanical stress.
- Pyroelectrics are those dielectric that get polarised when heated. All pyroelectrics are piezoelectrics, but not all piezoelectrics are pyroelectrics.
- Pyroelectrics are a special class of piezoelectric materials and ferroelectrics are a special class of pyroelectrics, which possess spontaneous and reversible polarisation.
- When a high frequency ac field is applied to a dielectric, there will be a dielectric loss and due to this the material gets heated up. So, when dielectrics are used as insulators in electric circuits, different types of dielectrics must be used for different temperature ranges.

Exercise

1. What is a dielectric? How it different from an insulator?
2. Explain electronic and ionic polarisability. Show that the polarisability of an atom of radius r is $4\ \pi\varepsilon_0 r^2$.
3. What is meant by ionic polarisation? How does it vary with temperature?
4. Explain polarisation in a dielectric material. Discuss its various kinds.
5. Distinguish between polar and non-polar dielectrics. Show that the polarisability of a non-polar dielectric is equal to $4\pi\varepsilon_0 r^2$.
6. Explain the behaviour of a dielectric in dc fields. Derive a relation between polarisation P and external electric field E.
7. State and prove Gauss's law in a dielectric material. Derive a relation between E, D and P.

8. Explain the following terms
   (i) Electric field
   (ii) Electric displacement vector D
   (iii) Polarisation vector P
   Also discuss the physical significance of these vectors.
9. Discuss Langevin–Debye theory of orientational polarisation. Also, prove that susceptibility of polar dielectric is inversely proportional to temperature.
10. Define relative permittivity and electric susceptibility and how these are related to each other.
11. Discuss dielectric polarisation in a dielectric material. Derive the following relation
$$P = \epsilon_0(\epsilon_r - 1)E$$
12. Derive an expression for electric field strength on a molecule within a dielectric. Also, obtain Clauuius-Mossotti equation. How can the diameter of atom be estimated with its help?
13. What is meant by Lorentz local field? Show that macroscopic field in the dielectric is given by
$$E_L = E + \frac{P}{3\varepsilon_0}$$
14. Discuss the behaviour of dielectric in alternating electric field. Explain the effect of frequency upon the dielectric constant of a material.
15. Draw a sketch of frequency dependence of different contributions to the polarisability of a dielectric.
16. Explain the terms dielectric loss and loss angle by a phasor diagram.
17. Explain the phenomenon of piezoelectricity and discuss some applications of piezoelectric crystals.
18. What is a dipolar relaxation? Show that the dipolar relaxation leads to complex dielectric constant of the material. Also, deduce an expression for it.
19. What are the ferroelectric materials? Discuss some applications of ferro-electric materials and their characteristics.
20. What is meant by ferroelectric hysteresis? Show that dielectric loss is given by the area of hysteresis loop.
21. What is meant by breakdown phenomenon in dielectric material? Discuss various kinds of breakdown mechanisms.
22. Compare the characteristics of piezoelectric, pyroelectric and ferroelectric crystals, and discuss some of their applications.
23. Write short notes on
   (i) Complex dielectric constant
   (ii) Dielectric loss
24. Prove that the energy dissipated in a dielectric material as dielectric loss is given by
$$P = \frac{\omega}{2} \cdot \epsilon_0 \epsilon_r' E_0^2$$

Numerical Problems

1. The dielectric strength of natural rubber is 40000 volts/mm at 50 Hz. Calculate the thickness of insulation on a wire carrying 44 kV to sustain the breakdown.
   Ans. (1.1 mm)
2. A paper capacitor of 0.03 μF capacitance is of 1 mm thickness, relative permittivity 2.6 and dielectric strength 1.6×10^7 V/m. Determine (a) the total area of dielectric needed for it, and (b) its breakdown voltage. **Ans. (1.29 m^2, 1.6×10^{-4} volts)**

3. Capacitance of a capacitor is 0.03 μf and the power factor is 0.0005. Calculate the dielectric loss of the capacitor when it carries a current of 250 ampere at a frequency of 25 kHz. **Ans. (14740 watts)**
4. Determine the polarization produced in steatite ($MgO.SiO_2$) by an electric field of 500 V/m, if its dielectric constant is 5.1? Ans. 18.1425×10^{-9} c/m^2
5. A simple parallel plate condenser is to be made to store 20 μC at a potential of 10 kV. The separation between the plates is to be 5×10^{-4} m. Calculate the area that the plates must have if the dielectric material between the plates is of alumina of dielectric constant 10. **Ans. 1.12×10^{-2} m^2**
6. A barium titanate crystal is inserted in a parallel plate condenser of plate dimensions 10 mm × 10 mm. A capacitance 10^{-9} F is noted when the plates are separated by 4 mm. Taking $\varepsilon_0 = 8.854 \times 10^{-12}$ F/m, determine the dielectric constant of the crystal. **Ans. 4.51×10^5**
7. Two capacitors are made of: one using glass plate ($\varepsilon_r = 6.0$) of thickness 0.30 and the other using a polymer film with ($\epsilon_r = 3.0$) and thickness 0.20 mm between the metal electrodes. Which one holds greater charge? **Ans. $C_g = 3\ C_p$**
8. Carbon tetrachloride contains 74 electrons in its molecule. Its relative permittivity is 2.26 and density 1.68×10^3 Kg/m^3. If the field acting on the liquid is 5.0×10^6 N/e. What is electronic polarizability, polarization and average electron displacement. **Ans. 7.1×10^{-16} m**
9. A capacitor of capacitance 0.2 μF has a loss factor of 0.004. What will be the power loss when it is connected across a 220 V, 60 Hz supply? **Ans. 1.4×10^{-2} watt**

Self-quiz Questions

1. What is the role of dielectric material in a photocopier machine?
2. Is the behaviour of dielectrics same under ac and dc fields? Explain.
3. What are polar and non-polar dielectrics?
4. What do you mean by dielectric and complex dielectric constant?
5. Discuss electric polarisation.
6. What is electronic polarisation?
7. What is ferroelectricity?
8. Discuss spontaneous polarisation in $BaTiO_3$.
9. What do you mean by piezoelectricity?
10. Discuss piezoelectric effect in quartz crystal.
11. What do you mean by dielectric losses?
12. Define Curie temperature.
13. What is a dielectric? How is it different from an insulator?
14. Explain what is meant by ionic polarisation. How does it vary with temperature?
15. What is meant by loss tangent when referred to polar dielectrics?
16. Define piezoelectricity. Explain piezoelectric effect.

Objective-type Questions

1. When dielectric is subjected to an electric field E_0, the macroscopic field inside the dielectric is
   (a) equal to E_0
   (b) zero
   (c) less than E_0
   (d) greater than E_0
2. The dielectric materials which lack centre of symmetry are
   (a) ferroelectric

(b) piezoelectric
(c) pyroelectric
(d) paraelectric

3. Which one of the following material does not exist?:
(a) Diaelectric
(b) Ferroelectric
(c) Paraelectric
(d) None of these

4. Which of the below given statement is true?
All ferroelectric materials
(a) exhibit piezoelectric effect
(b) possess centre of symmetry
(c) exhibit pyroelectric effect
(d) exhibit spontaneous polarisation

5. The unit of polarisation is
(a) coul-m (b) coul/m
(c) coul/m^2 (d) coul/m^3.

6. Water molecule is
(a) polar
(b) non-polar
(c) ionic
(d) none of these

7. The material that acquires spontaneous polarisation is
(a) ferroelectric
(b) piezoelectric
(c) pyroelectric
(d) paraelectric.

8. The transducers used in microphones are made of
(a) dielectrics
(b) piezoelectrics
(c) insulators
(d) none of these

9. Orientational polarisation
(a) varies exponentially with temperature
(b) varies linearly with temperature.
(c) varies inversely with temperature
(d) is independent of temperature

10. The dielectric constant at optical frequencies is
(a) nearly zero
(b) less than static dielectric constant
(c) greater than static dielectric constant
(d) equal to static dielectric constant.

11. The band gap energy of dielectric is
(a) more than 3 eV
(b) less than 1 eV
(c) zero
(d) none of these

12. Choose the correct relation for dielectric.
(a) $D = \in_0 E$
(b) $D = \in_0 E + P$
(c) $D = \in_0 P$
(d) $D = P \times E$

13. Relation between relative permittivity and electric susceptibility is
(a) $\in_r \in_0$ (b) $\in_r + \in_0$
(c) $\in_r - 1$ (d) $\in_r / \in_0$

14. The polarisation is dependent upon temperature in
(a) ionic materials
(b) non-polar material
(c) polar material
(d) none of these

15. In an isotopic dielectric medium
(a) D and E are perpendicular
(b) D and E are at 45°C
(c) D and E have the same direction
(d) D and P are perpendicular

16. Lorentz local field is given by
(a) $\text{E} + \dfrac{P}{\in_0}$ (b) $\text{E} + \dfrac{P}{3\in_0}$
(c) $\dfrac{P}{3\in_0}$ (d) $E + \dfrac{3\in_0 E}{P}$

17. Which of the following is a correct relation in dielectrics?
(a) $P = \in_0(\in_r + 2)$
(b) $P = \in_r E$
(c) $P = \in_0 E$
(d) $P = \in_0(\in_r - 1)E$

18. The loss factor in a dielectric depends upon
(a) conductivity
(b) RP of dielectric constant
(c) thickness of dielectric
(d) initial polarisation

19. For polar materials, dielectric constant is
    (a) unity (b) infinite
    (c) zero (d) very large

20. Unit of D is
    (a) V/m^2 (b) $Coul/m^2$
    (c) V/m (d) none of these

Answers

1. (a)	2. (b)	3. (a)	4. (a, b, d)	5. (c)
6. (b)	7. (a, c)	8. (b)	9. (c)	10. (b)
11. (a)	12. (b)	13. (c)	14. (b)	15. (c)
16. (b)	17. (d)	18. (d)	19. (d)	20. (b)

Fill in the Blanks

1. Clausius–Mossotti relation is valid for .... and non-polar dielectrics.
2. Non-polar molecules have .... structure.
3. Non-polar molecules can have only .... polarisability.
4. The dielectric polarisation is defined as ....
5. The common feature between ionic and electronic polarisation is ....
6. The relation between dielectric susceptibility and relative permittivity.................
7. Similar to the ϵ_r, the parameter in magnetism is ....
8. The relaxation time is smallest for ....
9. Quartz is an example of ....
10. In polar liquids the dielectric constant .... with temperature.
11. The dielectric constant of a dielectric .... with increasing frequency of applied electric field.
12. Optical absorption and IR absorption refer to ....
13. Line insulators are made up of ....
14. Capacitors for high frequency application require a material whose loss angle is ....
15. A ferroelectric material exhibits ....
16. Piezoelectric effect is observed in ....
17. Insulating material that can bear temperature up to 180°C is of .... types.

Answers

(1) gases (2) symmetrical (3) deformation
(4) induced dipole moment per unit volume
(5) polarisation in both is temperature independent (6) $\chi_e = \epsilon_r - 1$
(7) μ_r (8) electronic polarisation (9) piezoelectric crystal
(10) decreases (11) decreases (12) dielectric losses
(13) porcelain (14) low (15) magnetisation
(16) quartz (17) class c

Reading List

1. Raghavan, V (1990) *Material Science and Engineering*. New Delhi: PHI Learning Pvt Ltd.
2. Seth, S P *Electrical Engineering Materials*. New Delhi: Dhanpat Rai & Co.

3. Rajput, R K *Electric Engineering Material.* New Delhi: Laxmi Publications Pvt Ltd.
4. Vijaya, M S; Rangarajan, G *Material Science.* New Delhi: McGraw-Hill Education Pvt Ltd.
5. Shukla, R K *Fundamentals of Engineering Physics.* Kanpur: Ira Books.
6. Dekker, A J (1998) *Solid State Physics.* New Delhi: Macmillan India Ltd.
7. Gupta, Navneet (2000) *Engineering Practical Physics.* 2nd ed. New Delhi: Dhanpat Rai & Co.
8. Dekker, A J (2000) *Electrical Engineering Materials.* New Delhi: PHI Learning Pvt Ltd.
9. Kittle, Charles (1996) *Introduction to Solid State Physics.* 7th ed. New York: John Willey & Sons.

5

Conductivity of Metals and Superconductivity

Chapter Objectives

- To understand the conduction of electrons in metals.
- To provide explanation for the resistivity and high thermal conductivity in metals.
- To study various factors affecting electrical resistivity, like increase in temperature and alloying.
- To examine the limitations of free electron theory of metals.
- To get an idea about Fermi energy and Fermi function, and to explain some phenomena that use the concept of Fermi energy.
- To understand the Thermoelectric effect phenomenon and various related effects, like Seebeck effect, Peltier effect, Joule–Thomson effect, etc., and to discuss some practical applications of the heating effect.
- To understand superconductivity and transition temperature, and study the behaviour of materials at very low temperature.
- To explain the superconductivity phenomenon, development of high T_C materials and applications of superconducting materials.

5.1 INTRODUCTION

From the electrical point-of-view, materials are classified into three categories:

- Metals
- Insulators
- Semiconductors

Metals or conductors exhibit 'free electrons' in a very large number. They are responsible for high electrical and thermal conductivity. On the other hand, insulators exhibit no free electrons and hence, their conductivity is almost negligible. Semiconducting materials are intermediate in nature and therefore, exhibit very low conductivity at room temperature. Following are certain features of metals, insulators and semiconductors:

(i) Characteristic Properties of Metals

1. Metals exhibit a very high value of electrical and thermal conductivity, which is due to high free electron density.
2. Metals obey Ohm's law, according to which the current density varies directly to electric field strength,

i.e. $J = \sigma E$
where σ is known as electrical conductivity of metal.

3. The electrical conductivity of a metal decreases with the increases in temperature, i.e. they exhibit positive temperature coefficient of resistance.
4. Certain metals exhibit superconducting properties.
5. Metals obey Wiedemann–Franz law. The ratio of thermal to electrical conductivity remains constant at a given temperature, i.e. $\frac{K_T}{\sigma T}$ = Constant, where $K_T \rightarrow$ Thermal conductivity of metal.
6. Resistance of metals increases when a high conductivity material is mixed into it. This happens simply because such type of mixing leads to increase in the defects, which act like scattering centres. This is the reason why alloys are not made with the expectation to improve the electrical properties. Although they are made keeping in view an improvement in other properties, like strength.

(ii) Characteristic Properties of Insulators

1. The atoms of insulators normally have a saturated outermost shell and, therefore, practically no free electrons are available.
2. They exhibit a high value of electrical resistivity.
3. They undergo polarisation (separation of charges) on being subjected to an external electric field.
4. Every insulating material exhibits electrical capacitance which is expressed in terms of permittivity.
5. Permittivity is not constant, but varies with temperature and frequency. When subjected to ac fields, it decreases with increase in the frequency due to the inability of polarising charges to follow the electric field variations.
6. Dielectric strength of the material depends upon temperature and humidity. It decreases with increase in temperature in case of air. In case of liquid insulators, the effect varies with type of oil and its viscosity.
7. When subjected to alternating electric fields, they absorb some electric energy by the field which appear in the form of heat and is known as 'dielectric loss'.

(iii) Characteristic Properties of Semiconductors

Materials whose resistivity is high compared to conductors are known as semiconductors. These are extremely promising materials; they have drastically changed the earlier traditional electronic technology. Following are some features of semiconducting materials:

1. Semiconductors have negative temperature coefficients. Their conductivity increases with the rise in temperature.
2. The conductivity of these materials is high compared to insulators and is sensitive to the impurity addition, i.e. conductivity increases by many orders of magnitude with even a very little impurity addition. This is an additional advantage with semiconducting materials. Such flexibility is not possible in metals.
3. These material do not follow Ohm's law.
4. Band gap of these materials is nearly 1 eV.
5. The semiconducting material is a perfect insulator at 0 K.

Below, in Table 5.1 are given electrical conductivities of some selected materials at 0°C for comparison.

Table 5.1 *Electrical conductivities of some materials at 0°C*

Electrical Engineering Materials					
Metals		*Semiconductors*		*Insulators*	
Element	Electrical Conductivity at 0°C (in ohm^{-1} m^{-1})	Element	Electrical Conductivity at 0°C (in ohm^{-1} m^{-1})	Element	Electrical Conductivity at 0°C (in ohm^{-1} m^{-1})
Copper	5.9×10^7	Silicon	5×10^4	Ebonite	10^{-13}
Silver	6.3×10^7	Germanium	2×10^4	Silica Glass	10^{-11}
Gold	4.1×10^7	Gallium Arsenide	10^{-6}	Mica	10^{-15}
Aluminium	3.5×10^7	....		Bakelite	10^{-9}

In brief, the characteristic properties of electrical engineering materials can be compared on the basis of certain parameters as shown in Fig. 5.1.

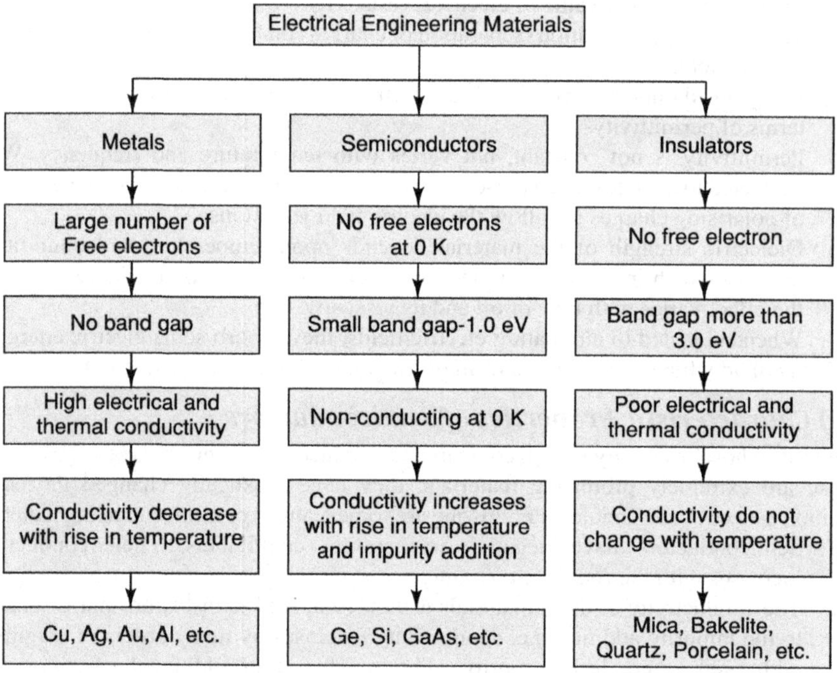

Fig. 5.1 *Parameters for comparing the characteristic properties of electrical engineering materials*

5.2 OHM'S LAW

For an electrical engineer, no other law is as fundamental as the Ohm's law, which is used almost every day. The law states that when an electrical field is applied to a conductor, the random motion of electrons inside converts into an average motion in a definite direction that constitutes an electric current. The current density is found proportional to electric field, E,

i.e. $$J = \sigma E \tag{5.1}$$

σ being a constant known as electrical conductivity and its reciprocal is named as resistivity, which is given by:

$$\rho = \frac{1}{\sigma} \tag{5.2}$$

One important feature of Ohm's law is that the current density remains constant as long as electric field E remains constant.

The motion of free electrons inside the metal may be considered like electron gas. Consider there are n free electrons per m^3. In the absence of external electric field, the velocity vectors associated with the electrons at a given instant add up to zero; hence, there is no resultant transport of charge. On applying the electric field in x-direction, the electrons acquire an acceleration given by

$$a_x = \left(\frac{e}{m}\right)E \tag{5.3}$$

In terms of velocity, it may be written as

$$\frac{\partial v_x}{\partial t} = -\frac{eE}{m} = -\frac{eV}{ml} \tag{5.4}$$

where l is length of the conductor–

or $$a_x = \frac{\partial v_x}{\partial t} = -\frac{eV}{ml} \tag{5.5}$$

While in motion, the electrons collide with other ions and neutral atoms within the conductor. The average time interval between two successive collisions of an electron with ions is known as relaxation time, τ. If v is the velocity gained by the electron in time τ then

Acceleration, $$a_x = \frac{v}{\tau} \tag{5.6}$$

The velocity of electron increases to v during two collisions and become zero after each collision. Thus, average drift velocity of the electrons is

$$v_d = \frac{0+v}{2} = \frac{v}{2} = \frac{a_x \tau}{2} \tag{5.7}$$

On substituting the value of a_x in above, we get

$$v_d = \frac{1}{2}\frac{eV\tau}{ml} \tag{5.8}$$

Also, $$v_d = \frac{J}{ne} = \frac{I/A}{ne} = \frac{I}{neA} \tag{5.9}$$

where J is current density, A is the area of cross-section. On comparing Eqs (5.5) and (5.6), we get

i.e. $$\frac{1}{2}\frac{eV}{ml}\tau = \frac{I}{neA}$$

$$I = \frac{ne^2\tau}{2m}\cdot\frac{A}{l}\cdot V$$

or $$\frac{V}{I} = \frac{2m}{ne^2\tau}\frac{l}{A} = \text{const. } (R) \quad (5.10)$$

The above used constant R denotes the resistance of conductor and the relation is known as Ohm's law.

Therefore, resistance of material $R = \frac{2m}{ne^2\tau}\cdot\frac{l}{A}$

and resistivity $$\rho = \frac{RA}{l} = \frac{2m}{ne^2\tau}$$

i.e. resistivity of material varies inversely with the relaxation time. But relaxation time, τ is inversely proportional to *rms* velocity, and *r.m.s.* velocity is inversely proportional to square root of absolute temperature, i.e.

$$\tau \alpha \frac{1}{v_{r.m.s.}} \alpha \frac{1}{\sqrt{T}}$$

Therefore, $$\boxed{\rho = \sqrt{T}}$$

On plotting resistivity ρ vs. temperature T, we find that some resistivity is still left at $T = 0$, which is known as residual resistivity, denoted by ρ_0. This is also called temperature independent part of resistivity, which is closely related with structural imperfections in the metal. The value of ρ_0 depends on the impurity concentration, as well as, presence of other defects in the metal. It can therefore be taken as a measure of purity of the specimen.

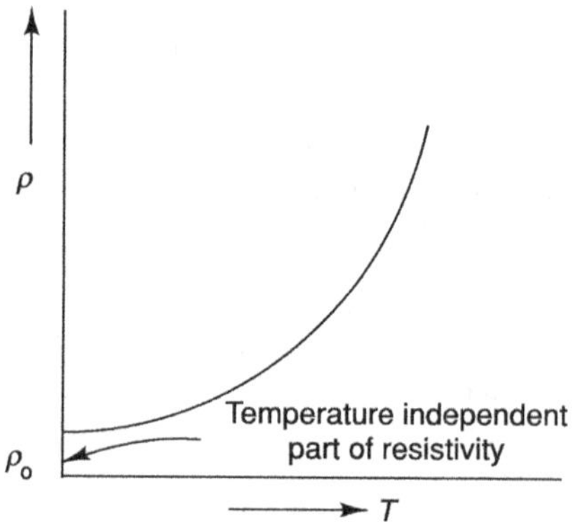

Fig. 5.2 *ρ vs. T for metals*

Equation (5.10) can also be expressed as

$$V = IR$$

$$E_x.l = J_x A.\rho.\frac{L}{A} = J_x.A.\frac{l}{\sigma A}$$

or $$E_x = \frac{J_x}{\sigma}$$ where σ is the conductivity of metal.

or $\boxed{J_x = \sigma E_x}$ which is Ohm's law in terms of conductivity of the metal.

where $\sigma = \frac{ne^2\tau}{m}$ is known as electrical conductivity of the metal.

Example 5.1

A sample of copper wire of 2 mm diameter carries a current of 5 mA. Calculate the average drift velocity of the free electrons. Copper is monovalent and crystallises in FCC structures with a lattice parameter of 3.61 Å.

Solution

$$J = nev_d$$

and $$v_d = \frac{J}{ne}$$

Copper is FCC, so the number of atoms per unit cell is = 4
Volume of the unit cell = $(3.61 \times 10^{-10})^3$ m^3

Therefore, the number of atoms per unit volume $= \dfrac{4}{(3.61\times10^{-10})^3}$

$= 8.5 \times 10^{28}$ atoms/m^3.

Since copper is monovalent, the number of free electrons per unit volume is equal to the number of atoms per unit volume = 8.5×10^{28} m^{-3}.
The current density

$$J = I/A = \frac{5\times10^{-3}}{\pi(1.0\times10^{-3})^2} = 1592.35 \text{ Amp/m}^2$$

$$v = \frac{J}{ne} = \frac{1592.35}{8.5\times10^{28}\times1.6\times10^{-19}} = 117.08\times10^{-9} \text{ m/s}$$ **Ans.**

Example 5.2

The drift velocity is the average velocity exhibited by the electrons in presence of electric field. Calculate the drift velocity of such electrons in an aluminium wire of diameter 1.0 mm carrying current of 6 A. Assume that 4.5 × 10^{28} electrons/m^3 are available for conduction.

Solution

Given: I = 6A, n = 4.5×10^{28} electrons/m^3 and radius

$$r = \frac{d}{2} = \frac{1.0\times10^{-3}}{2} = 5.0\times10^{-4} \text{ m}$$

Current density $J = \dfrac{I}{A} = \dfrac{6.0}{\pi\times(5.0\times10^{-4})^2}$

$$= \frac{6.0}{3.14\times(5.0\times10^{-4})^2}$$

$= 7.6 \times 10^6$ A/m^3

and drift velocity $v_d = \dfrac{J}{ne} = \dfrac{7.6\times10^6}{4.5\times10^{28}\times1.6\times10^{-19}}$

$= 1.05 \times 10^{-3}$ m/s **Ans.**

Example 5.3

The density of Cu is 8.92 × 10^3 kg/m^3 and its atomic weight is 63.5. Determine the current density, if the current of 10 A is maintained in Cu wire of radius 0.7 mm, assuming that only one electron of an atom takes part in conduction. Also, calculate the drift velocity of electrons.

Solution

Given:

Atomic weight = 63.5 kg
Density of copper = 8.92×10^3 kg/m^3, I = 10 A
Radius = 0.7×10^{-3} m

$$\text{Ratio } \frac{N}{V} = \frac{6.02\times10^{28}\times8.92\times10^{3}}{63.5}$$

$$= 8.456 \times 10^{30} \text{ electrons/m}^3$$

$$\text{Current density, } J = \frac{I}{A} = \frac{I}{\pi r^2}$$

$$= \frac{10}{3.14\times(0.7\times10^{-3})^2} = 03.25\times10^{6} \text{ m/s}$$

$$\text{Drift velocity } (v_d) = \frac{J}{ne} = \frac{3.25\times10^{6}}{8.456\times10^{30}\times1.6\times10^{-19}}$$

$$= 4.8 \times 10^{-6} \text{ m/s}$$ **Ans.**

Example 5.4

Resistivity is a characteristics property of the material. A copper wire is 1 m long and has a uniform cross-section of 0.1 mm²; the resistance of the wire at room temperature is found to be 0.182 ohm. What is the resistivity of the material?

Solution As the resistance of wire is given by

$$R = \rho\frac{l}{A}$$

Therefore, resistivity,

$$\rho = \frac{RA}{l} = \frac{0.182\times0.1\times10^{-6}\text{ m}^2}{1\text{ m}}$$

$$\rho = 1.82 \times 10^{-8} \text{ ohm m}$$ **Ans.**

Example 5.5

A uniform silver wire has a resistivity of 1.45 × 10⁻⁸ ohm m at room temperature. On applying electric field along the wire 1.0 volt cm⁻¹, compute the average drift velocity of the electrons, assuming there are 5.8 × 10²⁸ conduction electrons per m³. Also calculate the mobility and relaxation time of the electrons.

Solution Number of silver conduction electrons/ m³ = 5.8 × 10 ²⁸

$$n = 5.8 \times 10^{28}$$

$$\sigma = ne\mu$$

$$\mu = \frac{\sigma}{ne} = \frac{1}{\rho ne}$$

$$= \frac{1}{1.45\times10^{-8}\times5.8\times10^{28}\times1.6\times10^{-19}}$$

$$\mu = 7.0 \times 10^{-3} \text{ m}^2 \text{ volt}^{-1} \text{ s}^{-1}$$

and average drift velocity $v_d = \mu E$

Given that $E = 1$ V/cm $= 10^2$ V/m

$$v_d = 7.0 \times 10^{-3} \times 10^2$$

$$= 7.0 \times 10^{-1}$$

$$v_d = 0.7 \text{ m/s}$$ **Ans.**

Example 5.6

According to free electron theory, each atom contributes 1 electron for conduction process. Calculate the density of electrons in sample of silver which has density 10.5×10^3 kg/m^3 and conductivity 6.8×10^7 ohm^{-1} m^{-1} at 20°C. Also, calculate the mobility of conduction electrons in silver sample. (Atomic weight of silver is 107.9)

Solution The number of atoms per unit volume is given by

$$n = \frac{N\rho}{W} = \frac{6.02\times 10^{26}\times 10.5\times 10^{3}}{107.9} = 5.86\times 10^{28}$$

$$\sigma = n\mu e$$

$$\mu = \frac{\sigma}{ne} = \frac{6.8\times 10^{7}}{5.86\times 10^{28}\times 1.6\times 10^{-19}}$$

$$= 0.725 \times 10^{-2}\ \text{m}^2\ \text{V}^{-1}\ \text{s}^{-1}$$ **Ans.**

Example 5.7

Taking the required data for a common metal copper, show that the relaxation time is of the order of 10^{-14} s.

Solution

$$\sigma = ne^2\tau/m$$

$$\tau = m\sigma/ne^2$$

For copper, $n = 8.5 \times 10^{28}\ \text{m}^{-3}$

$$\sigma = 6 \times 10^7\ \text{ohm}^{-1}\ \text{m}^{-1}$$

Therefore, $$\tau = \frac{9.1\times 10^{-31}\times 6\times 10^{7}}{8.5\times 10^{28}\times (1.6\times 10^{-19})^2} = 2.5\times 10^{-14}\ \text{s}$$ **Ans.**

Example 5.8

Lorentz number is a universal constant which is identical for all the metals. Show that the unit of Lorenz number is watt-ohm / K^2.

Solution Lorenz number is given by

$$L = \frac{\pi^2}{3}\left(\frac{k}{e}\right)^2$$

unit of k is Joule/K. Unit of e is coulomb.

So, unit of L is $$\frac{(\text{joule})^2}{(K^2)(\text{coulomb})^2} = \left(\frac{\text{joule}}{\text{coulomb}}\right)\frac{\text{joule}}{(K^2)\text{Amp s}}$$

$$= (\text{volt})\left(\frac{\text{Joule}}{\text{s}}\right)\frac{1}{(K)^2\,\text{Amp}}$$

$$= \left(\frac{\text{Joule}}{\text{s}}\right)\left(\frac{\text{Volt}}{\text{Amp}}\right)\frac{1}{(K)^2}$$

$$= \text{watt–ohm}/(K)^2$$ **Ans.**

5.3 EXPLANATION FOR THE DEPENDENCE OF ELECTRICAL RESISTIVITY UPON TEMPERATURE

As resistivity of metal varies inversely with relaxation time, τ, therefore $\frac{1}{\tau}$ gives the probability of scattering of electrons in the metal. As the presence of imperfections and phonons (lattice vibrations) both contribute to the scattering of electrons, the probability of scattering can be obtained by considering the above two.
Therefore,

$$\frac{1}{\tau} = \frac{1}{\tau_{\text{phonon}}} + \frac{1}{\tau_{\text{imperfection}}} \tag{5.11}$$

or

$$\rho = \frac{m}{ne^2}\left[\frac{1}{\tau_{\text{phonon}}} + \frac{1}{\tau_{\text{imperfection}}}\right] \tag{5.12}$$

From the previous discussion, we can write

$$\rho = \rho_0 + \rho(T) \tag{5.13}$$

where

$$\rho_0 = \frac{m}{ne^2 \tau_{\text{imperfection}}} \tag{5.14}$$

and

$$\rho(T) = \frac{m}{ne^2 \tau_{\text{phenon}}}$$

Equation (5.12) is called the **Matthiessen's rule of resistivity**.

As ρ_0 depends upon imperfections and defects in metal, $\rho(T)$ strictly is decided by phonons and is known as lattice resistivity,

i.e. lattice resistivity $\rho(T) = \rho - \rho_0$

This is the parameter which is identical for different specimen of the same metal and is the inherent characteristic of metal whatever is its impurity level.

5.4 FREE ELECTRON THEORY OF METALS—REVIEW OF METALLIC CONDUCTION

Free electrons in a metal move inside in a random manner in all the directions, like molecules of an ideal gas. This is the reason why there is no net current in the absence of electric field. Free electrons which are present inside the conducting material are very large in number, therefore, constitute an electron gas. These free electrons were taken into consideration by Drude in 1900 to provide explanation for the distinctive character of metals. Drude theory was named as **Lorentz–Drude theory**. Its basic assumptions are as follows:

- Electrons are present in a very large number in metal; they are similar to molecules of an ideal gas.
- The assembly of free electrons in a metal is called 'electron gas', and this picture is known as 'electron gas model'.
- The free electrons move in a random manner with speed of order of 10^5 m/s at ordinary temperature.
- During their random motions, the electrons collide constantly with imperfections and with themselves. Such collisions have no practical contribution to conductivities.

- The electrical and thermal conductivities exhibited by the metals are only due to these free electrons.
- On applying the electric field, this random motion of electron change into a net motion in a direction opposite to the field, and therefore electrons acquire a net average velocity that constitute currents.
- In thermal equilibrium, free electrons follow Maxwell–Boltzmann (classical) statistics.

5.5 APPLICATION OF LORENTZ–DRUDE FREE ELECTRON THEORY

The Lorentz–Drude theory was applied to calculate the thermal and electrical conductivities and to derive Wiedemann–Franz law.

Electrical Conductivity

Drude used the electron gas model and kinetic theory to explain thermal and electrical conductivities. In general, the motion of electrons is random and the mean velocity $\bar{v}$ is given by kinetic theory of gases and law of equipartition of energy. Therefore,

$$\frac{1}{2}mv^2 = \frac{3}{2}kT \tag{5.15}$$

where m is mass of electron, k is Boltzmann constant and T is absolute temperature.

As soon as electric field is applied, these electron experience acceleration, $\frac{eE}{m}$ and a net velocity and undergo collision with positive ions of the metal. At each collision, the velocity of electron reduces to zero and again they get accelerated. If λ is the mean free path and t is the time taken between two successive collisions, numerically it can be calculated as follows:

$$t = \frac{\lambda}{v} = \frac{\lambda}{\sqrt{\frac{3kT}{m}}} \tag{5.16}$$

During this time, the velocity imparted by electric field $= \frac{eE}{m}t$

$$\text{Therefore, average velocity} = \frac{0 + \frac{eE}{m}t}{2} = \frac{1}{2}\frac{eE}{m}t \tag{5.17}$$

On substituting the value of t in the above equation, we get

$$u = \frac{eE}{2m}\frac{\lambda}{v} = \frac{eE\lambda}{\sqrt{12\,kTm}}$$

If n is the number of free electrons per unit volume of the metal, then the current density will be

$$J = neu = \frac{ne^2m\lambda}{\sqrt{12mkT}} \tag{5.18}$$

And, therefore, electrical conductivity of the metal will be

$$\boxed{\sigma = \frac{ne^2\lambda}{\sqrt{12\,mkT}}} \tag{5.19}$$

The above formula is known as Lorentz–Drude formula for electrical conductivity of a metal.

5.6 EFFECT OF VARIOUS PARAMETERS ON ELECTRICAL CONDUCTIVITY

(i) Variation of Electrical Conductivity of Metals with Temperature

From the above relation, the following conclusions could be drawn:

1. Electrical conductivity (σ) of metals is directly proportional to the number of free electrons per unit volume or electron density. Since n is very large for metals, therefore, electrical conductivity will also be large.
2. With an increase in the temperature of metal, scattering between electrons also increases. It reduces the mobility of charge carriers. This is the reason why conductivity also decreases.
3. The resistivity of a metal (e.g. copper) decreases linearly with decrease in the temperature until it reaches a finite value as shown in Fig. 5.3. According to the given empirical relation,

$$\rho_2 = \rho_1[1 + \alpha(T_2 - T_1)]$$

where α is the linear temperature coefficient of resistivity; ρ_2 and ρ_1 be the resistivites at temperatures T_2 and T_1, respectively.

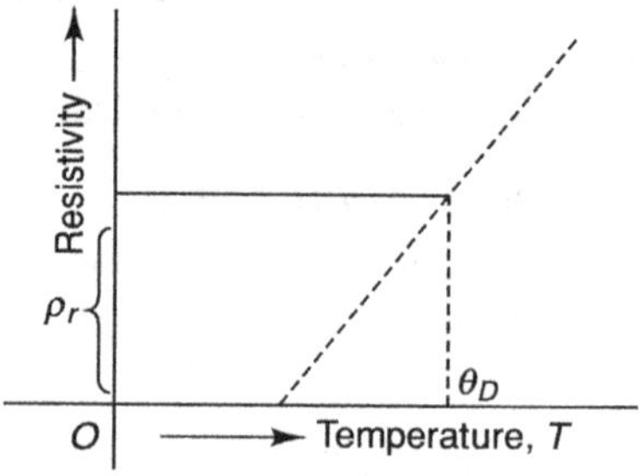

Fig. 5.3 *Variation of resistivity vs. temperature for metals*

(ii) Variation of Electrical Conductivity due to Defects

Departure from perfectness of crystal structure presence of impurity defects, points line defects and planar defects also affect the conductivity of metals.

Figure 5.3 shows that resistivity vs. temperature—although a linear function—gives an intercept which is represented by ρ_r and is known as residual resistivity, which can be interpreted due to the presence of various imperfections, such as impurity, vacancies, grain boundaries or dislocations. The residual resistivity ρ_r is independent of temperature.

According to Matthiessen's rule, the resistivity arises from independent scattering process are additive.

i.e. $$\rho = \rho_r + \rho_{th} \quad (5.20)$$

where $$\rho_r = \rho_{impurity} + \rho_{defect}$$

and ρ_{th} is the thermal resistivity.

The graph between resistivity ρ and temperature T shows that as the temperature increases above to a certain value (known as Debye temperature θ_D), the resistivity of the material increases linearly with temperature T. Debye temperature may be considered as a measure for the maximum frequency of the lattice vibrations; hard materials have high Debye temperature than soft materials. This may be seen in Table 5.2.

Table 5.2 *Debye temperatures of some metals*

Metal	*Ag*	*Al*	*Au*	*Cu*	*Fe*	*Pb*
Debye temp. (θ_D in K)	215	398	180	315	420	88

5.7 RESISTIVITY RATIO

It is defined as the ratio of resistivities of specimen material at room temperature to its resistivity at liquid Helium temperature,

i.e. $$RR = \frac{\rho(298)}{\rho_r(4.2K)}$$

$$\frac{\rho}{\rho_r} = \frac{\rho_r + \rho_{th}}{\rho_r} = \left[1 + \frac{\rho_{th}}{\rho_r}\right] \tag{5.21}$$

For high purity materials, this ratio is as high as 10^4 and for impurity materials this is nearly 10^2. Higher the purity of the material (metal), smaller will be the value of residual resistivity ρ_r and $\frac{\rho}{\rho_r}$ will be more.

5.8 VARIATION OF RESISTIVITY (CONDUCTIVITY) OF ALLOYS WITH TEMPERATURE

Figure 5.4 shows the variation of resistivity with temperature of Cu–Ni alloys having various contents of Ni, which shows that increasing the amount of Ni in Cu does not make any change in the slope, but the resistivity increase can be attributed to the following reasons:

1. Atoms of different sizes cause a vibration in the lattice parameter and hence, in electron scattering.
2. Atoms having different valences introduce a local charge difference, which increases the scattering.
3. Solutes (Ni in above case) having different electron concentration compared to parent metal Cu changes the position of Fermi level and this changes the population density $N(E_F)$ and hence, the conductivity

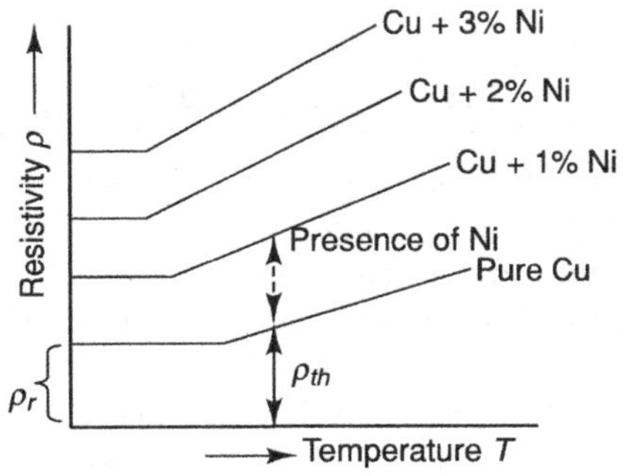

Fig. 5.4 *Resistivity (ρ) vs. T for Cu–Ni Alloys*

$$\sigma = \frac{1}{3}e^2 v_F^2 \tau N(E_F) \tag{5.22}$$

For dilute single phase alloys, resistivity ρ increases with square of the valence difference between the solute. The solvent constituents is

$$\rho \alpha (\Delta Z)^2$$

This is known as Linde's rule, as shown in Fig. 5.5.

In the case of concentrated single phase alloys, the isothermal resistivity has a maximum value near 50% solute (represented in Fig. 5.6 by a solid line). The figure shows the case of copper gold alloys. The residual resistivity of these alloys depend on the fractional atomic components X_A and X_B of the constituents.

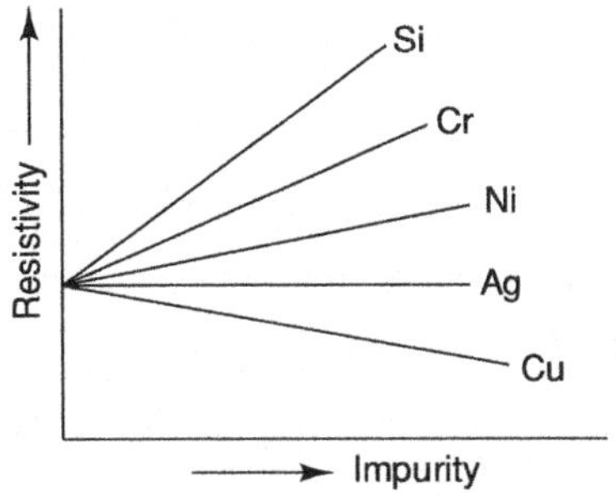

Fig. 5.5 *Change in resistivity vs. impurity atoms*

$$\rho = X_A\rho_A + X_B\rho_B + AX_AX_B$$

where $X_A + X_B = 1$ and A is a material constant.

This rule is known as **Nordheim's rule**, but it is only applicable for few binary systems as it does not take into consideration changes in the density of states with composition.

When solute atoms are periodically arranged then the electron wave scattering takes place coherently. This is why resistivity goes on decreasing. Some alloys such as Cu_3Au, Cu, Au, etc., show long-range ordering.

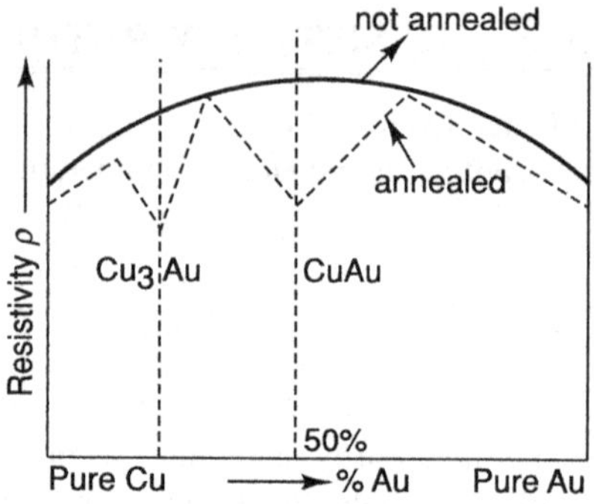

Fig. 5.6 *Resistivity ρ vs. Atomic per cent of annealed and unannealed Cu–Au alloy*

When annealing is done below order-disorder transition temperature then the ordered state can be achieved. But above the transition temperature, ordered state gets destroyed. In case of Cu and Au, the tendency towards ordering is so strong that even near the melting point some ordering remains.

5.9 THERMAL CONDUCTIVITY OF METALS K_T

Thermal conduction is a phenomenon by which the thermal energy transfer occurs from high temperature region to low temperature regions, when both regions are brought into contact. Free electrons do also assist in this process. Since free electrons behave like molecules of an ideal gas, they will have greater kinetic energy at the hot end of the metal bar than at the cold end. Property that characterises the ability of a material to transfer thermal energy (or heat) is known as 'thermal conductivity'.

Let us assume that n electrons are moving randomly in all directions in metal and their motion can be resolved along three axes. Along any one particular direction, only $\frac{n}{6}$ electrons will move. Let us consider three planes of unit area such that plane B and C are at the same distance of mean free path λ from plane A, as shown in Fig. 5.7. The temperatures of these planes are T_1 and T_2, respectively. If the temperature of plane B is greater than C, the transfer of energy will take place from plane B to plane C, i.e. $\frac{nv}{6}$ electrons will transfer from B to C. Since each electron has energy $\frac{1}{2}mv^2 = \frac{3}{2}kT_1$, the net energy transfer from B to C per unit area per second will be

$$= \frac{mv}{6} \times \frac{3}{2}kT_1$$

And similarly, the net energy transfer from C to $B = \frac{nv}{6}.\frac{3}{2}kT_2$

Thus, the net transfer of energy from plane B to C, through plane C per unit area per second.

$$Q = \frac{nv}{6}\left[\frac{3}{2}kT_1 - \frac{3}{2}kT_2\right]$$

$$= \frac{nvk}{4}[T_1 - T_2] \qquad (5.23)$$

But, $$Q = \frac{[T_1 - T_2].K_T}{2\lambda} \quad (5.24)$$

where K_T is thermal conductivity of metal.

From the above two equations,

$$\frac{[T_1 - T_2]K_T}{2\lambda} = \frac{nvk[T_1 - T_2]}{4}$$

Hence, $$\boxed{K_T = \frac{knv\,\lambda}{2}} \quad (5.25)$$

This equation represents the expression for thermal conductivity of a metal.

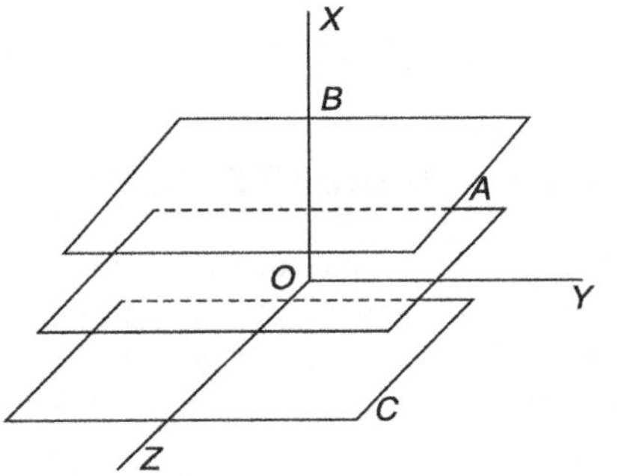

Fig. 5.7 *Planes at different temperature*

5.10 WIEDEMANN-FRANZ LAW

From the above two derived relations for thermal and electrical conductivities, we can now prove Wiedemann–Franz law.

$$\frac{K_T}{\sigma} = \frac{knv\,\lambda/2}{ne^2v\lambda/6kT}$$

$$\frac{K_T}{\sigma} = 3\left[\frac{k}{e}\right]^2 T$$

or $$\frac{K_T}{\sigma T} = 3\left[\frac{k}{e}\right]^2 = \text{Constant} \quad (5.26)$$

Since k is Boltzmann constant and e is electronic charge, therefore, $\frac{K_T}{\sigma T}$ has same value at all the temperatures for all the metals, or the ratio of K_T and σ is directly proportional to absolute temperature

$$\boxed{\frac{K_T}{\sigma T} = \text{Constant}}$$

This is called Wiedemann–Franz Law.

Example 5.9

Calculate the thermal conductivity for a metal assuming

$$\boldsymbol{\tau = 3.0 \times 10^{-14}\ \text{s} \quad T = 3.25\ \text{K}}$$
$$\boldsymbol{n = 2.5 \times 10^{22}\ \text{electrons per cm}^3}$$

Solution

$$\sigma = \frac{ne^2\tau}{m}$$

$$= \frac{2.5\times10^{22}\times(1.6\times10^{19})^2\times3.0\times10^{14}}{9.1\times10^{-31}}$$

$$= 21.09\times10^6 \qquad [1\ \text{cm}^3 = 10^{-6}\ \text{m}^3]$$

$$\frac{K}{\sigma T} = 2.44\times10^{-8}\ \Omega - \text{W/K}^2$$

$$K = 21.09 \times 325 \times 2.44 \times 10^{-8} \times 10^{6}\ \text{W/m-K}$$
$$= 167.24\ \text{W/m-K.}$$ **Ans.**

5.11 LORENTZ NUMBER (*L*)

The right-hand side of Eq. (5.26) is known as the Lorentz number; it is a universal constant. It does not involve any characteristic parameter of the metal and the ratio of thermal to electrical conductivity for a constant given value of temperature is identical for all metals. The value of Lorentz number could be calculated as:

$$L = 3.\frac{k^2}{e^2} = 3 \times \left[\frac{1.38 \times 10^{23}}{1.6 \times 10^{-19}}\right]^2$$

$$= 2.45 \times 10^{-8}\ \text{watt-ohm/k}^2$$

The experimental value of the Lorentz number for metals has been found to agree with the theoretical value, nearly at all temperatures except at very low ones. The experimental value of Lorentz number of some metals are given in Table 5.3

Table 5.3 *Value of Lorentz number of some metals at 273 K and 373 K*

Lorentz number L (in 10^{-8} watt-ohm/ K^2)		
Metal	*at 273 K*	*at 373 K*
Cu	2.23	2.33
Ag	2.31	2.37
Au	2.35	2.40
Pb	2.47	2.56

5.12 SUCCESS AND FAILURE OF CLASSICAL FREE ELECTRON THEORY

Table 5.4 gives a comparison of success and failure of classical Lorentz–Drude theory for free electrons.

Table 5.4 *Success and failure of classical Lorentz–Drude theory for free electrons*

S.No.	*Success*	*Failure*
1.	Explains well why the metals exhibit high value of electrical conductivity.	Theory does not provide any justification why metals exhibit low specific heat.
2.	Explains well why the metals exhibit high value of thermal conductivity.	Theory could not provide explanation for para magnetic susceptibility of conduction electrons.
3.	Explains well why metals obey Ohm's law.	Presence of long mean free path is also not justified in metals on the basis of this theory.
4.	Theory is quite successful at ordinary temperatures.	Theory could not explain why some substances are conductors and others not Unable to explain temperature dependence of electrical conductivity.

5.13 CALCULATION OF SPECIFIC HEAT OF METALS

Let us now calculate the specific heat of metals taking the Classical model into consideration. In a typical metal, each atom contributes one electron to **'electron gas'**. Therefore, in 1 K mol of the metal there are N_0 free electrons. If these electrons behave like molecules of an ideal gas, each would have $\frac{3}{2}kT$ or kinetic energy on the average, and the metal would have

$$Ue = \frac{3}{2}NkT = \frac{3}{2}RT$$

of electron energy per K mole.

The molar specific heat of the electrons, therefore, will be

$$C_v = \left[\frac{\partial Ue}{\partial T}\right]_V = \frac{3}{2}R$$

and the total specific heat of the metal at a temperature above the Debye temperature should be

$$C_v = 3R + \frac{3}{2}R = \frac{9}{2}R$$

But, in reality the value of C_v is only $3R$ at high temperatures. This departure from reality simply means that free electrons do not in fact contribute to the specific heat of metals. Again the scientists were in trouble to note this controversy. Why do the same free electrons which contribute to electronic conductivity and thermal conductivity don't do for specific heat? This was a big question.

To answer this unexpected behavior of electrons, we have to look and rethink the statistical behaviour of electrons. Actually this problem was first of all tackled by Sommerfeld, taking into consideration Pauli's exclusion principle, which is will valid and applicable for the electrons. This was the time when free electrons were given a special status and were treated by the Fermi–Dirac statistics along with principles of quantum mechanics. The theory was developed in the beginning of twentieth century.

5.14 QUANTUM MECHANICAL MODEL SOMMERFELD IDEA

Sommerfeld suspected that free electrons may not be behaving like molecules of an ideal gas. He assumed the assembly of electrons inside metal as Fermions and applied quantum mechanics to resolve the problem. The assumptions that were taken by him are as follows:

- Free electrons are valence electrons of the constituent atoms. Thus, the alkali metals are assumed to contain one free electron per atom while aluminium has three free electrons per atom.
- A valence electron inside the metal feels itself under the influence of the field of all ions and that of other electrons.
- The mutual repulsion between the electrons is neglected and the potential field representing the attractive interaction of ions is assumed to be completely uniform everywhere inside the solid.
- Since the emission of electrons is not possible at room temperature, the potential energy of an electron at rest inside the metal is assumed to be lower than that of an electron outside the metal.

- The free electron inside the metal at room temperature, therefore, can be treated as particle trapped in a box, such that it is totally free inside and cannot escape outside the box as the metal surfaces are the sites of a sharp potential barrier of height E_s.
- The free electron gas does not obey Maxwell–Boltzmann statistics.

Figure 5.8 represents the situation as though by Sommerfeld, taking into consideration the effect of Pauli's exclusion principle which permits only one electron per quantum state or max two (one with spin up and the other with spin down).

Figure 5.8 shows the distribution of electrons in various quantum states, as per the exclusion principle. E_F is energy of electrons which are at the topmost level at 0 K, known as Fermi energy and Φ is the work function of the metal.

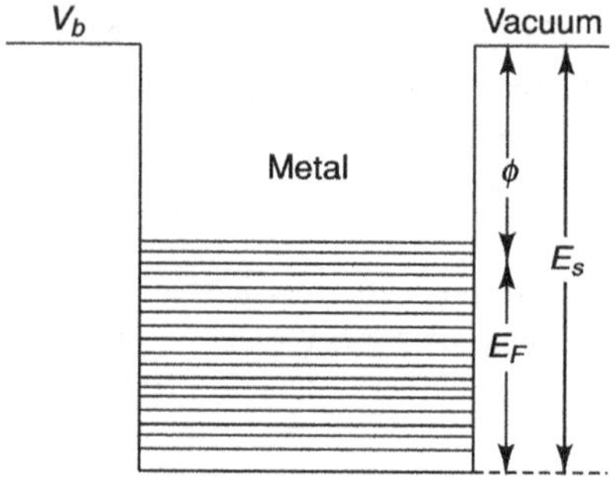

Fig. 5.8 *Electron assembly inside a metal*

5.15 ENERGY LEVEL AND DENSITY OF STATES FOR FREE ELECTRON GAS

Inside situation for electron may be assumed similar to a particle trapped in a three-dimensional box, where particle (electron) is completely free inside the box and cannot cross the height of the barrier, represented by V_b in Fig. 5.9.

Schrödinger's equation for such electron can be written as

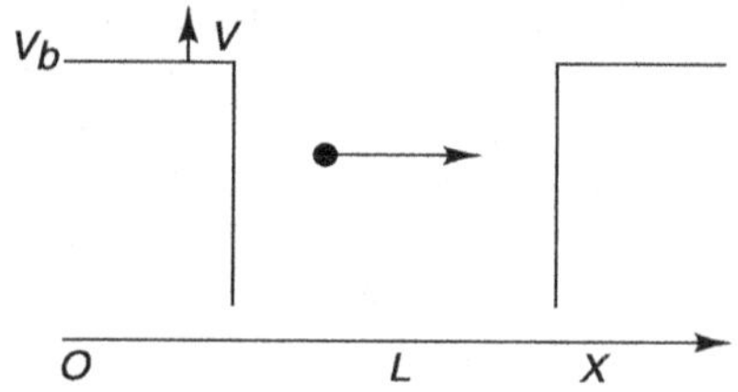

Fig. 5.9 *Situation of electron inside metal*

$$\frac{1}{X}\frac{\partial^2 X}{\partial x^2}+\frac{1}{Y}\frac{\partial^2 Y}{y\partial y^2}+\frac{1}{Z}\frac{\partial^2 Z}{\partial z^2}+\frac{2m}{\hbar^2}(E_x+E_y+E_z)=0 \tag{5.27}$$

The total energy of the trapped electron can be given by

or $$E_{n_x,n_y,n_z} = \frac{\pi^2\hbar^2}{2m}\left[\frac{n_x^2}{a^2}+\frac{n_y^2}{b^2}+\frac{n_z^2}{c^2}\right]$$

where $$n_x = 1, 2, 3\ldots$$
$$n_y = 1, 2, 3\ldots$$
$$n_z = 1, 2, 3\ldots$$

Zero value of n_x, n_y or n_z is not possible quantum mechanically.

For a cubical box of side L,

$$E_{n_x,n_yn_z} = \frac{\pi^2\hbar^2}{2mL^2}\left[n_x^2+n_y^2+n_z^2\right]$$

$$E_{n_x,n_yn_z} = \frac{h^2}{8mL^2}[n_x^2+n_y^2+n_z^2]$$

The number of allowed states between for such electrons energy range E and $E + dE$ can be given by

$$g(E)dE = 4\pi V\left[\frac{2m}{h^2}\right]^{3/2} E^{1/2}dE$$

or $$g(E)dE = CE^{1/2}\,dE \qquad (5.28)$$

where $$C = 4\pi V\left[\frac{2m}{h^2}\right]^{3/2}$$

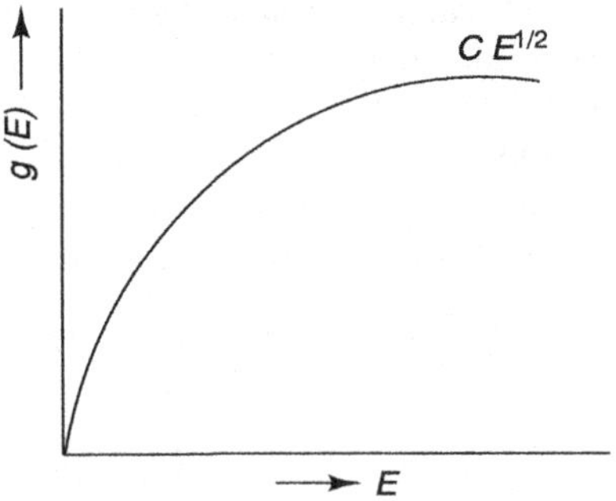

Fig. 5.10 *Plot of g(E) against E*

On plotting *g (E)* vs. *E*, we find that *g (E)* is a parabolic function of energy as shown in Fig. 5.10.

But, to find the actual states occupied by free electrons at temperature *T*, let us use Fermi–Dirac statistics, which is applicable to the free electron assembly. To know the probability that a particular quantum state is occupied, let us investigate the Fermi function.

$$f(E) = \frac{1}{e^{(E-E_F)/kT}+1} \qquad (5.29)$$

where E is the energy of allowed state and E_F is the Fermi energy (highest occupied level).

5.16 BEHAVIOUR OF FERMI FUNCTION AT ABSOLUTE ZERO

At $$T = 0 \text{ if } E < E_F$$

$$f(E) = \frac{1}{e^{-\infty}+1} = 1 = \frac{n_r}{g_r} \quad \text{one electron per quantum state}$$

if $$E > E_F\, f(E) = \frac{1}{e^{\infty}+1} = 0 \quad \text{all states above } E_F \text{ are empty}$$

and for any other temperature $T > 0$ when $E = E_F$

$$f(E) = \frac{1}{e^{0/KT}+1} = \frac{1}{2}$$

From the plot of Fermi function *f (E)* or occupation index it can be noticed that it shows a sharp cut-off at $E = E_F$. Using the above feature one can define Fermi energy.

Fermi Energy, E_F

Free electrons of a metal are particles exhibiting spin (either up or down) and are governed by Pauli's exclusion principle. Therefore, these are treated by Fermi–Dirac statistics.

The compulsion of Pauli's exclusion principle restricts their number at a particular energy level, even at absolute zero temperature. That is, even at 0 K only one electron or maximum two (one with spin up and the other with spin down) can be there at a particular level. Hence, the total population of electrons gets distributed among various energy levels beginning from zero (lowest) to a certain maximum. Therefore, only few electrons, which are located at the top level or near to it, become significant.

Energy of these electrons at the highest occupied level at 0 K is known as Fermi energy. From the nature of Fermi function it is evident that the energy levels above the Fermi energy level are always empty, and below it they are completely occupied (at 0 K).

So Fermi level can be defined as the highest occupied energy level (at $T = 0$) below which all the states are completely filled and above it all empty.

Taking this fact into consideration, $T = 0$ can be calculate by the below given method.

$$N = \int_0^\infty N(E)dE = \int_0^\infty f(E)g(E)dE$$

$$= \int_0^\infty E^{1/2}dE$$

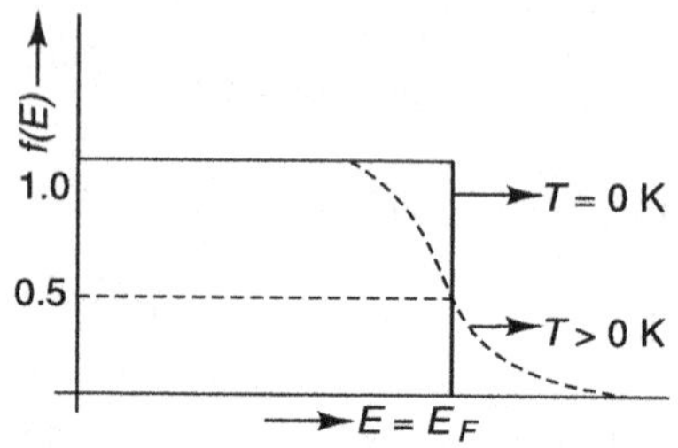

Fig. 5.11 *Fermi function at 0 K and above*

On substituting the value of C in the above, we get

$$E_{F_0} = \frac{h^2}{2m}\left[\frac{3n}{8\pi}\right]^{2/3} \tag{5.30}$$

where $n = \frac{N}{V}$ is the number of free electrons per unit volume.

Using the above formula, Fermi energy for various metals can be calculated, which comes out as

$$E_{F_0} \text{ for sodium } = 3.1 \text{ eV}$$

$$E_{F_0} \text{ for silver } = 5.5 \text{ eV}$$

$$E_{F_0} \text{ for copper } = 7.04 \text{ eV}$$

This implies that even at absolute zero there are some electrons that have energy of several electron volts. Naturally, these are the electrons of the topmost level. This is entirely different than what was expected classically. Classically all the electrons should have zero energy. This indicates the possibility of more than one energy state at 0 K as well, which is a consequence of the Pauli's exclusion principle. Hence, the electrons divide themselves among various energy levels beginning from zero up to a certain maximum known as Fermi energy. Therefore, as per Fermi–Dirac statistics, the scheme of distribution of electrons can be represented as given shown in Fig. 5.12.

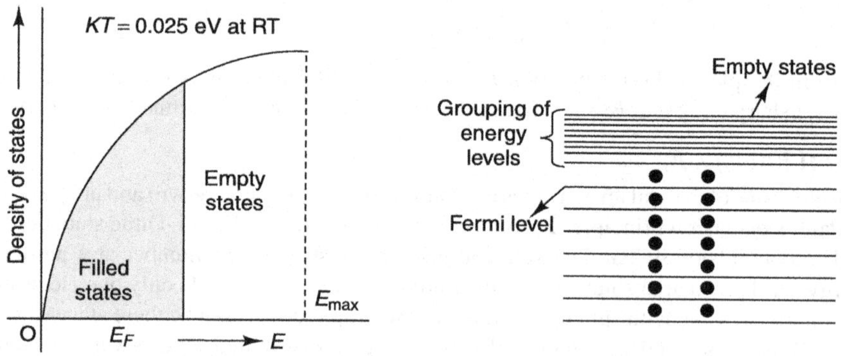

Fig. 5.12 *Scheme of electron distribution at 0 K*

It is now clear (from Fig. 5.12) that the number of electrons at the topmost level (Fermi level) that actually participate in various processes are not all, but only few. The electrons at the Fermi level or just near to the Fermi level pick-up energy and occupy the high energy level near to the Fermi level.

Example 5.10

Calculate the energy difference between the ground state and first excited state for an electron in a one-dimensional box of length 10^{-10} m, $V = 0$ for $0 \le x \le a$ and $V = \infty$ for $0 > x > a$.

Solution Under the given condition, the energy for the nth level is

$$E_n = \frac{\hbar^2 k^2}{2m} = \frac{n^2 h^2}{8ma^2} \text{ [where } a \text{ is the length of one-dimensional box]}$$

so,

$$E_1 = \frac{h^2}{8ma^2}(n=1)\left[\text{and } k = \frac{n\pi}{a}\right]$$

and

$$E_2 = \frac{4h^2}{8ma^2}(n=2)$$

The energy difference between first excited and ground state is

$$\Delta E = E_2 - E_1 = \frac{3h^2}{8ma^2} = \frac{3\times(6.62\times10^{-34})^2}{8\times9.1\times10^{-31}\times(10^{-10})^2}$$ **Ans.**

$$= 1.806\times10^{-17} J = 112.87 \text{ eV}$$

Example 5.11

Fermi energy is the energy exhibited by free electrons at 0 K and is characteristics of a metal. Compute the Fermi energy for sodium, assuming that each sodium atom contributes one free electron to the electron gas. The density of sodium is 0.971 g/cm³ and its atomic weight is 23. ($h = 6.63 \times 10^{-34}$ joule-s, $m = 9.11 \times 10^{-31}$ kg and Avogadro's number is 6.02×10^{23} atoms/mole).

Solution Let V be the volume of a piece of sodium and N the number of free electrons in it. The free electron density, N/V, in sodium is equal to the number of sodium atoms per unit volume—because each sodium atom contributes one free electron to the gas. This could also be found from the density and atomic weight of sodium.

The mass of 1 mole (or g-atom) of an element is M, where M is the molecular (or atomic) weight of the element. Its volume is M/ρ, where ρ is the density of the element. It contains N molecules (or atoms), where N is the Avogadro's number. Therefore, the number of molecules (or atoms) per unit volume is

$$\frac{N}{M/\rho} = N\rho/M$$

The electron density in sodium is thus

$$\frac{N}{V} = \frac{N\rho}{M}$$

$$= \frac{(6.02\times10^{23}\text{atoms/mole})(0.971\text{ gram/cm}^3)}{23\text{ gram/mole}}$$

$$= 2.54\times10^{22}\text{ atoms/cm}^3$$

$$= 2.54\times10^{28}\text{ atoms/m}^3$$

$$= 2.54\times10^{28}\text{ electrons/m}^3$$

The Fermi energy for sodium is

$$\varepsilon_F = \frac{h^2}{2m}\left(\frac{3N}{8\pi V}\right)^{2/3}$$

$$= \frac{(6.63\times10^{-34}\text{ joule}-\text{s})^2}{2(9.11\times10^{-31}\text{ kg})}\left(\frac{3\times2.54\times10^{28}\text{ m}^3}{8\times3.14}\right)^{2/3}$$

$$= 5.0\times10^{-19}\text{ joule}$$

$$= \frac{5.0\times10^{-19}}{1.6\times10^{-19}} = 3.1\text{ eV}$$ **Ans.**

Example 5.12

There are 2.54 × 10 ^{22} free electrons per cm^3 in sodium. Calculate its Fermi energy, Fermi velocity and Fermi temperature. (h = 6.63 × 10^{-34} joule-s, m = 9.11 × 10^{-31} kg, k = 1.38 × 10^{-23} joule/K, 1e V = 1.6 × 10^{-19} joule).

Solution The Fermi energy of a metal is given by

$$\varepsilon_F = \frac{h^2}{2m}\left(\frac{3N}{8\pi V}\right)^{2/3}$$

where m is the mass of electron and N is the number of free electrons in a volume V of the metal. Hence,

$$\frac{N}{V} = 2.54\times10^{22}/\text{cm}^3 = 25.4\times10^{28}/\text{metre}^3$$

Substituting this and the values of m and h, we get

$$\varepsilon_F = \frac{(6.63\times10^{-34}\text{ Joule}-\text{s})^2}{2\times(9.11\times10^{-31}\text{ kg})}\left(\frac{3\times2.54\times10^{28}/\text{m}^2}{8\times3.14}\right)^{2/3}$$

$$= 5.0\times10^{-19}\text{ joule} = 3.1\text{ eV}.$$

This is the (maximum) kinetic energy of the free electron at absolute zero. If v_F be the Fermi velocity, then

$$\frac{1}{2}mv_F^2 = 5.0\times10^{-19}\text{ joule}$$

$$v_F = \sqrt{\frac{2\times5.0\times10^{-19}\text{ joule}}{9.11\times10^{-31}}}$$

$$= 1.05\times10^6\text{ m/s}.$$

The Fermi temperature T_F is defined by

$$T_F = \frac{\varepsilon_F}{k} = \frac{5.0\times10^{-19}\text{ joule}}{1.38\times10^{-23}\text{ joule/K}}$$

$$= 3.6\times10^4\text{ K}$$ **Ans.**

Example 5.13

The electronic configuration of zinc is (1s^2 2s^2 2p^6 3s^2 3p^6 3d^{10} 4s^2), its density 7.13 g/cm^3 and its atomic weight is 6.54. Calculate its Fermi energy. The effective mass of a free electron in zinc crystal is 7.7 × 10^{-31} kg and the Avogadro's number is 6.02 × 10^{23} atoms/gram-atom.

Solution The electronic configuration of the ground state of Zn shows that each atom has 2 valence electrons which it contributes to the electron gas.

The mass of 1 gram-atom of Zn is 65.4 gram, so that the volume

$$V = \frac{\text{Mass}}{\text{Density}} = \frac{65.4}{7.13} = 9.17 \text{ cm}^3$$

The number of Zn atoms in this volume is 6.02×10^{23}. Therefore, the number of atoms per unit volume

$$= \frac{6.02 \times 10^{23}}{9.17} = 6.56 \times 10^{22}/\text{cm}^2$$

Since each contributes two valence electrons, the number of free electrons per unit volume is

$$\frac{N}{V} = 2 \times 6.56 \times 10^{22}$$

$$= 13.1 \times 10^{22}/\text{cm}^3 = 13.1 \times 10^{28}/\text{metre}^3$$

Now, the Fermi energy of the metal Zn is

$$\varepsilon_F = \frac{h^2}{2m}\left(\frac{3N}{8\pi V}\right)^{2/3}$$

where m is the (effective) mass of the electron in zinc. Substituting the values, we get

$$\varepsilon_F = \frac{(6.62 \times 10^{-34} \text{ joule} - \text{s})^2}{2 \times (7.7 \times 10^{-31} \text{ kg})}\left(\frac{3 \times 13.1 \times 10^{28}/\text{m}^3}{8 \times 3.14}\right)^{2/3}$$

$$= \frac{(6.62 \times 10^{-34})^2}{2 \times 7.7 \times 10^{-31}}(6.25 \times 10^{18})$$

$$= 17.78 \times 10^{-19} \text{ joule}$$

$$= \frac{17.78 \times 10^{-19}}{1.6 \times 10^{-19}} = 11 \text{ eV}$$ **Ans.**

Example 5.14

The Fermi energy for lithium is 4.72 eV. Calculate the number of conduction electrons per unit volume in lithium. ($h = 6.63 \times 10^{-34}$ J.-s, $m = 9.11 \times 10^{-31}$ kg).

Solution The Fermi energy is

$$\varepsilon_F = \frac{h^2}{2m}\left(\frac{3N}{8\pi V}\right)^{2/3}$$

where N is the number of conduction electrons in volume V of the lithium metal. The number of conduction electron per unit volume is therefore

$$\frac{N}{V} = \left(\frac{2m}{h^2}\right)^{3/2} \varepsilon_F^{3/2} \frac{8\pi}{3}$$

Then, with $\varepsilon_F = 4.72 \text{ eV} = 4.72 \times 1.6 \times 10^{-19}$ joule, we have

$$\frac{N}{V} = \left[\frac{2 \times 9.11 \times 10^{-31} \text{kg}}{(6.63 \times 10^{-34} \text{ joule–s})^2}\right]^{3/2} \times \frac{(4.72 \times 1.6 \times 10^{-19} \text{ joule})^{3/2} \times 8 \times 3.14}{3}$$

$$= 4.64 \times 10^{28}/\text{m}^3$$ **Ans.**

Example 5.15

For lithium, the Fermi energy is 4.70 eV and the density of electrons is $4.6 \times 10^{28}/m^3$. Find the electron density for a metal with Fermi energy 2.20 eV.

Solution The electron density for a metal with Fermi energy E_F is given by

$$\frac{N}{V} = \left(\frac{2m}{h^2}\right)^{3/2} \varepsilon_F^{3/2} \frac{8\pi}{3}$$

or

$$\frac{N}{V} \alpha \varepsilon_F^{3/2}$$

The electron density for the metal is

$$\left(\frac{N}{V}\right)_{\text{metal}} = 4.6\times10^{28}/m^3\left(\frac{2.20}{4.70}\right)^{3/2}$$

$$= 1.4 \times 10^{28}/\text{m}^3$$

Ans.

5.17 EXPLANATION FOR LOW SPECIFIC HEAT OF METALS

Low specific heat exhibited by metals, therefore, can be justified on the basis of the new scheme of distribution of electrons among different energy levels at 0 K. When a metal is heated, only those electrons which are at Fermi level or close to Fermi level absorb the energy and occupy the excited states; the remaining at lower energy levels never. We can estimate the number of such electrons by assuming that *N(E)* is directly proportional to *E* near the top of energy distribution. Therefore, the number *N* of electrons that contribute to specific heat of a metal at temperature *T* is

$$N = N_0 \frac{\Delta E}{E_F} = N_0\left(\frac{kT}{E_F}\right) \quad (5.31)$$

where E_F is the Fermi energy, N_0 is the number of free electrons per *k* mole (Avogadro's Number). Each of the *N* electrons within *kT* of E_F, acquired $\frac{3}{2}kT$ of energy. So, the total electronic contribution to the internal energy of the metal is

$$U_e = N_0\left(\frac{kT}{E_F}\right)\times\frac{3}{2}kT$$

$$= \frac{3}{2}\left(\frac{kT}{E_F}\right)RT$$

Therefore, electronic specific heat

$$C_{V_e} = \left(\frac{\partial U_e}{\partial t}\right)_V \approx \frac{3}{2}\left(\frac{kT}{E_F}\right)R$$

A more detailed calculation shows that

$$\boxed{C_{V_e} = \frac{\pi^2}{2}\left(\frac{kT}{E_F}\right)R} \quad (5.31)$$

At room temperature ($T = 300$ K), $kT = 0.025$ eV and E_F ranges from 1.53–1.8 eV for different metals. Thus, less than 1% of the electrons absorb energy. Their contribution to the specific heat, if calculated using the above formula, comes out 10^{-2} R. This is negligible compared to the atomic contribution of $3R$. If the temperature goes above 10^4 K, then only the electronic contribution could become comparable with the lattice contribution.

At very low temperatures, however, the electronic contribution exceeds the lattice contribution because the electronic specific heat remains proportional to T, but the atomic specific heat becomes proportional to T^3 (Debye theory). Close to absolute zero, the specific heat of metals is actually observed proportional to T, thus verifying the electronic contribution, and hence Fermi–Dirac distribution.

5.18 AVERAGE KE OF ELECTRONS

The average KE of an electron may be calculated from

$$\overline{E}_0 = \frac{1}{N}\int_0^{E_{F0}} E.g(E)dE$$

or

$$\overline{E}_0 = \frac{1}{N}\int_0^{E_{F0}} E.CE^{1/2}dE$$

$$= \frac{1}{N}\int_0^{E_{F0}} C.E^{3/2}dE = \frac{2}{5N}CE_{F_0}^{5/2}$$

On putting the value of N

$$\overline{E}_0 = \frac{2}{5}\cdot\frac{CE_{F_0}^{3/2}}{\frac{2}{3}CE_{F_0}^{3/2}} = \frac{3}{5}E_{F_0}$$

$$\boxed{\overline{E}_0 = \frac{3}{5}E_{F_0}}$$

Example 5.16

Silver is a highly conducting material having free electons in a large number. The Fermi energy of these electrons in silver metal is 5.4 eV. (a) Estimate the average energy of the free electrons in silver at 0 K? (b) At what temperature a classical free particle (such as ideal gas molecule) will have this kinetic energy? ($k = 1.38 \times 10^{-23}$ joule/K and 1 eV = 1.6×10^{-19} joule).

Solution

(a) The average energy of an electron in electron gas at 0 K is

$$\overline{\varepsilon}_0 = \frac{3}{5}\varepsilon_F$$

where ε_F is the Fermi energy. Now, $\varepsilon_F = 5.4$ eV. Therefore,

$$\overline{\varepsilon}_0 = \frac{3}{5}\times 5.4 = 3.25$$

(b) The kinetic energy of a classical particle at Kelvin temperature T is $\frac{3}{5}kT$. Also, if the particle has an energy 3.306 eV then we have

$$\frac{3}{5}kT = 3.256 \text{ eV} = 3.25\times(1.6\times 10^{-19}) \text{ joule}$$

where $$k = 1.38\times 10^{-23} \text{ joule/K}.$$

Therefore,

$$T = \frac{2}{3} \times \frac{3.25 \times 1.6 \times 10^{-19} \text{ joule}}{1.38 \times 10^{-23} \text{ joule/K}}$$

$$= 2.51 \times 10^4 \text{ K.}$$ **Ans.**

Example 5.17

The average energy of free electrons in a metal is linked with the Fermi energy which is the lowest energy exhibited by the electrons at 0 K. Determine the average energy and speed of electron and its mean energy at 0 K, if the Fermi energy is 15 eV.

Solution

Given $E_F = 15 \text{ eV}$

Average energy $\overline{E}_0 = \frac{3}{5} E_F$

$$= \frac{3}{5} \times 15 \text{ eV}$$

$$= 9.0 \text{ eV}$$

and $\frac{1}{2} mv^2 = \overline{E}_0 \text{ or } v = \sqrt{\frac{2\overline{E}_0}{m}}$

$$v = \sqrt{\frac{2 \times 9.0 \times 1.6 \times 10^{-19}}{9.1 \times 10^{-31}}}$$

$$v = 1.77 \times 10^6 \text{ m/s}$$ **Ans.**

Example 5.18

Fermi energy of a given substance is 7.5 eV. What is the average energy and speed of electron in this substance at 0 K?

Solution:

Given $E_F = 7.5 \text{ eV}$

Average energy $\overline{E}_0 = \frac{3}{2} E_F$

$$\overline{E}_0 = \frac{3}{5} \times 7.5 \text{ eV}$$

$$= 4.5 \text{ eV}$$

and $\frac{1}{2} mv^2 = \overline{E}_0 \text{ or } v = \sqrt{\frac{2E_0}{m}}$

or $v = \sqrt{\frac{2 \times 4.5 \times 1.6 \times 10^{-19}}{9.1 \times 10^{-31}}} = 1.25 \times 10^6 \text{ m/s}$ **Ans.**

Example 5.19

There are 2.5 × 10²⁸ free electrons per cubic metre of sodium. Calculate the Fermi energy and Fermi velocity.

Solution

Given $\frac{N}{V} = 2.5 \times 10^{28}$

Formula used is

$$E_F = \frac{1}{2}mv_F^2 = \frac{\hbar}{2m}\left(3\pi^2.\frac{N}{V}\right)^{2/3}$$

or $$E_F = \frac{\hbar^2}{8\pi^2 m}\left(3\pi^2.\frac{N}{V}\right)^{2/3}$$

$$= \frac{(6.62\times10^{-34})^2}{8\times(3.14)^2\times9.1\times10^{-31}}[3\times(3.14)^2\times2.5\times10^{28}]^{2/3}$$

$$= 4.99 \times 10^{-19}\text{ J}$$

$$E_F = 5.0 \times 10^{-19}\text{ J}$$

or $$E_F = 3.12\text{ eV}$$

and Fermi velocity $$v_F = \frac{\hbar}{m}\left(3\pi^2.\frac{N}{V}\right)^{1/3}$$

$$= \frac{6.62\times10^{-34}}{2\times3.14\times9.1\times10^{-31}}(3\times(3.14)^2\times2.5\times10^{28})^{1/3}$$

$$= 1.05 \times 10^6\text{ m/s}$$ **Ans.**

5.19 CONSTANCY OF WORK FUNCTION OVER A WIDE RANGE OF TEMPERATURE

The work function of a metal is found practically constant over a wide range of temperature. This fact can also be understood on the basis of quantum theory of free electrons. Work function of a metal is defined as the minimum energy needed to knockout an electron from the highest occupied level to a nearby empty level. Vast increase in temperature does not make any appreciable change in thermal energy kT. kT is found to be very less compared to Fermi energy, hence work function remains practically constant over a wide range of temperature.

5.20 SUCCESS OF QUANTUM FREE ELECTRON THEORY

The quantum free electron theory, despite much simpler, is highly successful in explaining most of the properties of metals, such as the low specific heat, electrical conductivity, electronic paramagnetism and thermionic emission. The most exciting feature of this theory is that even at very low temperatures the electrons have appreciable energy because they obey Fermi–Dirac distribution and are governed by Pauli's exclusion principle. Only those electrons which are close to the Fermi level take part in excitations—either due to thermal energy or due to external electric and magnetic fields.

5.21 HEAT PRODUCED IN CURRENT-CARRYING CONDUCTOR

If I be the current flowing in a conductor and R be its resistance then the total heat produced in the wire is given by

$$W_T = H_T = I^2 R = \frac{V^2}{R} = \frac{(E.l)^2}{\rho.\frac{l}{A}} \sigma E^2 lA \qquad (5.32)$$

This is known as Joule heating.

The amount of heat produced per unit volume in the current carrying conductor, therefore, will be

$$W = H = \sigma E^2 \qquad (5.33)$$

where σ is conductivity of the material of conductor and E is the electric field strength.

The above equation can also be expressed in term of current density J,

i.e.
$$H = JE \qquad (5.34)$$

On applying the electric field across the conductor, the drift motion of electron takes place and electrons are accelerated by increasing the field which is given by

$$a = \frac{e}{m} E \qquad (5.35)$$

where e is the electronic charge and m is the mass of electron.

Due to this acceleration, the velocity of electron increases and becomes $\frac{e}{m} E.t.$ Assuming that electric field is applied along the x-direction, the increase in energy of electron is given by

$$\Delta W = \frac{1}{2} mv^2 = \frac{1}{2} ma^2 t^2 = \frac{1}{2} m. \frac{e^2}{m^2} E^2 t^2 \qquad (5.36)$$

Average increase in the energy of electron

$$\overline{\Delta W} = \int_{t=0}^{\infty} \frac{e^2 E^2 \tau^2}{2m} \frac{dt}{\tau}$$

On integrating, we get

$$\overline{\Delta W} = \frac{e^2 E^2 \tau^2}{m}$$

Assuming that these electrons transfer their energy to crystal lattice, the total energy dissipated will be

$$W = \frac{n}{\tau} \frac{e^2 E^2 \tau^2}{m}$$

$$= \left[\frac{ne^2 \tau}{m} \right] E^2$$

or
$$\boxed{W = \sigma E^2} \qquad (5.37)$$

This is the energy transferred by the electrons during the collision process on applying the electric field across the conductor, which appears as heat in the conductor.

5.22 THERMOELECTRIC EFFECT

If a junction is made of two different metals and one end is kept cold and the other hot, a current flows through the circuit. This current is known as thermoelectric current and the

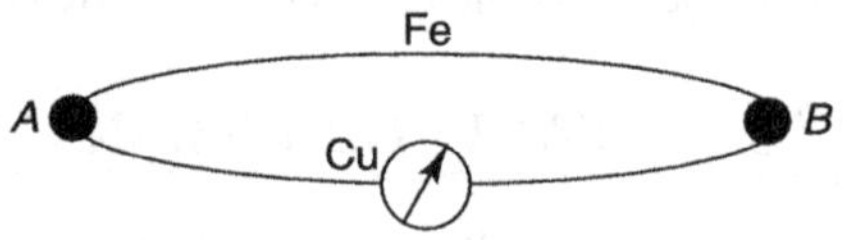

Fig. 5.13 *A Cu-Constantan thermocouple*

phenomenon is known as Seebeck effect or thermoelectric effect. The emf so developed is known as thermo-emf.

Thermocouple

The magnitude and the direction of emf depend upon the metals and the temperature of hot and cold function. Such a combination of two metals is known as 'thermocouple'. This fantastic device is very useful in measuring those temperatures that can not be measured by a thermometer. For example, the furnace temperature which is normally very high (or order of 1000°C or more) and the oven temperature. All these can be measured using a thermo couple. Figure 5.13 shows a simple Cu-Constantan thermocouple.

5.23 THERMOELECTRIC SERIES

For given temperatures of hot and cold junctions, the direction of current in a thermocouple depends on the metal chosen. Metals are arranged in a particular sequence, which may be used to predict the direction of the current in the temperature range 0°C to 100°C. The sequence of the metals is known as thermoelectric series. The following is a list of metals. Out of these if any two form a thermocouple, the current flows across the cold junction from the metal occurring first in the series to the one occurring later.

Selenium, antimony, iron, cadmium, zinc, silver, gold, (lead) mercury, copper, platinum, nickel, bismuth

The metals preceding lead are taken to be thermoelectrically negative and succeeding lead are thermoelectrically positive.

The farther the metals are apart in the above series, the greater will be in the thermo-emf produced in the circuit for a given difference of temperature between the two junctions. As Sb and Bi are quite distant, the thermo emf produced is large. Because of to this reason, the Sb-Bi thermocouples are preferred. Current flows from Sb to Bi across the cold junction.

5.24 SEEBECK'S EXPERIMENT

This remarkable experiment was performed in 1821 by Seebeck. He observed a flow of current in the junction circuit (containing two different metals) by simply maintaining a temperature difference across the junction. The experimental set-up is given below in Fig. 5.14.

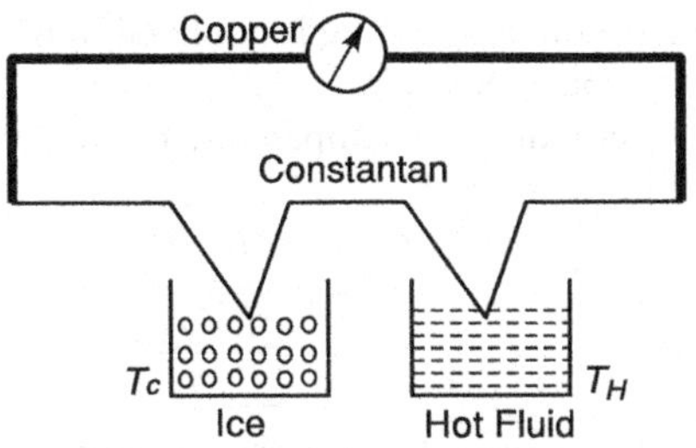

Fig. 5.14 *A thermocouple maintained at different temperatures*

No deflection in the galvanometer was observed if both of the junctions were at same temperature. When one end of junction was kept at melting temperature of ice and the other was heated gradually, current was observed in the circuit. The current increases until the hot junction is at a temperature of 270°C. If the heating was continued beyond 270°C, the current decreases, and finally at 540°C the current drops to zero.

It was discovered by Cumming in 1823 that on increasing the temperature of hot junction beyond 540°C, the direction of current is reversed. It flows from iron to copper through the hot junction and copper to iron at the cold junction.

5.25 NEUTRAL TEMPERATURE

The variation of thermo-emf vs. temperature is shown in Fig. 5.15. It is observed that on increasing the temperature difference, the thermo-emf rises and attains a saturation value. A further rise in the temperature decreases the thermo-emf. The temperature at which thermo-emf attains a peak-value is known as neutral temperature. Mathematically, the dependence of thermo-emf on temperature can be expressed as

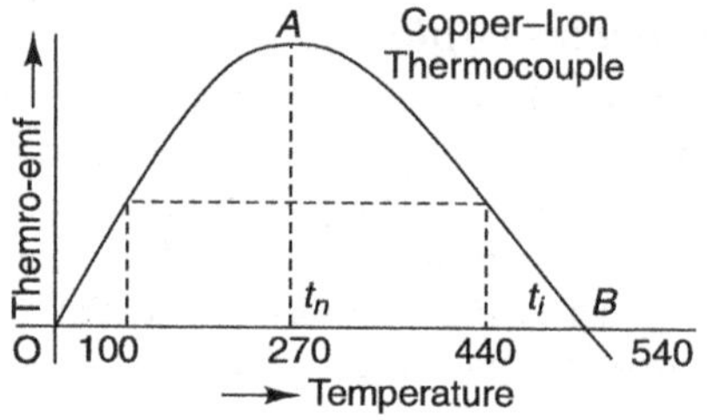

Fig. 5.15 *Variation of thermo-emf with temperature*

$$E = at + bt^2 + \ldots\ldots\ldots\ldots \qquad (5.38)$$

where a and b are constants.

5.26 THERMOELECTRIC POWER

From the above equation

$$\frac{dE}{dt} = a + bt \qquad (5.39)$$

The quantity $\frac{dE}{dt}$ is known as thermoelectric power.

The emf is maximum when $\frac{dE}{dt} = 0$

This means $t = -\frac{a}{b}$ which is neutral temperature.

The emf becomes zero when, $t = -\frac{2a}{b}$, which is known as inversion temperature.

Point A in the graph represents neutral temperature denoted by t_n and point B, where the thermo-emf changes its sign, represent inversion temperature denoted by t_i. The inversion temperature is not fixed. It is as much above the neutral temperature as the cold junction is below it.

Assuming the temperature of cold junction t_1, neutral temperature t_n and temperature of inversion t_2

$$t_n - t_1 = t_2 - t_n$$

or

$$\boxed{t_n = \frac{t_1 + t_2}{2}}$$

In the case of Cu-Fe thermocouple, if the cold junction is at 100°C, the temperature of inversion will be 440°C.

Example 5.20

A power generating station, producing 10 MW power, is connected to a town situated 20 km away by a copper feeder of resistance 0.1 ohm/km. The power is being transmitted at 33 kV; determine the (a) efficiency of transmission, and (b) percentage voltage drop in the feeder.

Solution

Since power $P = V \times I$ watt

$\therefore$ $I = \dfrac{P}{V} = \dfrac{10 \times 10^6}{33 \times 1000} = 303$ A

Total resistance of the feeder $R = 20 \times 0.1 = 2$ ohms

Hence, power lost in feeder $= I^2 R$ watt

$$= \frac{303^2 \times 2}{1000} \text{ kW} = 184 \text{ kW}$$

(a) Therfore, efficiency of transmission $= \dfrac{\text{Power delivered}}{\text{Power supplied}}$

$$= \frac{(10 \times 10^6) - 184}{10 \times 10^6} \times 100 = 99.99\%$$

(b) And, voltage drop in the feeder $= I \times R$

$$= 303 \times 2 = 606 \text{ V}$$

Therefore, percentage voltage drop $= \dfrac{606}{33 \times 1000} \times 100 = 1.8\%$ **Ans.**

5.27 REASON FOR THERMOELECTRIC EFFECT

When two metals are joined, a contact potential equal to difference in the work function of the two metals is established at the junction. The work function is defined as the energy difference between the topmost level and the Fermi level. The Fermi level is subjected to a small change in temperature which is of order of 10^{-5} to 10^{-4} eV/K. This causes a difference in the contact potentials at two junctions due to the difference in temperature at the two ends of thermocouple, and results in emf which drives the current in the circuit. The magnitude of emf is of the order of a few microvolts per degree temperature difference between the two junctions.

5.28 LAW OF INTERMEDIATE METAL

Let E_{AB} be the thermo-emf between metals A and B, E_{AC} between A and C, and E_{BC} between the metals B and C. Also, suppose the temperature of cold junctions is identical in the above three cases and the temperature of the hot junction is also the same.

Then,

$$E_{AB} = E_{AC} - E_{BC} \tag{5.40}$$

This law is known as the law of intermediate metal.

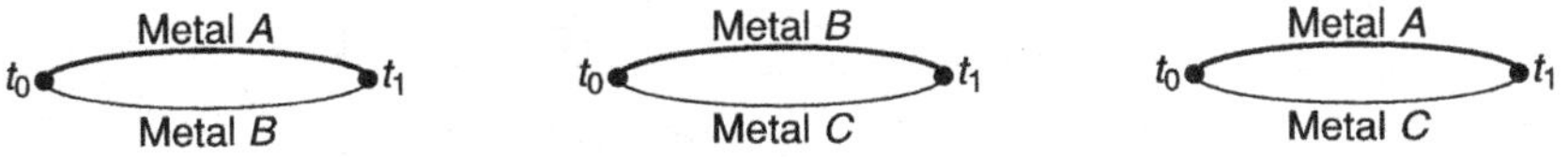

Fig. 5.16 *Thermocouples of different metals*

We know

$$E_{AB} = a_{AB}t + \frac{1}{2} b_{AB} t^2$$

$$E_{BC} = a_{BC}t + \frac{1}{2} b_{BC} t^2$$

$$E_{AC} = a_{AC}t + \frac{1}{2}b_{AC}t^2$$

On using these in the above equation, we get

$$a_{AB} = a_{AC} - a_{BC}$$

and $$b_{AB} = b_{AC} - b_{BC}$$

Table 5.5 shows the values of constants *a* and *b* with Pb metal.

Table 5.5 *Values of constants a and b for different metals*

Metal with Pb	*Constant a* (μv/°C)	*Constant b* (μv/°C)^2
Aluminium	– 0.047	0.003
Bismuth	– 43.7	– 0.47
Copper	2.76	0.012
Gold	2.90	0.0093
Iron	16.6	– 0.030
Nickel	19.1	– 0.030
Platinum	–1.79	– 0.035
Silver	2.50	0.012
Steel	10.8	– 0.016

This table, therefore may be used to find the values of *a* and *b* for any pair of metals in the list.

5.29 LAW OF INTERMEDIATE TEMPERATURES

Let E_{t_1, t_2} be the thermo-emf of a given thermocouple when the temperatures of the junctions are maintained at temperatures t_1 and t_2.

Then $$E_{t_1, t_2} = E_{t_1, t_3} + E_{t_3, t_2} \qquad (5.41)$$

This is known as the law of intermediate temperatures.

The above two laws show that we can include any metal wire or a galvanometer in the thermocouple circuit without changing the emf of the circuit.

The choice of metals for the thermocouples depends upon the accuracy and range of temperature measurement. Suitable materials for different ranges of temperature are given in Table 5.6.

Table 5.6 *Materials for thermocouples and thermopiles*

Temperature range	*Pair of metals*
Up to 1000°C	Fe-constantan
Up to 1400°C	Pt and Pt-Rh
Up to 1700°C	Mo-W
Thermopile:	
For very small	Sb-Bi

5.30 APPLICATIONS OF THERMOELECTRIC EFFECT

Below are some important applications of thermoelectric effect:

- It is useful in the measurement of temperatures of ovens and furnaces via thermocouple.
- A number of thermocouples are joined in a series to make a thermopile, which is also used for temperature measurement.
- It is useful in the detection of radiations by radio micrometre which is made by a series of thermocouples.
- It also comes handy in the measurement of ac and dc by thermo-millimetre.

5.31 PELTIER EFFECT

The reverse of Seebeck effect is also possible; in 1834 Peltier discovered it. Suppose, the two junctions of a thermocouple are initially at the same temperature and electric current is passed through the circuit by using a battery. It is observed that heat is evolved at one end of the junction, which becomes hot, and heat is absorbed at the other end, which becomes cold. If the direction of current is reversed, the colder junction becomes hot and *vice versa*. The heat absorbed or liberated at the junction is proportional to the charge passed through the it. If ΔH that is liberated or observed when a charge Δq is passed through junction, the Peltier emf is defined as

$$\pi_{AB} = \frac{\Delta H}{\Delta q} = \frac{\text{Peltier heat}}{\text{Charge transferred}} \tag{5.42}$$

A and B (Fig. 5.17) refers to the metals from which a thermocouple is made. π_{AB} (Peltier emf) is taken as positive if heat is absorbed by the junction when the electric current flows through the junction from metal A to B. When two different metals are joined at a point, this much emf is developed across the junction.

The heat developed in Peltier effect is not the heat as is observed in Joule heating when a current is passed through a resistor. Joule heat always warms up the resistor whatever be the direction of current, but in Peltier Effect, both heating and cooling are observed depending upon the direction of current in the junction.

Another difference between Joule heating and Peltier heating is

$$H_{\text{peltier}} \ \alpha \ \text{Current } (i)$$

while

$$H_{\text{joule}} \ \alpha \ i^2$$

Also Peltier heating or cooling is observed at the junction only, while in Joule heating the whole resistor gets warmed up. The practical application of Peltier effect is in refrigeration.

5.32 REASON FOR HEATING AND COOLING IN PELTIER EFFECT

Consider again the Cu–Fe thermocouple containing a battery which drives the current from iron to copper as shown in Fig. 5.17.

Contact potential difference between two different metals of thermocouple results in heating and cooling. Potential of iron is higher than that of copper. Hence, for a current flowing from Cu to Fe, i.e. from a lower to higher potential at B, energy is needed. This energy is absorbed from junction B and that's why junction B is cooled. At junction A, current that flows from Fe to Cu is from higher to lower potential, therefore heat energy is evolved and heats up junction A.

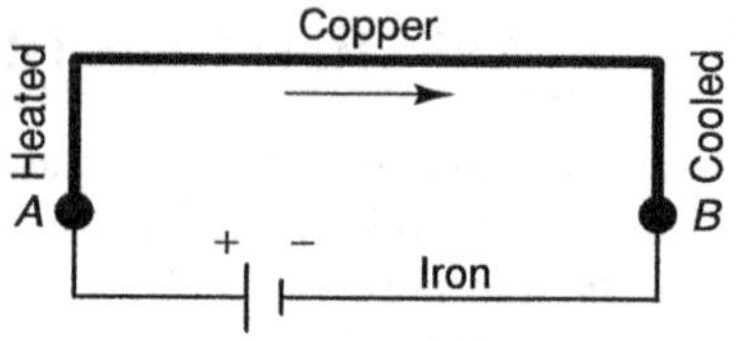

Fig. 5.17 *Cu–Fe thermocouple connected by a battery*

5.33 PELTIER COEFFICIENT

Peltier coefficient may be defined as—the amount of heat absorbed or liberated in Joules at any junction of a thermocouple, when a unit current flows through it for one second. If is denoted by π.

Following are some characteristics properties of Peltier coefficient:

- It is different for different pairs of metals of a thermocouple.
- It depends upon the temperature of junction of thermocouple.
- It is numerically equal to the difference of contact potential.

5.34 THOMSON EFFECT

In 1876, Thomson observed that if a piece of metal or a metallic wire has a temperature gradient and a current is passed through it, heat may be absorbed or liberated in different sections of the wire. This heat is over and above the Joule's heat $i^2 Rt$ and is known as Thomson heat. The effect is called Thomson effect.

Thomson heat produced or absorbed in a small section of a given wire is proportional to the charge passed through the section and the temperature difference between the ends of wire or metal piece.

If a charge ΔQ is passed through a small section of the wire having a temperature difference ΔT between the ends, the Thomson heat is

$$\Delta H = \sigma(\Delta Q).(\Delta T) \tag{5.43}$$

where σ is a constant for a given metal at a given temperature which is measured in volts per °C. For common metals, it is of order of a few micro volts per degree Celsius.

The quantity $$\sigma \Delta T = \frac{\Delta H}{\Delta Q} = \frac{\text{Thomson heat}}{\text{Charge difference}}$$

is known as Thomson emf.

In fact, this much amount of emf is produced when the ends of the section of the wire are maintained at different temperatures. The Thomson coefficient σ is taken to be positive if heat is absorbed when a current is passed from the low temperature end to the high temperature end. Copper, silver, zinc, antimony cadmium, etc., have positive Thomson coefficient. Iron, cobalt, nickel, bismuth and platinum have negative Thomson coefficient. In these metals, heat is absorbed when current is passed from hotter end to the colder end. In lead, σ is almost zero.

5.35 REASON FOR THOMSON EMF

If the temperature of a metal piece is not uniform throughout, the density of free electron therein also varies. The electrons tend to diffuse from higher concentration regions to lower concentration regions. This leads to emf between the hot and cold part of the metal. This emf is Thomson emf. If a current is forced through a wire having temperature difference, positive or negative work is done on the charge carriers depending on the direction of current, and hence, thermal energy is either liberated or absorbed.

The density of free electrons in a metal is dependent on temperature. Hence, Peltier emf developed across the junction is also dependent upon temperature of the junction. If in a thermocouple both junctions are kept at the same temperature, the Peltier emf developed across the two junctions balance each other and there is no net emf across the junction. But when temperature difference is there, we always have Peltier emf. The actual emf developed in a thermocouple is the algebraic sum of the net Peltier emf and the net Thomson emf developed in the thermocouple. With reference to Fig. 5.18, we can write

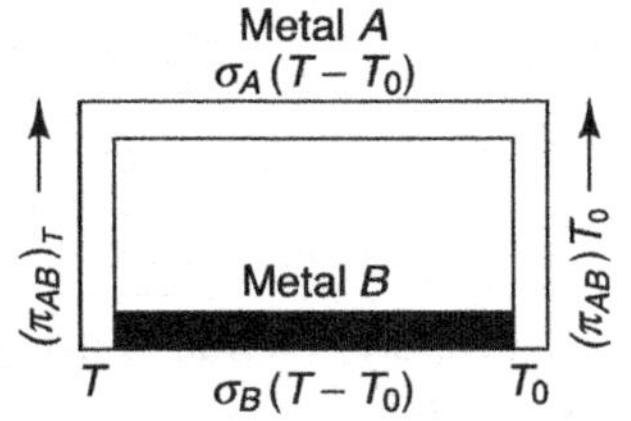

Fig. 5.18 *Seebeck emf*

$$E_{AB} = [\pi_{AB}]_T - [\pi_{AB}]_{T_0} + (T - T_0)(\sigma_A - \sigma_B)$$

Thus, Seebeck emf is a combination of two Peltier emfs and two Thomson emfs.

5.36 THERMODYNAMICS OF PELTIER EFFECT

Let us consider a thermocouple made of two difference metals A and B maintained at two different temperatures T_1 and T_2. This drives a current through the thermocouple. Due to this current, it absorbs heat at hot junction and evolves heat at cold junction. The heat absorbed at the hot junction is more than the heat evolved at the cold junction, and this difference of heat is the source of energy to which the current is due in the circuit. Therefore, a thermocouple acts like a heat engine taking heat from the source at higher temperature T_2 (hot junction), converting some part into work in driving the current through the loop and rejecting some to the junction at lower temperature (sink) T_2.

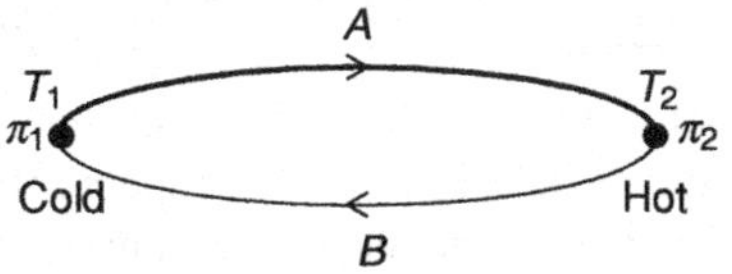

Fig. 5.19 *A thermocouple maintained at temperature difference.*

Let one ampere current flows for one second and the Peltier coefficients for the given thermocouple are π_2 at temperature T_2 and π_1 at temperature T_1, respectivily.

Heat energy absorbed at temperature T_2 at the hot junction

$$= \pi_2 I.t = \pi_2 \text{ Joule}$$

Heat energy liberated at temperature T_1 at the cold junction

$$= \pi_2 I.t = \pi_2 \text{ Joule}$$

Net heat absorbed = $(\pi_2 - \pi_1)$ Joule.

This energy is used in setting up a thermo-emf, E volts in the circuit.

That is

$$E = \pi_2 - \pi_1$$

Peltier effect is reversible, just like Carnot's reversible engine, i.e.

$$\frac{H_1}{T_1} = \frac{H_2}{T_2}$$

or $$\frac{\pi_2}{T_2} = \frac{\pi_1}{T_1}$$

or $$\frac{\pi_2}{\pi_1} = \frac{T_2}{T_1}$$

or $$\frac{\pi_2 - \pi_1}{\pi_1} = \frac{T_2 - T_1}{T_1}$$

Therefore, $$(\pi_2 - \pi_1) = \frac{\pi_1}{T_1}[T_2 - T_1]$$

or $$\boxed{E = \frac{\pi_1}{T_1}[T_2 - T_1]} \tag{5.44}$$

Therefore, thermo-emf is directly proportional to the temperature difference across the junction.

Experimentally, the graph between E and T is not a straight line, but a parabolic curve and E is not proportional to temperature difference. This shows that Peltier effect alone cannot explain the thermoelectric phenomenon, which lead to the discovery of Thomson effect in 1856. Thomson pointed out that there must be some additional source of emf which has not been taken into consideration in the Peltier effect. This additional emf is due to the different parts of the same metal being at different temperatures. The combined emf due to Peltier and Thomson effect showed agreement with experimental results.

5.37 POSITIVE AND NEGATIVE THOMSON EFFECT

As already discussed, when a current flows through a conductor which has a temperature gradient (i.e. different of temperatures), heat energy is evolved or absorbed not only at the junctions but also throughout the metals of a thermocouples. This effect is reversible. In case of copper, the hotter parts are at a higher potential than the colder ones and it is opposite in the case of iron. Consider a copper bar AB heated in the middle at the point P and a current is flowing in it, as shown below in Figs 5.20 and 5.21.

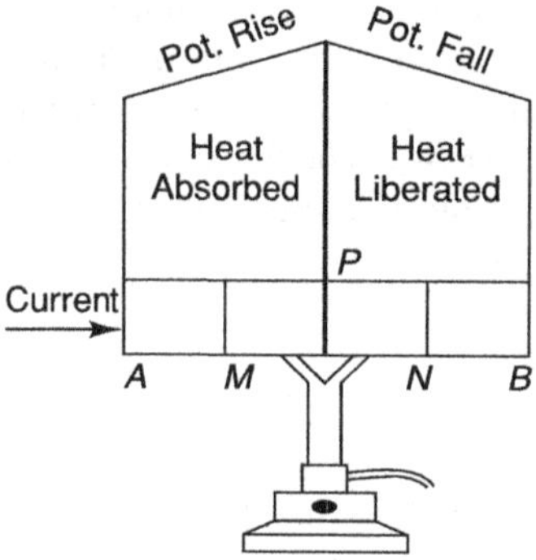

Fig. 5.20 *Positive Thomson effect (Cu, Sb, Ag, Zn, Cd) N at higher temperature at compared to M*

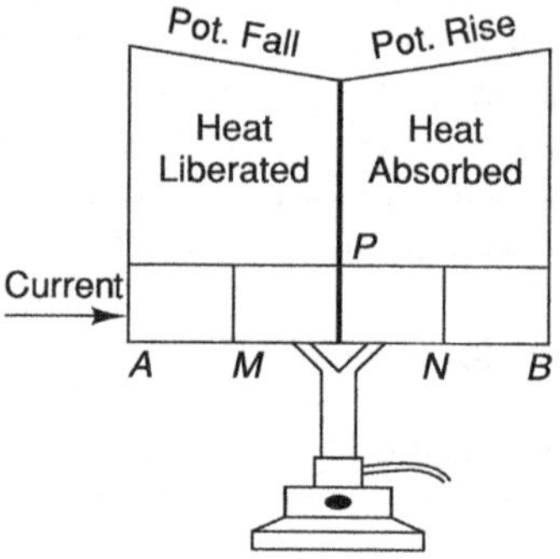

Fig. 5.21 *Negative Thomson effect (Fe, Pt, Bi, Co, Ni) M is at high or temperature as compared to N*

In the absence of a current, points M and N (equidistant from P) are at the same temperature. When current is passed from A to B, N shows a higher temperature than M (Fig. 5.20). The reverse is observed, if a bar is taken of Fe, Pt, Bi, Co or Ni. This means heat is absorbed from A to P and is given out from P to B. This is called positive Thomson effect.

Exactly reverse is observed, if the rod is replaced by Fe, Pt, Bi, Co or Ni, in which M shows a higher temperature compared to N and it is known as negative Thomson effect. But, if the bar is taken of Pb, the points equidistant from P will always be at the same temperature, i.e. heat is neither absorbed nor liberated. This means that in Pb **Thomson** effect is zero. Due to this reason, Pb is used as one of the metal to form a thermocouple with other metals for the purpose of drawing thermoelectric diagrams.

Example 5.21

Find the values of constants *a* and *b* for a Cu-Fe thermocouple.

Solution

$$\begin{aligned} a_{\text{Cu, Fe}} &= a_{\text{Cu, Pb}} - a_{\text{Fe, Pb}} \\ &= 2.76 - 16.6 \\ &= -13.8\ \mu\text{v}/^\circ\text{C} \\ b_{\text{Cu, Fe}} &= b_{\text{cu, Pb}} - b_{\text{Fe, Pb}} \\ &= 0.012 + 0.030 \\ &= 0.042\ \mu\text{v}/(^\circ\text{C})^2 \end{aligned}$$

Ans.

Example 5.22

The emf in a thermocouple, one junction of which is kept at 0°C, is given by $E = at + bt^2$. Determine the neutral temperature, Peltier coefficient and Thomson coefficients.

Solution Let the temperature of hot junction is t°C.

i.e. $t°\text{C} = (T - 273)$ where T is in Kelvin.

$$E = at + bt^2 = a(T-273) + (T-273)^2$$

$$\frac{dE}{dT} = a + 2b(T-273) \text{ and } \frac{d^2E}{dT} = 0$$

At neutral temperature $T = T_n$

$$\frac{dE}{dT} = 0$$

i.e. $a + 2b\,(T - 273) = 0;\quad T_n - 273 = \dfrac{a}{2b}$

Therefore, $$T_n = \left[\left(-\frac{a}{2b}\right) + 273\right]K = \left(-\frac{a}{2b}\right)^0 C$$

Peltier coefficient $$\pi = T.\frac{dE}{dT} = T[a + 2b(T-273)]$$

Thomson coefficient $$\sigma = T.\frac{d^2E}{dT^2} = 2Tb$$

Ans.

Example 5.23

The thermo-emf of certain thermocouple is related with the temperature by the equation $E = at + bt^2$ where *t* the temperature of hot junction is °C. The cold junction is at 0°C. The values of *a* and *b* are 15 $\mu v/°C$ and $-\frac{1}{30}\mu v/°C^2$. Estimate the neutral temperature and temperature of inversion.

Solution

$$E = at + bt^2$$

$$\frac{dE}{dt} = a + 2bt$$

At $t = t_n;\quad \frac{dE}{dt} = 0$

i.e. $a + 2bt_n = 0$

$$\Rightarrow \quad t_n = -\frac{a}{2b} = \frac{15}{2\times\frac{1}{30}} = 225°C$$

Also, $t_1 + t_2 = 2t_2$

$t_1 = 0$ and t_2 is the inversion temperature

$0 + t_2 = 2 \times 225$

$\Rightarrow \quad t_2 = 450°C$ **Ans.**

Example 5.24

The resistivity of pure copper is 1.42 $\mu\Omega$–cm. An alloy of copper containing 1 atomic per cent Ni has a resistivity of 2.75 $\mu\Omega$–cm. And, an alloy of copper containing 3 at % silver has resistivity of 1.98 $\mu\Omega$–cm. What is the resistivity of an alloy containing Z at % Ni and 2at%?

Solution

The resistivity of copper $\rho_{Cu} = 1.42$

$$\rho(C_u + N_i) = 2.75$$

$$\rho_{Ni} = 2.75 + 1.42 = 1.33$$

$$\rho_{Silver} = \frac{1.98 - 1.42}{3} = 0.186$$

$$\rho_{Alloy} = 1.42 + 2 \times 1.33 + 2 \times 0.186$$

$= 4.45\ \mu\Omega - cm$ **Ans.**

5.38 THOMSON COEFFICIENT

It is defined as the "amount of heat evolved or absorbed in Joules between two points of the same conductor which differ in temperature by 1°C, if a current of 1 amp flows for one second. It is denoted by σ.

Following are some of characteristics of Thomson coefficient:

- It is not uniform, but varies with temperature.
- It is numerically equal to potential difference per degree Celsius.
- It is also known as specific heat of electricity.

5.39 PRACTICAL APPLICATIONS OF HEATING EFFECT

On one hand, the heating effect of current is desirable, while under many other situations the cae is different.

Heating effect has an adverse result on the mechanical strength of the conducting material. Following are certain positive features of heating effect of current:

- Electrical furnaces utilise heating effect in metallurgical industries for melting the metals.
- It is used in various domestic appliances, like electric kettle, heater, boiler, immersion rods, ovens, etc.
- Heating effect is used in lamp filaments and arc lamps.
- It is also used for heating dc locomotive for industrial hauling uses.
- It is used in the calorie metre.
- It is used as a protection device (fuses) for excess flow of current.
- It is used in hot wire ammeter for measuring alternating current.

Table 5.7 *Difference between Joule and Peltier effects*

S.No.	*Joule Effect*	*Peltier Effect*
1.	It is an irreversible effect.	It is a reversible effect.
2.	Heat is always given out.	Heat is evolved at one junction and absorbed at the other.
3.	Heat is produced all along the conductor.	Heat is produced at one junction only.
4.	Heat produced is proportional to the square of the current.	Heat generated is proportional to the current itself.

5.40 DISCOVERY OF SUPERCONDUCTIVITY

The advent of superconductivity took place soon after the liquefactions of helium in the beginning of the twentieth century. Liquefaction of helium opened up new frontiers for the study of materials at very low temperatures. The behaviour of materials at very low temperatures (particularly electrical and magnetic behaviour) was found entirely different (rather extraordinary) at a temperature nearly equal to liquid helium temperature. The materials also exhibit perfect conductivity (almost zero resistance) and perfect diamagnetism (which means strong expulsion of magnetic flux or generation of strong forces against gravity). These extraordinary properties exhibited by the materials at such low temperatures provided them a tremendous potential and expectations. The only drawback of superconductivity phenomenon is the difficulty in maintaining such a low temperature. It forced scientists to think of materials that could show this remarkable characteristic at a high temperature or at room temperature. Such dream materials are known as high T_c superconductors.

5.41 SUPERCONDUCTIVITY AND TRANSITION TEMPERATURE

It is a well-known fact that resistivity of pure metals decreases with decrease in the temperature. The electrical resistivity of many metals and alloys drops suddenly to zero when they are cooled to a low temperature (nearly equal to liquid helium temperature).

This phenomenon was observed for the first time by Kamerlingh Onnes in 1911. The material exhibiting this remarkable characteristic is known as superconductor.

Figure 5.22 represents the case of mercury whose resistivity was fairly high at 4.2 K, but suddenly dropped to zero when the temperature reduced only slightly below 4.2 K. Naturally, the material undergoes a phase transition from a state of normal resistivity to a superconducting state. The temperature at which this drastic change in electrical resistivity occurs is known as critical temperature (*Tc*) or transition temperature. In a superconducting state, electrical resistivity is zero (or almost zero). Electric currents flow without any attenuation in superconducting rings for more than a year.

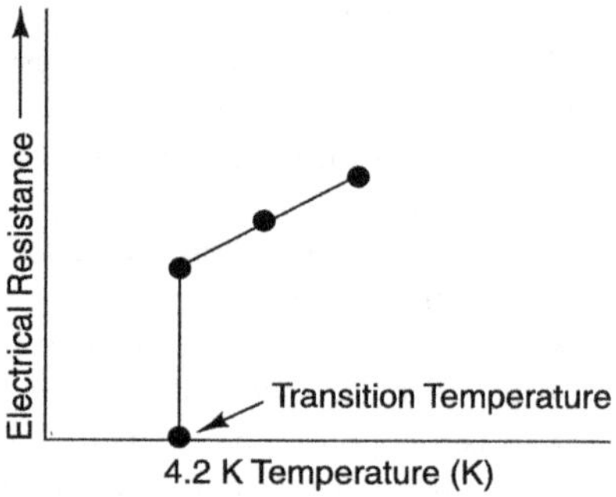

Fig. 5.22 *Resistance of Hg vs. absolute temperature*

Superconductivity is now well understood. It is an area with a large number of theoretical and practical interests. Attempts are going on to develop materials which could show this remarkable property even at room temperature, and some materials with high Tc have been developed. The decay of supercurrents in a solenoid was studied by File and Mills, and they concluded that the decay time of the supercurrent is not less than 10,000 years.

5.42 SUPERCONDUCTING MATERIALS

Many metals and compounds exhibit superconducting property at a very low temperature. This is found to occur in poorer metallic conductors, like Sn, Pb and tantalum rather than in good conductors.

Not only pure metals but various alloys and chemical compounds exhibit this remarkable property. At present, nearly 30 metals and 600 superconducting alloys are well known. The highest temperature superconductor has also been discovered which is a compound of niobium, aluminium and germanium, having a temperature of –253°C or 20 K. It is interesting to note that metals which are normally very good conductors of electricity are not superconductors.

Table 5.8 below gives a list of some inter-metallic compounds having a high transition temperature.

Table 5.8 *Transition temperature of some superconducting elements*

Element	*Critical temperature Tc (K)*
V	5.38
Nb	9.50
Te	7.77
Sn	3.72
Pb	7.19
Hg	4.15
La	6.00

Table 5.9 *Transition temperature of some superconducting compounds*

Alloy	*Critical temperature Tc (K)*
Nb_3 Sn	18.1
Nb_3 Ge	23.2
Nb_3 Al	17.5
V_3 Ga	16.5
V_3 Si	17.1
La In	10.4

Although the critical temperature of alloys is much higher than the elements, yet it is very difficult to attain these temperatures. Attempts are going on to develop room temperature superconductor, and some superconducting oxides have already been developed that exhibit *Tc* around 90 K. One example is $YBa_2Cu_3O_7$ which is popularly known as high temperature superconductor.

5.43 EXPLANATION OF SUPERCONDUCTING PHENOMENON

Superconductivity can be explained on the basis of free electron theory. The resistivity of a metal is given by

$$\rho = \frac{m}{ne^2\tau} \tag{5.45}$$

where n is the number of free electrons per unit volume and τ is the average collision time.

From the above equation it is very clear that, as the temperature decreases, the lattice-vibrations (phonons) begin to freeze which in turn decreases (diminishes) the electron scattering, and therefore, τ increases. At sufficient low temperatures, τ tends to infinite, and therefore, resistivity of the metal also approaches to zero.

Superconductivity is not a rare phenomenon. It is found in a large number of elements—in many alloys, in inter-metallic compounds and in semiconductors. The range of transition temperature extends from 23.2 K for the alloy of Nb_3Ge to 0.01 K for some semiconducting materials. The transition to superconducting state is not always sharp. It is usually sharp if the specimen is made up of a metallic element which is pure and has no structural imperfections.

Some experiments reveal—how readily a substance undergos a superconducting state depends upon how many electrons are there in its atom (electronic configuration). A substance exhibiting two to eight electrons per atom is found to exhibit superconductivity in general. Also, the material whose atoms exhibit an odd number of valence electrons (e.g., three, five, etc.) becomes superconducting most easily above absolute zero.

It has also been observed that superconducting phenomenon is favoured by certain kinds of crystal structures, and by the amount of empty space in the crystal.

Monovalent alkali and noble metals (e.g. gold) do not exhibit superconducting behaviour, even if they are cooled up to 0.01 K. This phenomenon is not detected in ferromagnetic materials.

5.44 ISOTOPE EFFECT

It has been observed that the critical temperature T_c of superconductor varies with isotopic mass. High T_c is found in samples with lighter nuclei. In mercury, T_c varies from 4.85 K to 4.146 K as the average atomic mass M varies from 199.5 amu to 203.4 amu. On mixing two different isotopes of the same element, T_c varies smoothly. The dependence of T_c on atomic mass shows lattice vibrations, and hence, electron-phonon interaction is deeply involved in superconductivity.

Experimental result show that

$$T_c \alpha M^{-\alpha}$$

or $$T_c M^{\alpha} = \text{constant} \quad (5.46)$$

where M is atomic mass, T_c is critical temperature and $\alpha = 0.49 \pm 0.01$. For most of the materials $\alpha = 0.5$.

Therefore,

$$T_c M^{1/2} = \text{constant}$$

$$\boxed{T_{c_1} M_1^{1/2} = T_{c_2} M_2^{1/2}} \quad (5.46)$$

The variation of transition with isotopic mass is given in Fig. 5.23.

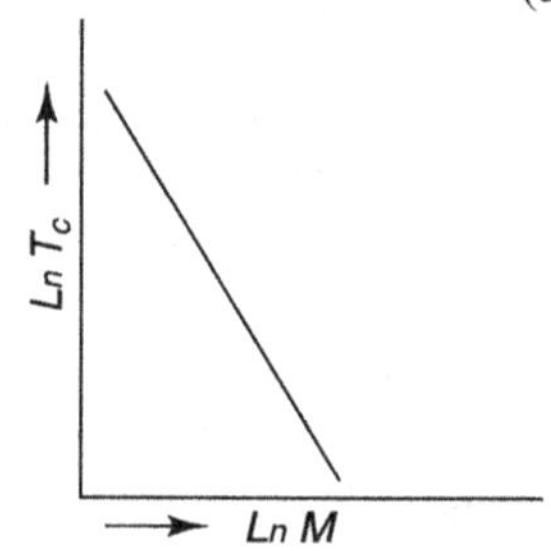

Fig. 5.23 *Variation of T_c with isotopic mass M*

Example 5.25

For mercury of mass number 202, the value of α is 0.50 and T_c is 4.153 K. Find the transition temperature T_c for the isolope of mercury of mass number 200.

Solution We know that

$$M^{\alpha} T_c = \text{constant}$$

$$(202)^{0.5} \times 4.153 = (200)^{0.5} . T_c$$

On solving, we get

$$T_c = \left[\frac{202}{200}\right]^{0.5} \times 4.153$$

$$= 4.174 \text{ K}$$

Ans.

Example 5.26

The critical temperature of a given superconducting sample is 2.1 K with mass 26.91. Determine the critical temperature when the isotope mass changes to 32.13.

Solution Given $T_{c_1} = 2.1$ K, $M_1 = 26.91$ and $M_2 = 32.13$, $T_{c_2} = ?$

Formula used is

$$T_{c_1} = M_1^{1/2} = T_{c_2} = M_2^{1/2}$$

$$T_{c_2} = \frac{T_{c_1} M_1^{1/2}}{M_2^{1/2}} = \frac{1.19 \times (26.91)^{1/2}}{(32.13)^{1/2}}$$

$$= \frac{2.1 \times 5.127}{5.668} = \frac{6.173}{5.668}$$

$$T_{c_2} = 1.89 \text{ K}$$

Ans.

5.45 EFFECT OF MAGNETIC FIELD (CRITICAL MAGNETIC FIELD)

Execution of zero-resistivity characteristic leads to a serious problem in practice. This is because when the critical current *Ic* exceeds a certain critical value, superconducting property of the material gets lost. In addition to current, magnetic field can also destroy the superconductivity. The magnetic field at which the superconducting property of the material is lost is known as critical magnetic field. Its value depends upon the nature of the material and its temperature. The variation of critical magnetic fields vs. temperature is shown in Fig. 5.24.

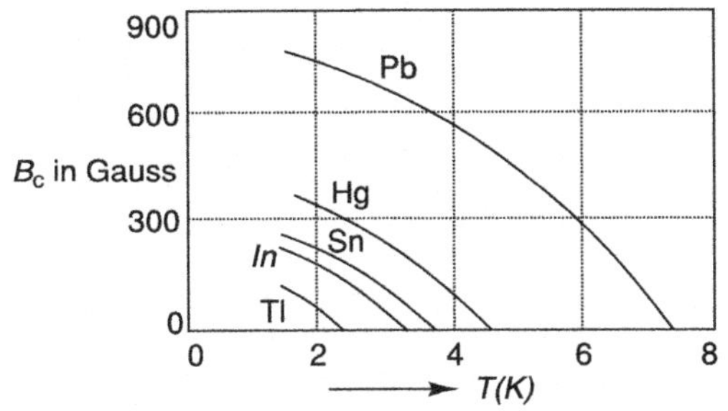

Fig. 5.24 *Variation of critical field as a function of temperature*

The temperature dependence of *Hc* may be given by the following relation:

$$Hc\ (T) = Hc\ (0)\left[1-\frac{T^2}{T_c^2}\right] \tag{5.47}$$

where, *Hc(T)* is critical magnetic field at temperature *T* and *Hc*(0) at 0 K.

Table 5.10 gives a list of some elemental superconductors along with their critical field at 0 K.

Table 5.10 *Critical magnetic field of some elemental superconductors*

Element	*Critical field at 0 k in (milli Tesla)*
V	142.0
Nb	198.0
Tc	141.0
In	29.3
La	83.0
Hg	41.2
Pb	80.3

Example 5.27

Determine the transition temperature and critical field at 4.2 K for a given specimen of a superconductor if the critical fields are 1.41 × 10⁵ and 4.205 × 10⁵ amp/m at 1.41 K and 12.9 K, respectively.

Solution

Given $H_{C_1} = 1.41 \times 10^5$ A/m at $T_1 = 14.1$ K

and $H_{C_2} = 10^5$A/m at $T_2 = 12.9$ K

Formula used is

$$H_c = H_0\left[1-\left(\frac{T}{T_c}\right)^2\right]$$

Thus, the critical fields at temperatures T_1 and T_2 can be written.

$$H_{c_1} = H_0\left[1-\left(\frac{T_1}{T_c}\right)^2\right] = H_0\left[\frac{T_c^2 - T_1^2}{T_c^2}\right]$$

$$H_{c_2} = H_0\left[\frac{T_c^2 - T_2^2}{T_c^2}\right]$$

From the above equations, we get

$$\frac{H_{c_1}}{H_{c_2}} = \frac{T_c^2 - T_1^2}{T_c^2 - T_2^2} \text{ or } \frac{1.41\times 10^5}{4.205\times 10^5} = \frac{T_c^2 - (14.1)^2}{T_c^2 - (12.9)^2}$$

$$T_c = 14.67 \text{ K}$$

Using the values,

$$H_c = 1.41 \times 10^5 \text{ A/m},\ T_1 = 14.1 \text{ K and } T_c = 14.67 \text{ K}$$

$$H_c = H_0\left[1-\left(\frac{T_1}{T_c}\right)^2\right]$$

$$1.41 \times 10^5 = H_0\left[1-\left(\frac{14.1}{14.67}\right)^2\right]$$

or $$H_0 = 18.504 \times 10^5 \text{ A/m}$$

The critical field at $T = 4.2$ K and $T_c = 14.67$ K

$$H_c = H_0\left[1-\left(\frac{T}{T_c}\right)^2\right] = 18.504\times 10^5 \times \left[1-\left(\frac{4.2}{14.67}\right)^2\right]$$

$$= 16.99 \times 10^5 \text{ A/m}$$ **Ans.**

Example 5.28

The material lead (Pb) behaves as a superconductor at a temperature of $T_c = 7.26$ K. If the value of critical magnetic field of lead at $T = 0$ K is $H_c = 7.0 \times 10^5$ A/m, find the critical magnetic field of lead at 4 K.

Solution Critical magnetic field at 0 K,

$$H_c\,(0) = 8 \times 10^5 = 7{,}00{,}000$$

Critical temperature, $T_c = 7.26$°C

From Eq. (2.53), the critical value of magnetic field at $T = 4$ K is given as

$$H_c(T) = H_c(0)\left[1-\left(\frac{T}{T_c}\right)^2\right]$$

$$= 7{,}00{,}000\left[1-\left(\frac{4}{7.26}\right)^2\right] = 6.37\times 10^5 \text{ A/m}$$ **Ans.**

Example 5.29

Determine the critical current density for 1 mm diameter wire of lead at 4.5 K. Given that for lead $T_c = 7.2$ K and $H_c = 8 \times 10^4$ A/m.

Solution

For lead, critical field, $H_c\,(0) = 8 \times 10^4$ A/m

For lead, critical temperature, $T_c = 7.2$ K

Critical field at $T = 4.5$ K,

$$H_c = H_c\,(0)\left[1-\left(\frac{T}{T_c}\right)^2\right] = 8\times10^4\left[1-\left(\frac{4.5}{7.2}\right)^2\right] = 4.875\times10^4 \text{ A/m}$$

Critical current, $I_c = 2\pi r\, H_c = 2\,\pi \times 0.5 \times 10^{-3} \times 4.875 \times 10^4 = 153.15$ A **Ans.**

Example 5.30

Superconducting tin has a critical temperature of 3.7 K in zero magnetic field and a critical field of 0.0306 T at 0 K. Find the critical field at 2 K.

Solution We known that the critical field H_c is given as

$$H_c = H_c(0)\left[1-\left(\frac{T}{T_c}\right)^2\right]$$

Hence, $H_c(0) = 0.0306\ T,\ T_c = 3.7$ K and $T = 2$ K

$$H_c = 0.0306\left[1-\left(\frac{2}{3.7}\right)^2\right] = 0.0306\,[1 - 0.292]$$

$$H_c = 0.0306 \times 0.708 = 0.0216 \text{ tesla}$$ **Ans.**

Example 5.31

The critical field for niobium is 1.5×10^5 A/m at 8 K and 2×10^5 A/m at 0 K. Calculate the transition temperature of the element.

Solution We know that

$$H_c(T) = H_c(0)\left[1-\left(\frac{T}{T_c}\right)^2\right]$$

or $$\frac{H_c(T)}{H_c(0)} = 1-\left(\frac{T}{T_c}\right)^2 \quad \text{or} \quad T_c = \frac{T}{\left[1-\dfrac{H_c(T)}{H_c(0)}\right]^{1/2}}$$

Here, $H_c\,(T) = 1 \times 10^5$A/m, $H_c(0) = 2 \times 10^5$ A/m and $T = 8$ K

$$T_c = \frac{8}{\left[1-\dfrac{1.5\times10^5}{2\times10^5}\right]^{1/2}} = 16$$

$$T_c = 16.0 \text{ K}$$ **Ans.**

Example 5.32

For a specimen of V_3Ga, the critical fields are respectively 0.176 T and 0.528 T for 14 K and 12 K, respectivily. Calculate the transition temperatures and critical fields at 0 K and 4.2 K.

Solution We know that,

$$H_c(T) = H_c(0)\left[1-\left(\frac{T}{T_c}\right)^2\right] \tag{1}$$

Given, At $T = 14$ K, $H_c(T) = 0.176\ T$ and at $T = 13$ K, $H_c(T) = 0.528$ T

Therefore,
$$0.176 = H_c(0)\left[1-\left(\frac{14}{T_c}\right)^2\right] \tag{2}$$

and
$$0.528 = H_c(0)\left[1-\left(\frac{13}{T_c}\right)^2\right] \tag{3}$$

Dividing Eq. (3) by Eq. (2), we get

$$\frac{1-(13/T_c)^2}{1-(14/T_c)^2} = \frac{0.528}{0.176} = 3$$

or
$$1-\frac{169}{T_c^2} = 3-\frac{588}{T_c^2} \quad \text{or} \quad \frac{419}{T_c^2} = 2$$

or
$$T_c = \left(\frac{419}{2}\right)^{1/2} = 14.5\ \text{K}$$

Substituting this value of T_c in Eq. (2), we get

$$0.176 = H_c(0)\left[1-\left(\frac{14}{14.5}\right)\right]^2 = H_c(0)[1-(0.9655)^2]$$

or
$$H_c(0) = \frac{0.176}{1-0.932} = 2.588\ \text{T}$$

For
$$T = 4.2\ \text{K}$$

$$H_c(T) = 2.588\left[1-\left(\frac{4.2}{14.5}\right)^2\right]$$
$$= 2.588\ [1 - 0.0839]$$
$$H_c(T) = 2.588 \times 0.9161 = 2.37\ \text{T}$$

Ans.

5.46 CRITICAL CURRENTS

The magnetic field by which the superconducting property of a material gets lost may not necessarily be an externally applied field; it can also be a magnetic field due to current flowing in the superconducting wire itself. It is defined as the maximum value of the current flowing through the superconductor at which the superconducting properly ceases; it is denoted by I_c.

If a superconducting wire of radius r carries a current I, then using Ampere's law

$$\oint_c H.dl = I \quad \text{or} \quad 2\pi r H = I$$

At $$H = H_c, \quad I = I_c$$

Therefore, $$\boxed{I_c = 2\pi r H_c}$$

If, in addition to the current, transverse magnetic field is also applied the critical current decreases.

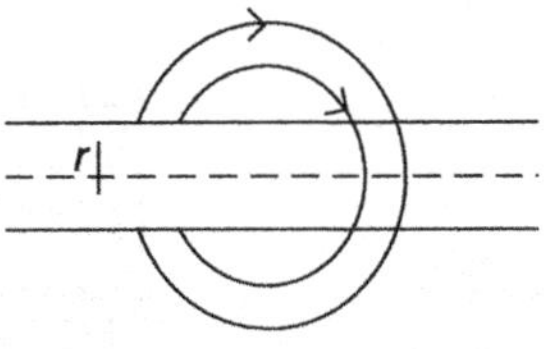

Fig. 5.25 *Superconducting wire carrying current*

Example 5.33

Determine the critical current that can flow through a long thin superconducting conductor of 2 mm radius. It is given that H_c = 7,900 A/m.

Solution Critical value of the magnetic field,

$$H_c = 7{,}900 \text{ A/m}$$

Radius of conductor, $$r = 2.0 \times 10^{-3} \text{ m}$$

According to Silsbee's rule,

critical current, $I_c = 2\pi r\, H_c = 2\pi \times 2.0 \times 10^{-3} \times 7{,}900 = 97.6$ A **Ans.**

Example 5.34

How much current can a Pb wire, 1.0 mm in diameter, carry in its superconducting state at 4.2 K? Given B_c (4.2) = 0.0548 T.

Solution Surface magnetic field associated with current I is

$$H_c = \frac{I_c}{2\pi r} \quad \text{or} \quad I_c = 2\pi r\, H_c$$

$$I_c = \frac{2\pi B_c(T)}{\mu_0} = \frac{\mu D B_c(T)}{\mu_0}$$

$$D = 2r = 1.0 \text{ mm} = 10^{-3} \text{ m and } \mu_0 = 4\pi \times 10^{-7}$$

$$I_c = \frac{\pi \times 10^{-3} \times 0.0548}{4\pi \times 10^{-7}} = \frac{548}{4} = 137$$

$$I_c = 137 \text{ Amp}$$ **Ans.**

5.47 THE MEISSNER EFFECT

This effect was discovered in 1933 by Meissner and Ochsenfeld. According to this, if a superconducting material is placed in a uniform magnetic field (below H_c), the magnetic flux lines that pass through the material above T_c get completely expelled out of the bulk material, when it is cooled below the transition temperature. In superconducting state, the magnetic flux lines inside the bulk material is zero. Fig. 5.26 shows how the flux is expelled at $T < T_c$.

Magnetic induction B inside the superconductor is always zero, as long as it is in the superconducting state. If M is the magnetisation vector and H be the magnetising field then

$$B = \mu_0(H + M) = 0$$

And magnetic susceptibility $\chi_m = M/H = I$
For normal diamagnets $\chi_m << I$. Therefore, a superconductor is an ideal diamagnet.

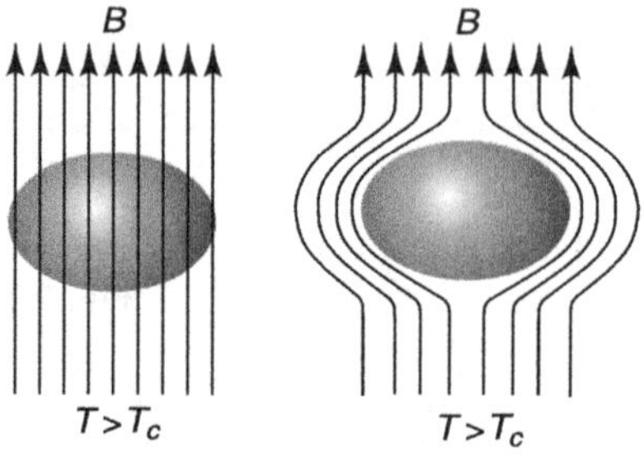

Fig. 5.26 *Expulsion of flux in superconducting state*

5.48 EXPERIMENTAL DEMONSTRATION OF MEISSNER EFFECT

Figure. 5.27 shows the circuit for experimental demonstration of Meissner effect. Two coils, one primary and the other secondary, are wound on a superconductor. The superconductor is kept above its critical temperature. In primary circuit a battery and a tapping key is connected, whereas in secondary circuit a ballistic galvanometer has been connected.

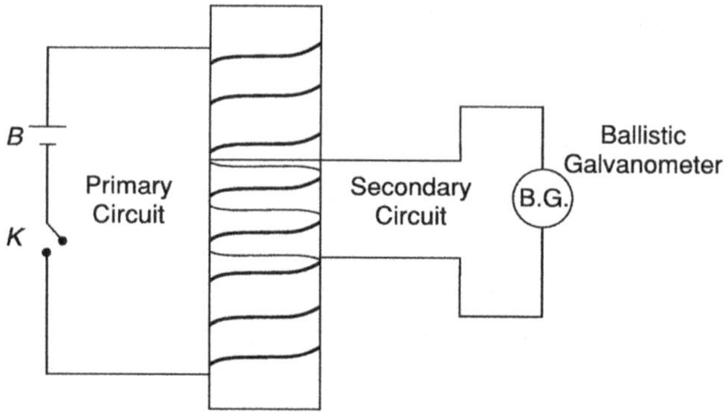

Fig. 5.27 *Experimental demonstration of Meissner effect*

Working

When the key is pressed, a current flows through the primary coil which sets up a magnetic field in the coil. The magnetic flux linked with the primary coil produces a flux in the secondary coil, as well. As a result, the ballistic galvanometer shows a kick. After that, the primary current becomes steady. Now, no current flows in the secondary coil. When the temperature of superconductor is gradually decreased, as soon as the temperature crosses below the transition temperature, the ballistic galvanometer shows a deflection. This indicates that the flux linkage with the secondary coil has changed. This change in the magnetic flux is due to the expulsion of magnetic flux from the superconducting material.

5.49 CLASSIFICATION OF SUPERCONDUCTORS

Superconductors are classified into two categories depending upon their behaviour in external magnetic field. There are given as

1. Type-I or Soft superconductors
2. Type-II or Hard superconductors

Type-I Superconductors

The superconductors in which the magnetic field is totally excluded from the interior below a certain magnetic field H_c; at H_c the material loses its superconducting property and magnetic field penetrates fully are known as Type-I or soft superconductors. The magnetisation curve of these superconductors is shown in Fig. 5.28. Low values of critical field are a disadvantage in practical applications because such superconductors cannot carry high currents.

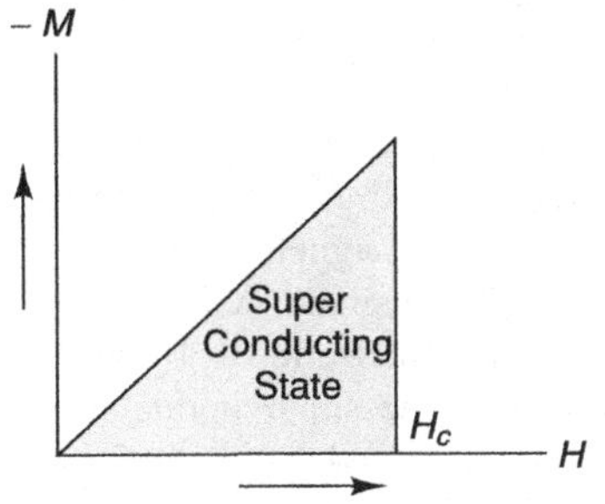

Fig. 5.28 *Variation of magnetisation vector with external field*

The important characteristics of Type-I superconductors are

1 They exhibit complete Meissner effect.

2 The critical values of magnetic field is low (roughly 0.1 tesla or 10^3 gauss).

The magnetisation current shows that transition at H_c is reversible, which means that below H_c the material again acquires superconducting property and the field is expelled. Table 5.11 lists some Type-I superconductors, their critical temperature T_c and critical magnetic field H_c.

Table 5.11 *Values of T_c and H_c for Type-I superconductor*

Element	*T_c (K)*	*H_c (Tesla)*
Al	1.18	0.0105
Hg	4.15	0.4011
In	3.4	0.0281
Pb	7.19	0.0803
Zn	0.85	0.0054
Sn	3.72	0.0305
Mo	0.92	0.0095

The variation of resistivity ρ vs. magnetic field H is shown in Fig. 5.29 for Type-I superconductors.

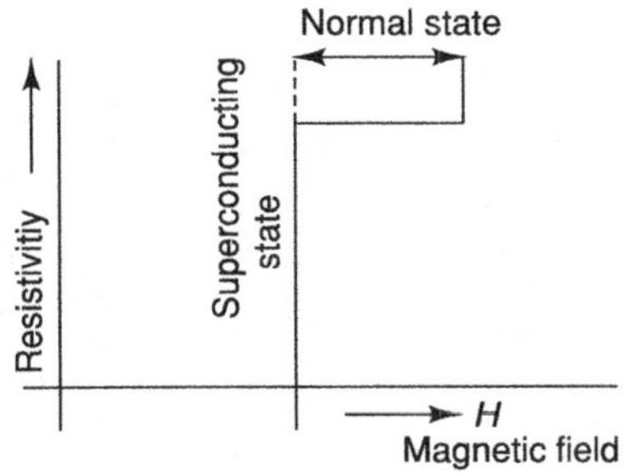

Fig. 5.29 *Variation of resistivity with magnetic field*

Type-II Superconductors

The superconductors in which the material loses magnetisation gradually rather than suddenly are known as Type-II or hard superconductors. That is, the transition from a perfect diamagnetic state to the paramagnetic state is not abrupt. The negative magnetisation increases linearly only up to a certain magnetic field, and beyond this the magnetisation falls slowly, becoming positive at a much higher magnetic field. These values are shown in the Fig. 5.30, as H_{c_1}, and H_{c_2}. Therefore, up to H_{c1} the material is perfect diamagnet and in between H_{c_1} and H_{c_2}, the material is partially diamagnetic. At H_{c_2}, it becomes paramagnetic with a small positive magnetisation. Between H_{c_1} and H_{c_2}, the superconductor is said to be in 'mixed state'. As for as electrical resistivity is concerned, it remains up to H_{c_2} as the superconducting regions shunt the normal regions into which flux has penetrated. H_{c_2} is the critical field at which the material goes from a superconducting state to normal state. The value of this critical field is much higher than Type-I conductors. Table 5.12 lists the values of T_c and H_c for Type-II superconductors.

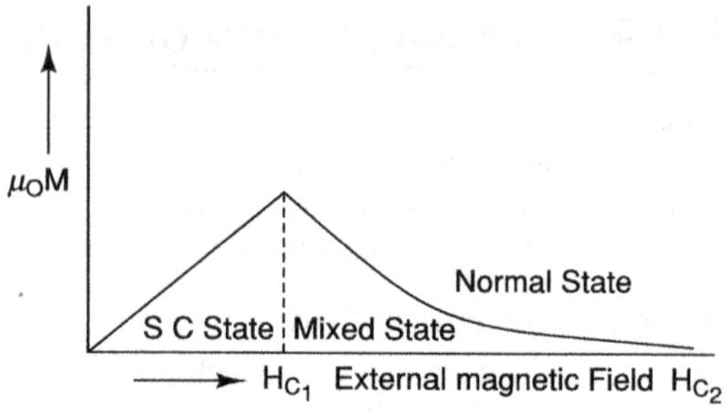

Fig. 5.30 *Variation of magnetisation vector with external field in Type-II superconductors*

Table 5.12 *Values of T_c and H_c for Type-I superconductor*

Compound	*T_c(K)*	*H_c(Tesla)*
NbTi	10.0	15.0
Nb_3Sn	18.0	24.5
Nb_3Al	18.7	32.4
Nb_3 (AlGe)	20.7	44.0
V_3Ge	14.8	2.08
V_3Si	16.9	2.35

The variation of resistivity vs. magnetic field for Type-II superconductors is shown in Fig. 5.31.

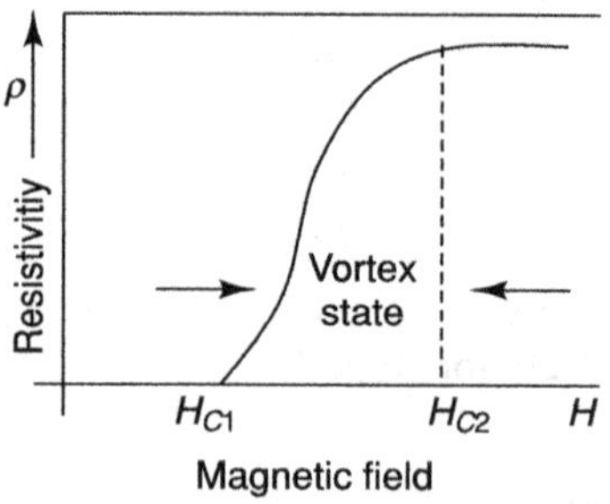

Fig. 5.31 *Variation of resistivity vs. magnetic field for Type-II superconductors*

5.50 ELECTRODYNAMICS OF SUPERCONDUCTORS (EXPLANATION OF MEISSNER EFFECT)

A simple explanation for the expulsion of flux below the transition temperature was provided by Fritz and Heinz Landon brothers in 1935. They proposed that a superconductor contains two kinds of electrons—the superelectrons and the normal electrons. At 0°K, a superconductor contains only superelectrons, but as the temperature increases, there is increase in normal electrons, the ratio of normal electrons to the superelectrons increases, and beyond the critical temperature all the electrons are normal. But at any temperature, the sum of the superelectrons and normal electron density is equal to the conduction electron density in the normal state. Superelectrons are not subjected to any lattice scattering, and therefore, are simply accelerated in an electric field.

Let us consider the normal electron density N_n and superelectron density N_s. Let V_n and V_s be their velocities. According to the above discussion, we can write

$$N = N_n + N_s \tag{5.48}$$

and total current density

$$J = J_n + J_s = eN_nV_n + eN_sV_s \tag{5.49}$$

As superelectrons have no resistance in their motion, these electrons are accelerated in a constant electric field E. The equation of motion is therefore given by

$$mdV_s/dt = eE \tag{5.50}$$

and

$$dJ_s/dt = e.Ns.dV_s/dt \tag{5.51}$$

or

$$dJ_s/dt = e^2N_sE/m \tag{5.52}$$

This is London's first equation which shows the absence of resistance, since no electric field is necessary unless the current changes or if the electric field E is zero, there is a constant supercurrent density Js. This is known as the superconducting phenomenon.

The corresponding equation for normal current is

$$J_n = \sigma E = e^2N_n\tau E/m$$

where σ is conductivity and τ the relaxation time. Therefore, one can conclude that electric field is necessary for steady current density in normal state, or if electric field is zero ($E = 0$) then the normal current density become zero, which does not happen with Js. Let us check the applicability of London's equation in Maxwell's equation.

$$\nabla \times E = -\frac{\partial B}{\partial t} = \mu_0 \frac{\partial H}{\partial t}$$

and

$$\nabla \times H = J_s + \frac{\partial D}{\partial t}$$

or

$$\nabla \times B = \mu_0 \left[J_s + \frac{\partial D}{\partial t} \right]$$

Now, if there is no variation in electric field then $\partial D/\partial t$ will be negligible, and can be ignored:

$$\nabla \times B = \mu_0 J_s$$

And using first equation (taking curl)

$$\frac{\partial}{\partial t}(\text{curl } J_s) = \frac{e^2 Ns}{m}(\text{curl } E)$$

or

$$\frac{\partial}{\partial t}(\text{curl } J_s) = -\frac{e^2 Ns}{m}\frac{\partial B}{\partial t}$$

$$\boxed{\text{curl } J_s = -\frac{e^2 Ns}{m} B} \tag{5.53}$$

This is London's second equation. The Meissner effect and magnetic flux penetration can be explained using this equation.

As previously discussed, we have

$$\nabla \times B = \mu_0 J_s$$

Again, taking curl on both sides

$$\text{Curl (curl } B) = \mu_0 (\text{curl } J_s)$$
$$\text{grad div } B - \nabla^2 B = \mu_0 \text{ (curl } J_s)$$

$$\boxed{\nabla^2 B = \mu_0 \frac{e^2 Ns}{m} B} \tag{5.54}$$

This equation shows that B is not uniform inside the superconductor, and the field gets exponentially dumped as we go into the bulk material from the external surface.

5.51 LONDON'S PENETRATION DEPTH

The above differential equation can also be written as

$$\nabla^2 B = \frac{1}{\lambda_l^2} B \tag{5.55}$$

where $\lambda_l = \left[\frac{m}{\mu_0 N_s e^2}\right]^{1/2}$

is known as London's penetration depth, which gives the depth of penetration insider the superconductor from the surface.

The above equation can easily be solved for one dimension.

$$\frac{\partial^2 B}{\partial x^2} = \frac{1}{\lambda_{l^2}} B \tag{5.56}$$

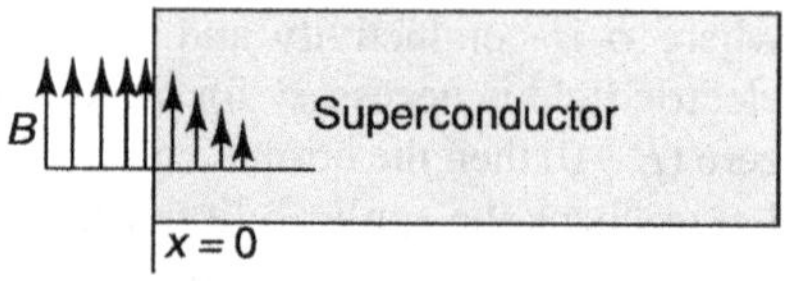

Fig. 5.32 *Penetration of B inside superconductor*

Figure 5.32 shows a semiinfinite superconductor occupying the positive side of the x-axis.

The surface of superconductor is at $\underline{x} = 0$. Let a magnetic field be applied parallel to the surface of the superconductor is in y-direction. The solution of the above equation can be given as,

$$B(x) = B(0)e^{-x/\lambda} \tag{5.57}$$

where $B(0)$ is the magnetic field at the surface and $B(x)$ at a distance x from the surface of the superconductor. The equation shows that magnetic field falls exponentially from the surface of the superconductor. The penetration depth for most of the superconductors is of the order of fraction of a micron.

A very important conclusion can be drawn from the above discussion that, which is—"Superconducting state is a macroscopic quantum mechanical state."

Let us look at some important features of London's penetration depth.

- The field decays below the surface of the specimen, which is brought about by the supercurrent flowing over the surface to shield the interior of the specimen from fields. Field is controlled by the equation given above.

- It does not allow the specimen to show complete Meissner effect. If the specimen is in the thin film form, there is no Meissner effect at all.
- Since London's penetration depth is dependent upon the density of the superelectron, which is dependent upon temperature, hence x also varies with temperature. Above T_c, N_s must be zero at 0 K, therefore it acquire some maximum value.
- The penetration depth is minimum at $T = 0$ and approaches infinity at $T = T_c$. Many experimental data support the below relation.

$$\lambda = \frac{\lambda_0}{\left[1-\left(\frac{T}{T_c}\right)^4\right]^{1/2}} \tag{5.58}$$

where λ_0 is the depth at 0 K.

- λ also varies with applied magnetic field, although this change is only very little.

Example 5.35

Niobium has a density of 8.5 × 10³ kg/m³. This has one condition electron per atom. Its atomic weight is 93. Calculate London's penetration depth in Nb.

Solution London's penetration depth is given by

$$\lambda_L = \left[\frac{m}{\mu_0 n_s e^2}\right]^{1/2}$$

$$n_s = \frac{\text{Density} \times N}{At.Wt} = \frac{8.5\times10^3\times6.023\times10^{26}}{93} = 5.505\times10^{28}/\text{m}^3$$

$$\lambda_L = \left[\frac{9.1\times10^{-31}}{4\pi\times10^{-7}\times5.505\times10^{28}\times(2\times1.6\times10^{-19})^2}\right]^{1/2} = 11.33 \text{ nm}$$

Ans.

Example 5.36

The critical temperature of lead is 7.2 K. Determine the penetration depth in lead at 5.5 K, if the penetration depth at 0 K is 380 Å.

Solution

Given $T_c = 7.2$ K and $\lambda_0 = 380$ Å, $\lambda(5.5\text{ K}) = 0$

The penetration depth is given by

$$\lambda(T) = \lambda_0\left[1-\left(\frac{T}{T_c}\right)^4\right]^{-1/2}$$

$$\lambda(5.5\text{ K}) = 380\left[1-\left(\frac{5.5}{7.2}\right)^4\right]^{-1/2}$$

$$= 567.16 \text{ Å}$$

Ans.

Example 5.37

Determine the critical temperature of aluminium if the penetration depth for aluminium is 16 nm and 96 nm at 2.18 K and 8.1 K, respectively.

Solution

Given $\lambda(2.18) = 16$ nm

and $\lambda(8.1) = 96$ nm

Formula used is

$$\lambda(T) = \lambda_0\left[1-\left(\frac{T}{T_c}\right)^4\right]^{-1/4}$$

$$\frac{96}{16} = \frac{\lambda_0[1-(8.1/T_c)^4]^{-1/2}}{\lambda_0[1-(2.18/T_c)^4]^{-1/2}}$$

$$6 = \sqrt{\frac{1-\left(\frac{2.18}{T_c}\right)^4}{1-\left(\frac{8.10}{T_c}\right)^4}}$$

or

$$35 = \frac{1}{T_c}4\{154968.19-22.58\}$$

$$T_c = 8.16 \text{ K}$$

Ans.

5.52 TEMPERATURE DEPENDENCE OF SUPERCONDUCTING ELECTRONS

We know that the penetration depth can be expressed as

$$\lambda_l = \sqrt{\frac{m}{\mu_0 N_s e^2}}$$

From the above expression it is clear that the penetration depth depends upon the number of superelectrons N_s. Since N_s varies with temperature, the penetration depth also strongly depend upon temperature. The corresponding variation of superelectrons with temperature can be expressed as

$$N_s(T) = N_s(0)\left[1-\left(\frac{T}{T_c}\right)^4\right]$$

5.53 BCS THEORY OF SUPERCONDUCTORS

The explanation to the superconducting phenomenon was given by Bardeen, Cooper and Schrieffer in 1957. This is also known as **microscopic theory of superconductors**. This theory is based on the formation of cooper pairs.

The electrical resistance in metals arises due to interaction of free electrons with the vibrating lattice ions. In normal state, the force between two electrons is repulsive. In superconducting state, the force between the two electrons become attractive.

5.54 FORMATION AND CHARACTERISTICS OF COOPER PAIRS

One electron interacts with positive ion of the lattice and deforms it; a second electron of compatible momentum also interacts with the ion in the distorted lattice to minimise its energy. The interacting electrons are said to form a pair. The electrons of a pair have equal and opposite momentum (one with $+k$ and the other with $-k$), one with the spin-up and other in spin-down state, so that the net spin of the pair become zero. Such electron pairs are known as cooper pairs**.** The formation of copper pairs is shown in Fig. 5.33.

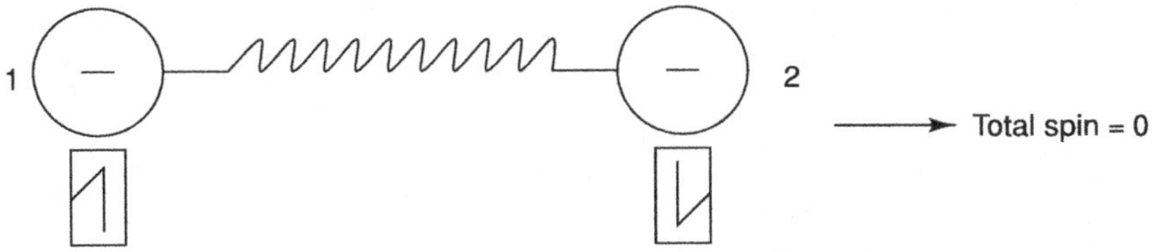

Fig. 5.33 *Representation of a cooper pair*

Cooper Pairs

The important characteristic of cooper pairs is that the excitation can take place only in the pairs. The cooper pairs so formed have net zero spin and are able to condense into a quantum mechanical ground state with long-range order. One consequence of this is the existence of a small energy gap near E_F. Many other characteristics of a superconductor can be understood in terms of the existence of small energy gap near E_F.

Energy gap E_g of a cooper pair is of the order of 10^{-3} eV. The energy gap E_g of a superconductor at 0 K is given by the formula

$$E_g(0) = 3.53\, kT_c$$

where k is Boltzmann constant.

At $T > 0$, some cooper pairs breaks up. The resulting electrons interact with the remaining cooper pairs and reduce the energy gap as 4hown in the Fig. 5.34.

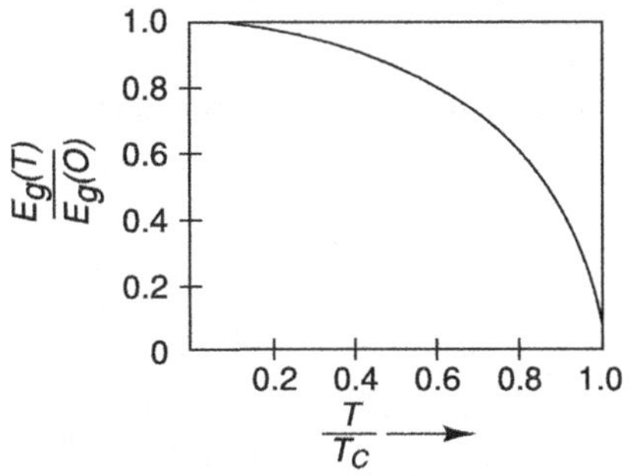

Fig. 5.34 *Variation of the energy gap with temperature*

At critical temperature T_c, the energy gap disappears; there are no more copper pairs and the material is therefore not superconducting.

5.55 EXPERIMENTAL EVIDENCE FOR THE ENERGY GAP

The energy gap in a superconductor corresponds to the long infrared or short microwave frequencies. The following experiments confirm this:

- Exponential decrease of electronic specific heat with decreasing temperature below $T_c/3$.
- Decrease in the absorption of ultrasonic waves below T_c.
- Transmission of microwaves and long IR waves through thin superconducting films.

Example 5.38

Calculate the upper frequency limit of the photon that is transmitted in a thin film of the Nb at 0 K. The energy gap of Nb at 0 K is 30.5×10^{-4} eV. Find the corresponding wave length.

Solution

The energy gap $E_g = h\nu = 30.5 \times 10^{-4} \times 1.6 \times 10^{-19}$

Therefore, $\nu = \dfrac{E_g}{h} = \dfrac{30.5 \times 10^{-4} \times 1.6 \times 10^{-19}}{6.6 \times 10^{-34}}$

The wavelength, $\lambda = \dfrac{c}{\nu} = \dfrac{3.0 \times 10^{8}}{7.39 \times 10^{11}}$

$$= 4.06 \times 10^{-4}\text{ m}$$

$$= 0.4\text{ mm}$$

Ans.

Example 5.39

Considering the critical temperature of mercury as 4.2 K, calculate the energy gap in eV at $T = 0$. Also find the wavelength of a photon whose energy is just sufficient to break up cooper pairs in mercury at $T = 0$. Find the region of the electromagnetic spectrum where such photons may be observed.

Solution

Copper pair binding energy or the energy gap is $E_g = 3\,\text{k}T_c$

$\Rightarrow$ $E_g = 3 \times 1.38 \times 10^{-23} \times 4.2 = 1.74 \times 10^{-22}$ J

or E_g in eV $= \dfrac{1.74 \times 10^{-22}}{1.6 \times 10^{-19}} = 1.08 \times 10^{-3}$ eV

The wavelength λ of a photon of energy E_g

$$\lambda = \frac{h_c}{E_g} \qquad \left(\because E_g = \frac{h_c}{\lambda}\right)$$

$$= \frac{6.6 \times 10^{-34} \times 3 \times 10^{8}}{1.74 \times 10^{-22}} = 1.14 \times 10^{-3}\text{ m}$$

From the value of λ, it is clear that these photons exist in a very short wavelength part of the microwave region.

5.56 FLUX QUANTISATION

This is a unique characteristic of superconductors. The total magnetic field that passes through a superconducting ring is quantised so that that it can acquire some fixed values only.

5.57 JOSEPHSON EFFECT

This is another interesting observation in superconducting materials. When two superconductors are separated by a very thin insulating layer of (10 – 50 Å), the junctions formed exhibit some striking properties which result from the quantum mechanical nature of the superconducting state and tunnelling. The junction is known as Josephson junction.

When two superconductors forming the Josephson junction are connected to each other, a direct current flows through the circuit. This effect is remarkable as the current is produced in a closed circuit in absence of any external electric or magnetic field. The effect is known as dc Josephson effect. This effect can be explained as due to tunnelling of the cooper pairs across the thin insulating layer.

Also, if a dc voltage is applied to Josephson junction an alternating current flows across the junction. This effect is known as ac Josephson effect. The current voltage characteristic of a Josephson junction is shown in Figs 5.35 and 5.36; Josephson shared the Nobel prize for this discovery in 1975.

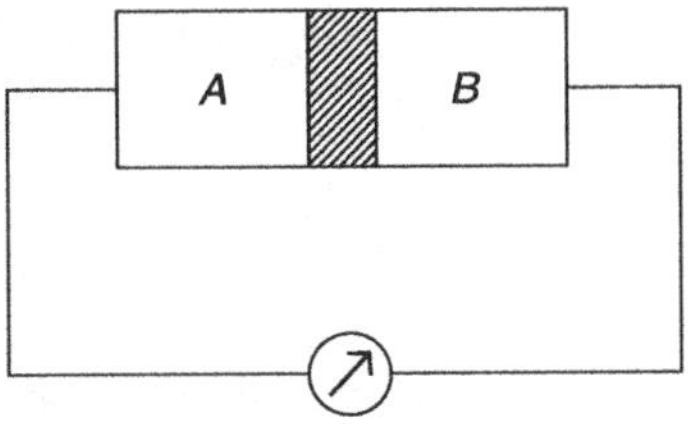

Fig. 5.35 *Showing a Josephson junction*

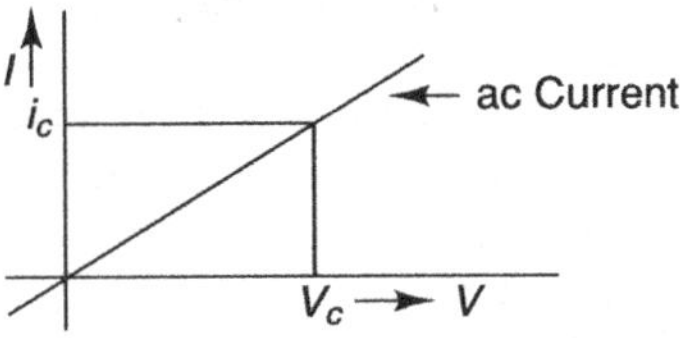

Fig. 5.36 *I–V characteristic of a Josephson junction*

5.58 CHARACTERISTICS OF SUPERCONDUCTORS

The characteristics of superconductors are described below from two angles:

- Properties that remain unchanged in superconducting state
- Properties that change in superconducting state

(i) Properties Unchanged in Superconducting State

- Crystal structure remains unchanged as confirmed by XRD.
- The photo-electric properties also remain unaffected in superconducting state.
- The thermal expansion and elastic properties also do not change in transition.
- In the absence of magnetic field, latent heat also remains unchanged.

(i) Properties that Change in Superconducting State

- At superconducting transition, there is total expulsion of magnetic flux and it behaves like a diamagnetic substance.
- Superconducting state ceases by applying a field equal to the critical field on the specimen.
- The specific heat varies exponentially with the temperature in superconducting state.
- The thermal conductivity of superconductors undergoes a continuous change between two phases. This is usually lower in superconducting state, as shown in Figs 5.37 and 5.38.
- The entropy of a superconductor decreases rapidly on cooling below the transition temperature. This shows that superconducting state is more ordered.

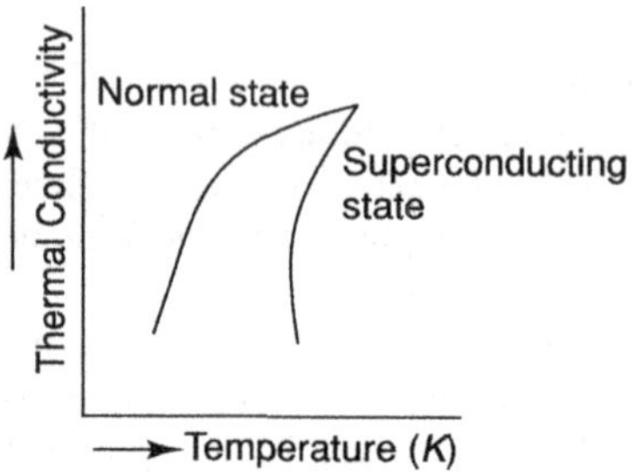

Fig. 5.37 *Thermal conductivities of a sample in normal and superconducting state*

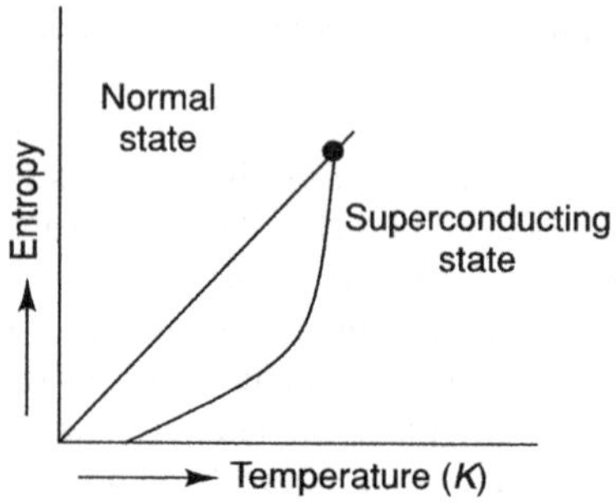

Fig. 5.38 *Change in entropy with temperature*

- All thermoelectric effects disappear in the superconducting state.
- There is an appreciable change in reflectivity of the metal either in visible or in infrared region.
- The magnetic characteristic in superconducting state is modified by the addition of impurities.
- Transition temperature of semiconductors also varies with isotopic mass, as shown in the Fig. 5.38
- At very high frequencies, the zero resistance of the superconductor is modified.

5.59 CHANGE IN THERMODYNAMIC PARAMETERS IN SUPERCONDUCTING STATE

Many thermodynamic parameters change when a superconducting material is brought into a magnetic field B. Some are described below:

(i) Decrease in Free Energy

Free energy of the system decreases when a system undergoes from normal state to superconducting state. Let a superconductor is subjected to a magnetic field B. The work done on the superconductor per unit volume is given by

$$W = -\int_0^B M dB \tag{5.59}$$

where $M \to$ magnetisation vector

This work done can be expressed in terms of change in free energy ΔF, i.e.

$$W = \Delta F = F_s(B) - F_s(0) = -\int_0^B M dB \tag{5.60}$$

Subscript 'S' denotes the superconducting state.

For a superconductor, we can write

$$M = -\frac{B}{\mu_0}$$

Using this in above equation, we get

$$F_s(B) - F_s(0) = -\int_0^B \left[-\frac{B}{\mu_0}\right] dB$$

i.e. $$F_s(B) = F_s(0) + \frac{B^2}{2\mu_0} \quad (5.61)$$

When a non-magnetic metal (non-ferromagnetic) is placed in a magnetic field, its magnetisation $M = 0$ (since it is paramagnetic). So the free energy F_N of the normal metal can be assumed to be independent of the applied field, i.e. for a normal metal, the free energy at the initial field B_C is same as the free energy at zero field, so that

$$F_N(B_C) = F_N(0) \quad (5.62)$$

When there is a transition from normal state to the superconducting state at critical field B_C, the free energy of the two states at B_C is equal, i.e.

$$F_N(B_C) = F_S(B_C) \quad (5.63)$$

From the above equation

$$F_S(B_C) = F_s(0) + \frac{B_C^2}{2\mu_0} \quad (5.64)$$

Using equations (5.63) and (5.64), in the above equation

$$F_s(B_C) = F_N(B_C) = F_N(0) = F_S(0) + \frac{B_C^2}{2\mu_0} \quad (5.65)$$

i.e. $$F_N(0) - F_S(0) = \frac{B_C^2}{2\mu_0} \quad (5.66)$$

The above equation shows that the free energy decreases when a system undergoes from normal state to superconducting state. The free energy of the system in superconducting state is lower compared to the normal state, which confirm that the superconducting state is an ordered state.

The free energy vs. temperature plot for normal and superconducting state, and the variation of free energy with magnetic field is given in Figs 5.39 and 5.40.

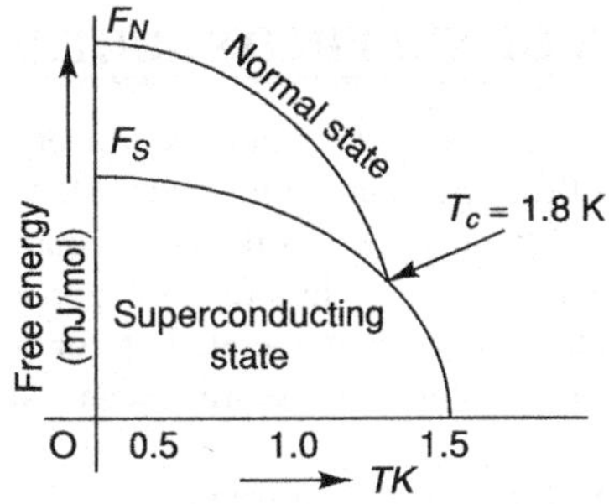

Fig. 5.39 *Free energy vs. temperature for normal and superconducting state*

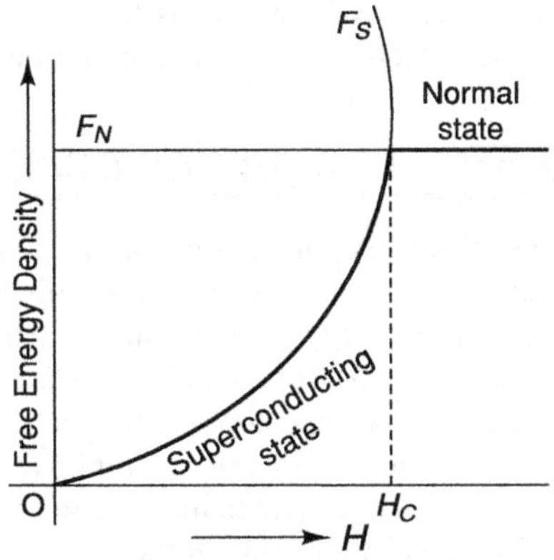

Fig. 5.40 *Variation of free energy with magnetic field*

(ii) Entropy

Entropy is a measure of degree of disorder in a system. This is also found much lower in superconducting state than in the normal state.

Change in entropy can be obtained by using Eq. (5.63)

$$S_N - S_S = -\frac{\partial}{\partial T}(F_N - F_S) = -\frac{\partial}{\partial T}\left[\frac{B_c^2}{2\mu_0}\right]$$

$$= -\frac{B_c}{\mu_0}\cdot\frac{\partial B_c}{\partial T}$$

We know

from equation, $B_c(T) = B_c(0)\left[1-\left(\frac{T}{T_C}\right)^2\right]$

Therefore, $\frac{2B_c}{\partial T} = \frac{2B_c(0)}{T_C^2}T$

Using this in the above equation

$$S_N - S_S = -\frac{B_c}{\mu_0}\cdot\frac{2B_c(0)}{T_C^2}$$

(iii) Change in Specific Heat

The change in specific heat when the material is brought from normal state to superconducting state is given by

$$C_N - C_S = T.\frac{\partial}{\partial T}(S_N - S_S)$$

On substituting the value of $S_N - S_S$ from the above equation and simplifying

$$[C_N - C_S]_{\text{at}T_C} = -\frac{4B_C^2(0)}{\mu_0 T_c}$$

The above relation shows that an increase in specific heat from normal to superconducting state is negative. This leads to the conclusion that at T_c, the transition from normal to superconducting state is a second order transition.

5.60 FREQUENCY DEPENDENCE OF SUPERCONDUCTIVITY

The existence of an energy gap in a superconductor suggests that photons of energy less than the gap energy can be transmitted through the superconducting material, and the photons of energy greater than the gap energy will be absorbed in the superconducting material (metal). It is found that nearly all photons with energy less than the gap energy incident on a superconducting metal are reflected as for any metal, but for a very thin film roughly of 20 Å. Significant transmission occurs with more transmission in superconducting state than in the normal state.

The absorption edge in the case of superconductors occurs at a frequency, $W = \frac{E_g}{\hbar}$, where E_g is the energy of gap. The absorption edge lies in the *IR* region between 5 cm–0.05 cm.

For photons of energy less than the gap energy, the resistance of a superconductor vanishes at absolute zero. Experimental evidences are there that at $T < T_C$, the resistance in superconducting state has a sharp threshold at the gap energy. Photons of lower energy find a resistance-less surface. Photons of higher energy find a resistance which approaches to that of normal state because such photons cause transitions to unoccupied normal energy levels above the gap. As temperature is increased, the gap decreases and the resistivity for photons below the gap no longer vanishes, except at zero frequency.

Superconductivity is not observed at optical frequencies. A superconductor is visibly opaque. It was found that superconductivity disappears at a frequency of more than 10^9 Hz.

5.61 CURRENT STATUS OF HIGH TEMPERATURE SUPERCONDUCTORS

Superconductor needs a very low temperature which is the biggest hurdle in using these materials for practical purposes to save the energy loss. Therefore, high T_c values are desirable in order to overcome the need of refrigeration to attain a superconducting state.

Two factors mainly decide the superconducting transition temperature:

- The electron phonon coupling constant
- The density of electron states

Some compounds containing chalcogen elements satisfy the condition of high T_c. The empirical formula of this compound is $A_xMO_6Z_{8-x}$, where A is a metal atom and Z is a chalcogen atoms–Se or Te. Experimentally, they have been found to have moderately high T_c–in the range of 14–20 k. These materials also show high H_{c_2} values.

The first high T_c ceramic superconductor was discovered in 1986 in a compound of barium, lanthanum, copper and oxygen. Since then a large number of metal oxide compounds with high *Tc* value have been reported. In this sequence, $YBa_2Cu_3O_7$ with a T_c of 90 K is the most studied superconductor. Some other compounds are given in Table 5.13.

Table 5.13 *The critical temperatures of some compounds*

Compound	*Symbol*	*Critical temperature(K)*
$La_{1.85}\,Ba_{0.15}\,CuO_4$	LBCO	36
$YBa_2\,Cu_3\,O_{609}$	YBCO	90
$Tl_2\,Ba_2\,Ca_2\,Cu_3O_{10}$	TBCO	125
$Hg\,Ba_2\,Ca\,Cu_2\,O^{6+}$	HBCO	133

Desirable Characteristics

Superconductors used for practical purposes need not only high T_c but also high H_c and J_c, and they must also exhibit favourable mechanical properties. They must be ductile, so that they may take shape of wires. Most of high T_c superconductors are brittle, which is the basic hurdle in converting them in the form of a wire.

5.62 PRACTICAL APPLICATIONS OF SUPERCONDUCTORS

The zero resistivity property of a superconductor has been explored for less power transmission. Many dc systems, like power transmission lines, electromagnets, motors, generators, etc., can be made more efficient by using superconductors.

(i) High-power Electromagnets

Magnetic Resonance Imaging (MRI) techniques are very useful for medical diagnosis and require very high magnetic field strengths. Electromagnets made of superconducting coils can be used in these instruments. Superconducting alloys of niobium and titanium are used for coil wires. High current can be passed through the coils without much heating effect. Such powerful magnets made out of normal metal coils would need huge cooling arrangements and the size would become too big. Although superconducting magnets need cryogenic cooling to maintain the superconducting state of the coils, they can be made smaller in size due to the high current carrying capacity of the superconducting wires. Type-II superconductors are used in electromagnets, as it is essential to have high critical fields.

Superconducting electromagnets are also used in many other applications where very high magnetic field strengths are required, like particle accelerators, controlled nuclear fusion reactors, etc.

(ii) Power Transmission Lines

DC power transmission through superconducting wires is the chief aim of electrical power engineers, as large current can be transmitted over very long distance with almost zero power loss. But this has practical difficulties, because high *Tc* superconducting require refrigeration and the cost of maintaining the superconducting wires at such low temperature is too high.

(iii) Frictionless Bearings and Motors

The diamagnetic characteristic of superconductors makes them suitable for achieving frictionless motion in bearings and motors.

The frictionless bearing consists of a superconducting coil of niobium wire above which a superconducting disc is mounted. The disc is capable of moving only vertically and not sideways. In frictionless motors the same principle of a magnetic air cushion between the stator and the rotor is used. This principle of magnetic levitation has also been successfully used in Japan, for friction-free super speed trains which have speeds up to 500 km/h.

(iv) Cryotron

Cryotron is a switching device which utilise the principle of destruction of superconductivity by the application of magnetic fields above B_c. The original cryotron consists of a thin wire of niobium wound over a thicker wire of tantalum (Fig. 5.41). The niobium winding is called the control coil and the thick tantalum wire is called the gate.

The critical magnetic field of the control coil wire must be much greater than that of the gate wire. The gate wire is in the superconducting state when no current is passing through the control coil. A current passed through the control coil, such that the magnetic field produced along its axis is greater than the critical field of the gate wire, switches the gate wire from superconducting state to normal resistive state. Thus, the gate can be switched between zero resistance state, by controlling the current through the control coil. The two states of the wire can be treated as binary digits 0 and 1 in computers.

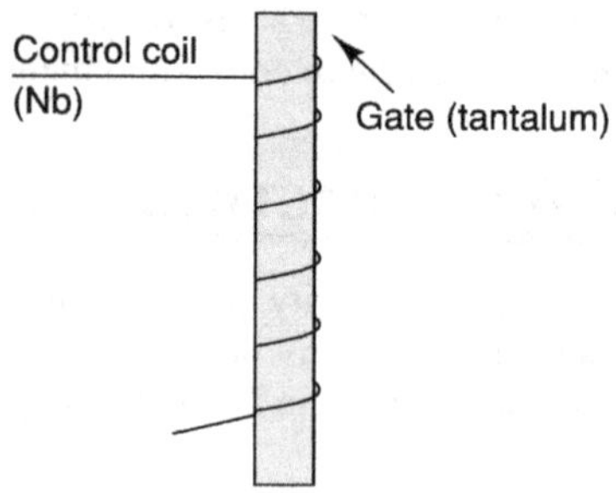

Fig. 5.41 *Cryotron*

(v) Josephson Junction Devices

❖ **SQUID** SQUID is a superconducting device based on the Josephson effect. It is also known as the superconducting quantum interference device. It consist of identical Josephson junctions connected in parallel. This is similar to the Young's double slit arrangement used for observing interference of light waves. The phase difference between the two junctions is varied by means of an applied magnetic field. The current through the circuit will have a periodicity which is very sensitive to the magnetic flux passing normally through the closed circuit. Extremely small magnetic flux can be detected with this device. The device can also detect voltages as small as 10^{-15} V and currents as small as 10^{-12} A. These devices are used in biomedical research to detect neural electrical activity in the brain. SQUID device are made of niobium and nitride thin films.

❖ **Microwave Generators and Detectors** The ac Josephson effect is used to generate microwaves. The power level from one single junction is too low (<1nW) and so arrays of junctions are used to generate microwaves. The device can also be used to detect microwaves.

Summary

- Materials are broadly classified as metals and non-metals. In metals, the valence electrons are free to move in the whole lattice, whereas in non-metals the valence electrons are strongly bound to the nuclei and are localised.
- Most of the properties of metals can be explained based on the classical free electron theory. In the Lorentz–Drude theory, the free electrons are treated as classical particles.
- In the quantum free electron theory (known as Sommerfeld's theory), the electrons are treated as quantum particles. The motion of the electrons is described by Schrodinger equation. The electrons occupy some definite energy levels, obeying Fermi–Dirac statistics.
- The highest occupied energy level at 0 K is called the Fermi energy. The Fermi energy is a characteristic parameter of the metal. It depends on the number of free electrons per unit volume in the metal.
- The quantum theory is successful in explaining many of the characteristic properties of the metals like electrical conductivity, thermionic emission, electronic paramagnetism, electronic specific heat, etc.
- In metals,, some resistivity is left at 0K which is known as residual resistivity (also called temperature independent part of resistivity). This is due to structural imperfections.
- Lattice resistivity of metal is defined as

  $$\rho(T) = \rho - \rho_0$$

  where ρ_0 is resistivity due to imperfection and ρ is total resistivity. This is identical for different specimens of some metal and is an inherent characteristic of metal.
- The above theory fails to explain low specific heat of metals and paramagnetic behaviour of conduction electrons.

- The above theory is very successful as it provides a logical interpretation of low specific heat of metals, spin paramagnetism and constancy of work function over a wide range of temperature.
- Thermoelectric effect is a phenomenon that arises due to a contact potential (equal to difference is work function of two metals) that is established across the junctions.
- Peltier effect is the reverse of Seebeck's effect in which the heat is evolved at one junction and absorbed at the other when a current is passed through it.
- The materials which exhibit zero resistivity below a certain temperature are known as the superconducting materials. The temperature is known as transition.
- T_c temperature is very low for elemental superconductors. Among the elements, Nb has the highest transition temperature of 9.5 K. Some alloys and intermetallic compounds have higher T_c of about 18–20 K.
- There is a limitation to the highest current that can pass through a superconductor. This current limitation is basically due to the effect of magnetic field on a superconductor. Application of sufficiently high magnetic field destroy the superconductivity.
- In the superconducting state, the material is a perfect diamagnet. This is known as Meissner effect.
- There are two types of superconductors—Type-I and Type-II. Type-I superconductors are perfect diamagnets below the critical field. Type-II superconductors have two critical fields B_{c1} and B_{c2}. Below B_{c1}, the magnetic lines of force are completely expelled, whereas between B_{c1} and B_{c2} the magnetic lines of force partially penetrate in to the material. Between the two critical fields the Type-II superconductor exists in the so-called mixed or vortex state (in which flux lines in the form of quantised vortices are arranged in a periodic two-dimensional array called the flux lattice). Beyond B_{c2} they become paramagnetic, i.e. they become normal metals.
- Type-II superconductors are more useful than Type-I in practical applications because the B_{c2} values are much higher than the B_c values of Type-I superconductors.
- Penetration depth is a measure of the depth of penetration of a magnetic field into the superconductors from its surface. It is of the order of fractions of microns for most superconductors.
- Superconducting state is an ordered state. There is a decrease in free energy of the system when there is a transition from the normal to the superconducting state. Correspondingly, the entropy in the superconducting state is lower than the entropy in the normal state.
- The isotope of a superconducting element is also a superconductor. The heavier isotope has lower T_c. The relation between atomic mass and T_c is given by

  $$Má\ T_c = \text{constant}$$

  A lies in the range of 0.15–0.5, for different elements.
- BCS theory is a microscopic theory that accounts for the superconducting behaviour.

Exercise

1. Explain free electron gas model of metals? Define free electron Fermi gas. Which properties of solids are explained by free electron gas theory?

2. Discuss the successes and failures of the free electron theory.
3. Obtain an expression for the electrical conductivity of a metal on the basis of free electron theory. Hence, prove Ohm's law.
4. (i) Obtain an expression for thermal conductivity of a metal on the basis of free electron theory.
   (ii) State the Wiedemann–Franz law.
5. Find an expression for Fermi energy and density of states of a system.
6. Give an account of the quantum theory of free electrons in metals. Derive an expression for the Fermi energy at absolute zero.
7. In terms of Fermi energy, calculate the kinetic energy at 0 K.
8. Derive an expression for Fermi energy of free electrons. Discuss briefly the effect of temperature.
9. Obtain an expression for energy levels in one dimensional free electron gas.
10. Write a note on
    (i) Fermi–Dirac distribution function
    (ii) Density of states in one-dimension
    (iii) Energy levels and wave function of free electrons in a box
11. Calculate the ground state energy and the first excited state energy for free electrons in a three-dimensional potential box of dimension $1 \times 1 \times 1$ cm^3.
12. Find the temperature at which thee is a 1% probability that an electron in a solid will have an energy 0.5 eV above the Fermi energy.
13. What is superconductivity? Give the main properties of a superconductor.
14. Explain Critical field in superconductivity and persistent current.
15. Explain the difference between Type-I and Type-II superconductors using Meissner effect.
16. What do you mean by superconductivity? Describe the effect of (i) magnetic field, (ii) frequency, and (iii) isotopes on superconductors.
17. Describe the effect of an external magnetic field on the superconducting state of a material. What do you mean by flux exclusion and what is Meissner effect?
18. What is the significance of critical temperature, critical magnetic field and critical current density for superconductors?
19. Derive the London's equations and discuss how they explain Meissner effect and flux penetration?
20. Give brief outline of BCS theory of superconductivity. Show that this theory provides adequate explanation of superconducting state.
21. Describe major uses and potentialities of superconductors.
22. Why superconductivity is a low temperature phenomenon?
23. What do you mean by high temperature superconductivity Give an account of high T_c superconductors.
24. Write a note on
    (i) Penetration of magnetic field in a superconductor and penetration depth
    (ii) Flux quantisation
25. Explain why the electrical resistivity of a solid goes down with the increase in temperature?
26. Explain superconductivity on the basis of free electron theory.
27. Mention some important changes that occur in substances when they undergo a change from normal state to superconducting state.

28. Explain the following terms:
    (i) Transition temperature
    (ii) Critical field
    (iii) Persistent current
29. Give a brief account of high-temperature superconductors and point out the hurdles in the Their development.
30. Discuss the formation of cooper pairs and energy gap in superconductors on the basis of BCS theory.

Numerical Problems

1. In a certain copper conductor, the current density is 2.4 A/mm^2 and electron density is 5×10^{28} free electrons per m^3 of the copper. Determine the drift velocity of the electrons. **[Ans. 0.3 mm/s]**
2. A conductor material has a free electron density of 10^{24} electrons per m^3. When a voltage is applied, a constant drift velocity of 1.5×10^{-2} m/s is attained by the electrons. If the cross-sectional area of the material is 1 cm^2 calculate the magnitude of current. Electronic charge is 1.6×10^{-19} coulomb. **[Ans. 0.24 A]**
3. The following data is known for a conductor,

   Mobility of electrons = 7.04×10^{-3} m^2/volt- second

   Electron density = 5.8×10^{28}/m^3

   Calculate (i) relaxation time, and (ii) resistivity of conductor. Given that $e = 1.6 \times 10^{-19}$ C, and $m = 9.1 \times 10^{-31}$ kg.

   [Ans. 40.04×10^{-15} second, $0.01531 \times 10^{-6}\ \Omega - $m] [IES 1994]
4. A conduction wire has a resistivity of $1.54 \times 10^{-8}\ \Omega$ m at room temperature. There are 5.8×10^{25} conduction electrons per m^2. Calculate the mobility and relaxation time of electrons. **[Ans. 3.979×10^{-14} second, 6.997×10^{-3} m^2/V – s] [IES 2000]**
5. A conducting wire has resistivity of $1.5 \times 10^{-8}\ \Omega$ m at room temperature. Its Fermi energy is 5 eV. Its volume of 1 m^3 contains 6×10^{28} conducting electrons. Calculate the relaxation time of conducting electrons. Hence or otherwise, obtain the value of drift velocity of electrons when the conductor is subjected an electric field of 100 V/m. Write the significance of relaxation time and drift velocity. Given: Mass of electron = 9.107×10^{-31} kg; Electron charge = 1.601×10^{-19} C.

   [Ans. 3.948×10^{-14} second, 0.694 m/s] [IES 2001]
6. A copper wire of 1.0 mm radius with conductivity of 5.8×10^7 s/m and electron mobility of 0.0032 m^2/Vs is subjected to an electric field of 20 mV/m. Find (i) the charge density of free electrons (ii) the current density (iii) the current flowing in the wire, and (iv) the electron drift velocity.

   [Ans. 1.133×10^{29} m^3, 1.16×10^6 A/m^2, 3.644 A, 6.4×10^{-5} m/s] [IES 2002]
7. A n-type germanium crystal has a current density of 100 A/m^2. The crystal has a resistivity of 0.5 Ω-m and electron mobility of 0.4 m^{2}Vs. Calculate the drift velocity and the time taken by the electron to travel 10 micro m in the crystal. Assume $e = 1.6 \times 10^{-19}$ C. **[Ans. 20 m/s, 0.5×10^{-6} second] [IES 2004]**
8. The critical temperature of mercury is 4.153 K for its one isotope of mass 200.59 amu. Calculate the critical temperature of mercury for its one isotope of mass 204 amu. **[Ans. 4.118 K]**
9. Determine the critical current that can flow through a long thin superconducting wire of aluminium of diameter 10^{-3} m. The critical magnetic field for aluminium is 8.9×10^3 A/m. **[Ans. 27.946 A]**

10. The specific resistance of platinum at 0°C is 10.3 $\mu\Omega$ – m. How long must a wire of 32 S W G radius = 0.0137 cm platinum be to have a resistance of 4 Ω at 0°C? What will be the resistance of the wire at 100°C, if the temperature coefficient of resistance is 0.0038 per °C? [Ans. 2.29 metres]
11. Find the current flowing at the instant of switching on a 100 watt metal filament lamp on a 220 V circuit. Given that the incandescent filament temperature is 2,000°C and the resistance temperature coefficient at room temperature of 20°C is 0.005. [Ans. 4.95 A]
12. The resistivity of copper rod as measured by Kelvin bridge is 1.74 micro-ohm-cm at 20°C. If the resistivity of the copper is 1.73 micro-ohm-cm at 20°C, calculate the impurity (%) content in the rod. [Ans. 0.5747%] [IES 2004]

Self-quiz Questions

1. What is free electron theory?
2. What do you mean by the drift velocity and mobility of a free electron?
3. What are the factors which affect the value of conductivity of material?
4. What is meant by temperature coefficient of resistance?
5. What is meant by thermal conductivity?
6. What is Seebeck effect?
7. State Peltier effect.
8. What is thermocouple and thermo-electric emf?
9. What is meant by a thermocouple?
10. On what factors does an emf of a thermocouple depend?
11. What is the effect of inserting a new metal into a circuit of a thermocouple on the thermo-emf?
12. Which junctions of a thermocouple get heated or cooled when an electric current is passed through them?
13. What is energy gap in superconductors?
14. Define critical magnetic field of a superconductor.
15. Why are Type-II superconductors more important?

Objective-type Questions

1. With increase in temperature, the resistance of a metal
   (a) remains constant
   (b) increases
   (c) decreases
   (d) becomes zero
2. Average kinetic energy ($\overline{E}_0$) of a free electron gas at 0K is
   (a) $\frac{2}{5}E_F$ (b) $\frac{5}{3}E_F$
   (c) $\frac{3}{5}E_F$ (d) E_F
3. The density of states of electrons between the energy range E and $E + dE$ is proportional to
   (a) $E^{1/2}$ (b) E^2
   (c) E (d) $E^{3/2}$
4. The phase space is a
   (a) two-dimensional space
   (b) one-dimensional space
   (c) three-dimensional space
   (d) six-dimensional space
5. At low temperature, the resistivity of a metal is proportional to
   (a) T^2 (b) T
   (c) T^5 (d) $T^{1/2}$

6. Which one of the following relations is correct for current density?

(a) $J = nev_d$ (b) $J = \frac{1}{nev_4}$

(c) $J = \frac{ne}{v_d}$ (d) $J = neAv_d$

7. Which one of the following relations is correct for the conductivity of metals?

(a) $\sigma = \frac{me^2}{2mt}$ (b) $\sigma = \frac{2m}{ne^2t}$

(c) $\sigma = \frac{ne^2t}{2m}$ (d) $\sigma = \frac{n}{2me^2t}$

8. The value of Fermi distribution function at absolute zero ($T = 0$ K) is 1, i.e. $F(E) = 1$, under the condition

(a) $E > E_F$ (b) $E < E_F$
(c) $E = E_F$ (d) $E >> E_F$

9. At any temperature T and for $E = E_F$, the Fermi distribution function becomes

(a) 0 (b) ∞

(c) 1 (d) $\frac{1}{2}$

10. The free electron theory of metals was initiated by
(a) Pauli
(b) Sommerfield
(c) Lorentz and Drude
(d) Fermi–Dirac

11. Which are correct for Ohm's Law? (J is current density, E is electric field, σ is electrical conductivity, ρ is electrical resistivity, I is current, R is resistance, L is length of the conductor, A is area of cross section of the conductor)

(a) $J = \sigma E$ (b) $\rho = J/E$
(c) $I = EL/R$ (d) $I = VA/\rho L$

12. Which of the following statements are true? The electrical conductivity of a metal is
(a) directly proportional to the probability of collision of the electrons with the lattice
(b) inversely proportional to the relaxation time
(c) directly proportional to the density of free electrons
(d) directly proportional to the mean free path

13. Which of the following processes increase the resistivity of a metal?
(a) The metal is heated
(b) The metal is purified
(c) The metal is cooled
(d) Imperfections are introduced into the metal.

14. The density of free electron states in a metal
(a) varies as $E^{1/2}$
(b) varies as $E^{-1/2}$
(c) varies as E
(d) varies as $1/E$

15. In an imperfection free crystal, the resistivity will be
(a) infinite (b) zero
(c) negative (d) unit

16. With increase in temperature, the mean free path
(a) decreases
(b) increases
(c) remains unchanged
(d) vanishes

17. Highest electrical resistivity exists in
(a) platinum wire
(b) nichrome wire
(c) silver wire
(d) kanthal wire

18. An example of a metal having zero Thomson effect is
(a) zinc (b) cobalt
(c) lead (d) rhodium

19. The transition temperature of mercury is
(a) 1 K (b) 1.14 K
(c) 4.12 K (d) 9.22 K

20. A superconducting material when placed in a magnetic field will
(a) attract the magnetic field toward its centre.
(b) repel all the magnetic lines of forces passing through it.

(c) attract the magnetic field but transfer it into a concentrated zone.
(d) not influence the magnetic field.

21. A superconducting material on being subjected to the critical field changes to
(a) critical conductivity
(b) superconducting which is independent of temperature
(c) normal state
(d) remains uninfluenced

22. The width of the energy gap of a superconductor is maximum at
(a) 0 K
(b) transition temperature
(c) room temperature
(d) none of these

23. The temperature at which a conductor superconductor is called
(a) superconducting temperature
(b) Curie temperature
(c) Neel temperature
(d) transition temperature

24. The resistance of a conductor in superconducting state is
(a) infinite (b) zero
(c) mega-ohm (d) arbitrary

25. Superconductivity is usually observed at
(a) very high temperatures
(b) room temperature
(c) very low temperatures
(d) any temperature

26. Meissner effect is related to
(a) optical properties of superconductors
(b) magnetic properties of superconductors
(c) thermal properties of superconductors
(d) electrical properties of superconductors

27. The specific heat of a superconducting material shows an abrupt change at $T = T_e$ jumping to a large value for
(a) $T < T_e$ (b) $T > T_e$
(c) $T = T_e$
(d) none of these

28. The transition temperature of most superconducting elements lies in the range
(a) Zero to 10 K
(b) 10 K to 20 K
(c) 20 K to 50 K
(d) above 50 K

29. Superconductors are generally
(a) ferromagnetic and anti ferromagnetic metals
(b) monovalent metals
(c) amorphous thin films of Be and Bi
(d) thin films of barium titanate

30. Hard superconductors observe
(a) breakdown of Silsbee's rule
(b) incomplete Meissner effect
(c) high critical field and transition temperature
(d) all of these

31. Soft superconductors observe
(a) Meissner effect
(b) Silsbee's rule
(c) both (a) and (b)
(d) none of these

32. A superconductor exhibits
(a) only zero resistance
(b) only diamagnetism
(c) zero resistance and paramagnetism
(d) zero resistance and diamagnetism

33. The property remains unchanged in superconducting state
(a) specific heat
(b) magnetisation
(c) internal energy
(d) volume

34. The critical field needed to destroy superconductivity
(a) $H_0\left(\frac{T}{T_c}\right)^2$
(b) $H_0\left(\frac{T_c}{T}\right)$

(c) $H_0\left[1-\left(\frac{T}{T_c}\right)^2\right]$

(d) $H_0\left[1-\left(\frac{T_c}{T}\right)^2\right]$

35. The electron pairs in a superconductors are called
(a) Bardeen pairs
(b) BCS pairs
(c) electron-hole pairs
(d) Cooper pairs

36. The cooper pairs are
(a) Bosons
(b) Fermions
(c) classical particles
(d) electron positron pairs

37. The magnetic susceptibility of a superconductor is (in SI units)
(a) zero (b) + 1
(c) –1 (d) infinity

38. The width of energy gap of a superconductor at 0 K is nearly
(a) K_BT_c (b) 35 K_BT_c
(c) 3.5 K_BT_c (d) 300 K_BT_c

39. The total charge on a Cooper pair is
(a) zero
(b) + 2e
(c) –2 e
(d) varies with T_c

40. The binding energy of a copper pairs is of the order of
(a) 10^{-12} eV (b) 10^{-3} eV
(c) 10^{-8} eV (d) 1 eV

41. In superconducting state entropy
(a) increase
(b) cecrease
(c) no effect on entropy
(d) none of these

42. The phenomenon of flux exclusion in a superconductor is called
(a) Meissner effect
(b) Joule effect
(c) Seebeck effect
(d) Cooper effect

43. The phenomenon of superconductivity was discovered by Kamerlingh Onnes in
(a) 1931 (b) 1911
(c) 1921 (d) 1811

44. The variation of critical temperature with increasing atomic mass is called
(a) isotope effect
(b) isobar effect
(c) isomeric effect
(d) none of these

45. The London's penetration depth is given by
(a) $\lambda=\left(\frac{m}{n\mu_0e^2}\right)^{1/2}$
(b) $\lambda=(nm\mu_0e^2)^{1/2}$
(c) $\lambda=\left(\frac{n}{m\mu_0e^2}\right)^{1/2}$
(d) none of these

Answers

1. (b)	2. (a)	3. (a)	4. (d)	5. (b)
6. (a)	7. (b)	8. (b)	9. (c)	10. (c)
11. (a)	12. (c)	13. (d)	14. (a)	15. (b)
16. (b)	17. (d)	18. (c)	19. (c)	20. (b)
21. (c)	22. (a)	23. (d)	24. (b)	25. (c)
26. (b)	27. (c)	28. (a)	29. (b)	30. (b)
31. (a)	32. (d)	33. (d)	34. (c)	35. (d)
36. (a)	37. (c)	38. (c)	39. (c)	40. (b)
41. (b)	42. (a)	43. (c)	44. (a)	45. (a)

Fill in the Blanks

1. Metals are characterised by .... value of conductivity.
2. The .... exhibit positive temperature coefficient.

3. Ratio of thermal to electrical conductivity is constant for .... which is known as Wiedemann–Franz law.
4. The resistance of metals................. on alloying.
5. The free electron theory is known as .... theory.
6. Good conductors of electricity .... show superconducting phenomenon.
7. Metals either exhibit no .... or partially-filled band gap.
8. Conductivity of metals decreases with .... of temperature.
9. Metals obey Ohm's law which tells that .... remains constant as long as electric field remains constant.
10. Resistivity ratio is defined as the ratio of resistivity of specimen material at room temperature to its resistivity at .... temperature.
11. Ratio of thermal to electrical conductivity has same valve at all temperatures for all metals and is called ....
12. The value of .... heat on the basis of free electron model comes out $\frac{9}{2}R$, but actually it is $3R$.
13. The .... obey Fermi–Dirac statistics which gives rise to Fermi energy.
14. Fermi energy is the energy of .... level at absolute zero.
15. The Fermi energy for .... is 7.04 eV.
16. Low specific heat of metals can be explained on the basis of............................. of electrons.
17. The temperature at which the thermo emf attains a peak value is known as .... temperature.
18. The reverse of .... is known as Peltier effect.
19. The resistivity of .... Suddenly drops to zero at 4°K.
20. The magnetic field at which the superconducting property of the material gets lost is known as ....
21. The minimum current that can be passed in a sample without destroying its superconducting property is known as ....
22. The magnetic flux expulsion from the material below critical temperature is known as ....
23. The materials in which the transition from superconducting state to normal state is sharp are known as .... superconductors.
24. The alloys exhibit .... of transition temperature compared to elements.
25. The depth inside the superconductor to which the magnetic field penetrate is known as ....
26. London's penetration depth is given by ....
27. The cooper pairs exhibit a total spin equal to ....
28. The copper pairs follow .... statistics.
29. Thermal conductivity is usually .... is superconducting state.
30. Josephson junction emits .... when a direct current is passed through the it.
31. The minimum magnetic flux trapped in a superconducting ring carrying current is given by ....
32. The structure of a high T_c superconductor is modified tetragonal perovskite.
33. The most suitable materials for making coils for an electromagnet of very high field strength are .... superconductors.

34. Type I superconductors have .... critical fields than type II.
35. The magnetisation of a superconductor just above B_C is ....
36. In a thin film of superconductor microwave radiation (1–4 mm wavelength) is ....
37. The critical magnetic field of a superconductor .... with increasing temperature.

Answers

1. high	2. conductivity materials	3. metals
4. increases	5. Lorentz–Drude	6. do not
7. band gap	8. increase	9. current density
10. liquid helium	11. Lorentz number	12. electronic specific
13. free electron gas	14. highest occupied	15. copper
16. Fermi–Dirac distribution	17. neutral	18. Seebeck effect
19. mercury	20. critical field	21. Meissner effect
22. type I	23. high value	24. penetration depth
25. $\lambda = \left[\dfrac{m}{\mu_0 nse^2}\right]^{9/2}$	26. zero	27. Bose–Einstein
28. lower	29. critical current	30. microwaves
31. *n*/2*e*	32. modified	33. type II
34. lower	35. small and positive	36. transmitted
37. decreases		

Reading List

1. Beiser, A (1977) *Concepts of Modern Physics*. New Delhi: McGraw-Hill Education Pvt Ltd.
2. Landau, L D; Lifshitz, E M (1985) *Statistical Physics*. Reading, Mass: Addison-Wesley.
3. Naik, P V (2000) *Principles of Physics*. New Delhi: PHI Learning Pvt Ltd.
4. Rajam, J B (1982) *Atomic Physics*. New Delhi: S Chand & Company Pvt Ltd.
5. Richtmyer, F K; Kennard, E H; Copper, J N (1966) *Introduction to Modern Physics*. New Delhi: McGraw-Hill Education Pvt Ltd.
6. Vijaya, M, S; Rangarajan G, *Material Science.* New Delhi: McGraw-Hill Education Pvt Ltd.
7. Baiser, Arthur *Perspectives of Modern Physics.* New Delhi: McGraw-Hill Education Pvt Ltd.
8. Dekker A J (1998) *Solid State Physics.* New Delhi: Macmillan India Ltd.
9. Gupta, Navneet (2000) *Engineering Practical Physics*. 2nd ed. New Delhi: Dhanpat Rai & Co.
10. Dekker AJ (2000) *Electrical Engineering Materials.* New Delhi: PHI Learning Pvt Ltd.
11. Kittel, Charles (1996) *Introduction to Solid State Physics*. 7th ed. New York: John Willey & Sons.

6

Electrical Conducting and Insulating Materials

Chapter Objectives

- Classification of electrical engineering materials into low- and high-resistivity materials.
- Knowing the materials used for various purposes, like overhead transmission lines, filaments and electrical contacts.
- Selection of insulating materials for various purposes.
- Popular electrical characteristics of the material to be considered on the basis of their uses.
- Protection of material against humidity, oxidation and other hazards.
- Various kinds of capacitors used in electrical engineering machines.
- Optimisation of electrical and mechanical characteristics of the materials used in electrical machines.

6.1 INTRODUCTION: PROPERTIES AND APPLICATIONS

Electrical engineering materials play a crucial role in almost all kinds of machines and electrical appliances. These are being used not only in electrical power generation, but also in distribution and in a wide range of electrical machine components. Some materials provide a very good insulation, which is the basic need for safe and risk-free transmission over long distances, while others are used for protection against moisture. The list of such materials is quite long and one cannot imagine the development of power infrastructure without using them. Some of these are given in Fig. 6.1 along with their characteristic property and main application.

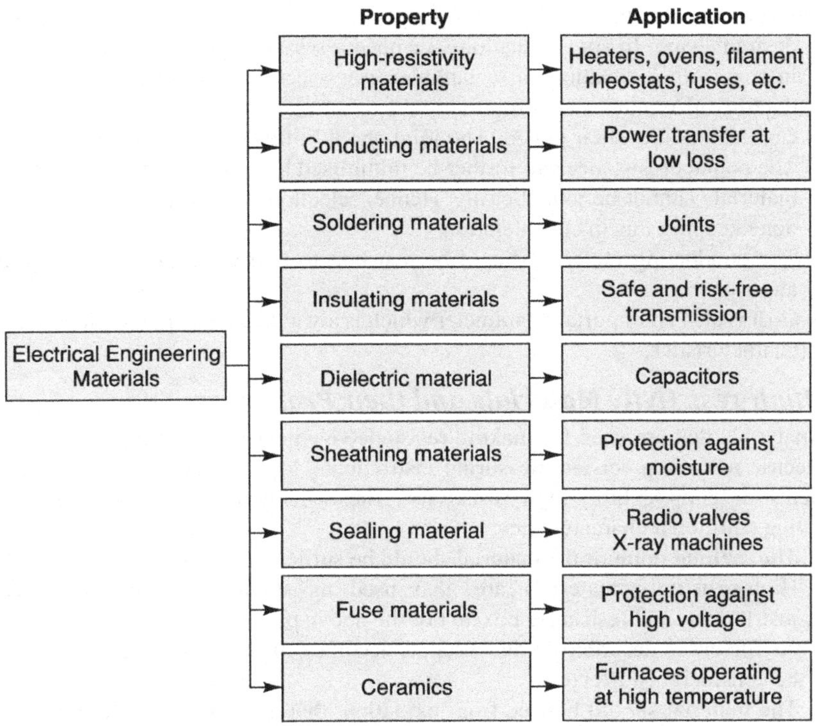

Fig. 6.1 *Electrical engineering materials*

6.2 CLASSIFICATION OF CONDUCTING MATERIALS

These materials are classified into two major categories on the basis of their resistivity:

- Low-resistivity materials
- High-resistivity materials

(i) Low-resistivity Materials and their Properties

The transmission and distribution of power requires a material which exhibits a low power loss. In fact, low-resistivity materials are used keeping in view the above features along with low voltage drop. Copper and aluminium are two materials that are commercially accepted. Silver is also a material which is suitable, but is not used because of high cost.

Some properties which are desirable for low-resistive materials are given below:

- The materials should exhibit a low temperature coefficient which means that the resistivity of such materials does not appreciably change with temperature. This is essential so as to avoid variation in voltage drop and power loss with the change of temperature, which normally changes in every season. Such materials are also desirable in windings of electrical machines and apparatus.
- Material should have sufficient mechanical strength. This is necessary because the overhead lines of the conductors used for transmission of power should be capable of bearing stresses due to wind and their own weight. Also, when coils are made for various equipments, the conducting material is subject to mechanical stresses. So, it should have sufficient strength.

- Ductility is another desirable property as it allows the material to be drawn in a desired shape. In some situations we need a wire of circular cross-section, while in some other situations a rectangular cross-section may be needed. Therefore, ductility also plays a vital role.
- Conductors are often joined. The joint should offer minimum contact resistance. The contact resistance can further be minimised by soldering the material. All the materials cannot be joined easily. Hence, selection of conducting material is also done keeping this in consideration.
- The conducting material should be resistant to corrosion when used in outdoor atmosphere.
- Cost is also an important parameter which is always looked upon besides the above characteristics.

(ii) High-resistivity Materials and their Properties

The materials that are used for making resistance elements for heating devices, starters for electric motors, precision measuring instruments, loading resistances, and rheostats filaments of lamp exhibit high resistivity. These materials should also exhibit the following additional characteristics:

- The melting point of the material should be sufficiently high.
- High-resistivity materials are also used as shunts in electrical measuring instruments and resistance boxes. For the above purposes a negligible temperature coefficient is desirable. In its absence, accuracy of the measuring instrument will be considerably affected.
- The material should be free from oxidation, that is, with rise in temperature and prolong usage it should not oxidise.
- High-resistance thin wires are normally made from these materials which in turn are given the shape of a coil. Ductility of the high-resistivity material, therefore, also is vital so that the material could be coiled.
- Mechanical strength should be high enough. This is necessary because the wires made from these materials are very thin, and in the absence of mechanical strength they may break during operation.

Some important properties of high resistivity of alloys are given in the Table 6.1.

Table 6.1 *Comparison of properties of some high-resistivity alloys*

S.No.	*Parameter*	*Nichrome*	*Constantan*	*Manganin*
1.	Resistivity (ohm-m at 20°C)	110×10^{-8}	52×10^{-8}	48×10^{-8}
2.	Melting point	1540°C	1300°C	1020°C
3.	Density (kg/m^3)	8250	8900	8190
4.	Permissible working range of temperature	1100°C	850°C	700°C
5.	Mechanical work ability	Can be stretched into thin wires	Can be stretched into thin wires	Can be stretched into wires
6.	Temperature coefficient of resistance (/°C at 20°C)	0.0001×10^{-4}	0.0003×10^{-4}	0.00002×10^{-4}

(Contd)

S.No.	*Parameter*	*Nichrome*	*Constantan*	*Manganin*
7.	Main applications	Heating lements for electric furnaces and ovens, room heaters, electric irons and resses	Resistance element for rheostats, starters of electric motors, etc.	Wire-wound shunts and precision resistances, coils for precision measuring instruments, etc.

6.3 DIFFERENCE IN PROPERTIES OF HARD-DRAWN AND ANNEALED COPPER

Low-resistivity materials that are very popular and universally accepted are copper, aluminium, silver and gold. The most widely used materials are copper and aluminium. Silver and gold are not used due to their high cost. Soldering and welding in copper wires is easily possible, while it is not the case with aluminium. Copper in pure its form is not much used as electrical contact material. When 10–30% Ni is added to it, the cost goes down and hardness also increases. Tensile strength of Cu or Al can be varied. These materials may be available in two forms:

- *Hard-drawn copper*: It is spongy, exhibits more tensile strength, and hence, can be used for over-head conductors and underground cables.
- *Annealed copper*: It is soft and exhibits less tensile strength, and is therefore, suitable for insulated conductors, low voltage power cable, winding wires, flexible wires, etc.

(i) Property-wise Applications of Copper

Some brilliant features of copper are given in Fig. 6.2; they make it suitable for a large number of applications.

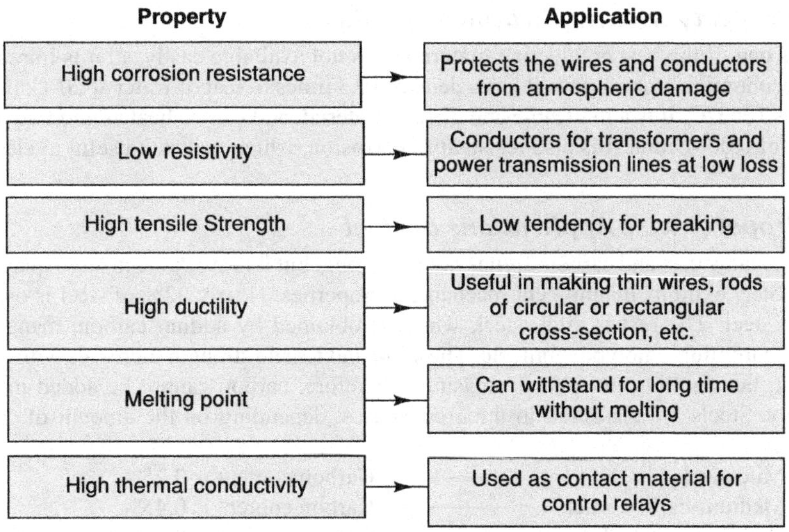

Fig. 6.2 *Features of copper*

(ii) Property-wise Applications of Aluminium

It is a another widely used material in electrical engineering; it acts as a substitute for copper due to its low cost and magnificent electrical and thermal characteristics. A low density makes it the most popular material to be used in aircrafts. Resistivity is 1.6 times more than copper and its melting point is less, i.e. 655°C. Compared to copper, heat produced in aluminium conductors is more, and therefore, chances of melting are also more which is its drawback. Aluminium is a soft metal compared to copper, but it can also be obtained in two kinds—hard-drawn and annealed. It also has a tendency to form a layer of oxide on it, which prevents the material from further oxidation and acts as a resistive layer to corrosion. Due to softness, chances of loose contact are more compared to copper, which is the greatest disadvantage with this material. Soldering is very difficult. As a substitute of copper, aluminium is quite extensively used for flexible wires, over-head transmission lines and in many other applications.

For economic reasons, aluminium has replaced copper in many fields. Similar to copper, it is available in abundance in India. Over-head transmission lines are now made from aluminium conductors with steel reinforcement. These are known as ACSR. This technique provides higher strength to over-head conductors.

Figure 6.3 shows an ACSR conductor.

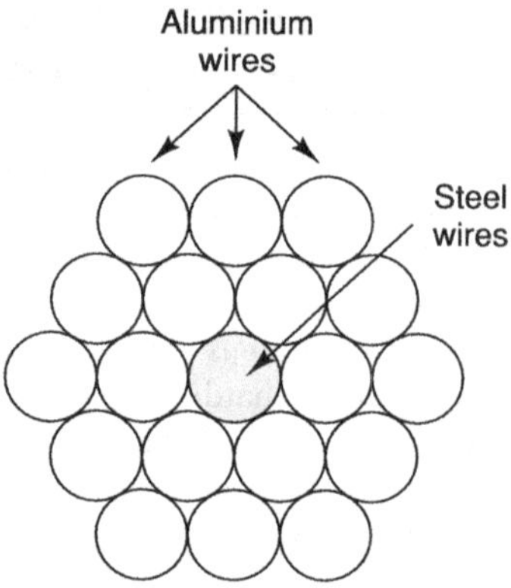

Fig. 6.3 *ACSR conductor*

(iii) Property wise Applications of Silver

Pure silver has high electrical conductivity and good corrosion resistance. In order to make it harder, 15% Cu is added to it. To make it more hard 40% copper is added and the alloy so obtained is used in commutator segments of small dc motors. For making brushes and collector rings a little graphite is added to it, which provides sliding lubrication.

(iv) Property wise Applications of Gold

This is one of the best conducting material. It is not available easily, so it is impractical from economic point of view. It has a density 19.3 times to that of water at 20°C; melting point is 1063°C. It is a malleable and ductile material, and can be beaten into translucent sheets of 0.0001 mm. It is also resistant to corrosion, which makes it useful in electrical fields.

(v) Property wise Applications of Steel

Steel contains iron and carbon. Iron is not too strong, but it is available in large quantity in India. Steel exhibits magnificent mechanical properties. Nearly 92% of steel is ordinary carbon steel. The rest is alloy steel, which is obtained by adding carbon, manganese, nickel, chromium, molybdenum, etc. The yield and tensile strength increases with carbon content, but ductility goes on decreasing. Therefore, carbon cannot be added in much quantity. Steels are classified in three categories, depending on the amount of carbon. These are

- Mild steel $\longrightarrow$ Carbon content is 0.25%
- Medium-carbon steel $\longrightarrow$ Carbon content is 0.45%
- High-carbon steel $\longrightarrow$ Carbon content is 0.7%

Tensile Strength of Steel is Increased by Heat Treatment and Quenching it in Water High resistivity does not allow this material to be used as a conductor. This is easily corroded by moisture and heat. Over-head steel conductors are galvanised to prevent from oxidation. Galvanised steel wires are used as over-head telephone wires and earth wires. Steel overhead conductors are also used for transferring small amounts of power.

6.4 STRANDED CONDUCTORS

To handle a single conductor of large radius, chances of breaking are always there, and its handling also becomes difficult. To overcome this, many conductors (bunch) are taken and twisted to get a high diameter. Such bunches of conductors are called 'strands'. Stranding provides flexibility and eliminates the breaking of conductors. Stranding is always done in opposite direction for successive layers. A standard stranding consist of six wires around one wire, then twelve wires around previous six, and then 18 wires around previous twelve, and so on. Instead of single wire at the centre, three or four can also be taken. If three stranded wires are put at the centre, nine wires will be put in the first layer, 15 in the second layer, and so on.

The formula for the above is given in Table 6.2.

Table 6.2 *Stranding of conductors*

Number of wires at the centre	*1 wire*	*3 wire*	*4 wire*
Number of wires in *n*th layer from the centre	$6n$	$3 + 6n$	$4 + 6n$
Total number of wires in a stranded conductor having *n* layers	$1 + 3n(1 + n)$	$3(1 + n)^2$	$(4 + 3n)(1 + n)$
Diameter of stranded conductor (overall)	$(1 + 2n)d$	$(2.155 + 2n)$	$d(2.414 + 2n)d$

where d is the diameter of each wire in cm and n is the number of layers.

Stranded conductor are expressed as,

7/2.24, 19/2.50, etc.

where first number 7, 19..., etc. indicate the total number of wires in the stranded conductor, and second number in the diameter of each wire in mm.

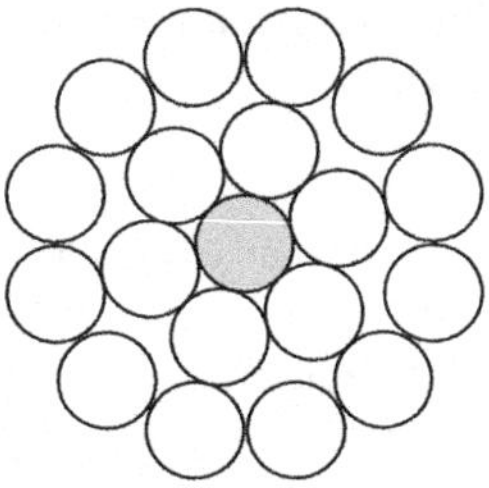

Fig. 6.4 *Stranded circular conductor containing two layers*

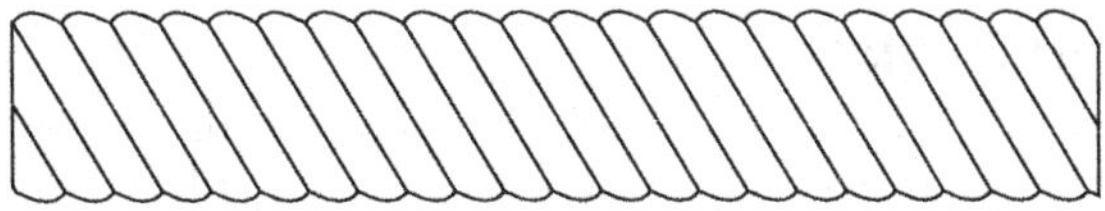

Fig. 6.5 *Stranded and twisted circular conductors*

Suppose we have a stranded conductor having specification 37/2.06. Therefore, such conductors will have 1 wire at the centre, six in the first layer, 12 in second layer and 15 in the third layer.

i.e. $37 = 1 + 3 \times 3(1 + 3)$

Therefore, number of layers = 3

Overall diameter = $(1 + 2n)d = (1 + 2 \times 3) \times 2.06 = 14.42$ mm

Circular stranded conductors and compact circular stranded conductors are normally used in single phase systems. This technique provides strength and economy by saving insulating and protective materials in the manufacture of cables.

Alloys of copper are also used as low-resistivity material. These are the following

- *Brass*: It is an alloy of copper and zinc (45%) which exhibits high tensile strength, but low conductivity than copper. It has good corrosion resistance, easy to solder and weld, and therefore widely used in plug-points socket outlets, switches, lamp holder fuse holders, etc.
- *Bronze*: It is another alloy of copper with many elements like aluminium, silicon and nickel. It is much superior to brass and is used in sliding contacts, fuse clips, etc.

6.5 ADDITIONAL REQUIREMENTS FOR HIGH-RESISTIVITY MATERIALS

High-resistivity materials play a key role in electrical engineering. Besides a high melting point, these materials should exhibit:

- Low temperature coefficient of resistance
- Anti-corrosive tendency
- Good mechanical strength
- High ductility
- Very low brittleness

Some popular resistance materials are described below:

(i) Platinum

This a greyish-white metal exhibiting high value of resistivity and high temperature coefficient. It is also a metal which is resistant to corrosion, and is malleable and ductile. Its melting temperature is 1775°C and resistivity 0.1 μ ohm-m. It can be drawn into thin wires and strips. It is normally used in elements of oven and in resistance thermometers. Resistance varies with temperature and this property is explored to measure very high temperatures. The devices made for the above purpose are named as thermocouples. Resistance thermometers, which are used to measure very high temperature, work on the above principle. Besides this, it is also used in vacuum tubes.

(ii) Tungsten

This is a silver-white, hard metal having a very high melting point—nearly 3400°C. This high value makes it useful in making bulb filaments. Its resistivity in 5.51×10^{-8} ohm-m and temperature coefficient of 0.0045 per ohm per °C. Tensile strength is also high. It is used in X-ray machines as targets. Tungsten is produced by a very complicated process from rare ores.

(iii) Molybdenum

This material has similar properties to tungsten. It has a melting point of 2620°C and resistivity about 4.8×10^{-8} ohm-m. The temperature coefficient is 0.0047 per degree Celsius. It is also used as target in x-ray tubes, and in vacuum tubes, as well.

(iv) Manganin

It is an alloy of 84% Cu, 12% Mn and 4% nickel. The colour is reddish brown. Its specific resistance lies in the range from 0.42–0.74 μ ohm-m. Temperature coefficient of resistance is 00003 ohm per degree Celsius. It can be used up to 60°C. Manganin is most widely used in the manufacture of resistance coils, shunts, etc.

(v) Constantan

It is an alloy of 60% Cu and 40% Ni. Resistivity is 49×10^{-8} ohm-m. Temperature coefficient is very small; the maximum working range of temperature is nearly 450°C. It is widely used in the manufacture of rheostats and thermocouples, starters, field regulator resistances, resistance boxes, etc.

(vi) Nichrome

Nichrome is an alloy of nickel and chromium. Its resistivity is nearly 1.0 μ ohm-m. Temperature coefficient is 0.00044 ohm/°C; working temperature range is 1000°C. It is available in the form of a wire or tape and is mainly used as heater element in furnaces in the above given temperature limit range.

(vii) Carbon

It is a tetravalent element consisting of a large band gap, and therefore exhibits a high resistance. It is used in making carbon resistors, brushes for electrical machines, instant boiler elements, dry cell anodes, etc. It is also used in automatic voltage regulators. It also comes handy in the manufacture of welding electrodes, fixed and variable resistors for light currents and contacts of certain classes of DC switch gear, which are subjected to arcing.

(viii) Lead

It is a poor conductor and exhibits anti-corrosion and softness. Its melting point is low—nearly 327°C. Due to this reason the alloy of lead with 60% Sn is used as a solder material. It is also used in storage batteries.

High-resistivity materials like mangnnin, nichrome and platinum are also used to make

- Standard resistances
- Rheostats
- Electric furnaces

Carbon, tantlum and tungsten materials are best suited for the lamp filaments. Elements of lamps require the following characteristics:

- Working temperatures should be high enough—1800°C for carbon, 2900°C for tantlum and 2000°C for tungsten.
- The efficiency of filaments should be highest—12 lumen/watt for tungsten, 1.6 for tantalum and nearly 1.2 for carbon.

Example 6.1

Determine the temperature coefficient of resistance of material used in a resistor, if the resistance at 25°C is 55 Ω and at 75°C is 57.2 Ω.

Solution Given: $R_{75} = 57.2\ \Omega$

and $R_{25} = 55\ \Omega$

As we know that $R_t = R_0(1 + \alpha t)$

Therefore, on writing above equation for temperature 25°C and 70°C, we get

$$R_{25} = R_0 (1 + 25\ \alpha) \quad \text{(i)}$$

$$R_{75} = R_0 (1 + 75\ \alpha) \quad \text{(ii)}$$

On dividing Eq. (ii) by (i), we get

$$\frac{R_{75}}{R_{25}} = \frac{1+75\alpha}{1+25\alpha}$$

or

$$\frac{57.2}{55} = \frac{1+75\alpha}{1+25\alpha}$$

i.e. $\quad 57.2 + (25 \times 57.2\alpha) = 55 + (75 \times 50\alpha) \quad$ (iii)

On solving Eq. (iii), we get

or $\quad \alpha = \dfrac{2.2}{2320} = 0.0009\ \text{K}^{-1} = 0 \quad$ **Ans.**

Example 6.2

A coil of standard annealed copper having a resistance of 50 ohms at 15°C is embedded in the core of a large transformer. After the transformer has been in operation for several hours the resistance of the coil is measured as 60 ohms. Determine its temperature. The temperature coefficient of resistance of the conductor materials is 0.00425 per °C at 0°C.

Solution Resistance at temperature 15°C (t_1), $\quad R_1 = 50\ \Omega$

Resistance at temperature t_2, $\quad R_2 = 60\ \Omega$

Final temperature t_2

Temperature coefficient of resistance, $\alpha = 0.00425$ per °C

Now, $\quad R_1 = R_0(1 + \alpha t_1)$

or $\quad 50 = R_0(1 + 0.00425 \times 15) \quad$ (i)

or $\quad 50 = 1.06375\ R_0$

and $\quad 60 = R_0(1 + 0.00425 t_2) \quad$ (ii)

Dividing (ii) by (i), we get

$$\frac{60}{50} = \frac{R_0(1+0.00425t_2)}{1.06375\ R_0}$$

$$1.2 = \frac{1+0.00425t_2}{1.06375}$$

$$t_2 = 65.83°\text{C}$$

Hence, final temperature = 65.83°C **Ans.**

Example 6.3

The resistance of a tungsten bulb is 140 Ω under normal working conditions. Find the temperature in degree assuming (i) cold resistance of bulb at 20°C is 8 Ω, and (ii) average temperature coefficient for tungsten wire at 20°C as 5 × 10^{-3}.

Solution Cold resistance of the bulb at $t_1 = 20°$C, $\quad R_1 = 8\ \Omega$

Resistance of bulb at t_2^0C, $\alpha_{20} = 140\ \Omega$

Average temperature coefficient at 20°C, $\alpha_{20} = 5 \times 10^{-3}$

Normal working temperature, t_2:

Now $$R_2 = R_1[1 + \alpha_1(t_2 - t_1)]$$
$$140 = 8\,[1 + 5 \times 10^{-3}(t_2 - 20)]$$
or $$\frac{14}{8} = 1 + 5 \times 10^{-3}(t_2 - 20)$$
or $$5 \times 10^{-3}(t_2 - 20) = 16.5$$
or $$t_2 - 20 = 3300$$
$$t_2 = 3320°\text{C}$$
Hence, temperature under normal working condition = 3320°C **Ans.**

Example 6.4

A resistance wire of 1 metre length and 0.08 mm diameter has a resistance of 95.5 Ω. Calculate the resistivity of the wire material.

Solution Length of wire, l = 1 m = 100 cm
Diameter of wire, d = 0.08 mm = 0.008 cm
Resistance of wire, R = 95.5 Ω
Resistivity of the wire material, ρ

Now, $$R = \frac{\rho l}{A}$$
or, $$\rho = \frac{RA}{l}$$ [where, A = cross-sectional area of the wire, l = length of the wire]

$$\rho = \frac{95.5 \times \frac{\pi}{4} \times (0.008)^2}{100} = 4.8 \times 10^{-5}\ \Omega\ \text{cm}$$
Hence, resistivity of the wire material = 4.8×10^{-5} Ω cm **Ans.**

Example 6.5

A copper conductor has a resistance of 17.5 ohms at 0°C. Find its percentage conductivity at 16°C. Assume the temperature coefficient of copper as 0.00428 per °C at 0°C.

Solution Resistance of copper conductor at 0°C, R_0 = 17.5 Ω
Temperature coefficient of copper at 0°C, α_0 = 0.00428 per °C
Percentage conductivity at 16°C:

Now, $$R_t = R_0\,(1 + \alpha t)$$
$$= 17.5\,(1 + 0.00428 \times 16) = 17.5 \times 1.06848$$
$$= 18.6984$$
Since conductance is reciprocal of resistance,
∴ percentage conductivity is at 16°C
$$= \frac{\text{Resistance at } 0°\text{ C}}{\text{Resistance at } 16°\text{ C}} \times 100$$
$$= \frac{17.5}{18.6984} \times 100 = 93.59\%$$
Hence, percentage conductivity at 16°C = 93.59% **Ans.**

Example 6.6

On adding a high conductivity material—copper—in small quantity to nickel of lower conductivity, why does the conductivity of the alloy decrease?

Solution It is due to the incresse in defects or imperfections in the parent material.

Example 6.7

How can the conductivity of carbon be increased?

Solution To increase the conductivity of carbon, different kinds of additives are used. These additives are mainly copper or bronze powder that is mixed with the carbon moulding compound.

Example 6.8

Permissible working temperature of manganin is low (about 60°C to 70°C), but that of the nichrome is high (about 350°C to 650°C). Why?

Solution It is because the resistance coefficient of temperature of manganin is high at higher temperature (about 0.00002/K) than that of nichrome, which is about 0.0000 1/K. Moreover, the electrical resistivity of manganin is 42×10^{-8} ohm-m as compared to 108×10^{-8} ohm-m for nichrome. Their other properties, which are also the deciding factor in this regard, are shown in Table 6.3.

Table 6.3

Material	*Density (kg / m³)*	*Tensile strength (MPa)*
Manganin	8200	420
Nichrome	8410	1000

Example 6.9

Nichrome is often designated as nichrome I, III, V, and Hastelloy by hastelloy A, B, C, D. What meaning do they convey?

Solution These designations specify different commercial grades of nichrome and hastelloy. Each grade differs in composition, properties and characteristics, and hence, is suitable for different specific applications. The details given in Table 6.4 are self-illustrious.

Table 6.4

Alloy	*Composition (%)*	*Application*
Nichrome I	Ni 80%, Cr 20%	Thermocouple, heating element of furnaces and ovens, and strain gauses
Hastelloy A	Ni 60%, Mo 30%, Fe 5%	Parts of chemical plants

6.6 COMPARISON BETWEEN SOME POPULAR LOW-RESISTIVITY MATERIALS

A comparison between copper, aluminium and steel on the basis of various parameters is given in Table 6.5

Table 6.5 *Comparison between copper, aluminium and steel*

S.No.	*Description*	*Copper*	*Aluminium*	*Steel*
1.	Resistivity (ohm-m)	17×10^{-9}	28×10^{-9}	–
2.	Coefficient of linear expansion (/K)	17.7×10^{-6}	23.7×10^{-6}	15.2×10^{-6}
3.	Temperature coefficient of resistance at 20°C	0.0039	0.0035	0.004–0.005
4.	Corrosion resistance	More	More	Less
5.	Tensile strength	More	Less	Maximum
6.	Ductility approximately	55%	50%	18–25%
7.	Density (kg/m^3)	8960	2700	7800
8.	Melting point (°C)	1083	660	1539
9.	Thermal conductivity at 25°C (W/m-K)	425	300	65
10.	Salient feature	Can be stretched into thin wires	Can be stretched into thick wires 0.1 mm diameter	Can be stretched into thick wires
11.	Comparative cost	12	8	1
12.	Working efficiency	Best	Better	Good
13.	Life	Highest	Higher	High

6.7 LOW-RESISTIVITY COPPER ALLOYS

A large number of alloys of copper fall in the category of low-resistivity materials. Central among them are brass and bronze. Brass is basically an alloy of copper and zinc, while bronze is an alloy of copper and tin. However, they are prepared with the addition of other elements, as well. Brass and bronze are available in different compositions and so possess varying properties. Consequently, they are of different types, viz. admiralty brass, muntz metal, delta metal, phosphor bronze, bell metal, etc. But among these, each type is not suitable for electrical applications. Various kinds of brass and bronze, and other non-ferrous alloys that find use in electrical industry are summarised below and tabulated in Table 6.6.

Table 6.6 *Compositions and applications of different types of brass and bronze along with other non-ferrous alloys*

Alloy	*Composition*	*Application*
Mg-alloy		
1. Electron	Al 3–12%, Mn 0.03%, Zn 2%, rest Mg	Parts of aircrafts and automobiles, collapsible tubes, coating on steel utensils solder, coating of petroleum containers and mild steel sheets
Al-alloy		
1. Duralumin	Cu 4%, Mg 0.5%, Si 0.7%, rest Al	Electrical cables, components of aeroplanes and automobiles
2. Y-alloy	Cu 4%, Mg 1.5%, Ni 2%, Fe 0.6%, rest Al	Casting of engine parts
Cu-alloy: Brass		
1. Muntz metal	Cu 60%, Zn 40%	Brass castings, stampings and extruded parts
2. Yellow brass	Cu 90%, Zn 10%	Screw, wires and hardwires
3. Admiralty bras	Cu 70%, Zn 29%, Sn 1%	Marine uses, pump parts and ship parts
4. Cartridge brass	Cu 70% Zn 30%	Bullet shots, cistern, storage batteries, military ammunition and foils
5. Arsenical copper	As 0.3%, rest Cu	Heat exchangers and condensers
Cu-alloy: Bronze		
1. Ounce metal low	Cu 85% Sn 5%, Zn 5%, Pb 5%	Ornamental fixtures, pipe fittings and pressure valves
2. Beryllium bronze	Be 1.5–2%, Sn 8%. rest Cu	Bellows, diaphragms and springs
3. Aluminium bronze	Al 10%, small quantities	Parts containing corrosive liquids
4. Alpha bronze	Cu 95%, Sn 5%	Coin and sprint making
5. Admiralty gunmetal	Cu 88%, Sn 10%, Zn 2%	Bearings and steam pipe fittings
6. Phosphor bronze	Cu 84–89%, Sn 10–15%, P 0.1 to 0.3% + Ni + Pb	Parts exposed to sea water, boiler fittings, etc.
7. Bell metal	Cu 70%, Sn 30%	Casting of bells

(i) Brass

It is an alloy of copper containing 40% Zn. Its conductivity is lower than that of copper. It has a high tensile strength and is fairly resistant to corrosion. It can be easily pressed into a desired shape and size, can be drawn into wires, and can be easily brazed. Different types of brass are widely used in the following applications:

- Plug points
- Socket-outlets

- Lamp holders
- Fuse holders
- Switches
- Knife switches
- Sliding contacts for rheostats and starters, etc.

(ii) Bronze

It has a composition of 10% Sn in 90% Cu. Its conductivity is lower than that of pure copper. Bronze components are generally made by forging process. It is corrosion resistant and possesses high strength. Different types of bronze are generally used in the following applications:

1. Beryllium bronze for making current carrying springs, sliding contacts, knife-switch blades, etc.
2. Phosphor bronze for making springs, bushings, etc.
3. Cadmium bronze for making commutator segments

6.8 MATERIALS FOR LAMP FILAMENTS

The following materials are used for lamp filaments:

(i) Carbon

The properties of carbon have been discussed earlier:

- Commercial efficiency of carbon filament lamp is 4.5 lumens per watt or 3.5 watts per candle power.
- To prevent the blackening of bulb, the working temperature is only 1800°C.

(ii) Tantalum

- Resistivity (ρ) = 12.4 $\mu\Omega$ cm
- Temperature coefficient of resistance (α) = 0.0036 per degree
- Melting point = 2900°C
- Specific gravity = 16.6

Its efficiency is about 1.6 watt per candle power. Due to low efficiency, it is not much used these days.

(iii) Tungsten

The efficiency of tungsten filaments, which are worked at 2000°C in an evacuated bulb, is 12 lumens per watt.

6.9 PREPARATION OF TUNGSTEN FILAMENTS

The following steps are undertaken in the preparation of tungsten filament:

- Chemically pure tungsten oxide is reduced on red heat in an atmosphere of hydrogen to metallic tungsten in the form of grey powder.
- The powder is pressed in steel moulds to form small bars in a hydraulic press.
- The bars are then heated in the furnace up to 1100°C in the presence of hydrogen, whereby the articles sinter together. This adds mechanical strength to the bars.
- The mechanical strength is improved further by heating (electrically) the bars almost to the melting point.

- The bars are then hammered or rolled in red heat to make them ductile, and finally drawn into filaments.

To improve efficiency of the bulb, it is filled with an inert gas, Argon with a small percentage of nitrogen. To minimise the convection currents produced by molecules of the gas in bulbs, the filament is wound into close spiral and suspended horizontally in the form of a circular arc. In high wattage bulbs, necks are provided. This is done to avoid blackening of the bulbs as the convection currents carrying particles from the filaments blacken the neck only, without impairing the candle power in the direction below the horizontal.

The efficiency of gas filled 'coiled coil' tungsten filament lamps is about 25 lumens per wattor 0.6 watts per candle power; thus, these lamps are called 'half-watt' lamp.

Figure 6.6 shows different types of filaments.

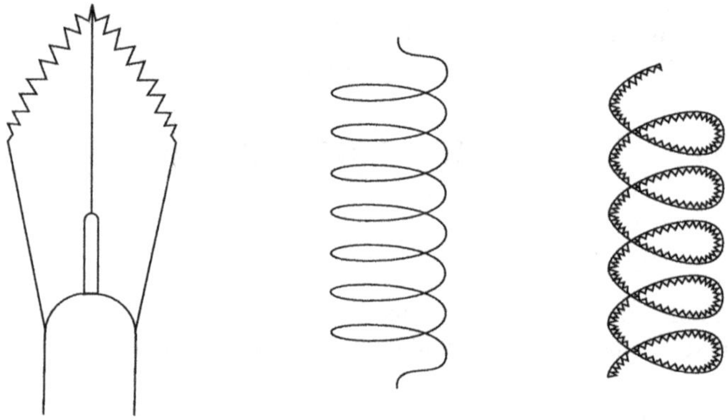

Fig. 6.6 *Filaments of different kinds*

6.10 MATERIALS USED FOR TRANSMISSION LINES

The high performing and most important material used for transmission lines is copper, since it has high conductivity and high tensile strength. Aluminium is used to a large extent, especially with a steel core, for a high voltage line. The choice of materials depends upon:

- Cost of materials,
- Required electrical properties,
- Local conditions, and
- Required mechanical strength.

Other materials employed for transmission lines are:

- Cadmium copper materials
- Copper weld materials
- Phosphor bronze materials
- Galvanised steel materials
- Galvanised iron
- Steel core copper
- Steel cored aluminium materials

6.11 ELECTRICAL CONTACT MATERIALS AND THEIR SELECTION

Contact materials play a vital role in electrical machines and appliances. The selection and details of various types of contact materials may be given as follows:

The successful operation of electrical contact depends upon following factors:

- Contact resistance
- Contact force
- Voltage and current

(i) Contact Resistance

The main function of nearly all contacts is to carry electric current. Therefore, they must offer least resistance to the flow of current. This factor is of greatest importance at very low voltages. The total resistance of a pair of contact is the sum of the resistance of the contact materials and the resistance at the interface of friction between the two contacts. The specific resistance of the contact materials is normally low compared to resistance at the interfaces.

The contact resistance may vary remarkably because of surface contamination caused by chemical compounds formed from the contact materials themselves. The most important and troublesome of these are the chemical compounds formed by heat and corrosion which result in oxide, chloride, carbonate, sulphide and sulphate films. The contact resistance is increased by the imposition of those relatively non-conducting films in the actual area of contact.

(ii) Contact Force

The contact resistance of a specific pair of contacts is directly related to the force applied to the contacts in a closed position. The total force should always be as high as possible.

(iii) Voltage and Current

The performance and life of contact materials are intimately related to the voltage and current the contacts must make and break.

Contacts operating in dc circuits are subjected to material transfer, i.e. transfer of metal from face of one contact to the face of its making contact. This results in a build up or mound on one contact and hole or crater in the making contact. The amount and directions of transfer, whether from positive to negative or *vice versa,* depends upon the particular phenomenon encountered in the electrical circuit. Material transfer in ac circuit is not usually encountered within the frequency range of about 25 to 400 cycles because the arc is extinguished as the current passes through the first zero point in the cycle and polarity of each contact is constantly changing.

In addition to the above factors, several others are given below:

- Contact bounce chatter
- Frequency of operation
- Speed of contact operation
- Type of electrical load
- Medium in which contacts operate

6.12 CLASSIFICATION OF CONTACT MATERIALS

Depending on the power level of the circuits, contact materials are classified in the following way:

- **Lightly and moderately loaded:** When the current does not exceed 1 ampere and the voltage causing acting is to the tune of 10 to 20 V.
- **Heavily loaded:** When the power levels are higher. The primary requirement of contact materials is probably stability of contact resistance.

Contacts operate more satisfactorily in a vacuum than in gaseous medium because vacuum eliminates any possible arcs between them.

1. Metals for Lightly-loaded Contacts

The following metals are normally used for making lightly loaded contacts:

(i) Platinum

The characteristics of this noble metal are quite suitable for the above purpose. Some of them are given below:

- It does not oxidise in air.
- It has no tendency to arc, but it may form bridges and needles at a low current.
- It is often alloyed with iridium for making light contacts of extra-high reliability.
- It is one of the most stable of metals under the combined action of corrosion and electrical erosion.
- It has a high melting point and does not corrode, and surfaces remain clean and low in resistance under most adverse atmospheric and electrical conditions.

(ii) Palladium

- It has properties similar to that of platinum.
- Being much cheaper, it is frequently used to replace platinum.
- It is less resistant to cathodic erosion and oxidation in air.

(iii) Silver

Below given properties of Ag are suitable for the above purpose:

- It has the highest thermal conductivity of any metal.
- It has low contact surface resistance, since its oxide decomposes at approximately 150°C.
- As silver oxidises easily dissociate at low temperature, the stability of contact resistance is slightly affected by them. Nevertheless, it is not recommended to use silver for contacts in light and precise devices with low contact pressures. Otherwise, silver is widely used as a contact material, either alone or in the alloy form.
- Silver easily combines with sulphur, therefore measures should be taken not to allow it to come in contact with sulphur-bearing substances, e.g. rubber.

(iv) Gold

- Although resistance to corrosion is more, but it has a much lower melting point.
- Gold and its alloys are ductile and easily formed into a variety of shapes.
- Because of its softness, it is usually alloyed.
- Gold alloys of silver and other metals are used to impart hardness and improve resistance to mechanical wear and electrical corrosion.

(v) Tungsten

- It is hard, dense and a slow-wearing metal.
- It is a good thermal and electrical conductor.
- It has a high melting point of 3380°C.
- It can be obtained in several grades having various grain sizes.
- It has no tendency to arc.
- Besides above positive features, it has a drawback that it oxidises when in contact with atmospheric oxygen.
- Tungsten contacts operating in an atmosphere of high relative humidity and temperature typical of moist tropical regions should not be located near phenol bearing materials, such as backlined paper or laminates bounded with phenolic resins because the gaseous product they evolve strongly corrode tungsten.
- Forged bars of tungsten have a longitudinal fibrous structure and contacts from them will have greater wear resistance than those from sheet tungsten.

(vi) Molybdenum

- It has contact surfaces midway between tungsten and fine silver; it often replaces either metals where greater wear resistance than that of silver is required or lower contact surface than that of tungsten is desired.
- It erodes faster than tungsten and corrodes in atmospheric condition.

(vii) Rhodium

It is an excellent contact material in light and precise devices.

(viii) Metal Ceramics (Cermets)

These materials are produced by powder metallurgy techniques. The blanks are pressed formed from a mixture of metal and non-metal powders and sintered at high temperature. Contacts, as compared to metal contacts, entail the advantages of greater resistance to fusion, welding and wear.

2. Metals for Heavily-loaded Contacts

The materials used for heavily-loaded contacts are given below:

1.
- Silver
- Silver-palladium
- Silver-copper
- Silver-cadmium

• **Operating Conditions** Voltage up to 500 V, arcing during opening and high contact pressure

• **Field of Application** These are used as 'circuit breakers of medium current rating'.

2. (i) Silver

(ii) Copper, cadmium

• **Operating Conditions** Main contacts

• **Field of Application** These are used in 'air circuit breakers'.

3. (i) Silver
(ii) Copper

• **Operating Conditions** Main contact

• **Field of Application** 'Oil circuit breakers'

4. (i) Silver
(ii) Silver-copper
(iii) Copper
(iv) Copper-cadmium
(v) Copper-silver

• **Operating Conditions** Medium and heavy alternating currents and voltage up to 500 V main contacts and arcing contacts (Cu–Cd).

• **Field of Application** 'Industrial-control contactors and magnetic starters'

5. Silver

• **Operating Conditions** Low direct current

• **Field of Application** 'dc contactors'

In addition to the above materials, following are also used in the making of heavily loaded contacts:

(i) Silver-nickel
(ii) Silver-graphite
(iii) Silver-cadmium oxide
(iv) Silver-molybdenum
(v) Silver-tungsten
(vi) Silver-tungsten carbide
(vii) Silver lead oxide
(viii) Silver cadmium-nickel
(ix) Copper-graphite
(x) Copper-tungsten

6.13 ELECTRICAL CARBON MATERIALS

Electrical carbon materials are manufactured from graphite and other forms of carbon, such as coal, etc. They all consist of practically pure carbon. Carbon claims the following fields of application:

1. For making brushes for electrical machines
2. For making contacts and resistors
3. For making carbon electrodes for electric-arc furnaces
4. For making battery cell elements
5. For making component for vacuum tubes
6. As arc light and welding carbons
7. For making microphone powders, membranes and other components for telecom equipments.

Electrical carbon materials are manufactured from (i) coke obtained from coal, coal tar residues, petroleum residues and peat, (ii) anthracites and so called thermo-anthracites obtained by heat-treating anthracites at 900 to 1200°C. Following operations are carried out in the manufacture of most of electrical carbon products:

1. Grinding the raw carbon materials.
2. Mixing the powdered carbon with a binding agent, such as coal tar pitch or coal tar.
3. Moulding the requisite articles from the resulting compound.
4. Baking the articles. It adds adequate mechanical strength and hardness to the product.

6.14 ARC LIGHT CARBONS

Cylindrical in shape, these are generally made from coke, graphite and carbon black. They are often made with a core of soften material which evaporates faster than the surrounding material, and thus provides a stabilised output of the arc.

Carbon Electrodes

Carbon electrodes for electric-arc furnaces and electro-chemical processes are of three kinds:

1. Baked
2. Graphitised
3. Self-baking

Baked electrodes are obtained from (i) foundry coke, (ii) anthracite, and (iii) thermoanthracite. If they are to be machined, cemented carbide-tipped tools shall be used, as they have high hardness.

Graphitised electrodes are not economic when compared with baked electrodes as they require an additional graphitising heat treatment in the course of manufacture. These electrodes exhibit the following advantages when used as electrical engineering materials:

1. Less brittleness
2. Lower resistivity
3. Greater chemical resistance

6.15 USE OF CARBON IN RESISTORS AND BRUSHES

These are produced with nominal resistance ratings ranging from 1 to 1×10^{12} ohms. The carbon resistors are of two types:

1. Film-type resistors
2. Solid-type resistors

In the film-type resistors, a film of carbon serves as the resistance element. In the solid-type resistors, the same purpose is served by a solid rod of special material consisting of carbon and inorganic binding agents.

Film-type resistors may be fixed or variable. The latter allows resistance to be varied within definite limits.

Solid-type resistors only serve as fixed resistors.

Carbon and Graphite Brushes

Commutator type machines utilise carbon brushes which are prepared in a number of grades for various kinds of requirements. The brushes play a leading hand in the reversal of the flow of current in the armature coils, i.e. in commutation. Some important types of brushes are discussed below:

(i) Carbon Brushes

The preparation of carbon brushes utilise the following steps: (i) reducing of amorphous carbon (available in various forms, such as gas carbon, lamp black or coal) to very fine powders, (ii) mixing with an appropriate binding material, such as pitch and (iii) carbonisation of the whole mass by baking the block at high temperature. The brushes obtained by this process are less expensive and have better mechanical and electrical properties.

Applications

- These are mainly used in small machines where rubbing speed, commutation or current density is of small magnitudes.
- These brushes can also be used in medium-size machines if a small amount of graphite is added to the original carbon before firing, because graphite addition improves lubrication adequately at the high speed running and current carrying capacity of the brushes.

(ii) Natural Graphite Brushes

Manufacture of these brushes is similar to carbon brushes. Graphite exhibits a high degree of lubricating property and thermal conductivity.

Uses

- They are employed in large, heavy-duty dc generators and motors.
- They are also used in turbo-alternators.

6.15(a) Soldering Materials

The common soldering materials are described below:

• **Tin-antimony** The composition of this solder material is 95% Sr and 5% Sb. Addition of Sb adds the strength.

• **Antimonial tin-lead** This solder is very good in the sense that it exhibits high mechanical strength and low flow ability.

• **Bismuth-lead-tin cadmium** This solder is also known as Wood's alloy solder. The ingredients are 50% Bi, 25% Pb, 12.5% Sn and 12.5% Cd. It has a very low melting point—nearly 60°C. It is quite useful in direct soldering of ICs.

6.16 ENCAPSULATING MATERIALS

All the electronic components require protection against environmental hazards. They should also be shock proof. Some protective jacket, therefore, is very essential for the above purpose. The material for this purpose should exhibit the following characteristics:

- It should be such that it prevents the component from mechanical abrasions.
- It should protect the electronic component from humidity.
- It should also prevent the component from greasing effect by touch.

The commonly used materials, for the above purpose, are aluminium Kobar (alloy of Ni, Fe, Co), ceramics-like alumina, plastic, epoxy, etc.

6.17 INSULATING GASES, LIQUIDS AND SOLIDS AND THEIR CHARACTERISTICS

Insulating materials play a major role in the satisfactory operations of electrical and electronic equipment where all current-carrying wires and surfaces are coated or covered with an insulating material. Such materials are available in a large number, some of which are described below. The factors that are looked upon for their effective use are as follows:

- Properties of the insulating material
- Ease of fabrication
- Availability
- Economy

1. Insulating Materials for Electrical Engineering

The insulating materials required for various applications in electrical engineering are classified in three categories:

- Insulating gases
- Liquid insulating material
- Solid insulating material

2. Insulating Gases

Many dielectric gases are used as insulators and for the medium of heat transfer. The major problems with these gases are the abnormalities in dielectric behaviour at high pressure temperature, instability and fire hazards. Normally, all known gasses are dielectric in pure form, but from electrical engineering point of view these are classified—as shown in Fig. 6.7 on the basis of different properties, like dielectric strength, dielectric loss, chemical instability and corrosion.

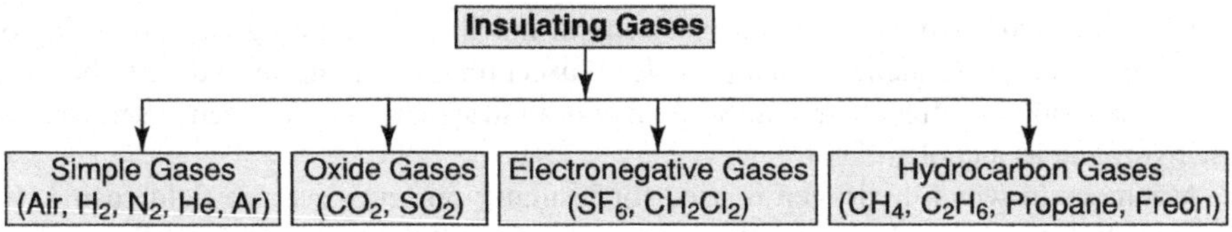

Fig. 6.7 *Insulating gases*

3. Simple Gases

(i) Air

The most important gift of nature is air, which serves as the insulation material between over-head transformers. This air provides insulation between the over-head transmission lines without any cost. It serves as a part of an electric device or machine whether we want it or not, acting as an insulator in addition to specially provided insulating materials (solids or liquids).

Air is an extreme insulating material when voltages are not very high. Current leakage through air is quite less than through solid or liquid insulating materials under the same conditions.

The dielectric strength of air is small—nearly 3 to 5 kV/ mm—for small distances between electrodes. The discharge voltage between two electrodes in air increases more slowly with the distance.

Variation of pressure has a considerable influence on the electric strength of air. The electric strength of gases increases considerably with pressure. Although the electrical strength of air and other gases becomes poorer at lower pressures, it exhibits a large value at high vacuum.

(ii) Nitrogen

Nitrogen is used as a dielectric in electrical apparatus. It is used in transformers to replace air (which causes oxidation). Oxidation reduces the life of equipment. In free breather-type oil-filled transformers, the oxidation of oil leads to sludge formation which degrades the mechanical stability of the apparatus. Transformers use chemically inert nitrogen to replace oxidising atmosphere and are sealed.

Nitrogen under pressure is used in certain types of electrical capacitors, as well. In gas pressure cables, nitrogen gas—along with oil treated paper—is used under pressure.

(iii) Hydrogen

Hydrogen is of special interest to electrical engineers as it is very light and exhibits nice properties favouring its use (instead of air) as a coolant in electrical machines. Table 6.7 lists the important properties of hydrogen, N_2 and CO_2, as compared to air, at identical temperatures and pressures; all the parameters for air taken equal to unity.

Table 6.7 *Comparison of properties of different gases*

Property	*Air*	*Nitrogen*	*Carbon dioxide*	*Hydrogen*
Density	1	0.97	1.52	0.07
Thermal conductivity	1	1.08	0.64	6.69
Thermal capacity	1	1.05	0.85	14.35
Electric strength	1	1.00	0.90	0.60

On using hydrogen as a coolant in electrical machines, the rating and efficiency of machine increases. Organic insulation is less susceptible to ageing in hydrogen because there is no oxidising effect. Large turbo-generators and synchronous condensers nowadays use hydrogen as a coolant.

Machines in which hydrogen is used for insulator purposes require tight casing. H_2 is kept slightly above the atmospheric pressure for safety purpose, as well, to avoid explosive mixture with air.

Hydrogen cooling also reduces windage losses and increases efficiency.

Inert gases, such as argon, neon and others, and also mercury and sodium vapours are used for filling various electronic tubes and appliances.

(iv) Oxide Gases (Carbon Dioxide)

Oxide gases have found little use in electrical equipment. Carbon dioxide has been used as an insulant in certain types of fixed capacitors and is used as a pre-impregnant for oil-filled high-voltage apparatus, such as cables and transformers. The relative dielectric constant of carbon dioxide is 1.000985 at 0°C. The dielectric strength of carbon dioxide compares favourably with that of air.

(v) Electronegative Gases

The electronegative gases have high dielectric strength compared to other traditional dielectric gases like nitrogen and air. The dielectric strength of SF_6 is 2.35 times more than air, and that of CCl_4 is 6.33 times. The electronegative gases are non-inflammable and non-explosive, which is essentially the additional advantage with them. The most important gas under this category is sulphur hexafluoride, while others are freon gases.

(vi) Sulphur Hexafluoride, SF_6

This gas is extremely popular for high-voltage applications and its use is most satisfactory in dielectric machines, like X-ray apparatus, Van de Graaff generators, voltage stabilisers, high-voltage switch gear, gas lasers, etc. Presently, it is used in metal cold substations. Following are some of the important properties of this gas:

- SF_6 is a colourless, non-toxic and non-inflammable gas. It is one of the heaviest known gases (five times than air) and has low solubility in water. The gas can be liquefied by compression. Its cooling characteristics are better than air and nitrogen.
- SF_6, under normal conditions, is chemically inert and completely stable. It does not react with quartz up to 500°C. Below 150°C, it does not react with metals, ceramics, rubber, resins, etc.
- It is inert, insulating gas, and is a coolant of high dielectric strength and thermal stability.
- Sulphur hexafluoride also increases the interrupting capacity of the circuit breakers.
- SF_6 is a gas with pronounced electronegative properties. Its relatively large molecules have a great affinity for free electrons, with which they combine making the gas-filled break much more resistant to dielectric breakdown.
- The excellent insulating properties of SF_6 make it possible to design circuit breakers with smaller overall dimensions and shorter break.
- SF_6 is the most commonly used gas in any compressed gas system due to its high dielectric strength.

(vii) Hydrocarbon Gases

Practically these are of little interest as they are explosive and flammable. Figure 6.8 shows the dielectric strength of various insulating gases at various values of pressure.

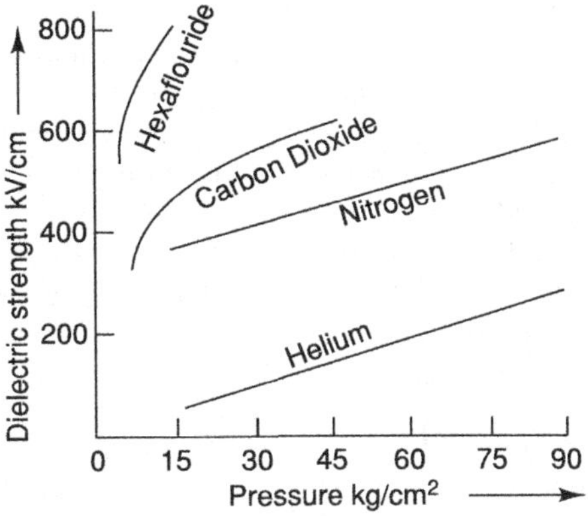

Fig. 6.8 *Dielectric strength of various insulating gases at various values of pressure*

4. Liquid Insulating Materials

These materials are used for dielectric purposes to eliminate the air and other gases. They are also as insulants and for heat transfer purposes. Liquid insulting materials are generally used along with solid insulants, and classified according to the temperature range in which they are used.

Insulating liquids are organic liquids which are used wholly as coolants. They are dielectric in nature and their presence, either alone or in conjunction with solid insulating and constructional materials, permits the design of electrical transmission and distribution equipment at minimum cost with an acceptable performance and life.

An ideal insulating liquid must have following properties:

- High electric strength, impulse strength and volume resistivity
- Low dielectric dissipation factor
- High or low dielectric constant (depending on application)
- High specific heat and thermal conductivity
- Excellent chemical stability and gas absorbing properties
- Low viscosity, density, volatility and solvent power, and high flash point
- Good arc quenching properties
- Non-flammable and non-toxic

The electrical equipment in which liquid-insulating materials are used are transformers, switchgear (circuit breakers), bushings, cables and capacitors. The liquid can be changed in transformers, switchgear, liquid-filled bushings and cables. However, it is not possible in liquid-impregnated equipment, such as capacitors, cables and electronic equipment which normally consists of sealed units.

(i) Mineral Oils

Mineral oils can generally be used in the temperature range of 50–110°C. The electrical properties and resistance to thermal expansion of these oils are ususally influenced by the presence of non-hydrocarbon compounds, oxygen, nitrogen, sulphur, etc. These hydrocarbon oils are widely used as insulating oils in transformers, circuit breakers, switch gears, capacitors, etc. In transformers, light fraction oil, such as transil oil, is used to allow convection cooling. This oil has a high flash point of 130°C to prevent fire hazard and loss caused by evaporation. Highly purified mineral oils have a dielectric strength as high as 180 kv/mm and if the oil contains polar or ionizing material, its dielectric loss increases. Clean hydrocarbons are among the best dielectric liquids due to their balanced structure and non-polar nature. They have a low dielectric constant of about 2.3, and therefore they are capable of dissolving only very few substances in them and produce the conducting ions. Hence, it is relatively easy to obtain and maintain low-power factor. They, however, undergo oxidation, particularly in the presence of catalysts such as copper, to form sludge and acids; proper refining and addition of antioxidants or inhibitors can remove this defect. Light oils, having Saybolt Viscosity of 100 seconds at 40°C, have been used under pressure in oil-filled high voltage cables. More viscous or tacky oils, having Saybolt viscosity of 2000 seconds at 40°C, are generally employed to impregnate the paper in solid type of cable.

(ii) Askarels

These are non-inflammable, synthetic insulating liquids, which are also used in the temperature range of 50–110°C. Chlorinated hydrocarbons are the most widely used among the askarels because of their excellent resistance, high dielectric strength, low dielectric constant (4 to 6) and small dielectric loss at ordinary temperature and power frequencies. They do not decompose under the influence of electric arc and have good thermal, chemical and electrical stability. Chlorinated hydrocarbons as askarles have widely been used as transformer fluids to reduce fire hazards and to obtain higher capacities in condensers because of their polar nature. Some special oils have widely been used as transformer fluids to reduce fire hazards and to obtain higher capacities in condensers because of their polar nature. Some special oils have also been developed for maintaining the dielectric constant at 4 to 5 over a wide range of temperatures. Chlorinated diphenyls belong to this class. Chlorinated diphenyl, penta chloro diphenyl, trichlorodiphenyl, hexa chloro diphenyl, trichloro benzene, etc., are the most widely used chlorinated hydrocarbons or askarles. Low chlorinated trichloro diphenyls are generally used for application at low temperatures because they retain their high dielectric constant and remain fluid at lower temperatures. At higher temperatures they are not very stable. In actual practice, askarles are generally used to impregnate a cellulosic insulating material, such as paper or press board, etc., which itself has a very high breakdown strength.

(iii) Silicone Fluids

Silicone liquids are generally used in the temperature range of 90–220°C and they are clear, water-like liquids. They are available in a wide range of viscosity. Silicone liquids are widely used as insulating materials because of their stability at higher temperatures. They are non-corrosive to metals up to 200°C. Because of excellent dielectric properties over a wide range of temperatures, they have widely been used as coolants in radio, pulse and aircraft transformers. Silicone fluids, however, have a tendency to breakdown and give considerable amounts of gases (such as hydrogen) and a residue which mainly contains carbon, silicon carbide and silicon dioxide (SiO_2).

(iv) Fluorinated Liquids

These liquids are used in the temperature range of 50–200°C. They are non-inflammable and highly stable chemically. They provide much better heat transfer from the windings and magnetic circuits in comparison to hydrocarbon oils or silicone liquids, and have widely been used in small electric and radio devices, and transformers, etc. The fluorinated liquids are, however, highly volatile and get deteriorated in their electrical properties in the presence of moisture.

(v) Synthetic Hydrocarbon Liquids

These are insulating oils, such as polybutylene and polypropylene, which are not better than mineral oils as dielectrics, but can be prepared as liquids having similar dielectric strength, thermal stability and susceptibility to oxidation as that of mineral oils. These are used in the temperature range of 50–110°C in high-pressure gas-filled cables and dc voltage capacitors.

(vi) Organic Esters

These organic fluids have been used in the temperature range of 50–110°C. They have a high dielectric constant and show very low dielectric losses. The dielectric constant of organic esters ranges from 2.0 to 3.5, but higher values such as 12.8 have also been found in tetra hydrofuryloxalate. Because of these properties, these fluids are well suited for use in high frequency capacitors.

(vii) Vegetable Oils

These are the oldest insulating liquids and used in the temperature range of 20–100°C. From dielectric point of view, both drying oils such as linseed and tung, and non-drying oils are important. Drying oils are generally suitable in the formation of insulating varnishes (which are used in transformers and motor coils, etc.), while non-drying oils can be used as plasticisers in insulating resin compositions.

Table 6.8 gives a comparison of various characteristics of liquid insulators.

Table 6.8 *Liquid insulating materials*

S.N	*Type of liquid*	*Temperature range*	*Relative cost*	*Use*
1.	Petroleum oils (Mineral oils)	– 50 to 110°C	1	All types
2.	Askarels	– 50 to 110°C	8	Transformers, capacitors, switchgear
3.	Silicon liquids	– 90 to 220°C	10	Transformers
4.	Halogenated hydrocarbons (other than askarels)	– 50 to 200°C	10–40	Electric equipment
5.	Synthetic hydrocarbons	– 50 to 110°C	2–3	Cables, capacitors
6.	Organic esters	– 50 to 110°C	4–8	Electronics
7.	Vegetable oils	– 20 to 100°C		Limited application

A good insulating material should possess the following characteristics.

- Large value of insulating resistance
- High value of dielectric strength
- Uniform viscosity; it gives uniform electrical and thermal properties
- Should be uniform throughout; it keeps the electric losses as low as possible and electric stresses uniform under high voltage difference
- Thermal expansion should be minimum
- When exposed to arcing, should be non-ignitable
- Should be resistant to oils or liquids, gas fumes, acids and alkalies
- Should have no deteriorating effect on the material in contact with it
- Low dissipation factor (loss tangent)
- High mechanical strength
- High thermal conductivity
- Low permittivity
- High thermal strength
- Free from gaseous insulation to avoid discharges (for solids and gases)
- Should be homogeneous to avoid local stress concentration
- Should be resistant to thermal and chemical deterioration

6.18 SELECTION OF THE INSULATING MATERIAL

Selection of an insulting material is done on the basis of following properties:

- Electrical
- Mechanical
- Thermal
- Chemical

(i) Electrical Properties

These materials are used either as dielectric or as insulators in electrical and electronic appliances. The properties taken into consideration are:

- Insulation resistance
- Dielectric constant or permittivity
- Breakdown voltage or dielectric strength
- Dielectric loss

Insulation Resistance and its Determination This is the ohmic resistance in an electric circuit—offered by an insulating coating, cover, material or support to an impressed voltage—tending to produce a leakage of current through the same.

Consider a cable of inner and outer radii r_1 and r_2, length l and of resistivity of insulating material ρ. Considering a very thin layer of radial thickness dr at a radius r; the length through which the leaking current flows is dr and area of cross-section provided to the flow of current is $2\pi rl$.

Hence, insulation resistance of the layer under consideration

$$= \frac{\rho dr}{2\pi rl}$$

Insulation resistance of the whole cable will be obtained by integrating the above expression between the limits r_1 and r_2.

$\therefore$ insulation resistance of the whole cable is

$$R = \int_{r_1}^{r_2} \frac{\rho dr}{2\pi rl} = \frac{\rho}{2\pi l}\int_{r_1}^{r_2} \frac{dr}{r} = \frac{\rho}{2\pi l}\log_e \frac{r_2}{r_1} \quad (6.1)$$

From the above expression, it is obvious that the insulation resistance of the cable decreases with the increase in length.

(ii) Dielectric Constant or Permittivity

The permittivity of dielectric materials is not constant, but varies with temperature and frequency. The variation of permittivity has been given in detail in the previous chapter (dielectric materials). The effect of temperature upon breakdown voltage for transformer oil is shown in Fig. 6.9.

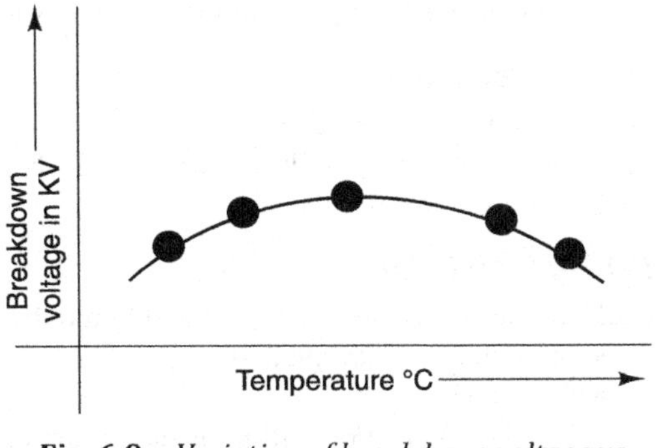

Fig. 6.9 *Variation of breakdown voltage vs temperature*

Approximate values of dielectric strength of different insulating materials are given in Table 6.9.

Table 6.9 *Approximate values of dielectric strength*

Material	*Dielectric Strength kV / mm*
Rubber	24
Asbestos	3.4 – 4.0
Mica	80
High-voltage porcelain	10 – 16

(Contd)

Low-voltage porcelain	1.5 – 4.0
Pure water	12 – 20
Paper	40 – 60
Glass	80 – 150

(iii) Dielectric Strength

- If the voltage across an insulating material is increased slowly, the way in which current increases depends upon the nature and condition of the material, as illustrated schematically in Fig. 6.10. For material *I*, the current increases very slowly and approximately linearly with voltage until a large, sharp increase results in what can be described disruptive dielectric breakdown. In contrast, for material *II*, the current increases more rapidly until current 'runway' occurs.

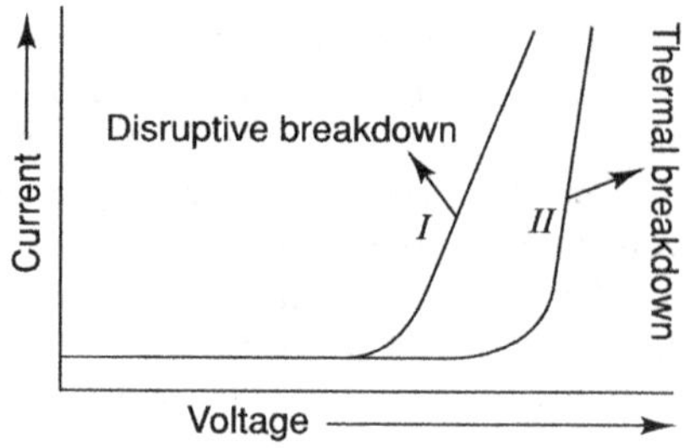

Fig 6.10 *Current flow in thermal and disruptive breakdown*

- The dielectric strength of an insulating material decreases with the length of time voltage is applied, moisture, contamination, elevated temperatures, heat ageing, mechanical stress and other factors may also markedly decrease dielectric strength to as little as 10% of the short-time values at standard laboratory conditions. At radio frequencies the dielectric strength may also be considerably less than 60 Hz.
- Dielectric failure that occurs along the interface, between a solid insulating material and air or a liquid insulating material, is termed 'surface breakdown'.

(iv) Power Factor

Power factor is a measure of power loss in the insulation and should be low. It varies with the temperature and usually increases with the rise in temperature of the insulation. A rapid increase indicates danger.

(v) Dielectric Loss

Dielectric losses occur in all solid and liquid dielectric due to:

- a conduction current, and
- hysteresis.
  - The conduction current is due to imperfect insulating qualities of the dielectric and is calculated by the application of Ohm's law. It is in phase with the voltage and results in a power loss (I^2R) in the material, which is dissipated as heat.
  - Dielectric hysteresis is defined as the lagging of electric flux behind the electric force producing it so that under varying electric forces a dissipation of energy occurs. The energy loss due to these causes is called the dielectric hysteresis loss.

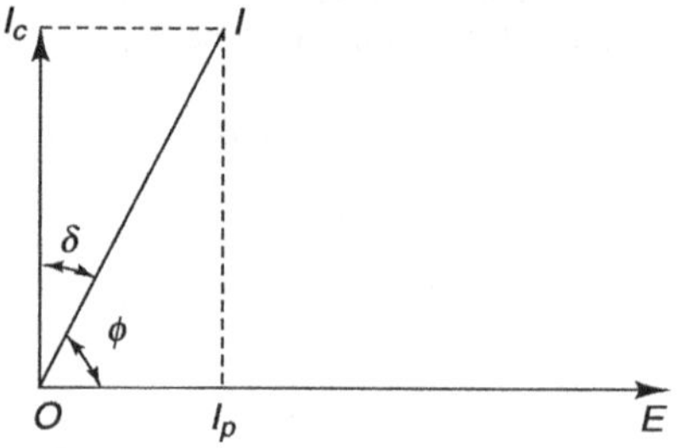

Fig. 6.11 *Plot of E against Ic*

The energy is dissipated as heat. Dielectric hysteresis is somewhat analogous to magnetic hysteresis.

Dielectric loss is also important as it gives an indication of the amount of energy absorbed by the material, when subjected to ac fields. The factors affecting the dielectric loss are

- Dielectric loss increases up to the increase in frequency of applied voltage
- Dielectric loss increases with humidity
- Dielectric loss increases with temperature
- Dielectric loss increases with increase in voltage

(vi) Effect of Moisture on Insulation

In an insulating material, moisture diffusion takes place when the electrical equipment is in operation. The effect of moisture on insulating materials changes the following properties:

- The volume resistivity, surface resistivity and dielectric strength decrease as a result of moisture absorption.
- The moisture also increases dissipation factor and dielectric constant.
- High humidity often leads to hydrolysis.
- High humidity favours the growth of fungi in some insulating materials, which degrade the insulating material.
- Swelling in some materials is also observed in the presence of high humidity.
- Mechanical strength of insulating material is reduced in the presence of moisture.

Example 6.10

Find the resistance of a liquid resistor consisting of two concentric metal cylinders of diameter $D = 38$ cm and $d = 18$ cm, respectively with water of specific resistance $\rho = 8{,}000\ \Omega$-cm between them. The length of both cylinders is 60 cm.

Solution Length of liquid resistor, $l = 60 \text{ cm} = 0.6 \text{ cm}$

Radius of outer cylinder, $r_2 = \dfrac{D}{2} = \dfrac{38}{2} = 19.0 \text{ cm or } 0.0191 \text{ m}$

Radius of inner cylinder, $r_1 = \dfrac{d}{2} = \dfrac{18}{2} = 9.0 \text{ or } 0.009 \text{ m}$

$$\log_e \frac{r_2}{r_1} = \log_e \frac{0.0191}{0.009} = 0.7514$$

Resistivity of water, $\rho = 8{,}000\ \Omega\text{m}$

Insulation resistance of the liquid resistor

$$= \frac{\rho}{2\pi l} \log_e \frac{r_2}{r_1} = \frac{80}{2\pi \times 0.6} \times 0.7514 = 16.21\ \Omega$$ **Ans.**

Example 6.11

Calculate the resistivity of the insulation material in a cable, the core diameter of which is 1.8 cm and sheath diameter is 5 cm. For a length of 3,000 m of cable, the insulation resistance is 1,820 MΩ.

Solution: Insulation resistance, $R' = 1{,}820\ \text{M}\Omega$

Core radius, $r_1 = \dfrac{1.8}{2} \text{ cm} = 0.009 \text{ cm}$

Sheath radius, $r_2 = \frac{5}{2} = 2.5 \text{ cm} = 0.025 \text{ m}$

Length of cable, $l = 3{,}000$ metres

Resistivity of dielectric,

$$\rho = \frac{2\pi l \times R'}{\log_e \frac{r_2}{r_1}} = \frac{2\pi \times 3.000 \times 1{,}820 \times 10^6}{\log_e \frac{0.025}{0.009}} = 33.1 \times 10^{12}\ \Omega\text{m} \quad \textbf{Ans.}$$

Example 6.12

Determine the insulation resistance of a single core cable of inside diameter 0.05 metre, outside diameter 0.070 m and length 2 km; the resistivity of the insulating material being 6 × 10¹² ohm metre.

Solution Radius, $r_1 = \frac{0.05}{2} = 0.025 \text{ m}$

Radius, $r_2 = \frac{0.070}{2} = 0.035 \text{ m}$

Length, $l = 2 \text{ km} = 2000 \text{ m}$

Specific resistance, $\rho = 6 \times 10^{12}\ \Omega\text{m}$

Now insulation resistance

$$= \frac{\rho}{2\pi l} \log_e \frac{r_2}{r_1}$$

$$= \frac{6 \times 10^{12}}{2\pi \times 2000} \log_e \frac{0.035}{0.025}$$

$$= 3.14 \times 10^8\ \Omega$$ **Ans.**

Example 6.13

A capacitor is fabricated from two square metal plates, each of side 120 mm. The plates are separated by a dielectric and are thick on 2 mm and relative permittivity 5. Calculate the capacitance.

If the electric field strength of the dielectric is 1.25 kV/mm, calculate the total charge on each plate.

Solution Area of plate $A = 110 \times 10^{-3} \times 110 \times 10^{-3} = 121 \times 10^{-4} \text{ m}^2$

Separation $d = 2 \text{ mm} = 2 \times 10^{-3} \text{ m}$

$\varepsilon_r = 5$

$$C = \varepsilon_0 \frac{\varepsilon A}{d} = 8.854 \times 10^{12} \times 5 \times \frac{121 \times 10^{-4}}{2 \times 10^{-3}} = 267.7 \times 10^{-12} \text{ F}$$

$E = 12.5 \text{ kV/mm} = 12.5 \times 10^3 \text{ V/mm}.$

$V = Ed = 12.5 \times 10^3 \times 2 = 25 \times 10^3 \text{ V}$

$Q = CV = 267 \times 10^{-12} \times 25 \times 10^3 = 6.67 \times 10^{-6} \text{ C}$ **Ans.**

Example 6.14

A capacitor consists of two parallel rectangular plates, each of 120 mm square separated by 1 mm in air. When a voltage of 1000 V is applied between the plate, an

average current of 15 mA flows in 5S. Determine (i) charge on capacitor, (ii) electric flux, (ii) electric flux density, and (iv) electric field strength in dielectric

Solution

(i) Charge on capacitor

$$Q = I.\,t = 15 \times 10^{-3} \times 5 = 75 \times 10^{-3}\text{ C}$$

(ii) Since charge and electric field are equal to each other

$$\phi = Q = 75\text{ mc}$$

(iii) Area $A = 120 \times 10^{-3} \times 120 \times 10^{-3} = 144 \times 10^{-4}\text{ m}^2$

Electric flux density $D = \frac{Q}{A} = \frac{75 \times 10^{-3}}{144 \times 10^{-4}} = 5.20\text{ c/m}^2$

(iv) Electric field strength in dielectric

$$E = \frac{V}{d} = \frac{1000}{1 \times 10^{-3}} = 1.0 \times 10^{6}\text{ V/m}$$ **Ans.**

Example 6.15

A multiple capacitor has 12 rectangular plates 120 mm × 150 mm. The plates are separated by mica sheet 1 mm thick. Calculate the capacitance, assuming that the relative permittivity of medium is 4.

Solution

$$n = 12, \quad \varepsilon_r = 4.0, \quad d = 1.0 \times 10^{-3}\text{ m}$$
$$A = 120 \times 150 \times 10^{-6} = 18 \times 10^{-3}\text{ m}^2$$
$$\varepsilon_0 = 8.854 \times 10^{-12}\text{ F/m}$$
$$c = (n-1)\varepsilon_0\varepsilon_r.\frac{A}{d}$$
$$= (12-1) \times 8.854 \times 10^{-12} \times 4 \times \frac{18 \times 10^{-3}}{1 \times 10^{-3}}$$
$$= 7009 \times 10^{-12}\text{ F}$$ **Ans.**

Example 6.16

The dielectric strength of a rubber is 40000 volts/mm at frequency 60 Hz. Calculate the thickness of insulation on a wire carrying 33 kV to sustain the breakdown.

Solution The required thickness of insulation will be

$$= \frac{33\text{ kV}}{40000\text{ volts/mm}} = \frac{33 \times 1000\text{ mm}}{40000}$$
$$= 0.82\text{ mm}$$ **Ans.**

Example 6.17

A paper capacitor of 0.03 μF capacitance is of 1 mm thickness, dielectric constant 2.6 and dielectric strength 1.8 × 10^7 V/m. Determine (a) the total area of dielectric needed for it, and (b) its breakdown voltage.

Solution The capacitance of parallel plate capacitor given as

$$C = \frac{\varepsilon_0\varepsilon_r A}{d}$$, we obtain

(a) $$A = \frac{Cd}{\varepsilon_0 \varepsilon_r} = \frac{(0.03 \times 10^{-6}) \text{ farad} \times 0.001 \text{ m}}{(8.85 \times 10^{-12}) \times 2.6}$$

$$= 1.302 \text{ m}^2$$

(b) Since the dielectric strength is equal to the breakdown voltage $V_{\text{breakdown}}$, divided by the thickness of dielectric, therefore

$$V_{\text{breakdown}} = (1.8 \times 107) \times 0.001 \text{ m}$$

$$= 1.8 \times 10^4 \text{ V}$$ **Ans.**

Example 6.18

Capacitance of a capacitor is 0.035 μF and the power factor is 0.005. Estimate the dielectric loss of the capacitor when it carries a current of 250 ampere at a frequency of 25 kHz.

Solution The given values are

$$C = 0.035 \ \mu\text{F} = 0.025 \times 10^{-6} \text{ F}$$

$$\tan \delta = 0.0005 = 5 \times 10^{-4}$$

$$f = 25 \text{ kHz} = 25 \times 10^{-4}$$

and $$I = 250 \text{ A}$$

The voltage across capacitor is given by

$$V = \frac{1}{2\pi fC} = \frac{250}{2\pi \times 25 \times 10^3 \times 0.035 \times 10^{-6}}$$

$$= 45495 \text{ volt}$$

Dielectric loss is given by

$$P = \text{V.I. } \tan \delta = 45495 \times 250 \times 5 \times 10^{-4}$$

$$= 5686 \text{ watts}$$ **Ans.**

Example 6.19

A parallel plate capacitor is to be made to store 20 μC at a potential of 10 kV. The separation between the plates is 5 × 10^{-9} m. Calculate the area that the plates must have if the dielectric material between the plates is of alumina of dielectric constant 10.

Solution Given are: $Q = 20 \times 10^{-6}$ coulomb, $V = 10 \times 10^3$ V, $\varepsilon_r = 10$,

$$\varepsilon_0 = 8.854 \times 10^{-12}, d = 5 \times 10^{-4} \text{ m}$$

As $$Q = CV$$

$\therefore$ $$C = \frac{Q}{V} = \frac{20 \times 10^{-6}}{10 \times 10^3} = 2 \times 10^{-9} \text{ farad}$$

$\because$ $$C = \frac{\varepsilon_r \varepsilon_0 A}{d}$$

$\therefore$ $$A = \frac{Cd}{\varepsilon_r \varepsilon_0} = \frac{2 \times 10^{-9} \times 5 \times 10^{-4}}{10 \times 8.854 \times 10^{-2}}$$

$$= 10.294 \times 10^{-3} \text{ m}^2$$ **Ans.**

6.19 OTHER IMPORTANT PROPERTIES OF INSULATING MATERIALS

Mechanical Properties

The strength of material is also a property which is practically important. Following properties related with the material are always taken into consideration:

- Tensile strength
- Compressive strength
- Flexural strength

6.20 THERMAL CHARACTERISTICS

The important thermal properties which are looked upon are

- Thermal conductivity
- Frost resistance
- Heat resistance

(i) Frost Resistance

It is the ability of insulating material to function in low temperature range of nearly 60°C – 70°C. It is a very important property and is utilised in aircraft equipments, outdoor substations and in radio and telephone field equipments.

(ii) Thermal Conductivity

When one part of a material is heated, heat energy gets transferred to the other parts as well. The rate of this heat flow depends on the nature of the material, the cross-section area normal to the direction of heat flow and the temperature gradient. It is defined as the amount of thermal energy transferred per unit area of cross-section per second per unit temperature gradient. Its unit is $wm^{-1} K^{-1}$.

$$K = \frac{Q}{A.\dfrac{\Delta\theta}{\Delta x}}$$

Heat conduction in metals takes place through free-electrons and lattice vibrations (phonons), as well. But, in non-metals it takes place only due to lattice vibrations. This is why conduction of heat is easily possible in metals as compared to non-metals. Thermal conductivities are given in Table 6.10.

Table 6.10 *Thermal conductivities of some popular materials*

Material	*Thermal Conductivity $Wm^{-1} K^{-1}$*
Aluminium	247.0
Copper	398.0
Silver	428.0
Alumina	35–39
Glass	1.70
Perspex	0.15–0.21
Wood	0.14–0.16

(iii) Thermal Classification of Insulating Materials

Materials should have the ability to work properly at a high operating temperature without losing their main characteristics. Therefore, insulating materials can be classified on the basis of operating temperature. Table 6.11 shows the various classes of insulating materials and their operating temperature along with some examples.

Table 6.11 *Various classes of insulating materials*

Class	*Operating Temperature (0°C)*	*Examples*
Class Y	Up to 90	Cotton, silk, paper, thermoplastics like PVC
Class A	Up to 105	Cotton, silk, paper impregnated with lacquer or polyamide resins
Class B	Up to 130	Mica, fibre-glass, asbestos
Class E	Up to 120	Enameled wire, polyurethane
Class F	Up to 155	Mica, fibre-glass, asbestos coated with insulating varnish
Class C	More than 180	Quartz, ceramics, teflon

Example 6.20

Find the thermal conductivity of a metal containing 3.0×10^{22} electrons/cm^3 at 300 K. The average collision time between electrons and the obstacles is 3×10^{-14} s. Take Lorentz number = 2.44×10^{-8} ohm W/K^2; the mass of electron is 9.1×10^{-31} kg.

Solution Given are

$$n = 3.0 \times 10^{22}/\text{cm}^3 = 3.0 \times 10^{28}/\text{m}^3$$
$$t = 3 \times 10^{-14}\ \text{s}$$
$$m = 9.1 \times 10^{-31}\ \text{kg}$$
$$L = 2.44 \times 10^{-8}\ \text{ohm W/K}^2\text{, and}$$
$$T = 300\ \text{K. We take } e = 1.6 \times 10^{-19}\ \text{C.}$$

The electrical conductivity given as

$$\sigma = \frac{ne^2t}{m}$$

On substituting the appropriate values, we get

$$\sigma = \frac{(3.0\times10^{28})\times(1.6\times10^{-19})^2\times(3\times10^{-14})}{9.1\times10^{-31}}$$
$$= 25.3 \times 10^6\ (\text{ohm-m})^{-1}$$

The Lorentz number is

$$\frac{K_t}{\sigma T} = 2.44 \times 10^{-8}\ \text{ohm W/K}^2$$

or

$$K = \sigma T \times (2.44 \times 10^{-8})$$
$$= (25.3 \times 10^6) \times (300) \times (2.44 \times 10^{-8})$$
$$= 185.19\ \text{W/mK}$$

Ans.

6.21 CHEMICAL PROPERTIES OF INSULATING MATERIALS

Certain chemical properties are also required to be considered for the insulating materials. They are described below:

- Chemical resistance
- Hygroscopicity
- Moisture permeability

(i) Chemical Resistance

It is the ability of the insulating material to fight against corrosion in the presence of gases, water, acids and alkalies. For materials which are subjected to high voltage, high chemical resistance is also necessary.

(ii) Hygroscopicity

Many insulation materials are hygroscopic. Sometimes the insulation may come in direct contact with water. The porous materials are more hygroscopic than dense ones. Small amount of moisture absorbed by an insulated material affects its electrical properties drastically.

(iii) Moisture Permeability

The tendency of an insulating material to pass moisture through them is known as 'moisture permeability'. This property is always kept in mind while appraising the quality of materials used for filling, insulating varnish coating, protective coatings, cable sheaths, etc. Moisture can penetrate through very small pores as the size of water molecule is very small.

6.22 CLASSIFICATION OF INSULATING MATERIALS ON THE BASIS OF STRUCTURE

Classification of insulating material is also done on the basis of their physical and chemical structure (see Table 6.12).

Table 6.12 *Classification of materials on the basis of structure of material*

Fibrous material	Wood, paper, cotton, adhesive tapes.
Insulating liquids	Transformer oil, cable oil, silicone fluids.
Non-resinous materials	Bitumen's, waxes
Glass and ceramics	Glass, porcelain, etc.
Plastics	Molding powder, rubber laminations
Mineral	Mica, micanites
Gaseous	Air, H_2, N_2, Ne, CO_2 SF_6, Hg and Na vapour

Table 6.13 lists a comparison of some popular electrical engineering materials.

Table 6.13 *Comparison between popular insulting electrical engineering materials*

S.No.	*Material*	*Properties*	*Uses*
1.	Paper and pressboard	Low dielectric loss, discharge current is lower	High frequency capacitors
2.	Cotton	Hygroscopic, low dielectric strength, properties can be improved by impregnation	Winding of small magnet coils, armature windings of coils and chokes
3.	Wood	High dielectric constant, highly hygroscopic, dry wood can bear a voltage gradient of 10 KV/inch	Terminal blocks, wedges for armature winding, operating rods in high voltage switch gears

(Contd)

4.	Bitumen	Hydrocarbons of jet black colour, highly soluble in mineral oil, poor insulating property, low hygroscopic, acid and alkali resistant	Underground cables
5.	Waxes	Complex organic substance, high insulating property, low hygroscopic	Used as impregnated material for paper and cloth insulation, dipping medium coating on conductors
6.	Glass	Organic material containing oxides silicates and borates, etc., best insulating properties, high-resistivity and dielectric strength	Insulating material to form envelope for internal support in bulbs, valves X-ray tubes fuse bodies, etc.
7.	Ceramics	Hard, strong, dense, unaffected by chemical action, stable at high temperature, excellent dielectric properties, weak impact strength	High voltage insulation at elevated temperatures in ovens
8.	Asbestos	Exhibit fibre structure, can work at high temperatures, good tensile strength	Capacitor dielectric, transist ors, hybrid circuit substrates electro-mechanical transducers; not useful for high voltage used as thermal insulator in wires and cables at high temperature
9.	Rubber	Stretchable, moisture repellent, good insulating properties, good corrosion resistance. Can be obtained as hard rubber synthetic rubber, butadiene rubber, butyl rubbers, chloroprene rubber and silicone rubber.	Used as protective clothing such as boots and gloves. Also used as insulation covering for wires and cables, hard rubber is used in housing for storage batteries, panel board jacketing material chloroprene (Neoprene) is extensively used as jacketing for conductors

Table 6.14 compares the various properties of all kinds of rubbers described above.

Table 6.14 *Comparison of various parameters of some insulating materials*

Property	*Natural Rubber*	*Hard Rubber*	*Neoprene*	*Nitrile Butadiene*	*Silicon*	*Hypalon*	*Butyl Rubber*
Specific gravity	0.93–1.6	1.1–1.4	1.09–1.6	1.00	1.2 to 1.5	–	0.9–1.3
Water absorbtion (Percentage)	1–2	0.2–1	3–4	–	0.5 to 2.4	–	0.3–0.5
Operating temp. (°C)	75	0	90	–	175–200	150	80–90
Tensile strength (kg /cm^2 × 10^{-3})	0.21–0.32	0.28–0.72	0.14–.26	0.19–0.32	0.09–0.12	0.11–0.26	0.05–0.21
Dielectric constant (MHz)	2.7–5.0	2–5	7.5–14	3.9–10	2.8 to 7	5 to 11	2.11–4
Volume resistivity (Ω -m)	10^{15}–10^{17}	10^{14}	10^{11}–10^{12}	10^{12}–10^{17}	10^{13}–10^{17}	10^{13}–10^{15}	10^{14}–10^{16}

(Contd)

Surface resistivity (Ω)	10^{14}–10^{15}	–	10^{11}–10^{12}	10^{12}–10^{15}	10^{13}	10^{14}	10^{13}–10^{14}
Dielectric strength (kV / m.m.)	18–24	12–28	4–20	16–20	12–28	16–24	16–32
Power factor at MHz (× 10^{-2})	0. 5	0.5–2.0	1.0 to 6.0	3–5	0.1 to 1	2–9	0.3–8.0

6.23 MICA AND ITS VARIOUS KINDS

Mica is one of the best neutral insulating material. Normally, two kinds of mica are used for electrical purposes. They are:

- Muscovite mica
- Phlogophite mica

(i) Muscovite Mica

It is found in India, Brazil and the US. The chemical composition of muscovite mica is $KAl_3SiO_3O_{10}(OH)_2$. It is translucent green, ruby, silver or brown, and is strong, tough and flexible. Muscovite mica exhibits good corrosion resistance and is not affected by alkalis. It is used under sever situations like capacitors and commutators.

(ii) Phlogophite Mica

Also known as magnesium mica, the chemical composition of phlogophite mica is $KMg_3AlSiO_3O_{10}(OH)_2$.

It is less flexible as compared to muscovite mica. It is amber, yellow green or grey in colour. Phlogophite mica is more stable, but its electrical properties are poorer compared to muscovite mica. It is used where thermal stabilities are required as in domestic appliances like inn, hotplates, etc.

6.24 MICA PRODUCTS

Mica, in its natural form, has limited industrial applications because of its poor chemical and insulating properties. Some of important mica products are given below.

(i) Synthetic Mica

Made synthetically, its structure is similar to phlogophite mica with hydroxil groups (OH) replaced by fluorine. Its dielectric constant, tan δ and resistivity are similar to those of muscovite mica. The main advantage is its thermal stability; it can bear heating up to 1200°. It is not available in sizeable splitting, but only as a compacted aggregate.

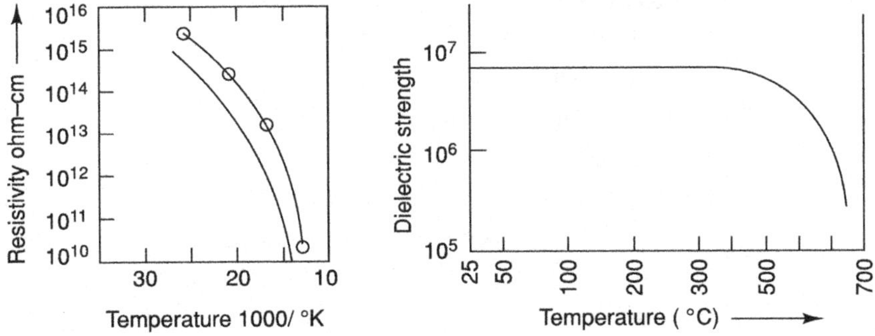

Fig. 6.12 *Resistivity and dielectric strength variation of mica with temperature*

(ii) Mica Paper

Mica, being crystalline in nature, has many structural limitations for use in electrical applications. It is broken into small particles of aqueous solution. Out of these particles sheets of mica paper are produced by a process similar to the one used in manufacturing paper. Its thickness varies from 0.05 mm to 0.1 mm. It is common used in armature and field coils of rotating machines.

(iii) Manufactured Mica

When mica flakes are held together with adhesives, the product is called manufactured mica. The binding material is about 20%. Binding resins are shellac, epoxy, silicon resin, etc. It is used in commutators, electrical heating devices, transformers, etc.

Table 6.15 *Comparison of properties of mica*

Property	*Muscovite Mica*	*Phlogophite Mica*
Specific gravity	2.7–3.2	2.6–2.8
Moisture absorption (%)	Low	Low
Tensile strength ($kg/cm^2 \times 10^{-3}$)	0.35 to 0.7	–
Shear strength (kg/mm)	23–26	10–13
Resistivity ohm-cm at 20°C	10^{15}–10^{16}	10^{12}–10^{14}
Power factor ($\times 10^{-4}$)	1–3	10–50
Operating temperature (max)°C	500–600	800–900
Relative permitivity	6–7	5–6
Dielectric strength (kV/mm)	80	60

6.25 POLYSTYRENE

Polystyrene is a chain polymer which is obtained by polymerising the organic liquid styrene. The lower polymers of styrene are liquids, but high polymers are hard, rigid and transparent solids. It is a good insulator having low dielectric loss. Table 6.16 shows various parameters related to polystyrene.

Table 6.16 *Properties of polystyrene*

Material	*Value*
Dielectric constant	2.55
Dielectric strength (kV/mm)	20–28
Power factor (60 Hz)	$10–30 \times 10^{-5}$
Insulation resistance (Ω - cm)	10^{17}–10^{19}
Specific gravity	10.6
Tensile strength (kg/cm^2)	$0.35 \times 0.65 \times 10^{-3}$
Compressive strength (kg/cm^2)	$0.78–1.12 \times 10^{-3}$
Elongation (%)	1–3.5
Water absorption (%)	0.03–0.6

(Contd)

Specific heat	0.32
Max operating Temperature	65–80°C

Applications of Polystyrene

- It is used as dielectric material in capacitors.
- It is used as tape wrapper in cables.
- It is used as a base material for audio and video tapes.

6.26 POLYTHENE, TEFLON, PVC, EPOXY GLASS AND BAKELITE

(i) Polyethylene

Also known as polythene, it is obtained by polymerisation of ethylene. The polymerisation is performed in the presence of a catalyst at atmospheric temperature and pressure around 100°C. These are of two kinds:

- Low-density polythene
- High-density polythene

To obtain heat resistance property, polyethylene is subjected to ionising radiation. Some parameters of polythene are given in Table 6.17.

Table 6.17 *Properties of polythene*

Property	*Low-density Polythene*	*High-density Polythene*
Dielectric constant (60 Hz)	2.28–2.32	2.25–2.32
Dielectric strength (kV/mm)	20–160	20–60
Power factor at (60 Hz)	$1–5 \times 10^{-4}$	$2–6 \times 10^{-4}$
Volume resistivity (Ω/cm)	$10^{17}–10^{19}$	$10^{15}–10^{16}$
Surface resistance (Ω)	$10^{14}–10^{17}$	$10^{14}–10^{15}$
Specific heat	0.55	–
Softening Temperature (°C)	94–110	–
Melting point (°C)	110–116	126–136
Tensile strength (kg/cm^2)	0.09 to 000.16×10^{-3}	$0.16–0.365 \times 10^3$
Water absorption (%)	< 0.01	< 0.01
Elongation (%)	110–600	40–300

❖ **Properties** Polyethylene exhibits good electrical and mechanical properties. It is moisture resistant and is not soluble in many solvents except benzene and petroleum at high temperature. The dielectric constant and power factor remains steady over a wide range of temperature.

❖ **Applications**

- Used as general-purpose insulation
- Used in the insulation of wires and cable conductors
- Used in high-frequency cables and in television circuits
- Used as a jacketing material for cables
- Polyurathene films are also used as dielectric material in capacitors

(ii) Teflon

The chemical name of Teflon is polytetra fluoro-ethylene. This is synthesised by polymerisation of tetrafluoro ethylene. It is a valuable insulator having very good electrical, mechanical and thermal properties. It can tolerate very high temperatures without damage. Its dielectric constant does not change with time, frequency, and temperature. Teflon's insulation resistance is also very high. It is also water resistant. Table 6.18 gives some important parameters of Teflon.

Table 6.18 *Properties of Teflon*

Teflon	*Value*
Dielectric constant (60 Hz)	2–2.2
Dielectric strength (kV/mm)	16–20
Power factor (60 Hz)	$2–8 \times 10^{-4}$
Volume resistivity (Ω-cm)	10^{13}
Specific gravity	2.1–2.3
Melting point (°C)	327
Decomposition temperature (°C)	350
Moisture absorption (%)	Nil
Maximum useable temperature (°C)	300
Flammability	Nil
Compressive strength (kg/cm^2)	$1.6–1.7 \times 10^{-3}$
Tensile strength (kg/cm^2)	$1.8–4 \times 10^{-4}$

Applications

- It is used as dielectric material in capacitors.
- It is used as covering for conductors and cables.
- It is also used as the base material for PCBs.

(iii) Polyvinyal Chloride (PVC)

PVC is another important dielectric material which is obtained by polymerisation of vinyl chloride (gaseous) in the presence of a catalyst at 50°C. Its composition can be varied depending upon the requirements. The properties of PVC (mechanical strength, porosity, flexibility, mixture absorption capacity and electrical properties) can be changed by adding some additives. It has been observed that PVC exhibits good electrical and mechanical properties. It is hard, brittle and has higher loss tangent. It is non-hygroscopic and can resist flame and sunlight.

Applications

- It is used as jacketing material for wires and cables.
- It is also used as an insulation material for dry batteries.

(iv) Epoxy Glass

Epoxy glasses are made by bonding two or more layers of material. The layers used reinforcing glass fibres impregnated with an epoxy resin. The main characteristics of epoxy glass are given in Table 6.19.

Table 6.19 *Main characteristic parameters of epoxy glass*

Epoxy Glass	*Value*
Volume resistivity	10^{14} ohm cm
Dielectric constant	5.0
Dielectric strength	400 V/mm
Loos factor (at 60 Hz)	0.1

Epoxy glass is water resistant and in not affected by alkalies and acids.

Applications

- It is used as a base material for copper-clad sheets used for PCBs.
- It is also used for terminal ports, instrument case, etc.

(v) Bakelite

It is a hard, dark-coloured thermosetting material, which is a type of phenol formaldehyde. Some parameters of Bakelite are presented in Table 6.20.

Table 6.20 *Values of some parameters of Bakelite*

Bakelite	*Value*
Dielectric constant	4.1
Loss factor	0.001
Dielectric strength	400 V/mm
Volume resistivity	10^3 ohm cm
Max. operating temperature	300° F
Water absorption	Low

6.27 TRANSFORMER OIL

It is a liquid insulator in which transformer cores, chokes, etc. are dipped. Besides insulation, it also works as a coolant. Transformer oil is also used in other apparatus like switches, circuit breakers, etc. It is obtained by distilling the higher fractions (gasoline, petroleum, naptha, etc.) in the process of crude oil distillation. The oil so obtained is treated with H_2SO_4, and then treated with alkali to neutralise the acid. It is washed with water, filtered and dried. Table 6.21 gives some specific parameters related with transformer oil.

Table 6.21 *Values of some products of transformer oil*

Transformer oil	*Value*
Specific gravity	0.885
Flash point	135°C
Resistivity	1.0×10^{12}–10×10^{12} Ohms
Relativity permitivity at RT	2.2

Specific heat	0.42
Dielectric strength	30 KV/mm
Viscosity	Low

Purification of Transformer Oil

Several impurities such as water, resins, fibre and carbon particles, etc., may be present in the transformer oil. Such impurities decrease the breakdown voltage of this oil. Water in the oil is normally present in the form of tiny bubbles suspended or in a dissolved state. The oil is purified in the following ways:

- It is boiled, preferably in vacuum, to evaporate the water.
- Since the impurities including water are heavier than oil these can be removed by quickly rotating the oil in the centrifuge. The impurities are pressed against the sides of the vessel by centrifugal force.
- The solid impurities are also removed by filtering the oil through a special filter cardboard under pressure.

Adsorbants whose particles readily absorb the products of ageing and moisture on the surface are added. The heated oil can be thoroughly mixed with the crushed adsorbent and then filtered, or can be passed through a thick layer of the absorbent. The following artificial products are mostly used as adsorbants for the transformer oil: silica gel, alumina or special type of moulding clays, etc.

6.28 SILICONE FLUIDS

These are organic fluids containing Si–O chains having normally methyl group. They exhibit high thermal stability, low temperature coefficient, low dielectric loss and high dielectric strength. These fluids can be obtained over a wide range of viscosity. Operating temperature is quite wide, from 65°C to 300°C. Their dielectric constant is nearly around 2.6. These fluids are used as cooling liquids for capacitors and transformers. They are also used with rubber to make silicon rubbers.

6.29 VARNISH

It is the liquid form of resinous matter in oil or a volatile liquid which—after applying—dries either by evaporation or chemical action to form hard, lustrous coating which is resistant to air and water. Varnish does not contain colour pigments. The dried film is normally transparent or translucent. It is used to improve insulation properties, mechanical strength and to reduce degradation caused by oxidation and adverse atmospheric conditions. It also provides a fireproof finish.

6.30 ENAMELS FOR WINDINGS

It is a fusible insulated coating material which is generally applied on conducting surfaces. The thickness of enamel coating is of order of 0.05 mm. Enamels are used for coating wires and windings of transformers, motors, etc. These are also used to furnish a heavy protective coating on electronic equipments.

6.31 SUMMARY OF SOME WIDELY USED INSULATING MATERIAL

Table 6.22 describes—in a nutshell—the characteristics of some important and widely used insulating materials along with their applications.

Table 6.22 *Widely used insulating materials*

S.No.	*Insulating Material*	*Characteristics*	*Applications*
1.	Polyethylene	Electrically insulating can be made into thin sheets	Used as general purpose insulation Material; jacket material for cables and also as dielectric material
2.	PVC	Electrically insulating; non-corrosive	Insulating layers on copper wires; tubes for carrying wires; jacket material for wires and cables
3.	Thermocole (Polystyrene)	Thermally insulating and light weight	Can be used as insulating material to protect from heat
4.	Teflon	Thermally and electrically insulating Material; High MP	Insulators at high temperature non-stick pans; sheets and rods used at high temperatures
5.	PUF (Polyurathene foam)	Thermally insulating material	Used in inner walls of refrigerators

6.32 CAPACITORS AND THEIR WORKING

Capacitor is well-known component and is the backbone of electrical and electronic equipment. It is a device to store electrical energy and is also used as a filter in electric circuits.

A capacitor is composed of two conducting plates filled in between an insulating material known as 'dielectric material'. The capacitance of a capacitor is given by the relation

$$C = \frac{\epsilon_r \epsilon_0 A}{d} \tag{6.2}$$

where ϵ_r is the dielectric constant of the material also known as permittivity of dielectric medium, ϵ_0 is the permittivity of free space, A is the surface area of the plates and d, the separation between them. Dielectric material plays an important role, as the capacity of the capacitor depends upon it along with the separation. Another parameter which is always taken into consideration is the operating voltage of the capacitor. It is the maximum voltage that can be applied between the plates of a capacitor without causing any damage to it. If the voltage exceeds this value, the electric field on electrons of the dielectric material become so large that they break up their bonding and render them free. The large number of free electrons, thus produced, move towards the positive plate constituting a heavy current through the dielectric. This is known as electrical breakdown of dielectric; in this situation it no longer remains an insulator and the capacitor action is destroyed.

A number of dielectric materials are used in the commercially available capacitors; the important ones being air, paper, mica, plastic film, ceramics and electrolytic films.

6.33 VARIOUS KINDS OF CAPACITORS

(i) Variable Air Capacitor

Variable air capacitor consists of a semicircular metal plate fixed in a position. Another set of plates move parallel to the fixed plate by turning a shaft. By doing so, effective area of plates changes and capacitance varies. The actual variable air capacitor, therefore, onsist of a parallel combination of plates. The plates are usually made up of aluminium. The capacitors are limited to maximum capacitance of nearly 1000 PF. These are used in radio receivers for tuning different wavelength of commercial broadcasting stations.

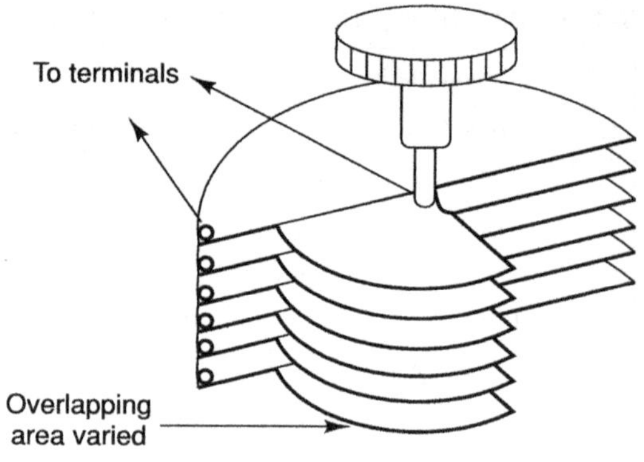

Fig. 6.13 *Variable air capacitor*

(ii) Paper Capacitor

Paper capacitors have the advantage of being compact and relatively inexpensive. The dielectric constant of paper is 2.0. There are two kinds of paper capacitors:

(1) Impregnated paper capacitor
(2) Metalised paper capacitor

Impregnated paper consists of two thin sheets of paper between metal foils compactly rolled and impregnated with an impregnate oil, moral wax, petroleum jelly, castor oil, etc. Increasing the number of paper sheets increases the voltage rating of the capacitor, but decreases its capacitance. Fig. 6.14 shows a paper capacitor.

In metallised paper capacitor, a metal layer is sprayed on to the paper so as to avoid a separate foil. Paper capacitors are usually made in the range of 0.001 to 1.0 microfarads having a typical voltage rating of 500 volts, but higher voltage ratings are also fairly common. To increase the stability and to reduce the power loss, the paper is now replaced by a thin layer of polystyrene.

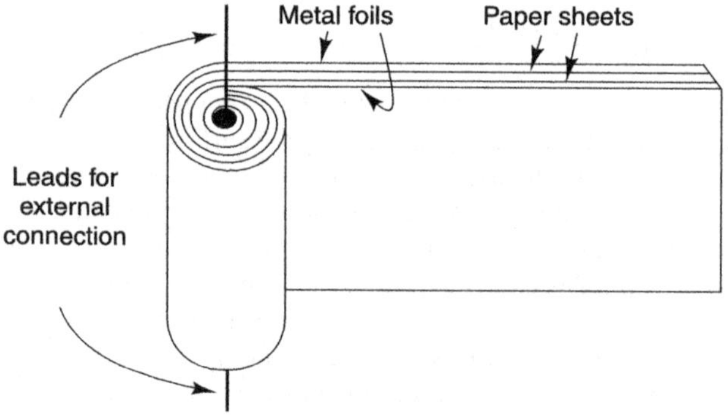

Fig. 6.14 *Paper capacitor*

(iii) Mica Capacitor

Mica has still higher value of dielectric constant, which is 6.0. Therefore, it gives larger capacitance for same physical dimensions. Further, a mica sheet can withstand higher voltage than a paper sheet of same thickness. However, mica capacitors are relatively expensive. In mica capacitor the sheets of mica are put in-between metal foils. Figure 6.15. shows a mica capacitor.

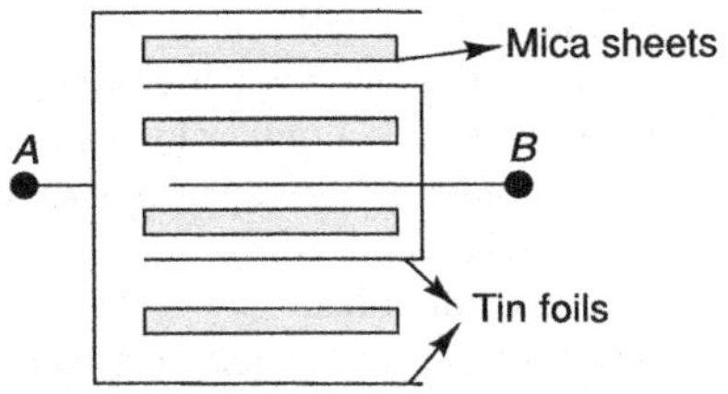

Fig. 6.15 *Mica capacitor*

Alternate plates are connected together to form the two terminal (*A* and *B*) of the capacitor. If *n* is the number of spaces between the plates, the system is equivalent to *n* capacitors in parallel, the total capacity of which is given by

$$C = \frac{n \in_0 KA}{d}$$

By making *n* large, it is possible to increase the capacity of the capacitor without appreciably increasing its physical dimensions. Tin foil is often used for conducting plates.

(iv) Ceramic Capacitors

Ceramic materials are very common in the manufacture of capacitors. Some special ceramics have a very high value of dielectric constant (~100). Therefore, ceramic capacitors constructed by depositing silver or copper coating on two sides of a ceramic base offer a large value of capacitance in very small sizes. In addition, they provide the advantage of a very high voltage rating. Figure 6.16 shows a ceramic capacitor.

Some materials which are used in a ceramic capacitor are given in Table 6.23.

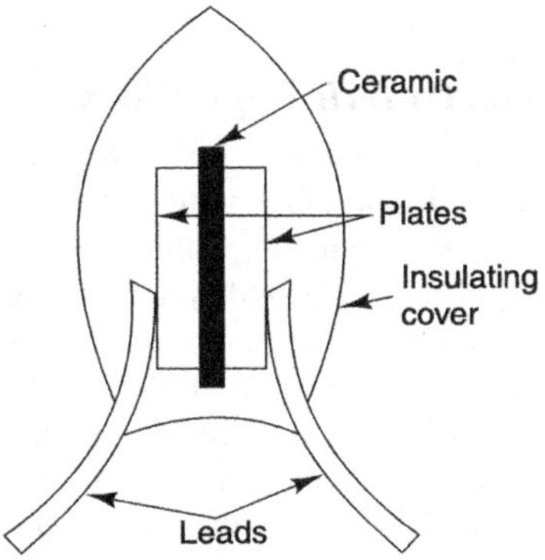

Fig. 6.16 *Ceramic capacitor*

Table 6.23 *Values of dielectric constant in some materials*

Material	*Dielectric Constant (at 25°C)*
Teflon	2.1
PVC	4.6
Al_2O_3	10
$MgTiO_3$	20
TiO_2	100
$BaTiO_3$	1000–2000

6.34 MATERIALS FOR CAPACITORS: DESIRABLE CHARACTERISTICS

1. Capacitors for dc Applications

The dielectric for a capacitor should have properties so that:

(i) the size is small
(ii) it could operate at a high temperature
(iii) it must have a higher insulation resistance
(iv) it should exhibit lower power factor
(v) it has a longer life
(vi) it is economical

Several types of capacitors are now used, especially for dc applications. These are in addition to paper capacitors (paper with impregnants like castor oil, chlorinated diphenyls, mineral oils, polyesters, etc.), electrolytic capacitors with aluminium or tantalum ceramic capacitors with variety of ceramic bodies (tubular, disc, blocking, etc.), and film dielectric (Teflon, polystyrene, etc.) capacitors.

2. Power Capacitors

Most power capacitors consist of aluminium foils interleaved with multiple sheets of special kraft cellulose paper. Depending on the application, the paper is impregnated with mineral oils or askarels. Askarels have a higher dielectric constant that gives a better electric stress distribution between paper and askarel. With askerals, the capacitor size reduces.

Electrolytic Capacitors

Certain metals, for example aluminium, have the property that if they are placed in a suitable conducting solution (usually a solution of boric acid) and a current is passed with the metal plates as anode, and the solution as cathode, a thin film of metal oxide is deposited on the plate as a result of electrolysis. This film is non-conducting and acts as dielectric material between the conducting plates. The whole thing then acts as a capacitor and is known as electrolytic capacitor. Since the thin film is only a few microns thick, electrolytic capacitors have a very large capacitance. They are commercially available in the range of 1 μF to 1000 μF. Their working voltage arc is, however, low—from about 12 V to 500 volts. Their polarity in the circuit also cannot be reversed. It this is done, the oxide is reduced into metal and oxygen. The dielectric is thus destroyed and the capacitor action is lost. Further, the released oxygen gas buds up sufficient pressure inside the sealed capacitor which might then explode. Figure 6.17 shows an electrolytic capacitor. In spite of the above drawbacks, for large capacitance they are the least expensive and most compact capacitors available. The film of dielectric which gets deposited has a thickness of order of 10^{-7} m and its dielectric strength is hardly 10^9 volts/m, which is extremely high.

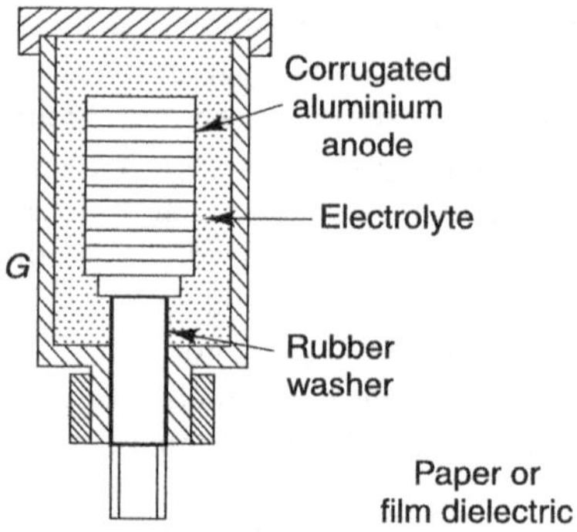

Fig. 6.17 *Electrolytic capacitor*

6.35 HIGH-TENSION UNDERGROUND CABLES

Underground cables are required for transmission of high voltages. These are least effected by meteorological effects and lighting, but a detection of underground fault is not easily possible. They are more beneficial than opened transmission lines from the point of view of safe transmission of high voltage. In such cables the insulation material that is normally used is impregnated paper insulation. All paper insulated cables must be provided with a lead sheath because the cable operation entirely depends upon the capability of the sheath.

Figure 6.18 shows the cross-section of a three core cable suitable for 33 kV. There are seven small diameter copper conductors in each core. These conductors are stranded together and covered with paper insulation. The three paper-covered cores are enclosed in a belt insulation of impregnated paper. The jute is filled in the space between the outer insulated and covered cores. Lead sheath covering the belt insulation provides proper protection to the cable from the external agencies. Eddy currents in the sheath are negligible because the inductive effects of the three cores enclosed in one sheath practically neutralise each another. For exceptionally high voltages to the tune of 132 kV to 220 kV, special hollow conductors, oil filled, single-core cables are used.

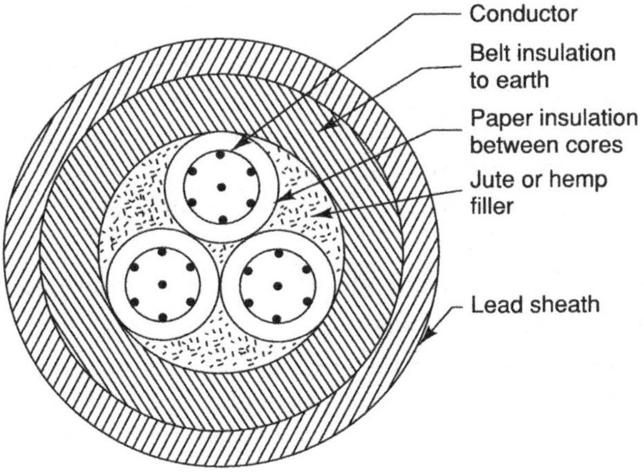

Fig. 6.18 *High voltage underground cable*

Summary

- A conductor of electricity is any substance or material which can carry an electric current when subjected to a difference of potential; the greater the density of current for a given potential difference, the more efficient the conductor is said to be.
- Materials of high conductivity are employed for making conductors for all kinds of windings required in electrical machines, apparatus and devices, as well as for transmission and distribution of electric energy. Important materials of this class are: copper-, aluminium- and copper-based alloys.
- Important materials of high-resistivity are tungsten, carbon, nichrome V or brightway C, manganin, constant or eureka, German or nickel silver or electrum, nirosta, fechral, chromal, etc.
- Materials used for lamps filaments are carbon, tantalum and tungsten.
- Material used for transmission lines incude cadmium copper materials, copper weld materials, phosphor bronze materials, galvanised steel materials, galvanised steel core copper and steel cored aluminium materials.

- Some of the commonly used materials for making bimetallic strips are:

  Iron, nickel and constantan—high coefficient of expansion

  Alloy of iron and nickel—low coefficient of expansion
- Metals suitable for lightly loaded contacts are platinum, palladium, gold, tungsten, molybdenum, rhodium and cermets.
- Metals and alloy for fuses include lead and tin, copper, silver, rose's alloys, and wood's alloy.
- The possible thermocouples are constantan-iron couple, constantan-copper couple, copel, alumel cromel, copper cope couple, platnium and platinum-rhodium copel, iron-copel, thermocouples, cromel-alumel, couple-cromel and copel thermocouple.
- Electrical insulating materials are those which offer a very large resistance to flow of current, and hence they are used to keep the current in its proper path along the conductor.
- Thermoplastic materials are those which soften on the application of heat, with or without pressure, but they require cooling to set them to shape.
- Insulating varnishes are generally categorised according to their composition as 'oil varnishes' and 'spirit varnishes'.
- The electrical insulating films comprise a group of thin flexible insulating materials made from different polymers.
- The function of insulating liquids is to provide electrical insulation and heat transfer.
- Transformer oil works in two ways:

  (i) It transfers heat by convection from winding and core to the cooling surfaces.

  (ii) It maintains the insulation of the windings.
- Presence of even a trace of water in transformer oil reduces its insulation strength considerably.
- The various tests which are carried out an transformer oil are given below:

  (i) Moisture test (ii) Acidity test

  (iii) Sludge resistance test (iv) Electric strength test
- Insulation resistance is the resistance between two conductors (or systems of conductors) usually separated by insulating materials. It is the total resistance in respect of two parallel paths, one through the body and other over the surface of the body.
- The potential gradient at which breakdown occurs is known as dielectric strength.
- Dielectric constant or relative permittivity is the ratio of the electric flux density in the material to that produced in free space by the same electric force.
- Dielectric hysteresis is defined as the lagging of the electric flux behind the electric force producing it so that under varying electric forces a dissipation of energy occurs; the energy loss due to this cause is called dielectric hysteresis loss.
- Ageing is, in effect, the wearing out of an insulating material by reducing its resistance to mechanical injury.

Exercise

1. What are the properties that high-conducting materials should possess? What are

the materials used for (i) electrical contacts, (ii) rheostats, and (iii) fuses? Give reasons.

2. Discuss the important electrical and mechanical properties of copper. How do its electrical and mechanical properties compare with that of aluminium for use as a conductor?
3. Discuss the properties and uses of the following materials:
   (i) Tungsten (ii) Nichrome (iii) Constantan (iv) Chromel
4. What are the advantages of ACSR conductors? What is a solder? Explain the different uses of solder. Explain the various uses of high-resistivity materials used in the electrical industry.
5. On adding a higher conductivity material (copper) in small quantity to nickel of lower conductivity, why does the conductivity of the alloy decrease? Explain the phenomenon involved.
6. Write the properties and applications of any two noble metals used in electrical contacts.
7. Name some of the commonly used materials as fuse element. Why silver is preferred as fuse element inspite of its higher cost?
8. Describe the applications of following materials in electrical equipment:
   (i) Mica (ii) Paper (iii) PVC (iv) Asbestos (v) Glass
9. Explain the properties and uses of the following insulating materials in electrical engineering:
   (i) Rubber (ii) Mica (iii) Polyvinyl chlorides
10. Describe the characteristics of good insulating materials and discuss briefly the properties and use of mica, porcelain and rubber.
11. What are the materials used in (i) electrical contacts, (ii) rheostats, and (iii) fuses?
12. Discuss in detail any two of the following properties of insulating materials:
    (i) Electrical properties (ii) Thermal properties
    (iii) Chemical properties (iv) Mechanical properties
13. Explain the classification of insulating materials. What is the effect of moisture on insulation? Briefly discuss the methods used to purify transfer oil.
14. Discuss the role of mica as an insulating material. Briefly explain the properties and application of micanite, mica paper, mica tape and moulding micanities.
15. Make a list of ten materials used as insulating materials in electrical engineering. Give the characteristics and applications of each of these ten materials.
16. (i) What insulating materials would you suggest for the following? Mention the reasons.
    (a) Flexible wire
    (b) Heating elements in an oven
    (c) Commutator in dc machine
    (d) Electric iron
    (e) Fuse holder
    (f) Switch for domestic purpose
    (g) Distribution board
    (h) Cable joining box
    (i) High-voltage cable
    (j) Low-voltage cable.

    (ii) What do you understand by the terms 'dielectric loss' and 'loss angle'? Explain your answer with the help of phasor diagram. What are the factors which affect the dielectric loss of an insulating material?

17. Write notes on any three of the following:
    (i) Transformer oil
    (ii) Gaseous dielectrics
    (iii) Materials used in arc lamps
    (iv) Different types of solders
18. Explain the following terms: dielectric constant, polarisability, frequency response of polarisation and electrical breakdown of insulation.
19. What are thermoplastic and thermosetting types of insulating materials? Name any two thermoplastic materials and discuss their electrical characteristics.
20. (i) What is the effect of moisture on insulators?
    (ii) What is meant by the term dielectric strength of an insulating material and how it is expressed?
21. (i) Classify the insulating materials on the basis of maximum working temperature. Name different materials belonging to such a classification along with their applications in electrical instruments, accessories and equipment.
    (ii) What are the causes of breakdown of dielectric solids? How is the permissible operating voltage determined for a solid dielectric?

Numerical Problems

1. The insulation resistance of a kilometre of the cable having a conductor diameter of 1.5 cm and insulation thickness of 1.5 cm is 500 M Ω. What would be the insulation resistance if the thickness of insulation were increased to 2.5 cm?
   [Ans. 667.35 MΩ]
2. The conductor of single-core cable has a radius of 30 mm; the radius over the insulation is 12 mm. If the insulation resistance of the cable is 16000 ohms per km, calculate the specific resistance of the dielectric used. **[Ans. 72.517 MΩ]**
3. A parallel plate capacitor has an area of 20 cm^2 and separation between the plates is 0.2 mm. The space between the plates is filled with dielectric having the real part of the dielectric constant $\varepsilon_r' = 2.5$ when subjected to 2.0 volt alternating voltage at 1 MHz. The loss tangent at this frequency is 4×10^{-4}. Find the elements of parallel R-C circuit. **[Ans. 221.35 μ μF.]**
4. Calculate the diameter of a copper wire of 100 metres length used as winding material in a transformer such that resistance of the whole winding is 2 ohms. Calculate the diameter of the wire if aluminium is to be used for the above winding, resistance remaining the same. **[Ans. 1.28]**
5. Two coils, A and B, connected in series have resistances of 600 Ω and 300 Ω and temperature co-efficient of 0.1% and 0.4%, respectively at 20°C. Find the resistance of the combination at a temperature of 50°C. What is the effective temperature co-efficient of combination? **[Ans. 0.2%]**
6. A field coil at 15°C takes 5 A when a voltage of 250 V is impressed across it. Find the mean temperature of the coil if the current reduces to 4A for the same impressed voltage. Assume that the coil is made of copper conductor.
   [Ans. 84.5°C]
7. A certain load is to be carried by an overhead line. For the same power loss in the line, compare the diameter, weight and cost of aluminium conductor with that of copper, assuming aluminium to have a specific resistance of 2.85 μΩ cm, specific gravity 2.71 and cost Rs 15000 per tonne. The corresponding figures for copper may be taken as 1.7, 8.89 and Rs 30000, respectively. **[Ans. $\frac{0.2556}{1}$]**

Self-quiz Questions

1. What should a good insulator possess?
2. What is the effect of humidity on dielectric strength?
3. Mention some of the factors that influence the dielectric strength of an insulator.
4. Mention some gaseous insulating materials.
5. Mention some liquid insulating materials.
6. What are askarels?
7. Name some materials that can be used as insulators up to about 105°C?
8. Mention some insulators used up to 90°C.
9. Mention some insulators used up to 180°C.
10. Why are electrical properties of paper greatly influenced by the presence of moisture?

Objective type Questions

1. The material normally used in cable shields is
   (a) copper (b) aluminium
   (c) cast iron (d) lead
2. The material which is commonly used as filaments of electric lamps is
   (a) aluminium (b) nichrome
   (c) tungsten (d) brass
3. The material with lowest resistivity is
   (a) constantan
   (b) German silver
   (c) manganin
   (d) nichrome
4. The ratio of percentage of copper and zinc in brass is
   (a) 3 : 2 (b) 4 : 1
   (c) 1 : 1 (d) 2 : 3
5. Bronze is an alloy of copper and
   (a) zinc (b) tin
   (c) lead (d) carbon
6. The material with highest resistivity is
   (a) nichrome (b) constantan
   (c) moneal meal (d) maganin
7. The heating elements of iron are made of
   (a) copper (b) nichrome
   (c) constantan (d) aluminium
8. The material which is used for making overhead telephone wires is
   (a) galvanized steel
   (b) copper
   (c) aluminium
   (d) platinum
9. The material which is used for sliding contacts for starter and rheostats is
   (a) bronze (b) gold
   (c) hydrogen (d) lead
10. A platinum wire senses temperature because of its
    (a) high resistivity
    (b) high coefficient of resistance
    (c) high melting point
    (d) non-corroding texture
11. Copper is completely miscible with
    (a) nickel (b) gold
    (c) hydrogen (d) lead
12. Titanium alloys are
    (a) magnetic
    (b) prone to corrosion
    (c) cheap and easy to machine
    (d) alloys with high strength
13. The material which is used as liquid contact material in electrical switches is
    (a) copper
    (b) vegetable oil
    (c) fluxes
    (d) mercury
14. Resistivity of electrical conductors is not affected by
    (a) pressure (b) temperature
    (c) composition (d) ageing

15. Glass is basically
    (a) anhydride (b) oxide
    (c) arsenic (d) chloride
16. Mica and micanites fall under which class of insulating material?
    (a) Class Y (b) Class E
    (c) Class C (d) Class H
17. Polystyrene is a
    (a) brittle plastic at room temperature
    (b) malleable plastic at room temperature
    (c) elastic at room temperature
    (d) none of these
18. Which of the following materials possesses highest resistivity?
    (a) Mica
    (b) Paraffin wax
    (c) Air
    (d) Mineral oil
19. Eddy currents in a core will be reduced considerably by
    (a) reducing flux density
    (b) laminating core
    (c) reducing core volume
    (d) increasing the number of turns
20. Which of the following materials is used in making spark plugs?
    (a) Porcelain (b) Steatite
    (c) Alumina (d) Zicronia
21. Which insulating material is used in underground cables to protect lead and steel armour?
    (a) Bitumen (b) Cotton
    (c) Quartz glass (d) Polystyrene
22. The density of gold is closer to density of
    (a) platinum (b) nichrome
    (c) tungsten (d) kanthal
23. The insulating materials to be used for most of electronic equipments have
    (a) arc resistance
    (b) thermal stability
    (c) high electric and mechanical strength
    (d) all of these
24. Dielectric materials are essentially
    (a) insulating materials
    (b) conducting materials
    (c) semiconducting materials
    (d) ferro-electric materials
25. Breakdown in a dielectric may occur due to
    (a) electrical breakdown
    (b) thermal breakdown
    (c) electro-chemical breakdown
    (d) any of these
26. Dielectric strength of paper is
    (a) 2 to 4 kV/mm thickness
    (b) 4 to 10 k V / mm thickness
    (c) 12 to 15 kV / mm thickness
    (d) 15 to 20 k V / mm thickness
27. Loss tangent is lowest for
    (a) ebonite
    (b) quartz
    (c) rubber
    (d) paraffin wax
28. Loss in a dielectric may occur due to
    (a) polarisation
    (b) conductivity
    (c) ionisation
    (d) any of these
29. Highest electrical resistivity exists in
    (a) platinum wire
    (b) nichrome wire
    (c) silver wire
    (d) kanthal wire
30. Condensers to be used at high frequencies require a material whose loss angle is
    (a) low (b) high
    (c) uniform (d) one
31. Phosphor bronze contains phosphorus
    (a) 0.1 to 1.5% (b) 2%
    (c) 3 to 4% (d) 9–10%
32. Brass is an alloy of
    (a) copper and zinc
    (b) copper and tin
    (c) copper and aluminium
    (d) copper and iron

33. Melting point of nickel is
    (a) 1452°C
    (b) 1300°C
    (c) 1100°C
    (d) none of these
34. Tungsten has a
    (a) BCC structure
    (b) FCC structure
    (c) HCP structure
    (d) SC structure
35. Babbit metal contains tin nearly
    (a) 93% (b) 40%
    (c) 60% (d) 75%
36. Bakelite is made by condensation reaction between phenol and
    (a) urea
    (b) formaldehyde
    (c) furfural
    (d) polyamine
37. A dielectric should possess?
    (a) High humidity
    (b) High resistivity
    (c) Medium conductivity
    (d) High loss tangent
38. A *p*-type conductor possesses
    (a) electrical dipoles
    (b) conductivity almost that of a metal
    (c) conductivity due to movement of excess electrons
    (d) conductivity due to movement of excess positively charged holes
39. At absolute zero, an intrinsic semiconductor is
    (a) an insulator
    (b) conductor
    (c) a *p*-type semiconductor
    (d) an *n*-type semiconductor
40. Which is true about insulators?
    (a) They possess high conductivity
    (b) They possess low resistivity
    (c) They possess low conductivity
    (d) They permit the flow of sound or heat or electricity
41. The resistance of a semiconductor
    (a) increases with temperature
    (b) decreases with increase in temperature
    (c) becomes zero at 0°C
    (d) remains constant and does not vary with temperature
42. Which of the following thermal insulators is suitable above 180°C?
    (a) Cotton (b) Silk
    (c) Mica (d) Paper
43. A semiconductor should possess resistivity of
    (a) About 10^{-6} ohm cm
    (b) 10^{-6}–10^{3} ohm cm
    (c) 10^{13} ohm cm
    (d) 10^{6} ohm cm
44. Germanium doped with gallium is an example of
    (a) *n*-type semiconductor
    (b) *p*-type semiconductor
    (c) intrinsic semiconductor
    (d) *n-p-n* type semiconductor
45. A perfect insulator should possess
    (a) phase angle 0°C
    (b) phase angle of 45°C
    (c) phase angle of 90°C and loss angle zero
    (d) phase angle of 125°C
46. The limiting electrical field density of an insulator above which breakdown occurs is called
    (a) dielectric strength
    (b) capacitance
    (c) ageing strength
    (d) dielectric constant
47. Which of the following liquid insulating materials can be used between 90°C and 220°C?
    (a) Mineral oils
    (b) Vegetable oils
    (c) Ageing strength
    (d) Dielectric constant
48. An ideal thermal insulator
    (a) should possess low thermal conductivity
    (b) should absorb moisture from the surroundings
    (c) should be chemically stable to surrounding conditions
    (d) should be fireproof

49. Silicon doped with arsenic is an example of
    (a) *n*-type semiconductor
    (b) *p*-type semiconductor
    (c) *n-p* type semiconductor
    (d) *n-p-n* type semiconductor

50. A *p*-type semiconductor can be obtained by doping silicon or germanium with
    (a) As (b) B
    (c) Sb (d) P

Answers

1. (d)	2. (c)	3. (b)	4. (a)	5. (b)
6. (a)	7. (b)	8. (a)	9. (c)	10. (b)
11. (b)	12. (d)	13. (d)	14. (b)	15. (b)
16. (c)	17. (a)	18. (d)	19. (b)	20. (c)
21. (a)	22. (b)	23. (d)	24. (a)	25. (d)
26. (b)	27. (d)	28. (b)	29. (d)	30. (a)
31. (a)	32. (a)	33. (a)	34. (a)	35. (a)
36. (b)	37. (b)	38. (d)	39. (a)	40. (c)
41. (c)	42. (c)	43. (b)	44. (b)	45. (c)
46. (a)	47. (c)	48. (b)	49. (a)	50. (b)

Fill in the Blanks

1. Materials which are used to prevent the loss of electricity through certain parts in an electrical system are known as ....
2. A good insulator is that which has .... thermal ageing.
3. Dielectrics having little porosity are generally used because moisture holding ability................. with porosity.
4. Porosity of insulating material .... affect the electrical properties and gives .... surface resistivity and .... dielectric constant.
5. The surface tension of dielectrics can be .... by using wetting agents.
6. A good insulator possesses .... or high resistivity.
7. A good insulator possesses .... constant.
8. Dielectric strength decreases with an .... in temperature in case of air as insulator.
9. .... decreases as the frequency of applied electrical field is increased.
10. .... of an insulating material is badly affected by ageing or weathering.
11. Air is an important .... insulating material.
12. SF_6 possesses .... property than air and nitrogen.
13. The dielectric strength of .... is about 2.35 times greater than that of air.
14. Chlorinated hydrocarbons are mostly used as ....
15. .... are not suitable to be used as switch gear oils.

Answers

1. electrical insulating materials
2. least
3. increases
4. adversely, reduced, increased
5. improved
6. low conductivity
7. low
8. increase
9. dielectric strength
10. dielectric strength

11. gaseous
12. better cooling
13. SF_6
14. askarels
15. silicon fluids

Reading List

1. Indulkar, C S; Thiruvengadan *An Introduction to Electrical Engineering*. New Delhi: S.Chand & Company Limited
2. Solymer, L; Walsh, D *Electrical Properties of Materials*. 8ed, New York: Oxford University Press.
3. Dekker, A J (2000) *Electrical Engineering Materials*. New Delhi: PHI Learning Pvt Ltd.
4. Gupta, M L *Electronic Engineering Materials*. New Delhi: Dhanpat Rai & Sons.
5. Seth, S P *A course in Electrical Engineering Materials*. New Delhi: Dhanpat Rai & Sons.
6. Allison, John *Electrical Engineering Materials and Devices*. New Delhi: McGraw-Hill Education Pvt Ltd.
7. Raghvan, V *Material Science and Engineering*. New Delhi: PHI Learning Pvt Ltd.
8. Vijaya, M S, Rangarajan G *Material Science*. New Delhi: McGraw-Hill Education Pvt Ltd.

7

Junction Rectifier, Transistors and Devices

Chapter Objectives

- Classification of electronic devices, their preparation and the conduction mechanism
- Current–voltage relationship in forward and reverse bias *p-n* junction
- To understand breakdown mechanism in *p-n* junctions
- Applications of *p-n* junction as voltage regulator and rectifier
- Properties, characteristics and applications of tunnel diode and field effect transistor; various parameters like trans conductance, amplification factor, and ac and dc drain resistance
- Bi-polar junction transistor and its characteristics
- Use of transistor in various modes like common emitter, common base and common collector
- IC and its fabrication

7.1 INTRODUCTION TO SEMICONDUCTOR DEVICES

A device comes into existence when things are put together. In semiconductor electronics there is one such device which is simplest of its kind i.e. the *p-n* junction. This junction is obtained by taking a *p*-type semiconductor material and an *n*-type semiconductor and putting these in contact. The contact between these two materials is not simple, but it is done by some special technique so that the two materials are metallurgically joined. Conduction of charge carriers takes place on either side of the junction due to concentration gradient in space on either side of the junction. This will be discussed in the forthcoming section. The reason to discuss and to analyse such a simplest semiconductor device is to learn the basic operating mechanism and the related physics, so as to develop a background in understanding the other semiconductor devices. Almost all the devices and the conduction mechanisms taking place therein are similar—more or less—to *p-n* junction. So, it forms the fundamental basis of semiconductor devices.

Semiconductor devices are put in two categories:

- Electronic devices
- Photonic devices

The electronic devices are those in which the movement of charge carriers (electron or holes or both) plays an important role—either under the influence of electric field or due to concentration gradient. Such devices can also be classified into two categories:

- Bi-polar devices
- Uni-polar devices

In bi-polar devices, both electrons and holes participate in the conduction process and discharge their duties well. The ordinary *p-n* junction diode, the transistors (*p-n-p* or *n-p-n*) are simple examples of this kind.

Apart from the above, there are devices where only one kind of charge carrier, either electron or hole, plays an active role. Field-Effect Transistor (FET) is a device of this kind. Details about the FET will be discussed in the forthcoming section of this chapter.

The photonic devices are an important class of devices in which photon plays an important role. In some devices, the photon is emitted as a result of recombination process like in LEDs and in semiconductor lasers. Some devices detect the incidence of photons over them, which serve as photo detector. In some other devices a voltage is developed due to incidence of light photon like the solar cell or photocell. The above devices will also be discussed in detail in this chapter.

7.2 ELECTRONIC DEVICES

A detailed description about some semiconductor devices is given below:

- *p-n* junction
- Bi-polar junction transistor
- Uni-polar junction transistor

(i) p-n Junction

The *p-n* junction is of great importance in modern electronics. It is one of the simplest devices and the basic building block of modern electronics. Its theory serves as the foundation of the physics of semiconductor devices. The theory which deals with the current–voltage characteristics was developed by Shockley.

When *p*-type semiconductor materials are metallurgically joined with a *n*-type semiconductor, a *p-n* junction is formed. The technique of getting a *p-n* junction has been described in forth comming section. Simplified diagrams related with the junction are given below.

It can be noticed that during junction formation there is an abrupt change in concentration of charge carriers on either side of the junction. The *p*-type side is rich with holes and *n*-type side is rich with electrons. This leads to migration of charge carriers across the junction, which is known as diffusion. The electrons enter *p*-side from *n*-side, leaving immobilised positive ions there and holes on *p*-side pick up these electrons. The atoms on *p*-side (near to the junction) therefore become immobilised. Hence, a space-charge region is created near the junction on both the sides. This migration of charge carriers cannot continue indefinitely. The net positive and negative charges in *n*- and *p*-regions induce an electric field in the region near the metallurgical junction. It is in the direction from the positive to negative charge or from *n*- to *p*-region. Since the space charge region is depleted of any mobile charge, this region is also referred as the depletion region.

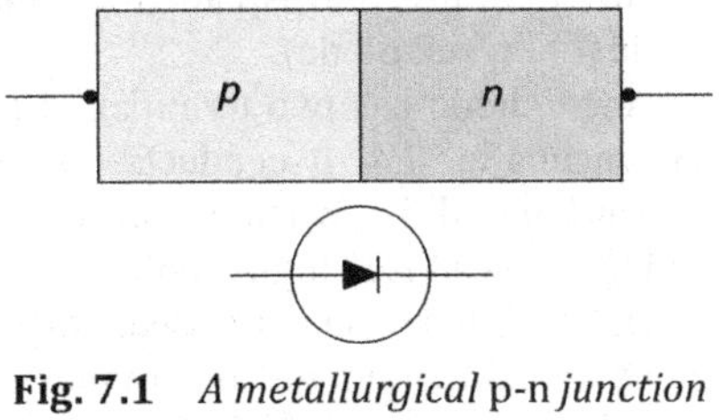

Fig. 7.1 *A metallurgical* p-n *junction and its symbol*

The thickness of depletion region is of the order of one tenth of a micrometer. Density gradient still exist in the majority carrier concentrations at each edge of the space charge region. One can think of a density gradient as producing a diffusion force that acts on the majority carriers. The same has been shown in Fig. 7.2. The electric field in the space-charge region produces another force on the electrons and holes, which is in the opposite direction to the diffusion force for each type of particle. In thermal equilibrium, the diffusion force and the electric field force exactly balance each other.

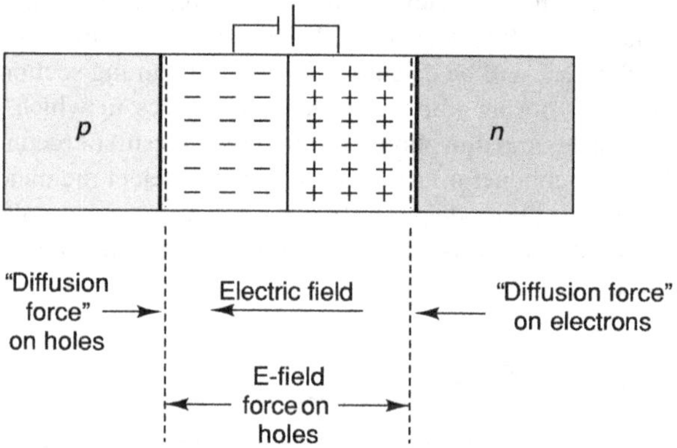

Fig. 7.2 *Formation of potential barrier in a p-n junction*

The diffusion process discussed above is a self-limiting process due to the fact that it disturbs electrical neutrality of both sides of the junction. The voltage developed across the junction is called potential barrier or junction barrier V_0. The potential barrier is of the order of 0.1 to 0.3 volt. The potential distribution diagram is shown in the Fig. 7.3. The electric field, so developed near the junction, prevents the further entry of charge carriers in either of the regions.

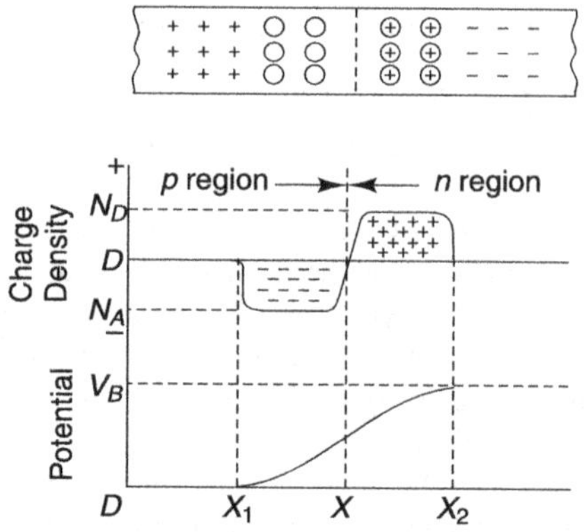

Fig. 7.3 *Potential distribution diagram in p-n junction*

(ii) Method of Growing a p-n Junction

When a *p*-type semiconductor metallurgically joined to an *n*-type semiconductor, the contact surface so formed is called *p-n* junction. A *p-n* junction that conducts current in only one direction is called a semiconductor diode (or crystal diode).

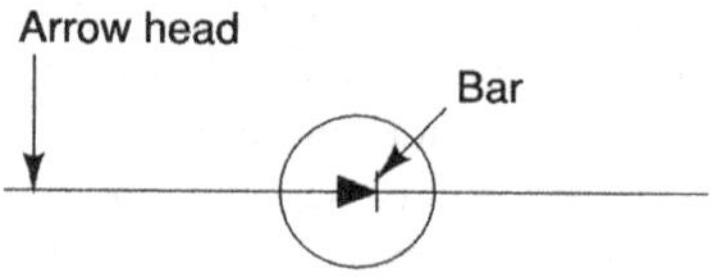

Fig. 7.4 *Symbol of p-n diode*

A *p-n* diode has two terminals. Its symbol is shown in Fig. 7.4. It conducts only when it is forward biased, i.e. when the terminal connected with arrowhead is at higher potential than the terminal connected to the bar. On the other hand, when it is reversed biased, practically it does conduct.

Fabrication of *p-n* junction require some special techniques, namely growing, alloying and diffusion.

The most common technique is alloying. An alloyed junction is made from an *n*-type slice of semiconductor (germanium or silicon) by melting a pellet of trivalent indium placed on the slice. This is done by heating the system to about 500°C. The indium is absorbed into germanium or silicon to produce a *p*-region and therefore a *p-n* junction is formed. The same is shown in Fig. 7.5.

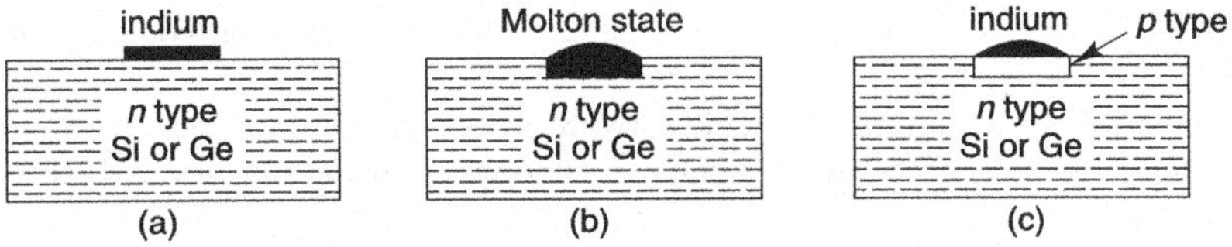

Fig. 7.5 *Growing a p-n junction: different stages*

In some cases, diffusion methods are also used to prepare a *p-n* junction. In this process, the semiconductor wafers of one conductivity (either *p*-type or *n*-type) are placed in a vessel that contains an oxide of impurity to be added. The combination is passed slowly through a furnace with precisely known and controlled temperature of 800°C to 1200°C depending upon the type of junction desired. At such a high temperature, a gas of impurity atoms diffuses into the semiconductor material and forms a thin layer of opposite conductivity; thus, resulting in the formation of *p-n* junction.

The quality of the semiconductor material is very important to the production of good diodes. Germanium may be grown in conventional vertical pulling furnace, or it may be zone-refined and doped in a horizontal furnace. Silicon is usually doped and grown in a crystal-pulling furnace. Since germanium and silicon both are group IV members, group III and V elements are used to dope the crystals. Most group III and V elements have been used, but arsenic and antimony for *n*-type, germanium and gallium for *p*-type, germanium and phosphorus for *n*-type, and silicon and boron or gallium for *p*-type silicon are most widely used.

There are at least two different types of diodes. One type is produced by alloying and the other by diffusion. Actually, combinations of these techniques are used to make diodes with several layers, such as the *p-i-n* or *p-n-p-n* diodes.

(iii) Biasing of a p-n Junction

When *p*-type material of *p-n* junction is connected to the positive terminal of the battery, the junction is said to be forward biased, and when it is connected with negative terminal it is said to be reverse biased.

When forward voltage exceeds the potential barrier under the forward-biased condition, the width of depletion region decreases. Under reverse-biased condition, the width of depletion region increases. The corresponding variation in the width of depletion layer is shown below in Fig. 7.6.

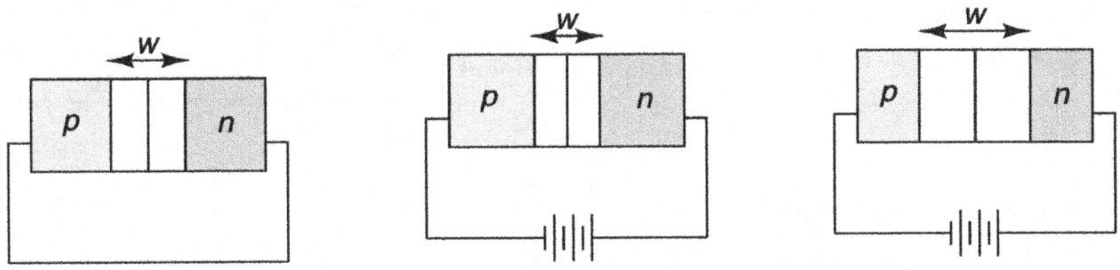

Fig. 7.6(a) *p-n junction with zero bias* **Fig. 7.6(b)** *p-n junction with forward bias* **Fig. 6.(c)** *p-n junction with reverse bias*

Thickness of the depletion region also depends upon impurity concentration. It is found inversely proportional to square root of impurity concentration.

$$\text{Width of depletion region } \alpha \frac{1}{\sqrt{\text{impurity concentration}}}$$

The variation of thickness of depletion region with external applied voltage across the junction has been already discussed above. In the absence of external voltage, thickness is typically of the order of 1 micron.

An increase in the current under the forward-biased condition is almost exponential for the most part and it follows the equation known as diode equation, which will be derived later.

$$I = I_s(e^{eV/kT} - 1) \tag{7.1}$$

where I_s is equal to saturation current under reverse bias condition. The term kT/e in the equation has the dimensions of voltage and is known as the voltage equivalent of temperature, denoted by V_T.

$$I = I_s\{e^{V/V_T} - 1\} \tag{7.2}$$

At room temperature, V_T comes nearly 26 mV. As shown in the characteristic curve given in Fig. 7.8, the diode current is negligibly small until the voltage exceeds a critical value known as the cut-in voltage or the threshold voltage. The cut-in voltage is about 0.2 volts for Ge diodes and about 0.6 volts for Si diodes. Another important phenomenon occurs in the forward-biased condition. As the movement of charge carriers is resumed, the depletion layer narrows down and finally vanishes as the bias exceeds the cut-in voltage.

The variation in the current under forward- and reversed-biased junction is shown in Fig. 7.7.

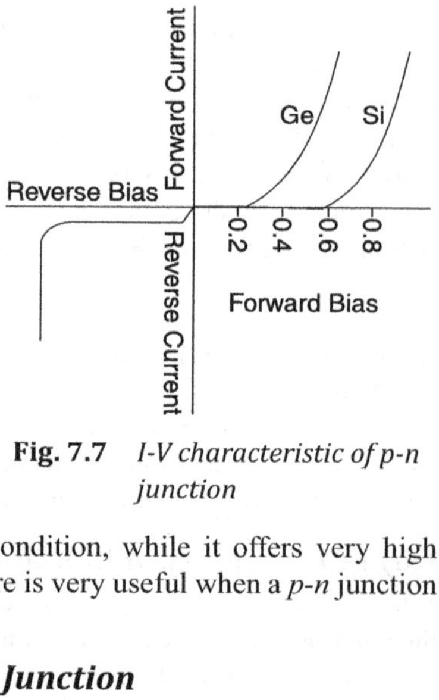

Fig. 7.7 *I-V characteristic of p-n junction*

It is clear from the figure that *p-n* junction offers negligible resistance under forward-biased condition, while it offers very high resistance under reverse biased condition. This feature is very useful when a *p-n* junction is used to convert ac signal into dc signal.

(iv) Current-voltage Relationship for p-n *Junction*

Figure 7.8 shows a *p-n* junction under reverse-biased condition.

In a reverse-biased *p-n* junction, the current can only be due to intrinsic charges appearing in the barrier region by thermal generation. The expression for the electron densities in *n*- and *p*-region as a function of temperature is given by

$$n = n_i e^{-eV_B/KT} \tag{7.3}$$

Similarly,

$$p = p_i e^{-eV_B/KT} \tag{7.4}$$

where n = number of donor electrons

n_i = number of intrinsic electrons

T = junction temperature

V_B = barrier potential

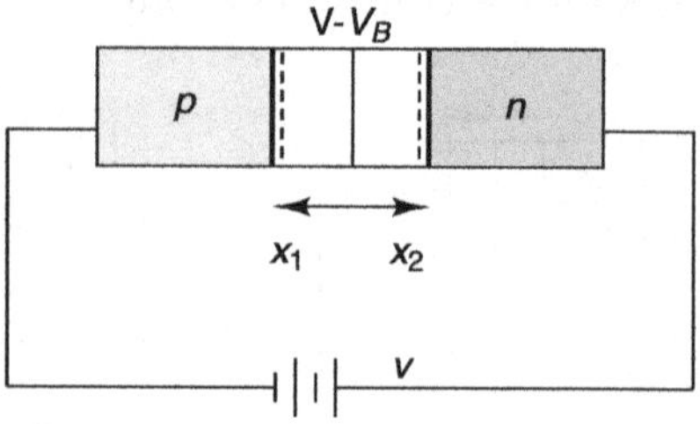

Fig. 7.8 *A p-n junction under reverse-biased condition*

Intrinsic reverse current

$$I_s = A[n_i \cdot e \cdot v_n + p_i e v_p]e^{-eV_B/kT}$$

In semiconductors, drift and diffusion currents are present simultaneously. Therefore, total current at *p-n* junction at low potential is the sum of both components. Therefore,

$$J_n = ne\mu_n E + eD_n \frac{dn}{dx} \quad (7.5)$$

$$J_h = ep\mu_h E - e.D_h \frac{dp}{dx} \quad (7.6)$$

$$-Edx = -\frac{J_n dx}{n_e\mu_n} + \frac{e.D_n}{n_e\mu_n} dn \quad (7.7)$$

$$-Edx = \frac{-J_n dx}{n_e\mu_n} + \frac{kT}{e} \cdot \frac{dn}{n} \quad (7.8)$$

$$Edx = \frac{J_h dx}{pe\mu_p} + \frac{kT}{e}\frac{dp}{p} \quad (7.9)$$

Therefore, junction potential V_0 will be

$$V_0 = V - V_B = -\int_{x_1}^{x_2} Edx = -\frac{1}{en\mu_n} + \int_{x_1}^{x_2} J_n dx + \frac{kT}{e}\int_{n}^{ni} \frac{dn}{n} \quad (7.10)$$

The first term is the conductivity term which create ohmic resistance drop being small it can be neglected.

The remaining term gives the relation between the electron densities and potential across the depletion layer in forward direction.

$$V - V_B = \frac{kT}{e}\log\frac{n}{n_i}$$

$$n = n_i e^{(V-V_B)e/kT} \quad (7.11)$$

$$p = p_i e^{(V-V_B)e/kT} \quad (7.12)$$

Therefore,

$$I_n = A.\ n_i\, e\, v_n = An_i\, ev_h\ e^{(V-V_B)e/kT}$$
$$I_p = A\, p_i\, e\, v_h$$

Total junction current

$$I = I_n + I_p - I_s = A(n_i ev_n + p_i ev_p)e^{-eV_B/kT}$$

Therefore, $$I = I_s\,[e^{eV/kT-1}] \quad (7.13)$$

The above equation is the ideal diode equation and gives a good description of the current–voltage characteristics of the *p-n* junction over a wide range of currents and voltages. If voltage V becomes negative, (i.e. under reverse bias), by few $\frac{kT}{eV}$, then the reverse-biased current becomes independent of reverse-biased voltage. The parameter I_s is then known as **reverse saturation current.**

The reverse saturation current is small due to low rate of pair generation at ambient temperatures, but I_s is strongly temperature dependent. It is usually considered as doubling its value for every 100°C rise of junction temperature. The increase in I_s place a limit of about 1000°C on germanium devices and 2000°C for silicon. The reverse current is independent of voltage.

The reverse saturation current is also written as

$$I = I_s\left[\frac{eV}{e^{nkT}} - 1\right] = I_s\left[\frac{V}{e^{nkT}} - 1\right] \tag{7.14}$$

when the voltage V is positive and greater than V_T then diode equation becomes

$$\boxed{I = I_s\, e^{V/nkT}} \tag{7.15}$$

Equation 7.15 indicates that except for a small range in the neighbourhood of the origin, the current increases exponentially with voltage, as discussed earlier.

When the diode is reverse-biased and V is several times V_T, $I \approx -Is$. The reverse current is therefore constant independent of the applied reverse-biased voltage. That is why Is is known as reverse saturation current.

Example 7.1

Derive an expression for the ratio of diode currents resulting from two different forward voltage drops if $V >> n\, V_T$.

Solution If $V >> n\, V_T$ then diode equation becomes

$$I = I_s\, e^{\frac{V}{nV_T}}$$

$$\frac{I_2}{I_1} = \frac{I_s e^{\frac{V_2}{nV_T}}}{I_s e^{\frac{V_1}{nV_T}}} = e^{\frac{(V_2 - V_1)}{nV_T}}$$ **Ans.**

Example 7.2

Find the increase in temperature ΔT necessary to increase Is by a factor of 150.

Solution As the reverse saturation current I_0 of a Si diode and Ge diode approximately doubles for every 100°C rise in temperature. Therefore,

$$\frac{I_{s2}}{I_{s1}} = 150 = 2^{\frac{(T_2 - T_1)}{10}} = 2^{\frac{\Delta T}{10}}$$

or $$\Delta T = \frac{10\, In\,(150)}{In\,(2)} = \frac{10 \times 5.01}{0.6931} = 72.20°\text{C}$$ **Ans.**

Example 7.3

The current flowing through a certain germanium p-n junction at room temperature when reverse-biased (bias voltage being very large in comparison to V_T) is 0.25 μA. Determine the current flowing through the germanium diode when the applied voltage is 0.12 V.

Solution The diode current is given as $I = I_s^{(e^{V/V_T} - 1)}$

From the given data, large reverse-biased current, $I \approx I_0 = 0.25 \times 10^{-6}$ A

Applied voltage, $V = 0.12$ V

Volt-equivalent of temperature, $V_T = 26$ mV or 0.026 V at room temperature

So, $$I = 0.25 \times 10^{-6}\,(e^{0.12/0.026} - 1) = 24.8\ \mu\text{A}$$ **Ans.**

7.3 EINSTEIN'S RELATION

There exists an important relationship between the diffusion coefficient and the mobility of charge carrier. This is known as Einstein's relation. This states that the ratio of diffusion constant to mobility of charge carrier is constant and is equal to, $\frac{kT}{e}$, i.e.

$$\frac{D_n}{\mu_n} = \frac{D_h}{\mu_h} = \frac{kT}{e} \tag{7.16}$$

This relationship may be derived by considering a situation in a semiconductor in which there exist an electric field E and concentration gradient such that the resultant current is zero. Naturally, this will be a condition when no current will be flowing and system will be in thermal equilibrium.

$$E(x) = -\frac{dV}{dx} \tag{7.17}$$

In thermal equilibrium, the density of holes as a function of x is given by Maxwell–Boltzmann statistics

$$n_n(x) = n_o e^{-\frac{eV}{kT}} \tag{7.18}$$

where n_o is a constant. Hence, there will be a gradient of hole density given by

$$\frac{dn_n}{dx} = -\frac{e}{kT} n_n \frac{dV}{dx} = \frac{e}{kT} n_n E_x \tag{7.19}$$

In equilibrium

$$0 = n_n e\mu_n E - e.D_n.\frac{e}{kT} n_n E$$

$$= n_n e\mu_n E - \frac{e^2}{kT} D_n n_n E$$

$$D_n = \left(\frac{kT}{e}\right)\mu_n \tag{7.20}$$

Hence, diffusion constant, $D_n \alpha \mu_h$

Similarly, $$D_h = \left(\frac{kT}{e}\right)\mu_h \tag{7.21}$$

For Ge $\quad D_n = 0.0044 \text{ m}^2 \text{ sec}^{-2}$
at RT $\quad D_h = 0.0093 \text{ m}^2 \text{ sec}^{-2}$

7.4 *p-n* JUNCTION DIODE AS A RECTIFIER

p-n junction diode exhibits a magnificent property. It offers a negligible resistance during forward bias, while a very high resistance under reverse-biased condition. This property is utilised to get the dc signal from ac signal. When the diode is used in this way, it is known as rectifier. For the first half cycle of alternating voltage signal, the diode conducts and for the remaining half it does not. So, a pulsating dc is obtained. To get a continuous dc, one has to use two diodes, so that one conducts for first half cycle and the other for second half cycle. The rectification using single diode, therefore, is known as half-wave rectification and with two diodes in circuit is known as full-wave rectification. The corresponding circuit diagrams and waveforms are given in Figs 7.9 and 7.10.

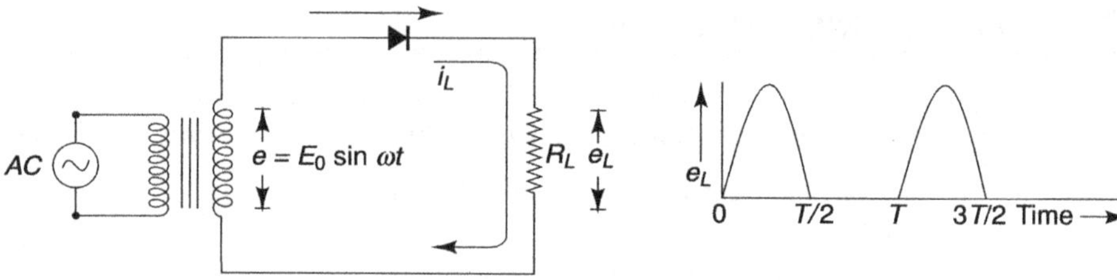

Fig. 7.9 *A half-wave rectifier circuit waveforms for voltage*

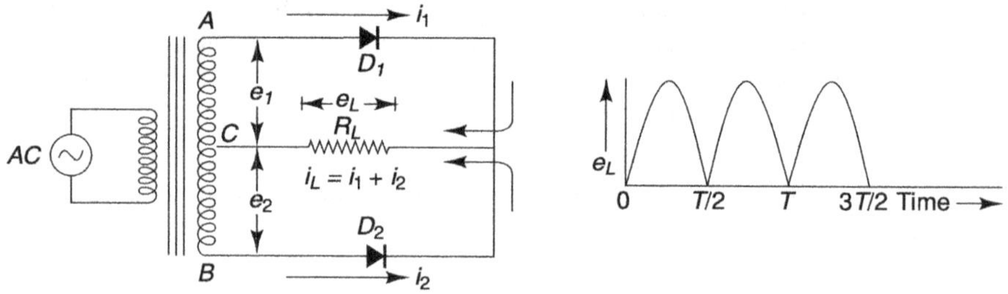

Fig. 7.10 *A full-wave rectifier circuit and waveform for voltage*

7.5 JUNCTION CAPACITANCE

Under reverse bias condition, the transition region of *p-n* junction is depleted of mobile charge carriers and behaves like a dielectric, with only fixed charges present. The normal *n*- and *p*-conducting-regions, outside the transition, serve as electrodes of the capacitor.

On increasing the reverse voltage from zero, more mobile charge is removed and the width of depletion region increases. Therefore, with increased dielectric thickness, the capacitance decreases. The impurity density variation across a junction is designed to maximise the change in depletion thickness per volt applied. The dynamic transition capacitance with reverse voltage is given by

$$C_T = \frac{\partial Q}{\partial (V_B + V)} = C_0 (V_B + V)^{-n} \text{ farad/m}^2 \tag{7.22}$$

The capacitance C_0 is measured at zero-applied voltage. The exponent will vary from $\frac{1}{3}$ to $\frac{1}{2}$ depending upon the diode construction.

Such reverse-biased diodes are known as **Voltage-Variable Capacitance abbreviated as VVC.**

7.6 BREAKDOWN IN *p-n* JUNCTION

If the polarity of the external bias is reversed, it has two effects. First, the reverse bias pulls the majority carriers further away from the junction so that the depletion layer widens. The other effect of the reverse bias is that it effectively raises the potential barrier at the junction. Therefore, the current through the diode is expected to be zero for all reverse bias voltages. However, the reverse bias has the right polarity for the movement of the minority charge carriers, which will carry a small but non-zero reverse current through the diode. As the reverse bias is increased, the reverse current increases and quickly saturates because the minority carriers are in short supply. This is shown in Fig. 7.11. It can be verified from the equation, as well. For negative V with $|V| > V_T$, the exponential term will be small compared to 1 and can be ignored so that $I = -I_s$. The reverse saturation current I_s is of the order of microamperes for Ge diodes and nanoamperes for Si diodes. The reverse current is constant at the value I_s until the reverse bias is so large that the junction undergoes an electrical breakdown. It is characterised by sharp increase of reverse current to very large values. The breakdown mechanism is of two types:

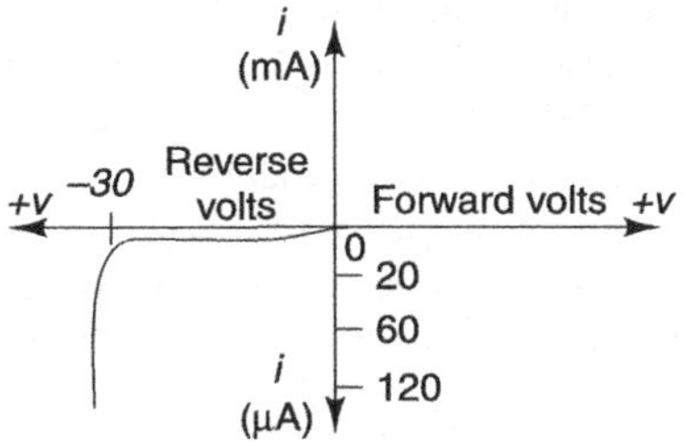

Fig. 7.11 *Breakdown in a diode (IN5545) under reverse bias*

- Zener breakdown
- Avalanche breakdown

7.7 ZENER BREAKDOWN

This breakdown occurs in highly doped *p-n* junctions through tunnelling mechanism. In a highly doped junction under reverse bias, the valence band and conduction band on opposite sides of the junction are sufficiently close, and therefore, the electrons may tunnel directly from *p* region into the conduction band on *n* side. This is shown in Fig. 7.12.

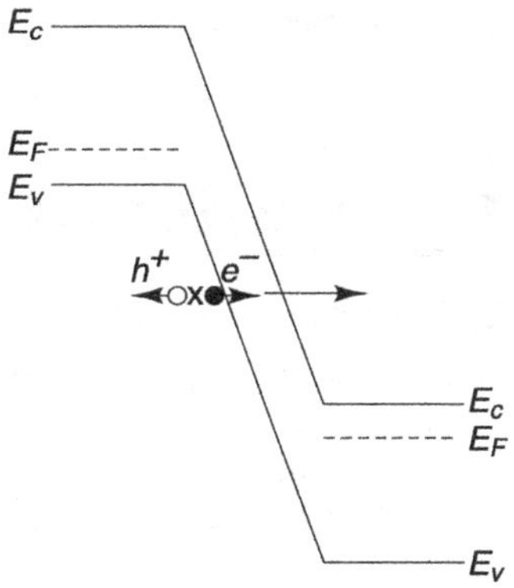

Fig. 7.12 *Zener breakdown in a p-n junction*

This breakdown is obtained at very low reverse bias voltages (of the order of about 5 volts). This happens in diodes with very thin depletion layers—about 100–200 A in thickness. The electric field across the depletion layer is approximately equal to the bias voltage divided by the thickness of the layer. Thus, a thin layer results in strong electric fields of the order of 10^6 volts/cm even for low-bias voltages. This field is so strong that it is able to pull out more electrons from the fixed ions in the depletion region by the process of field emission, as discussed. A very large number of new carriers are thus created that carry a large reverse current through the diode. This is known as the Zener effect. Since, the thickness of the depletion layer depends on the impurity concentration in the semiconductor. The larger this concentration, the thinner is the layer. Thus, zener effect is observed only in heavily doped diodes. In others, the mechanism of breakdown is the Avalanche effect. Both mechanisms are reversible in the sense that if the voltage is reduced, the excess carriers recombine, the breakdown vanishes and the diode comes back to its original condition.

7.8 AVALANCHE BREAKDOWN

In the reverse-biased condition, the velocity of minority carriers increases as the bias voltage is increased. If this is sufficiently large (of the order of 100 volts for a typical *p-n* junction), the minority carriers acquire high velocities and are able to further ionise the fixed ions of the crystal with which they collide. This generates new electron-hole pairs. They also move in opposite directions and acquire sufficient velocities to ionise other ions fixed in the crystal. The process is thus regenerative and very soon, a large number of new carriers are generated which carry a large current through the junction. This is known as avalanche breakdown of the junction. This current will increase excessively and might damage the diode by the excessive heat produced unless it is limited by the external circuit to safe values. Figure 7.13 shows the avalanche breakdown.

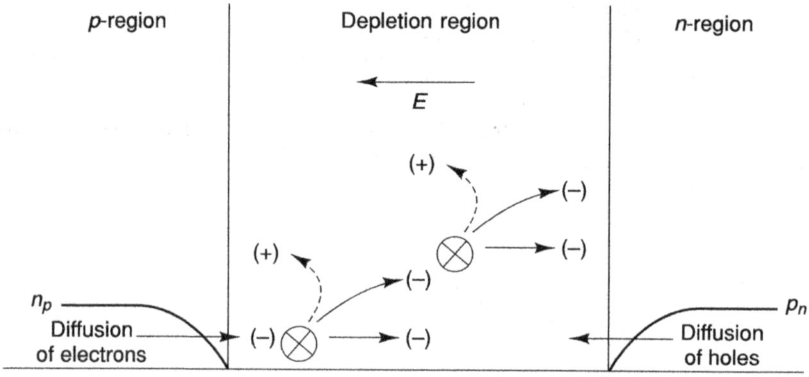

Fig. 7.13 *Avalanche breakdown in a p-n junction*

7.9 SPECIAL PURPOSE DIODES

(i) Zener Diode

A zener diode is a *p-n* diode operated under reverse bias large enough to cause breakdown of the junction. Its circuit symbol is slightly different from that of an ordinary *pn* diode as shown in Fig. 7.14. The voltage V_z, at which breakdown occurs, is known as the zener voltage. By varying the impurity concentration, zener diodes having zener voltage from a few volts to a few hundred volts are commercially made. Zener effect is the only mechanism of breakdown in the diodes having low zener voltage; in high V_z diodes it is avalanche effect. Yet, the name zener diode is conventionally used for all diodes having a sharp breakdown.

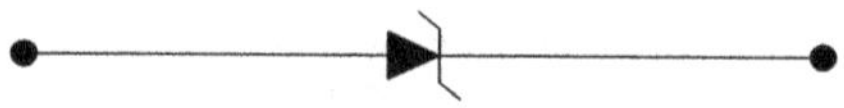

Fig. 7.14 *Circuit symbol of a zener diode*

The characteristic curve of a zener diode is shown in the Fig. 7.15. It is seen that once the breakdown occurs, the diode maintains a nearly constant voltage across it while the current through it changes over a large range. This property can be exploited to design a voltage regulator circuit for maintaining the constant output voltage in a power supply. An important precaution in using a zener diode in any circuit is to ensure that the zener current I_z is limited by the external circuit, so that the power dissipation I_zV_z in the zener is within the maximum rating specified by the manufacturer.

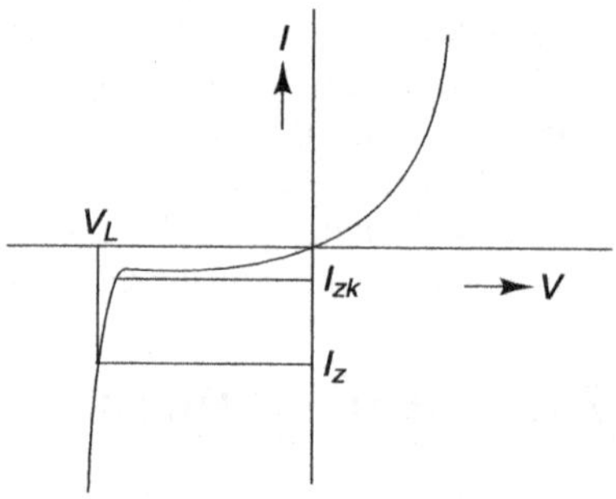

Fig. 7.15 *Characteristics of a zener diode*

(ii) Zener Diode as a Voltage Regulator

The circuit of a voltage regulator using a zener diode is given in Fig. 7.16.

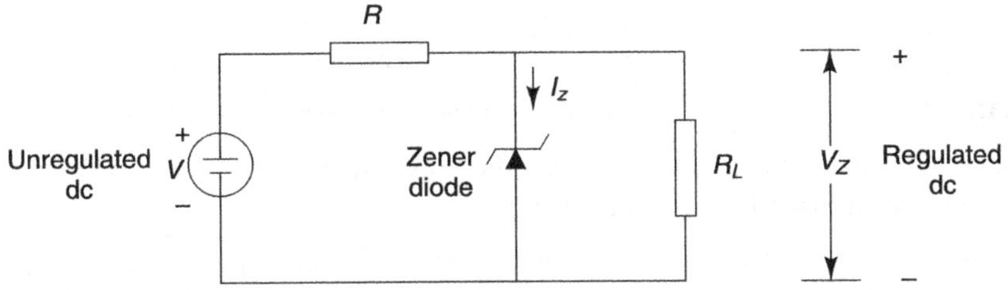

Fig. 7.16 *Zener diode as a voltage regulator*

If we study the *I-V* characteristic of zener diode, it is found that the reverse current increase abruptly near zener voltage (break voltage down). In other words, one can expect approximately no variation in the output voltage against abnormal variation in the reverse current, if this diode is used in the breakdown region. This simple principle is used in a voltage regulator.

The source voltage and resistor *R* are selected so that—initially—the diode is operating in the breakdown region. The diode will now regulate the load voltage against variations in load current and against variations in supply voltage *v*, because in the breakdown region large changes in diode current produce only small changes in diode voltage. Moreover as load current or supply voltage changes, the diode current will accommodate itself to the changes and maintain a nearly constant load voltage.

Example 7.4

Find the diffusion coefficients of electrons and holes of a silicon single crystal at 300 K if the mobilities of electrons and holes are 0.19 and 0.027 m^2V^{-1} s^{-1}, respectively at 3000 K.

Solution Diffusion coefficient is given by

$$D = \left(\frac{kT}{e}\right)\mu$$

Given

$$k = 1.38 \times 10^{-23} \text{ J/k}$$
$$e = 1.6 \times 10^{-19} \text{ C}$$
$$\mu_e = 0.19 \text{ m}^2\text{V}^{-1}\text{ s}^{-1}$$
$$\mu_h = 0.027 \text{ m}^2\text{V}^{-1}\text{s}^{-1}$$

For electron,

$$D_n = \left(\frac{kT}{e}\right)\mu_n$$

$$= \frac{1.38 \times 10^{-23} \times 300 \times 0.19}{1.6 \times 10^{-19}}$$

$$= 49.1 \times 10^{-6} \text{ m}^2\text{s}^{-1}$$

For hole,

$$D_h = \left(\frac{kT}{e}\right)\mu_h$$

$$= \frac{1.38 \times 10^{-23} \times 300 \times 0.027}{1.6 \times 10^{-19}}$$

$$= 6.98 \times 10^{-4}\ \text{m}^2\ \text{s}^{-1}$$ Ans.

Example 7.5

Can we take one slab of *p*-type semiconductor and physically join it to another *n*-type semiconductor to get a *p-n* junction?

Solution No! Any slab, howsoever flat, will have roughness much larger than the inter-atomic crystal spacing (~ 2 to 3Å) and hence, continuous contact at the atomic level will not be possible. The junction between *p*-type and *n*-type material should not be an ordinary junction, but it should be a metallurgical junction. The junction made by physically joining the materials will behave as a discontinuous connection for the flowing of charge carriers.

Example 7.6

The *V-I* characteristic of a silicon diode is shown in Fig. 7.17. Calculate the resistance of the diode at (a) $I_D = 15$ mA, and (b) $V_D = -10$ V.

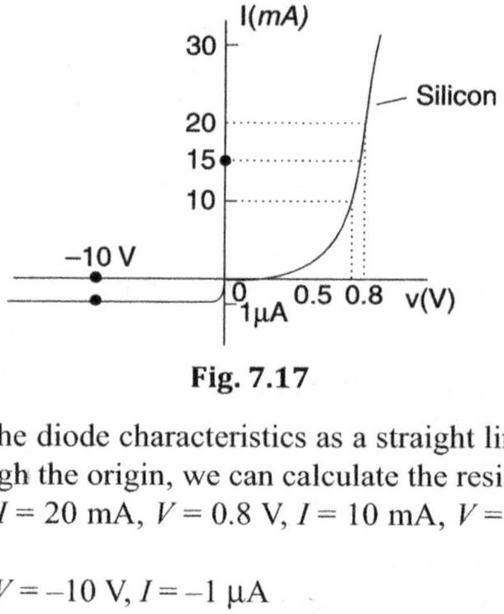

Fig. 7.17

Solution Considering the diode characteristics as a straight line between $I = 10$ mA to $I = 20$ mA, passing through the origin, we can calculate the resistance using Ohm's law.

(a) From the curve at $I = 20$ mA, $V = 0.8$ V, $I = 10$ mA, $V = 0.7$ V, $r_{fb} = \Delta V/\Delta I = 0.1$ V/10 mA = 10 Ω

(b) From the curve at $V = -10$ V, $I = -1$ μA
Therefore,

$$V_{reb} = 10\ \text{V}/1\ \mu\text{A} = 1.0 \times 10^7\ \Omega$$ **Ans.**

Example 7.7

Find the diffusion coefficients of holes and electrons for germanium at 300 K. The carrier mobilities in cm² / V–s at 300 K for electrons and holes are 3,650 and 1,720, respectively. Density of carriers is 2.5×10^{13}/cm². Boltzmann constant $k' = 1.38 \times 10^{-23}$ J degree^{-1} and $e = 1.602 \times 10^{-19}$ C.

Solution We know that

diffusion constant, $$D_n = \mu_n \frac{k'T}{e} = \frac{3650 \times 1.38 \times 10^{-23} \times 300}{1.602 \times 10^{-19}} = 94.4\ \text{cm}^2/\text{s}$$ **Ans.**

Diffusion constant, $D_h = \mu_h \dfrac{k'T}{e} = \dfrac{1720 \times 1.38 \times 10^{-23} \times 300}{1.602 \times 10^{-19}} = 44.5 \text{ cm}^2/\text{s}$ **Ans.**

(ii) Tunnel Diode

The tunnel diode is a non-degenerate *p-n* junction diode in which the concentration of impurities is very high, in both *n*- and *p*-region–about 1 part in 10^3. In normal diodes, the concentration is about 1 part in 10^8. The *I-V* characteristics of these diodes are quite different from that of ordinary diodes.

Principle As the width of the depletion region, decreases with increasing impurity concentration. The width depletion region in a tunnel diode is of the order of 100Å. If the width of the barrier is very small and if occupied energy states exist on one side of the junction and allowed empty states exist on the other side of the junction at the same energy levels, the electrons penetrate or tunnel through the barrier. In a diode which is heavily doped, these conditions exist and so the tunnelling mechanism determines the *I-V* characteristic.

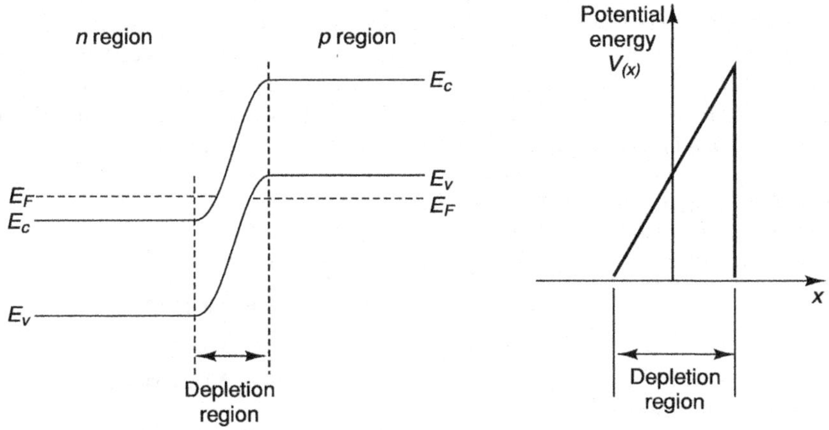

Fig. 7.18 *Energy band diagram of tunnel diode*

Due to tunnelling effect, the tunnel diode has a very distinct I-V characteristic under forward bias. The characteristics are shown Fig. 7.19.

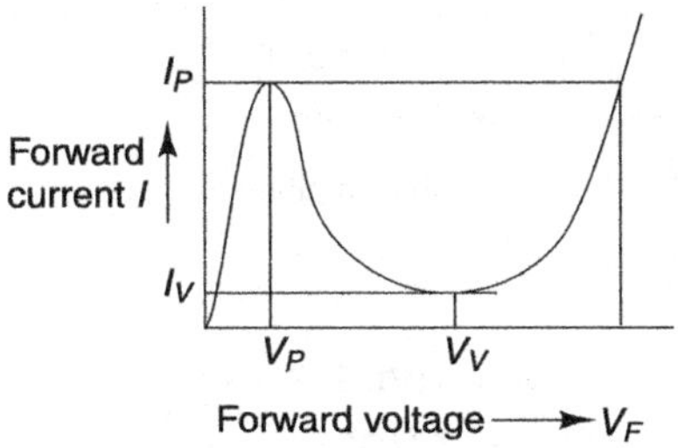

Fig. 7.19 *I-V characteristic of a tunnel diode*

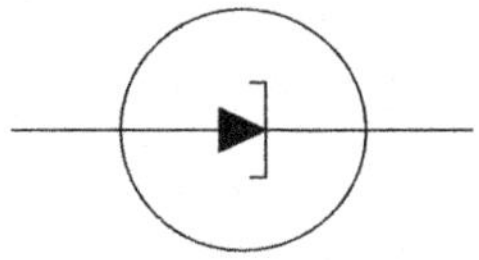

Fig. 7.20 *A symbol of tunnel diode*

For a small forward voltage, the resistance remains very small. At the peak current *Ip*, corresponding to the voltage V_p, the slope of the *I-V* curve becomes zero. If *V* is increased beyond V_p, the current starts decreasing! It decreases up to a certain voltage V_V, called the valley voltage, at which $I = I_V$.

Between the current I_p and I_V, the diode has a negative resistance (i.e. current decreases with increasing voltage). This characteristic of tunnel diode was earlier used in oscillator circuits. Beyond the valley voltage V_V, resistance becomes positive and the current increases with increasing voltage. It can be seen from the curve that a current value lying between *Ip* and I_V can be obtained at three different applied voltages. It is this multi-valued feature that makes tunnel diode a special device useful in pulse and digital circuits. The diode is also used in high frequency (microwave) oscillators.

Tunnel diodes are normally made from germanium and gallium arsenide. The switching times in tunnel diodes are of the order of nano seconds. The suitability of a tunnel diode for applications is measured by the ratio I_p/I_V. This ratio is highest (= 15) for GaAs and lowest (= 3.5) for Si.

7.10 FIELD EFFECT TRANSISTOR

The junction field effect transistor was first of all analysed by Shockley in 1952. It is basically a voltage control resistor. It is a uni-polar device, as it utilises only one type of charge carrier either or holes electrons. The conduction takes place in a channel of semiconductor material. Current through this channel can be controlled by applying electric field. Therefore, it is a voltage control device. This electric field is applied across either reverse band *p-n* junction or a thin film layer of an insulating material. Depending upon this it is classified in two categories:

- JFET—Junction Field Effect Transistor
- IGFET—Insulated Gate Field Effect Transistor

A family tree of field effect transistor is shown in Fig. 7.21.

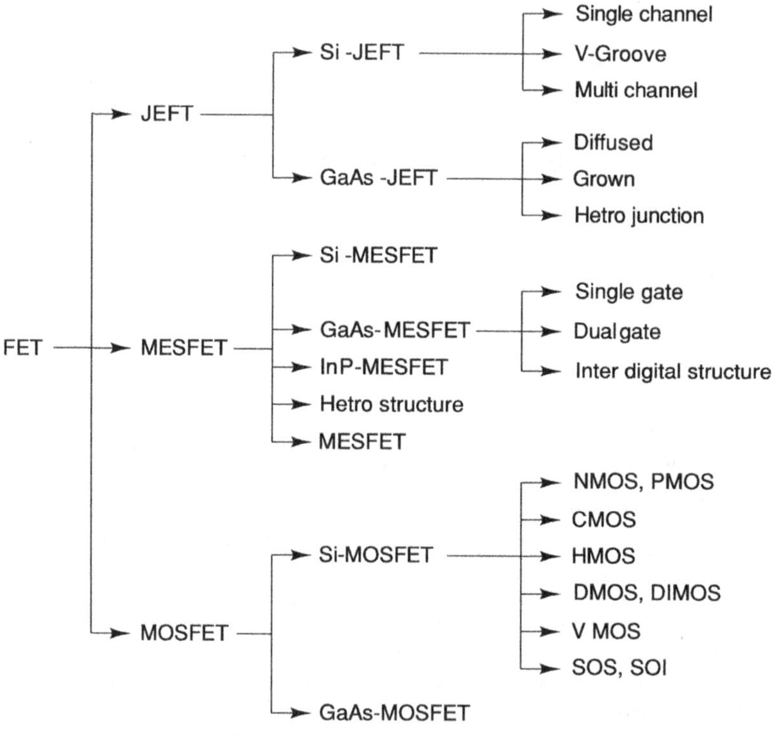

Fig. 7.21 *Family tree of FET*

7.11 FEATURES OF FET

FET exhibits many attractive features for applications in analog switching, high-input impedance amplifiers, microwave amplifications and integrated circuits. The FETs have considerably high-input impedance than BJTs, which allows the input of FET to be more readily matched to the standard microwave systems. The FET has a negative temperature co-efficient at high-current levels, that is, the original current decreases with increase in temperature. This characteristic leads to a more uniform temperature distribution over the device area and prevents FETs from thermal runaway or second breakdown which occurs in BJTs. The device is thermally stable, even when the active area is large or when many devices are connected in parallel which is an additional advantage with this. Since FETs are uni-polar devices, they do not suffer from minority carrier storage effects and consequently, have higher switching speeds and higher cut-off frequencies.

Figure 7.22 illustrates a junction field effect transistor. A small bar of *n*-type semiconducting material has a negative terminal known as source *S*. Heavily doped gate electrodes, *G* of *p*-type material from the junction on either side of the bar.

The thin region in JEFT consists of a conductive channel provided between two ohmic contacts—one acting as source and the other as the drain. When a positive voltage is applied to the drain with respect to the source, electrons flow from source to drain. Hence, the source acts as origin of charge carriers and the drain as sink. The third electrode, the gate, forms a rectifying junction with the channel. As the device is basically a voltage-controlled resistor, its resistance can be varied by varying the width of layer extending into the channel region. The device could be made of *p*-type material as well, in which the conduction takes place due to holes. Therefore, FETs can be classified as:

- *n*-channel FETs
- *p*-channel FETs

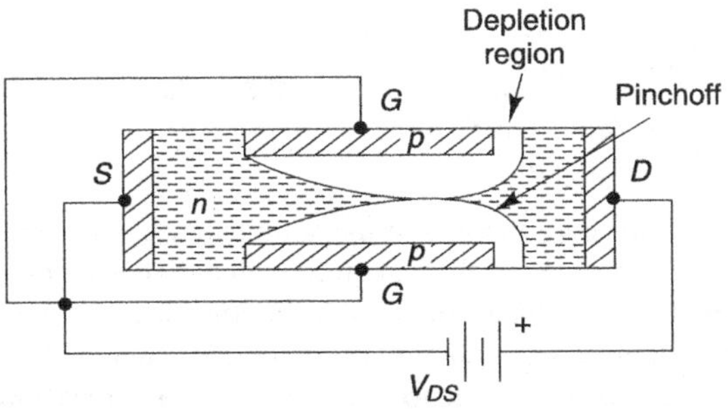

Fig. 7.22 *JFET in pinch off condition*

7.12 BASIC DEVICE CHARACTERISTICS

(i) Drain or Output Characteristics

When $V_{GS} = 0$, then $V_{DS} = 0$ and $I_D = 0$

As V_{DS} increases, I_D also increases. The channel than acts like a simple semiconductor.

When V_{DS} is applied, electrons move in the channel. There is a voltage drop along the bar and with the gates at source voltage, the channel will be positive to the *p*-gates. The junctions at the gate channel interfaces are reverse biased and the depletion regions form in the lightly doped channel.

Since the channel material has non-zero resistance, the potential gradually rises as we move for source to drain along the length of the channel.

Because of this distributed drop along the channel, the reverse voltage at the right end of the gates is more compared to left, and the depletion layer will be thicker at the right end. With no mobile charges in the depletion zone, the channel current will be to the wedge-shaped portion of the channel.

At low V_{DS}, the depletion regions are thin and the current increases as shown in (1) in Fig. 7.23. As V_{DS} increases, the cross-section of the channel is reduced and we reach to (2) on the graph. The channel nears closure or pinch-off and for higher V_{DS} the pinched-off region lengthens as shown by the flat portion on the curve. The pinch-off voltage, therefore, can be defined as the drain source voltage (V_{DS}) above which the drain current becomes almost constant. It is denoted by V_p and numerically it is given by (for *n*-channel FET).

$$V_p = \frac{q\alpha^2 N_D}{2 \in} \tag{7.23}$$

where $\alpha \rightarrow$ Channel width

$q \rightarrow$ Charge of the carrier

$N_D \rightarrow$ Concentration of donor

$\in \rightarrow$ Permittivity of channel material

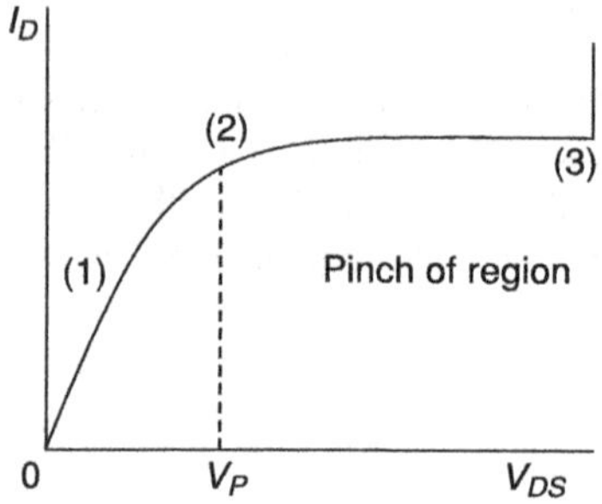

Fig. 7.23 *Variation of current with drain source voltage (V_{GS} = 0 or shorted gate)*

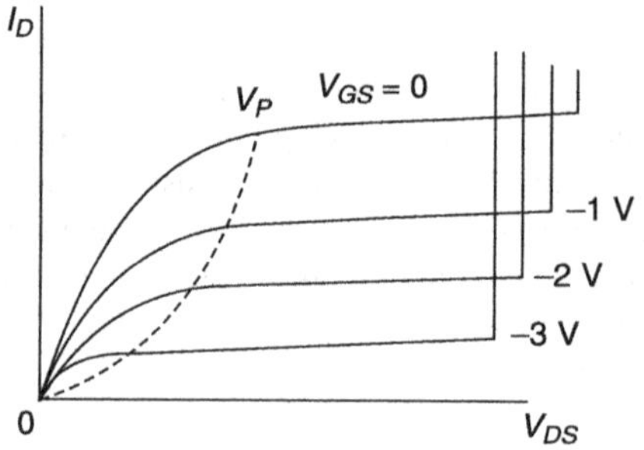

Fig. 7.24 **I_D vs. V_{DS} with different V_{GS}**

Placing the gates at negative voltage to the source drives the depletion regions more deeply into the channel. Above V_p, the pinch off voltage and the current is sensitive to V_{GS}. The characteristics are suitable for use as a control device. The operation is in the depletion mode with negative V_{GS} reducing the channel current.

The abrupt rise in the current is due to avalanche voltage breakdown in the depletion region, between gate and drain.

7.13 SIGNIFICANCE OF THE PINCH-OFF REGION

The region of the characteristic curve in which drain current I_D remains fairly constant is called the pinch-off region. It is sometimes also called the saturation region or amplifier region. In this region the JFET operates as a constant current device, since drain current (or output current) remains almost constant. It is the normal operating region of the JFET when used as an amplifier. The drain current in the pinch-off region with $V_{GS} = 0$ is referred to the drain-source saturation current, I_{DSS}.

It is to be noted that in the pinch-off (or saturation) region the channel resistance increases in proportion to increase in V_{DS}. So, it keeps the drain current almost constant and the reverse bias required by the gate-channel junction is supplied entirely by the voltage drop across the channel resistance, which is due to the flow of I_{DSS} and not by the external bias because $V_{GS} = 0$.

Drain current in the pinch-off region is given by Shockley's equation

$$I_D = I_{DSS}\left(1 - \frac{V_{GS}}{V_P}\right)^2 = I_{DSS}\left(1 - \frac{V_{GS}}{V_{GS(OFF)}}\right)^2 \tag{7.24}$$

where I_D is the drain current at a given gate-source voltage V_{GS}, I_{DSS} is the drain-current with gate shorted to source and $V_{GS}(O_{FF})$ is the gate-source cut-off voltage.

Therefore, it is very interesting to observe that JFET behaves as an ordinary resistor in the ohmic region, as a constant current source in pinch-off (or saturation) region and as a constant voltage source in breakdown region.

7.14 FET CONFIGURATION

FETs (JFETs and MOSFETs) have three configurations just as like BJTs. The three configurations of FETs—the common gate, common source, and common drain configurations—are given in Fig. 7.25. Input and output terminals of different configurations are given in Table 7.1.

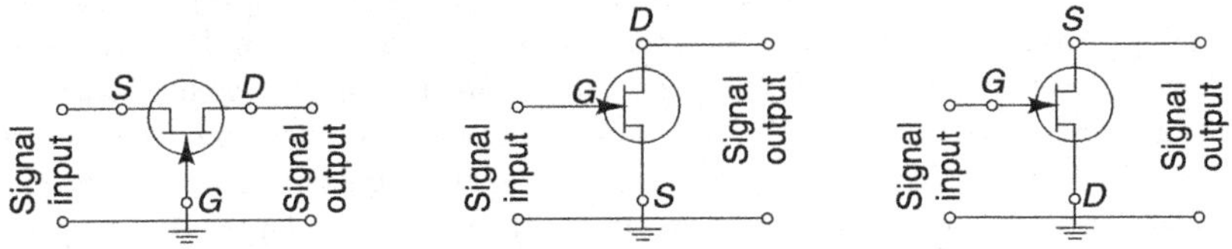

(a) Common Gate Configuration (b) Common Source Configuration (c) Common Drain Configuration

Fig. 7.25 *N-Channel FET configurations*

Table 7.1 *FET configurations*

Configuration	*Input Terminal*	*Output Terminal*
Common gate	source	drain
Common source	gate	drain
Common drain	gate	source

The three FET configurations have characteristics that are similar to the three corresponding BJT configurations, as given below in Fig. 7.26.

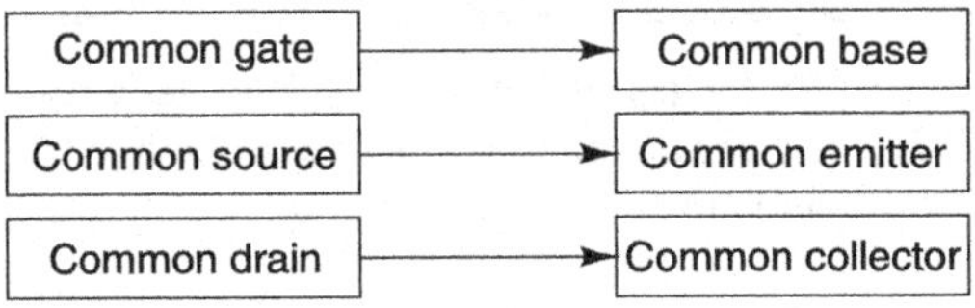

Fig. 7.26 *Characteristics of FET configuration*

7.15 JFET AS AMPLIFIER

One of the interesting features of JFET is the amplification of input signal. The circuit diagram is shown in Fig. 7.27.

In FET the weak signal is applied between gate and source and an amplified output is taken across the drain-source circuit, having load just as in a vacuum tube where signal is connected between grid and cathode and output is taken across a load connected in plate circuit.

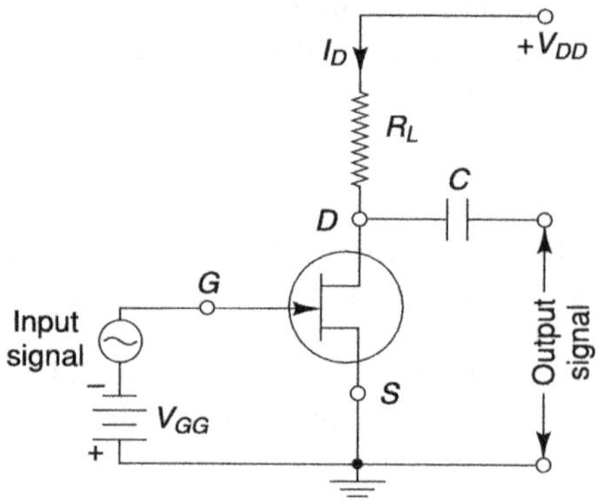

Fig. 7.27 *JFET as an amplifier*

For proper working of JFET, it is necessary that the gate is made negative with respect to source (i.e. the input circuit is always reverse biased). This is achieved either by inserting a battery V_{GG} in the gate circuit with the polarity, as shown in Fig. 7.27, by providing a biasing circuit.

On applying an alternating weak signal, the fixed reverse bias of the gate changes and that causes a large change in drain current, which gives rise to change in the width of the channel. This fact makes the FET capable of amplifying the strength of weak signals. During the positive half cycle of the input signal, the reverse bias on the gate decreases causing increases in channel width and hence, the drain current. During the negative half cycle of the input signal, the reverse bias on the gate increases causing drain current to decreases. The result is that a small change in voltage at the gate produces a large change in the drain current. These large variations in drain current give large output across the load R_L and JFET, thus, operates as an amplifier.

7.16 JFET PARAMETERS

JFET, like vacuum tubes, has certain parameters which determine its performances. These parameters are given below:

- ac drain resistance
- Transconductance
- Amplification factor
- dc drain resistance

7.17 AC DRAIN RESISTANCE (r_d)

The ac drain resistance of a FET (r_d) is similar to the ac plate resistance (r_p) of a tube. It may be defined as the ratio of change in drain-source voltage (ΔV_{DS}) to the change in drain current (ΔI_D at constant gate-source voltage (V_{GS}),
i.e. ac drain resistance,

$$r_d = \frac{\Delta V_{DS}}{\Delta I_D} \text{ at constant } V_{GS} \tag{7.25}$$

Referring to the FET drain characteristics, it is clear that beyond the pinch-off voltage, the change in drain current is very small for a change in drain-source voltage because the curve is almost flat. Therefore, the value of drain resistance (r_d) is very large ranging from 10 kΩ to hundreds of megaohms.

For instance, if a change in drain current of 0.01 mA is produced by a change in drain-source voltage of 4 V, then

$$r_d = \frac{4\text{V}}{0.01\text{ mA}} = 400\ k\Omega$$

Typical value of *rd* is about 400 kΩ.

7.18 TRANSCONDUCTANCE (g_m)

Similar to the transconductance of the tube, FET also has transconductance which determines the control that the gate-source voltage (V_{GS}) has over the drain current (I_D). The prefix 'trans' in the terminology applied to g_m reveals that it establishes a relationship between an output and input quality. The root word conductance was chosen because gm is determined by a voltage-current ratio similar to the ratio that defines the conductance of a resistor $G = \frac{1}{R} = \frac{1}{V}$. It may be defined as the ratio of change in drain current (ΔI_D) to the change in gate-source voltage (ΔV_{GS}) at constant drain source voltage (V_{DS}), i.e.

$$\text{transconductance, } g_m = \frac{\Delta I_D}{\Delta V_{GS}} \text{ at constant } V_{DS} \tag{7.26}$$

Transconductance is usually expressed either in mA/volt or micromhos. For instance, if a change in gate voltage of 0.1 V causes a change in drain current of 0.35 mA, then

$$g_m = \frac{0.35\text{ mA}}{0.1\text{ V}} = 3.5\text{ mA/V}$$

$$= 3500\ \mu\text{ mhos}$$

7.19 MATHEMATICAL EXPRESSION FOR TRANSCONDUCTANCE

Differentiating both sides of Shockley equation $I_D = I_{DS}\left(1 - \frac{V_{GS}}{V_P}\right)^2$, we get

$$dI_D = 2I_{DSS}\left(1 - \frac{V_{GS}}{V_p}\right)\left(-\frac{1}{V_p}\right)dV_{GS}$$

$$\frac{dI_D}{dV_{GS}} = 2I_{DSS}\left(1 - \frac{V_{GS}}{V_p}\right)\left(-\frac{1}{V_P}\right)$$

or

$$g_m = \frac{2I_{DSS}}{V_p}\left(1 - \frac{V_{GS}}{V_p}\right) \tag{7.27}$$

Substituting $V_{GS} = 0$ in above expression, we get

$$g_{mo} = -\frac{2I_{DSS}}{V_p} \tag{7.28}$$

From Eqs (7.27) and (7.28), we have

$$g_m = g_{mo}\left(1 - \frac{V_{GS}}{V_p}\right) \tag{7.29}$$

7.20 AMPLIFICATION FACTOR (μ)

The amplification factor (μ) of FET may be defined as the ratio of change in drain-source voltage (ΔV_{DS}) to the (ΔV_{DS}) at constant drain current I_D.

$$\text{Amplification factor, } \mu = \frac{\Delta V_{DS}}{\Delta V_{GS}} \text{ at constant } I_D \tag{7.30}$$

Application factor has no unit; it is just a number. It indicates how much more control the gate voltage has over drain current.

For instance, if the amplification factor of a FET is 90, it indicates that gate voltage is 90 times as effective as the drain voltage in controlling the drain current in the given FET.

7.21 RELATION BETWEEN r_d, g_m AND μ

The relationship among the three parameters of FET can be established as under:

We know,
$$\mu = \frac{\Delta V_{DS}}{\Delta V_{GS}} \text{ at constant } I_D$$

Multiplying and dividing the RHS by ΔI_D, we get

$$\mu = \frac{\Delta V_{DS}}{\Delta V_{GS}} \times \frac{\Delta I_D}{\Delta I_D}$$

$$= \frac{\Delta V_{DS}}{\Delta I_D} \times \frac{\Delta I_D}{\Delta V_{GS}}$$

or
$$\mu = r_d \times g_m \tag{7.31}$$

i.e. amplification factor = ac drain resistance × transconductance

7.22 DC DRAIN RESISTANCE

It is also called the static or ohmic resistance of the channel and is defined as ratio of drain-source voltage and drain current. It is denoted by R_{DS}.

i.e.
$$R_{DS}^* = \frac{V_{DS}}{I_D} \tag{7.32}$$

The FET exhibits some interesting characteristics given below:

- Its operation depends only upon one kind of charge carriers, hence it is uni-polar.
- It is simpler to fabricate and occupies less space in integrated form.
- High input resistance—typically many mega ohms.
- Less noisy than BJT.
- Exhibits no offset voltage at zero drain current, and hence makes an excellent signal chopper.

7.23 DIFFERENCE BETWEEN FET AND ORDINARY TRANSISTOR (BJT)

The field effect transistor differs from an ordinary or bi-polar transistor in the many ways:

- The operation of FET is uni-polar whereas, ordinary transistor utilises two types of charges, carriers.

- FET has high input impedance (of order of megaohms) since its input circuit (i.e. gate to source) is reverse biased. Whereas, the input circuit of BJT is forward biased, and hence has low input impedance.
- FET is essentially a voltage driven device whereas, BJT is basically a current operated device.
- For current conduction in a field effect transistor, no junction is involved. The conduction is through an *n*-type or *p*-type semiconductor. Therefore, the noise level in FET is very small.
- The gain of a bi-polar transistor is characterised as current gain since the current between collector and emitter is controlled by the base current. Whereas, the FET gain is characterised as a transconductance (i.e. the ratio of change in output current or drain current to the gate voltage) because in this case the drain current is controlled by gate voltage.
- FET is simpler to fabricate and it occupies less space in integrated form.

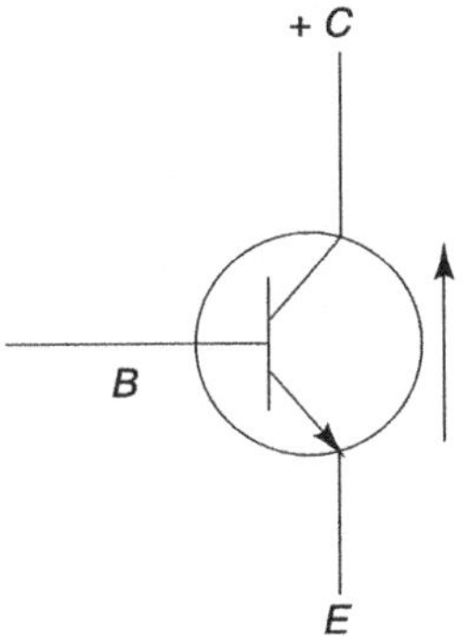

Fig. 7.28(a) *n-p-n transistor*

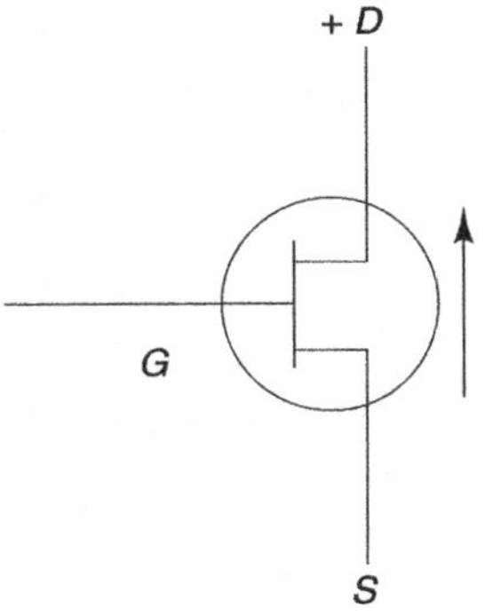

Fig. 7.28(b) *Field effect transistor*

- The performance of BJT is degraded by neutron radiation because of the reduction in minority-carrier lifetime. Whereas, FET can tolerate a much high level of radiation since they do not rely on minority carriers for their operation.
- The performance of FET is relativity unaffected by ambient temperature changes as it has a negative temperature coefficient at high current levels. It prevents the FET from thermal breakdown. The BJT has a positive temperature co-efficient at high current levels, which leads to thermal breakdown.
- FET amplifiers have low gain bandwidth product due to the junction capacitive effects and produce more signal distortion except for small signal operation.

7.24 ADVANTAGES AND DISADVANTAGES OF FET

❖ Advantages

A FET is a voltage controlled device (similar to a vacuum tube) in which the output current (drain current) is controlled by the input (gate) voltage. Therefore, it has the following important advantages:

(i) FET has a very high input impedance (of the order to 100 magaohms) which shows a high degree of isolation between the input and output circuit.

(ii) The operation of FET depends upon the majority carriers (i.e. electron in *N*-channel and holes in *P*-channel FET), which do not cross junctions. Therefore, the inherent noise of tubes (because of high-temperature operation) and those of ordinary transistors (because of junctions) are not present in FET.

(iii) In FET, there is no risk of thermal runway since it has a negative temperature co-efficient of resistance.

(iv) A FET has a smaller size, longer life and high efficiency.

(v) The power gain of FET is very high which eliminates the necessity of using driver stage while applying it as a power amplifier.

❖ **Disadvantage** Since FET has high input impedance, the gate voltage has less control over the drain current. Therefore, FET amplifier has much less voltage gain than a bi-polar amplifier.

Example 7.8

Determine the pinch-off voltage of a Ge *P*-channel FET with half channel height 2 microns and resistivity of 2 Ω cm. The relative dielectric constant of Ge is 16 and the mobility of holes is 1800 cm²/ s.

Solution

$$qN_A = \frac{\sigma}{\mu_p} = \frac{1}{2 \times 1800} = 2.777 \times 10^{-4}$$

$$\in = 16 \times 8.85 \times 10^{-12}$$

$$= 14.16 \times 10^{-12} \text{ F/cm}$$

Pinch-off voltage, $V_p = \dfrac{qa^2 N_A}{2 \in}$

$$V_p = \frac{2.777 \times 10^{-4} \times (2 \times 10^{-4})^2}{2 \times 1.416 \times 10^{-12}}$$

$V_p = 3.92$ V **Ans.**

Example 7.9

An *n*-channel silicon (dielectric constant = 12) FET with a channel width *a* = 3.5 × 10⁻⁶ m is doped with 10²¹ electrons/m³. Determine the pinch-off voltage.

Solution Pinch-off voltage

$$Vp = \frac{qa^2 N_D}{2 \in}$$

$$\in = \in_0 \in_r = 8.85 \times 10^{-12} \text{ F/m} \times 12$$

$$= 106.2 \times 10^{-12} \text{ F/m}$$

$$a = 3.5 \times 10^{-6} \text{ m}$$

$$V_p = \frac{1.6 \times 10^{-19} \times (3.5 \times 10^{-6} \text{m})^2 \times 10^{21}}{2 \times 106.2 \times 10^{-12}}$$

$$= \frac{1.6 \times 9 \times 10^2}{212.4}$$

$= 9.2$ V **Ans.**

Example 7.10

A JFET has $Vp = -4.0$ V, $I_{DSS} = 10$ mA and $I_{DS} = 2.2$ mA. Determine the transconductance.

Solution Drain-source saturation current, $I_{DSS} = 10$ mA
Pinch-off voltage, $Vp = -4.0$ V
Drain-source current, $I_{DS} = 2.2$ mA

From Shockley's equation, drain-source current, $I_{DS} = I_{DSS}\left[1 - \frac{V_{GS}}{V_p}\right]^2$

or
$$V_{GS} = V_p\left[1 - \sqrt{\frac{I_{DS}}{I_{DSS}}}\right] = -4.0\left[1 - \sqrt{\frac{2.2}{10}}\right] = -2.16 \text{ V}$$

Then,
$$g_m = \frac{-2I_{DSS}}{V_p}\left[1 - \frac{V_{GS}}{V_p}\right] = \frac{-2 \times 10 \times 10^{-3}}{-4.0}\left[1 - \frac{-2.25}{-4.0}\right]$$
$$= 2.19 \text{ m A/V}$$
Ans.

Example 7.11

An *n*-channel JFET has $I_{DSS} = 10$ mA and $V_p = -5$ V. Determine the minimum value of V_{DS} for pinch-off region and drain current I_D for $V_{GS} = -3$ V pinch-off region.

Solution Pinch-off voltage, $V_p = -5$ V
Gate-source voltage, $V_{GS} = -2$ V
Drain source saturation current, $I_{DSS} = 10 \text{ mA} = 10 \times 10^{-3}$ A

Drain current,
$$I_D = I_{DSS}\left[1 - \frac{V_{GS}}{V_p}\right] = 10 \times 10^{-3}\left[1 - \frac{-3}{-5}\right]^2$$
$$= 1.6 \text{ mA}$$
Ans.

Example 7.12

An *n*-channel JFET has a pinch-off voltage of – 4.5 V and $I_{DSS} = 8$ mA. At what value of V_{GS} will I_{DS} equal to 2 mA? What is its g_m at this I_{DS}?

Solution Pinch-off voltage, $V_p = -4.5$ V
Drain-source saturation current, $I_{DSS} = 8 \text{ mA} = 8 \times 10^{-3}$ A
Drain-source current, $I_{DS} = 2 \text{ mA} = 2 \times 10^{-3}$ A.
From Shockley's equation (Eq. 5.1),
$$V_{GS} = Vp\left[1 - \sqrt{\frac{I_{DS}}{I_{DSS}}}\right] = -4.5 \times \left[1 - \sqrt{\frac{2 \times 10^{-3}}{8 \times 10^{-3}}}\right]$$
$$= -2.25 \text{ V}$$
Ans.

Transconductance g_m for $I_{DS} = 3$ mA for which $V_{GS} = -1.902$ V
$$g_m = \frac{-2I_{DSS}}{V_p}\left[1 - \frac{V_{GS}}{V_p}\right]$$
$$= \frac{-2 \times 8 \times 10^{-3}}{-4.5}\left(1 - \frac{-1.902}{-4.5}\right)$$
$$= 2.06 \text{ mA/V}$$
or
$$= 2.06 \text{ m S}$$
Ans.

Example 7.13

When a reverse gate voltage of 26 V is applied to a FET, the gate current is 1.6 × 10⁻³ μA. Determine the resistance between gate and source.

Solution Gate-source voltage,

$$V_{GS} = 26 \text{ V}$$

Gate current, $I_G = 1.6 \times 10^{-3}$ mA $= 1.6 \times 10^{-9}$A

∴ gate to source resistance

$$= \frac{V_{GS}}{I_G}$$

$$= \frac{26}{1.6 \times 10^{-9}} = 16.25 \times 10^{9} \text{ ohms} \qquad \textbf{Ans.}$$

From this example it could be noticed that a FET has a high impedance (usually hundreds of megaohms), which permits high degree of isolation between the input and output.

Example 7.14

Determine the value of trans-conductance of a FET, when the drain current changes from 1 mA to 2.1 mA with a change in gate-source voltage from – 3.5 V to – 3 V.

Solution Change in drain current,

$$\Delta I_D = 2.1 - 1.0 = 1.1 \text{ mA}$$

Change in gate-source voltage.

$$\Delta V_{GS} = 3.5 - 3.0 = 0.5 \text{ V}$$

Transconductance,

$$g_m = \frac{\Delta I_D}{\Delta V_{GS}} = \frac{1.1 \text{ mA}}{0.5 \text{ V}} = 2.2\ V = 220\ \mu\Omega \qquad \textbf{Ans.}$$

Example 7.15

The following readings were obtained experimentally from a FET:

V_{GS}	**– 0.1 V**	**– 0.1 V**	**– 0.4 V**
V_{DS}	**5 V**	**14 V**	**14 V**
I_D	**8 mA**	**8.3 mA**	**7.1 mA**

Determine (i) ac drain resistance, (ii) transconductance, and (iii) amplification factor.

Solution:

(i) At constant V_{GS} (i.e. at $V_{GS} = -0.1$ V), when V_{DS} is increased from 5 V to 14 V, the drain current I_D increases from 8 mA to 8.3 mA.

∴ change in drain-source voltage,

$$\Delta V_{DS} = 14 - 5 = 9 \text{ V}$$

Change in drain current,

$$\Delta I_D = 8.3 - 8 = 0.3 \text{ mA}$$

∴ ac drain resistance,

$$r_d = \frac{\Delta V_{DS}}{\Delta I_D} \text{ at constant } V_{GS}$$

$$= \frac{9 \text{ V}}{0.3 \text{ mA}} = 30 \text{ k-ohms} \qquad \text{Ans.}$$

(ii) At constant V_{DS} (i.e. at $V_{DS} = 14$ V), when V_{GS} is changed from –0.1 V to –0.4 V, the drain current I_D changes from 8.3 mA to 7.1 mA.

$$\therefore \quad \Delta V_{GS} = 0.4 - 0.1 = 0.3 \text{ V}$$

Change in drain current,

$$\Delta I_D = 8.3 - 7.1 = 1.2 \text{ mA}$$

∴ transconductance,

$$g_m = \frac{\Delta I_D}{\Delta V_{GS}} \text{ at constant } V_{DS}$$

$$= \frac{1.2 \text{ mA}}{0.3 \text{ V}} = 4 \text{ mA/V} = 4000\ \mu \text{ mhos}$$ **Ans.**

(iii) Amplification factor,

$$\mu = r_d \times g_m = (30 \times 10^3) \times (4000 \times 10^{-6}) = 120$$ **Ans.**

Example 7.16

Determine the transconductance of a FET if its amplification factor is 100 and drain resistance is 33 kilo-ohms.

Solution Amplification factor,

$$\mu = 100$$

Drain resistance, $r_d = 33$ kilo ohms $= 33 \times 10^3$ ohms

∴ transconductance,

$$g_m = \frac{\mu}{r_d} = \frac{100}{33 \times 10^3}$$

$$= 3.03 \times 10^{-3} \text{ mhos} = 303\ \mu\text{mhos}$$ **Ans.**

7.25 BI-POLAR JUNCTION TRANSISTOR (BJT)

The transistor was invented by J Bardeen in 1947 at Bell Labs USA. The first junction transistor consisting of two back-to-back *p-n* junctions was invented by William Shockley in 1951. A junction transistor consists of two *p-n* junctions created in the same single crystal and their symbols are given in Figs 7.29 and 7.30.

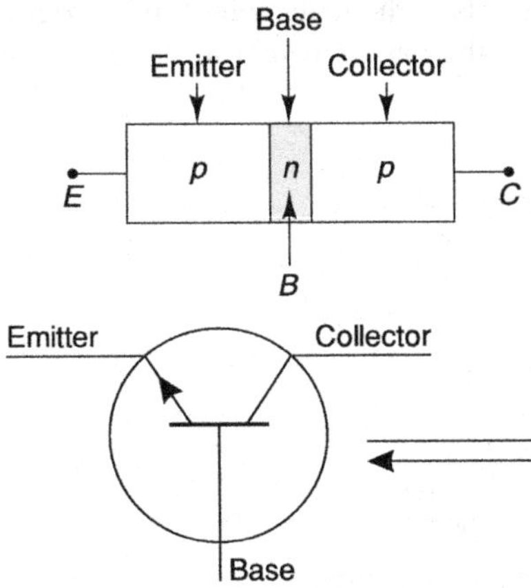

Fig. 7.29 *p-n-p transistor and its symbol*

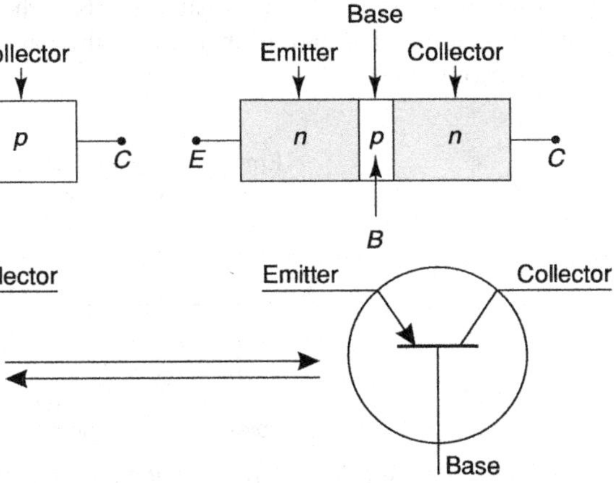

Fig. 7.30 *n-p-n transistor and its symbol*

A transistor has three doped regions with entirely different doping profiles. There are two kinds of transistors.

- *n-p-n* transistor
- *p-n-p* transistor

All the three segments of a transistor have different thickness and their doping levels are also different. The direction of the arrow in the symbols (Figs 7.29 and 7.30) represent the direction of conventional current in the transistor. A brief description of the three segments of a transistor is given below:

Emitter : This is the segment on the side of the transistor shown in the figure. It is of moderate size and heavily doped. It supplies a large number of majority carriers for the current flow through the transistor.

Base : This is the central segment. It is very thin and lightly doped.

Collector : This segment collects a major portion of the majority carriers supplied by the emitter. The collector side is moderately doped and larger in size as compared to the emitter.

We have seen earlier, in the case of a *p-n* junction, there is a formation of depletion region across the junction. In case of a transistor, depletion regions are formed at the emitter-base junction and at the base collector junction. Therefore, to understand the action of a transistor, we have to consider the nature of depletion regions formed at these junctions. The charge carriers move across different regions of the transistor when proper voltages are applied across its terminals.

The biasing of the transistor is done differently for different uses. The transistor can be used in two distinct ways. Basically, it was invented to function as an amplifier—a device that produces enlarged copy of a signal. But, later its use as a switch acquired equal importance.

In operation, the transistor is biased in such a way as to make the emitter-base junction forward biased and the base-collector junction reverse biased as shown in Figs 7.31 and 7.32. The energy band diagram of the biased transistor is shown in Fig. 7.33. The biasing results in the reduction of the barrier height at the emitter base junction and an increase in the barrier height at the base-collector junction, as shown in Fig. 7.34.

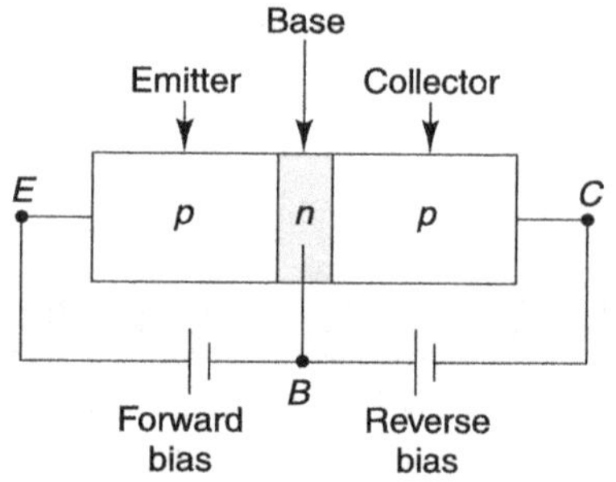

Fig. 7.31 *Biasing of a p-n-p transistor*

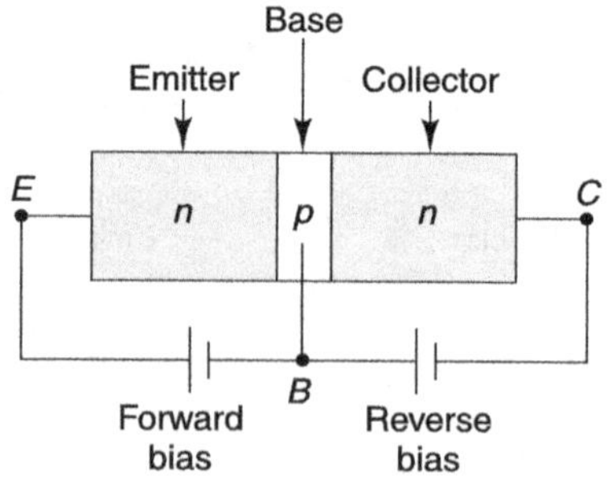

Fig. 7.32 *Biasing of an n-p-n transistor*

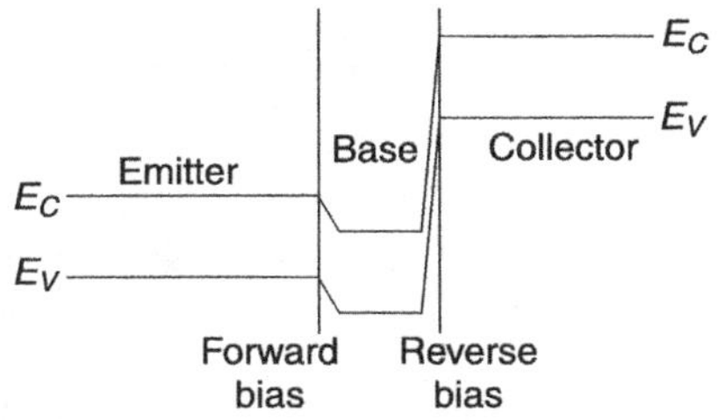

Fig. 7.33 *Energy-band diagram of a biased transistor*

7.26 WORKING MECHANISM OF A *p-n-p* TRANSISTOR

Let us consider the flow of charge carriers in a *p-n-p* transistor. Holes (majority carriers) flow from the emitter into the n-type base region and diffuse across the thin base region into the collector region, i.e. the emitter emits holes and the collector collects them. The hole current in the collector region is only slightly less than the hole current in the emitter region. The resistance of the emitter-base junction is low as it is forward biased and the resistance of the base-collector region is high as it is reverse biased. So, in a transistor we are forcing a certain current I to flow from a low-resistance region into a high-resistance one, which results in power gain. In fact, a transistor by virtue of its construction and biasing, is an inherent power amplifier. The name transistor has been coined to indicate the transfer of resistor from low to high across the device.

The amplification factor of transistor depends on how much of the hole current emitted by the emitter is effectively collected by the collector. This can be controlled by the proper design of the transistor. In order to make a large hole current flow from the emitter into the collector, across the base, the *p*-type emitter region is heavily doped and the *n*-type base region is lightly doped. The base region is made very thin ($\cong$,10^{-5} m), so that it would be less than the diffusion length of the carrier. The *p*-type collector region is lightly doped to increase the junction width so that the capacitance of the collector region would be low and the reverse breakdown voltage would be high.

7.27 TRANSISTOR CONFIGURATION

A transistor has three terminals, namely, emitter, base and collector. We require four terminals when a transistor is to be connected in a circuit; two for the input and two for the output. This is achieved by making one terminal of the transistor common to both input and output terminals. The input is fed to this common terminal and any one of the other two terminals. The output is obtained between the common terminal and the remaining terminal. Transistor is configured based on the common terminal.

- Common-base configuration
- Common-emitter configuration
- Common-collector configuration

7.28 COMMON-BASE CONFIGURATION

In common-base (CB) configuration, the base terminal of a transistor is made common to both input and output. Figure 7.34 shows this common base configuration of *npn* transistor. The input is applied between the emitter and base of the transistor. The output is obtained between collector and base terminal of the transistor.

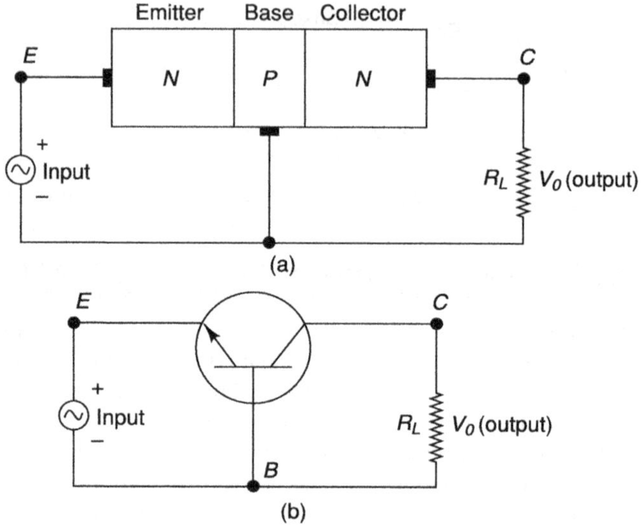

Fig. 7.34 *Common-base transistor amplifier*

7.29 COMMON-EMITTER CONFIGURATION

In common-emitter (CE) configuration, the emitter terminal of a transistor is made common to both input and output. Figure 7.35 shows the common-emitter configuration of n_{pn} transistor. The input is applied between base and emitter and the output is obtained between collector and emitter. This configuration is widely used for many applications.

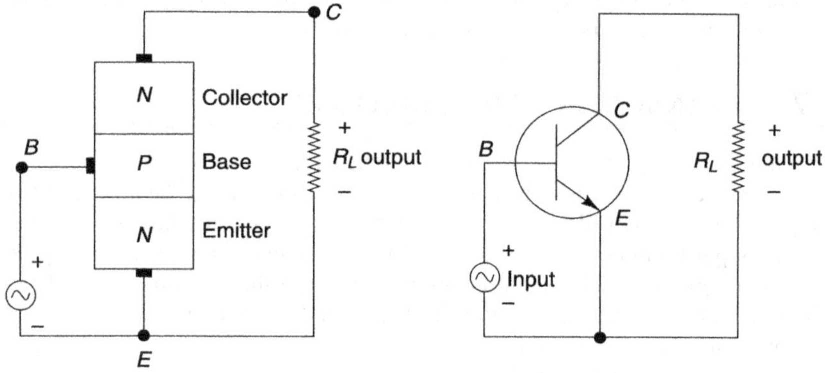

Fig. 7.35 *Common-emitter configuration*

7.30 COMMON-COLLECTOR CONFIGURATION

In common-collector (CC) configuration, the collector terminal is made common to both input and output. The input is applied between the terminals emitter and collector. Figure 7.36 shows the common collector configuration.

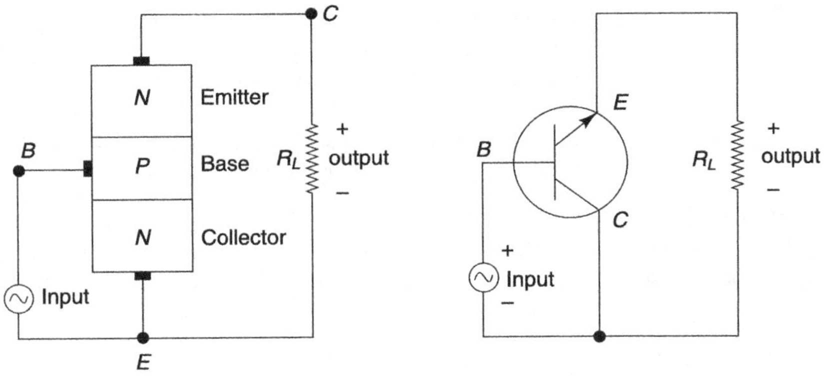

Fig. 7.36 *Common-collector configuration*

7.31 TRANSISTOR AS AN AMPLIFIER

Transistor is used for many applications, but is mainly used as an amplifier. Amplifier raises the strength of a weak signal applied as input signal. Figure 7.37 shows circuit of a common-base transistor amplifier. The weak signal is applied between emitter and base terminals. The output is taken across the load R_C connected in the collector circuit.

In order to achieve faithful amplification, the input circuit (emitter-base junction) should always remain forward biased. To do so, a dc voltage V_{EE} is applied at input circuit in addition to the input signal. The voltage V_{CC} is used to maintain the reverse bias in the collector-base junction.

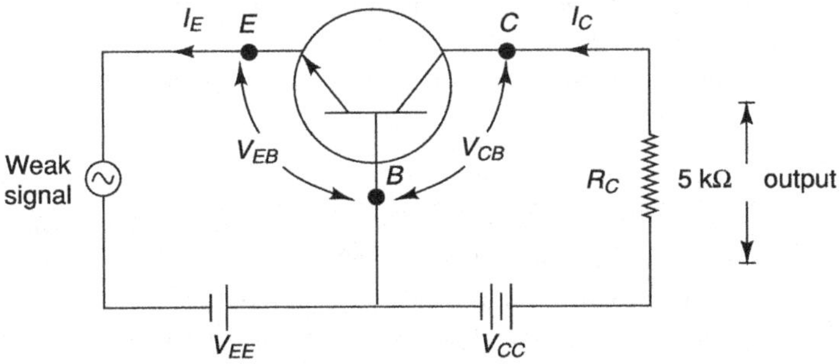

Fig. 7.37 *Transistor as an amplifier*

7.32 COMPARISON OF VARIOUS TRANSISTOR CONFIGURATIONS

Comparison of various characteristics of the three configurations is shown in Table 7.2.

Table 7.2 *A comparison of CE, CB and CC transistors*

S.No.	*Characteristic*	*Common base*	*Common emitter*	*Common collector*
1.	Input resistance	Low (about 100 Ω)	Low (about 750 Ω)	Very high (about 750 kΩ)
2.	Output resistance	Very high (about 450 kΩ)	(about 45 kΩ)	(about 50 Ω)

Contd

3.	Voltage gain	Low (about 150)	(about 500)	(Less than 1)
4.	Leakage current	Very small (5 μA for Ge 1 μA for Si)	Very large (500 μA for Ge 20 μA for Si)	Very large (500 μA for Ge 20 μA for Si)
5.	Application	For high frequency application	For audio frequency applications	For impedance matching

Conclusion

From the three configuration, the common emitter circuit is the most efficient (it is used about 90% to 95% of all transistor applications). The reasons for use of this CE configuration are:

- High current gain
- High voltage and power gain
- Moderate output to input ratio

7.33 MICROELECTRONIC CIRCUITS AND ICs

The latest techniques of producing semiconductor devices are so fantastic that one can get very small electronic circuits, known as micro-electronic circuits, out of these. These micro-electronic circuits can be interconnected to get integrated circuits popularly known as ICs. Today IC is so common now that it is on the tongue of every layman.

An integrated circuit chip contains millions of electronic components interconnected on a very small single crystal of silicon. The quality of this silicon must be of 'electronic grade'. No other technology has witnessed such a revolutionary change in modern electronics as the IC technology. The basic objective of IC technologists is to put as many components as possible on a single small chip of silicon, and to make the device to respond very fast and efficient.

Since its inception, the various stages to achieve IC technology are described in the Table 7.3.

Table 7.3 *Various stages of IC development*

Year	*Technology*	*Number of components on a chip*
1960–65	Small Scale Integration (SSI)	10
1965–70	Medium Scale Integration (MSI)	100–1000
1970–80	Large Scale Integration (LSI)	100–20,000
1980–85	Very Large Scale Integration (VLSI)	20,000–500,000
1985–95	Ultra Large Scale Integration (UVLSI)	>500.000
1995	Giant Scale Integration (GSI)	>10.00000

7.34 BASIC STEPS IN IC FABRICATION

In brief the several steps involved in IC fabrication, some main procedures are listed below:

Metallurgical-Grade Silicon The starting material for silicon wafer fabrication is sand (SiO_2). It is abundant in the earth's crust; 28% of the earth's crust consists of silicon in one way or the other. SiO_2 is heated with carbon in an arc furnace to reduce it. According to the reaction.

$$SiO_2 + 2C \longrightarrow Si + 2CO$$

This silicon is 98% pure (called metallurgical grade silicon). To make it useful for the manufacturing of ICs, it is further purified to get semiconductor grade silicon or electronic-grade silicon.

- **Electronic-Grade Silicon** This powdered raw material is then reacted with hydrogen chloride to form trichlorosilane gas ($SiHCl_3$).

$$Si + 3HCl \longrightarrow SiHCl_3 + H_2$$

$SiHCl_3$ is fractionally distilled for purification and subsequently reduced with hydrogen to polycrystalline silicon.

$$SiHCl_3 + H_2 \longrightarrow Si + 3HCl$$

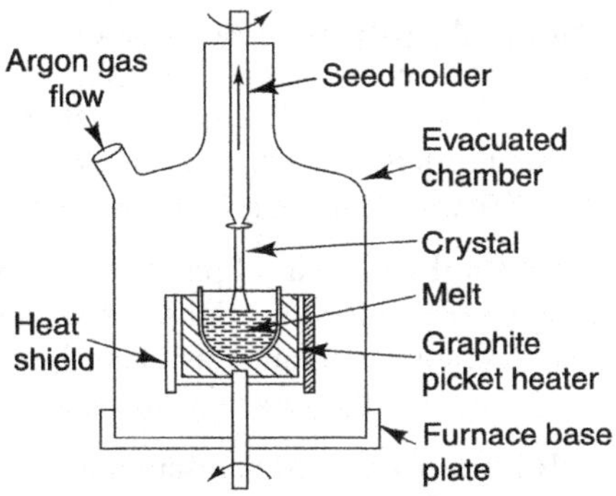

Fig. 7.38 *Czochralski method for single-crystal growth*

- **Single-crystal Growth** There are many ways for single-crystal growth of electronic grade silicon. We have Czochralski (CZ) method, floating zone (FZ) method and Bridgman method. Mostly crystals are grown by using CZ method. Crystal diameters of up to 150–200 mm are made. The necessary apparatus is shown in Fig. 7.47. The details of this technique have been discussed in the previous chapter.

- **Wafer Manufacture** Silicon rods once obtained, are sliced, lapped, etched and polished to obtain the 0.3–0.4 mm thick wafers. Sharp edges of the wafer are rounded off to avoid the damage of wafer.

- **Oxidation** SiO_2 (silica) forms readily on silicon by placing the wafer into a tube furnace, heated between 900°C and 1300°C, and exposed to water vapour. This type of oxidation is called wet or steam oxidation. This oxidation is much faster than dry oxidation for which oxygen without steam is reacted with the silicon slice.

The details about the above steps will be describe the chaper 8. Besides the above many more steps are important which are described below.

Photolithography

Photolithography means 'printing through light'. It is done to transfer the geometric pattern that produces the desired electrical characteristics into the Si-wafer. In this a protective layer (photo-resist) is first coated on the SiO_2 layer and then exposed to UV light from a mercury lamp. Photo resist is applied to the wafer by first applying a small amount of the material at the centre of the wafer and spinning it—mounted on a shaft—at a particular speed, so that the photo resist spreads outwards uniformly coating the wafer. The spin speed and the viscosity of the photo resist determine the thickness of the coating.

The purpose of photo resist is to transfer the image from the mask to the SiO_2 layer on the surface of the vapour. After the exposure, the unexposed photo resist is dissolved at the places where the SiO_2 needs to be removed. The remaining photo resist protects SiO_2 from the etchant as it is etch resistant.

- **Mask** Masks are geometric patterns comprising opaque and transparent regions which are desired to be designed according to the electrical circuit required to be integrated on a wafer. Several masks are used and they have to be precisely aligned on the wafer.

• **Impurity Doping** Addition of impurity is done using ion implantation technique. This technique involves the ionisation of the species to be implanted and their subsequent acceleration towards the substrate in an electric field. There is a possibility that due to collision of ions, the lattice symmetry may damage. For this, annealing between 700°C and 1000°C is done for a short time, which restores the original lattice symmetry.

• **Epitaxial Growth** Epitaxial growth refers to the growth of a layer of silicon on a substrate. This growth is done in such a way that the crystal structure of the layer must have the same orientation as that of substrate. Silane (SiH_4), dichlorosilane (SiH_2Cl_2), etc., are used as the vapour source and hydrogen as carrier gas.

• **Metallisation** Metallisation is the process of providing electrical connections between different parts of the ICs. For electrical connections aluminium is used, as it possesses high electrical conductivity and low melting point. A thin coating of aluminium is deposited over the entire wafer by vacuum evaporation and the undesired aluminium area is then etched away.

• **Mounting and Packing** Individual chips are very small and brittle in nature hence, each chip is mounted on the gold-plated header of a packing. Connections between the IC and the package leads are made by aluminium wires from the bonding pads on the chip to the leads on the header, and the IC is sealed in an inert atmosphere.

ICs are very useful components in microelectronics because of their small size, light weight, low cost, low power consumption and high reliability. ICs are of two types—linear and non-linear. Linear ICs are used in operational amplifiers, power amplifiers, voltage regulators and in analog computer circuits. Non-linear ICs are used in digital computers to perform switching functions in logic gates and memory units.

7.35 IC CHIP FABRICATION

The main steps involved in the fabrication of IC are described below taking an example of a n-MOS device. The following patterns are given in the diagrams, which will be used to indicate various materials:

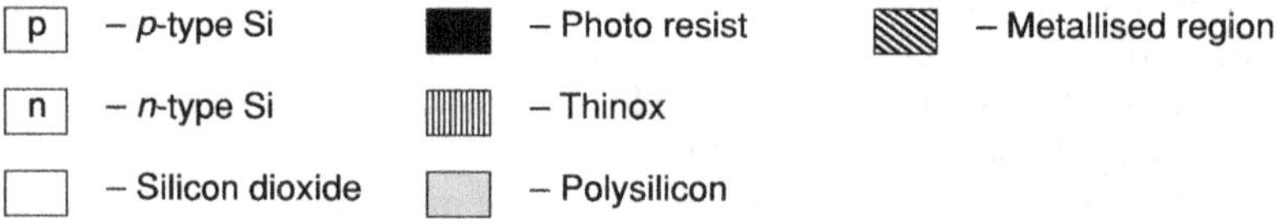

Step 1

A thin wafer (0.4 mm thick and 7.51 mm diameter), cut from a single crystal of silicon of high purity and doped with p-type impurity, is taken.

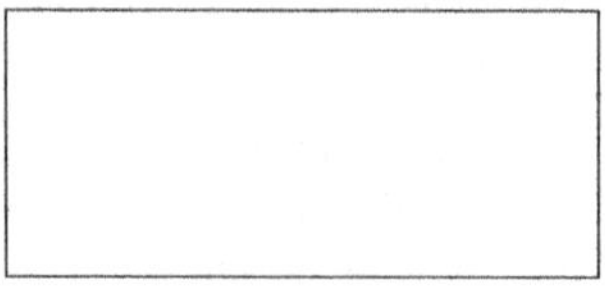

Fig. 7.39(a)

Step 2

A layer of SiO_2 of 1 μm thickness is grown on the surface by wet or dry oxidation process.

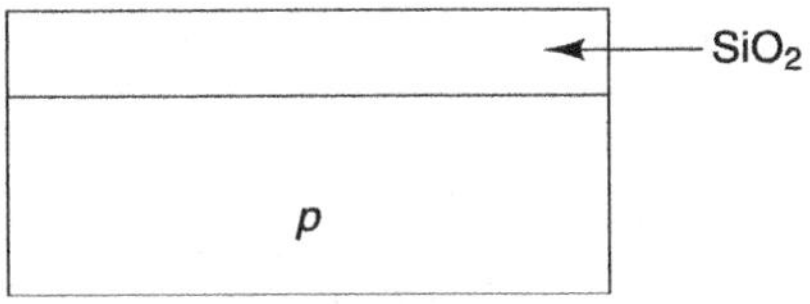

Fig. 7.39(b)

Step 3

The surface is covered with a photo resist which is coated uniformly.

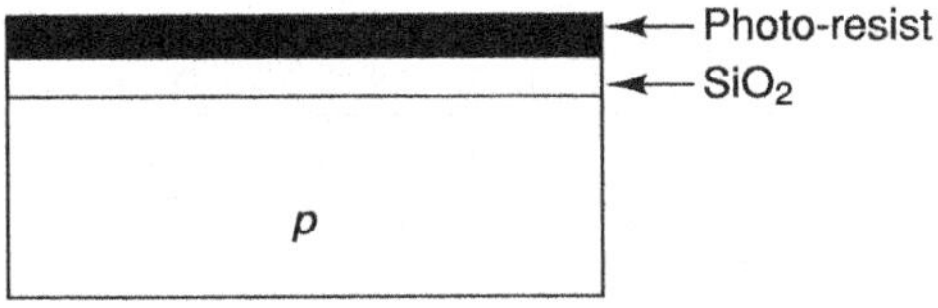

Fig. 7.39(c)

Step 4

A mask (suitably designed for the required circuit design) is placed on the photo resist. In Fig. 7.39(d) a simple mask with a small opaque portion at the centre and transparent elsewhere is shown. Ultraviolet radiation is allowed to incident on the photo resist through the mask. Those portions of the photo resist which get exposed to the UV radiation, get polymerised and hardened and the other unexposed areas (centre portion) remain unaffected.

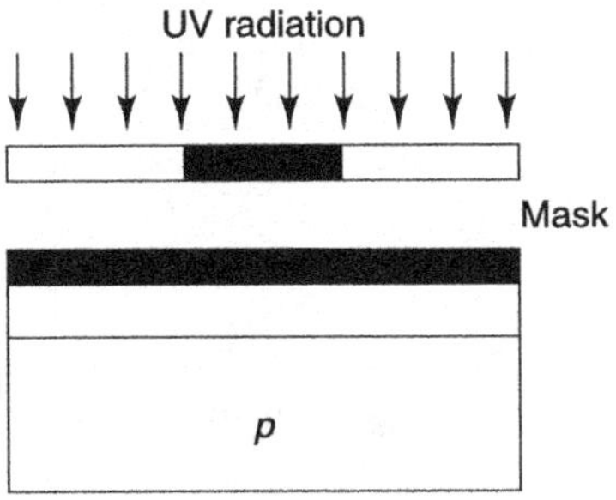

Fig. 7.39(d)

Step 5

The unpolymerised portion (the central portion) is readily etched away together with the SiO_2 layer, using suitable etchant solutions. Subsequently, the polymerised photo resist is also removed in the other portions, leaving the SiO_2 layer intact in those regions. In this way, windows are created on the SiO_2 coated Si, for diffusion of desired impurities, as shown in Fig. 7.39(e).

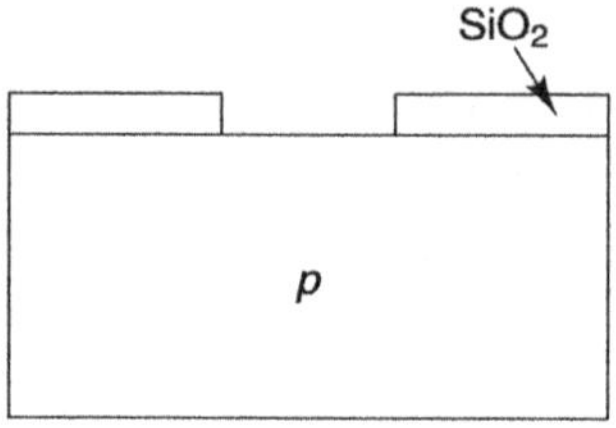

Fig. 7.39(e)

Step 6

A very thin layer of SiO_2 (0.1 μm thickness) called thinox is grown upon the entire chip surface by the oxidation process described above.

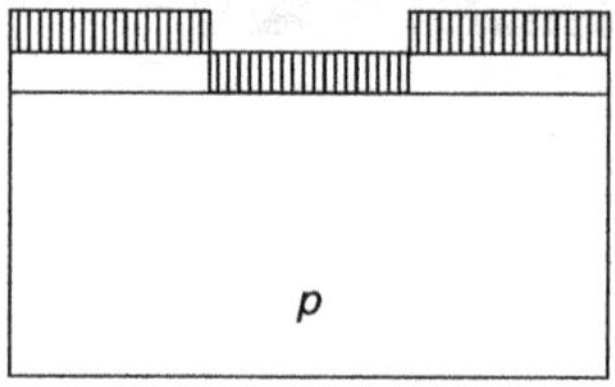

Fig. 7.39(f)

Step 7

Photo resist coating, masking and etching process is utilised again to deposit polysilicon on the thin oxide (thinox) layer by chemical vapour deposition technique at the region in the window, as shown in Fig. 7.39(g).

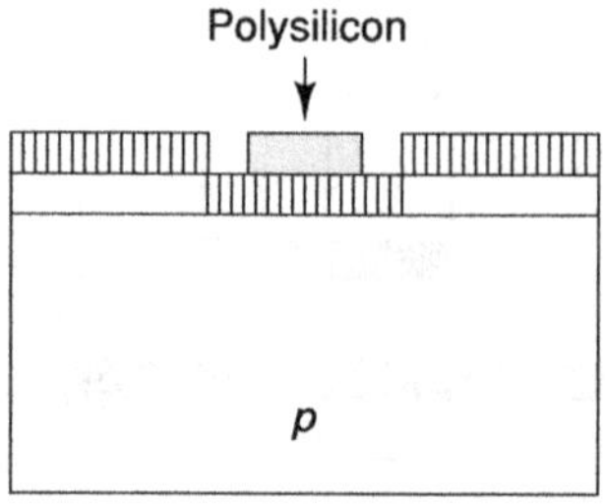

Fig. 7.39(g)

Step 8

The thinox layer is removed by a suitable etching procedure.

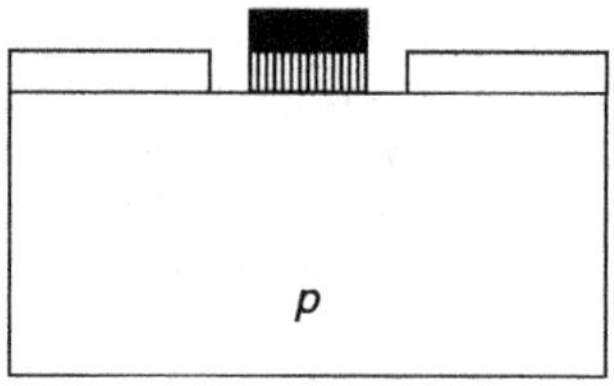

Fig. 7.39(h)

Step 9

n-type impurities are diffused into the exposed areas of *p*-type Si by heating the wafer to a high temperature and passing a gas containing the desired impurity (phosphorus). *n*-type Si is formed in the two regions for source (S) and drain (D), as shown In Fig. 7.39(i).

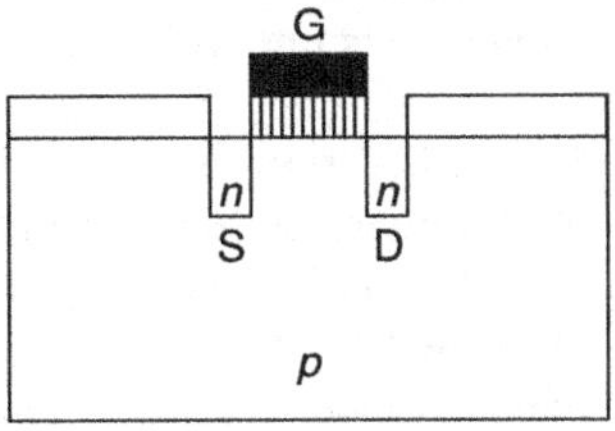

Fig. 7.39(i)

Step 10

Thick oxide layer is formed again on the whole surface. It is than coated with photo resist and suitable mask is used to select areas of polysilicon gate (G), source and drain where connections are to be made. Exposing to UV radiation and etching is done to expose those areas where contacts have to be made.

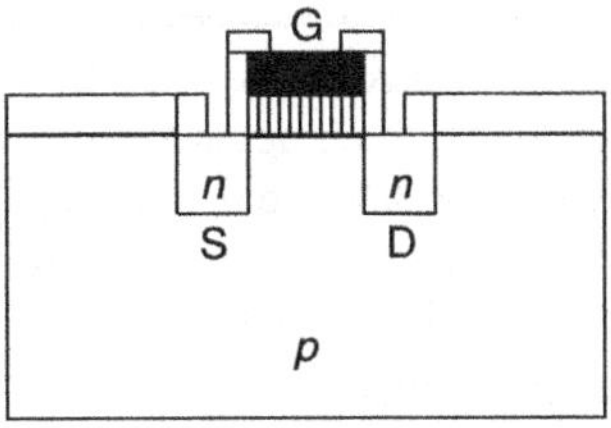

Fig. 7.39(j)

Step 11

A thin film (1 μm thick) of aluminium is coated over the surface (metallisation). The metal layer is then masked and etched to get contact points for S, D and G.

The IC technology described above is only a summary of the steps involved in the fabrication of ICs. The technology is developing so fast that new and modern techniques are adopted at each stage.

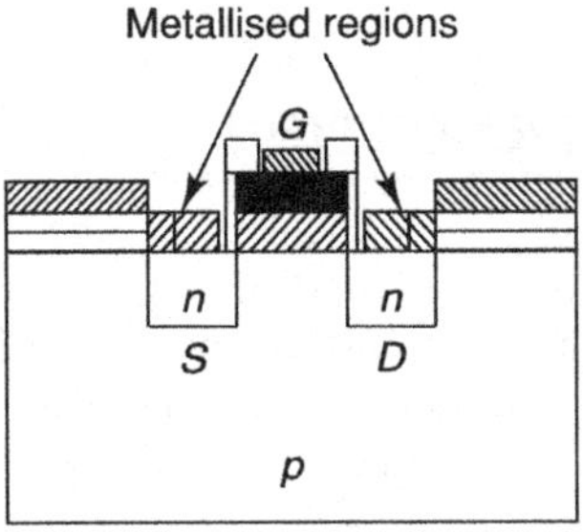

Fig. 7.39(k)

As an example, the impurity doping—instead of the diffusion technique—ion implantation technique is used. In this technique, the Si substrate is subjected to highly energised donor or acceptor atoms. When these atoms impinge on the Si surface, they travel below the surface, forming regions with varying dopant concentrations. At high temperature (>8000°C), diffusion occurs from regions of high to that of low concentration. Once the doped areas are put in place, it is important that the temperature be kept as low as possible.

In the photolithographic technique described above, UV radiation is made incident on the photo resist coated Si surface through the mask. To prepare the mask, the required pattern is first drawn on a thin glass plate—on a bigger scale—by using stand and photographic techniques. The mask will have extremely small transparent and opaque regions intricately patterned, and so when UV radiation is incident on it, diffraction occurs around the edges of the patterns. This limits the line widths in the pattern to around 0.8 μm. To avoid this, electron beam lithography technique has been developed recently; it get line widths of the order of 0.5 μm. This is a direct method in which no mask is needed. An electron beam-sensitive resist is used. The patterns are obtained by scanning the surface with an electron beam, according to the required pattern. The drawbacks of this technique are the cost of the equipment and the large amount of time required.

7.36 PHOTONIC DEVICES

Till now we have concentrated on electronic devices. There is a wide variety of devices where the photon plays an active role. Such devices are known as optoelectronic devices or photonic devices. These devices provide the optical source and the detectors that allow broadband telecommunications and data transmission through optical fibres. This section is mainly devoted to those devices that either detect or emit photons. Devices which convert the photon (optical energy) into electrical energy will also be discussed. Photon-emitting devices include the incoherent sources like LED and highly coherent sources like semiconductor lasers. Since the modern electronics often involve optical as well as electrical signals, photo diodes have become as important as electronic devices.

Photonic devices are classified in three categories:

- The devices that convert electrical signal into optical radiation commonly known as LED (light emitting diodes) and diode lasers.
- The devices that detect optical signals through electronic processes known as photo detectors.
- The devices that convert optical radiation into electrical energy known as photo-voltaic device or solar cell.

7.37 RADIATIVE TRANSITION

The LED and semiconductor laser belong to the luminescent devices family. Luminescence can be defined as emission of light radiation as a result of electronic excitation of the material. The light emitted by a LED is incoherent, while the light emitted by semiconductor laser is highly coherent. This is entirely different from the radiation emitted as a result of temperature (incandescence). Although different process are used to excite radiations of different wavelengths, all the radiation are fundamentally identical. Figure 7.40 gives a classification of electromagnetic radiations.

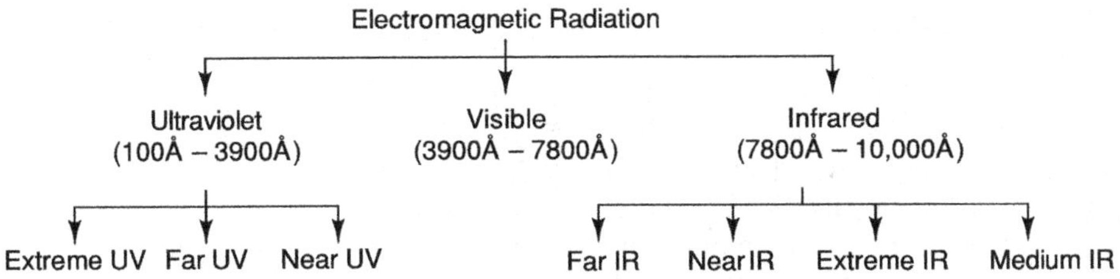

Fig. 7.40 *Classification of electromagnetic radiations*

7.38 PHOTO-TRANSISTOR

A very important application for which transistor is receiving considerable attention is its use in photo-electric equipment such as flame detectors, automatic door openers, light dimmers, etc. These devices work upon the principle that when semiconductor material is exposed to the light, its electrical conductivity increases. A transistor specially fabricated for absorption of electromagnetic radiation is by its very nature a sensitive photo-detector. Such a transistor is called a photo-transistor. The photo-transistor must have a large base-collector junction to collect the incident photons. An *n-p-n* transistor is shown in Fig. 7.41.

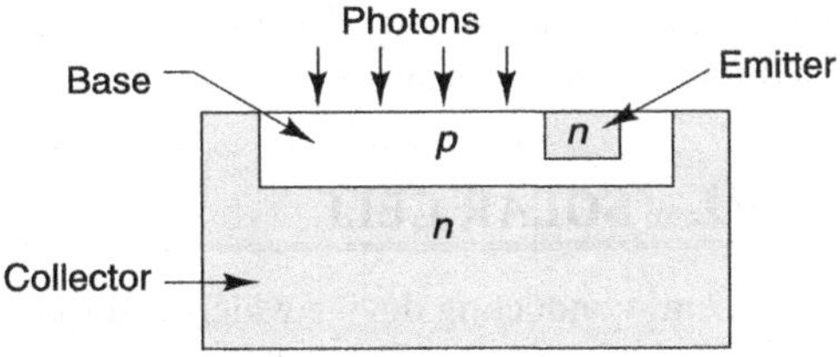

Fig. 7.41 *Schematic diagram of* n-p-n *transistor with large base collector region*

In a photo-transistor, the base collector junction (which is reverse biased) acts as a photo diode. When photons are incident on this junction, electron-hole pairs are generated.

The photo-transistor in above devices is preferred against photo-cell. This is because the photo-cells deliver only microamperes of current, but photo-transistors deliver milliamperes. Thus, they often eliminate the use of amplifiers for increasing the current to usable proportions. The photo-transistors may be point-contact type or junction type. One such *p-n* junction assembly is shown in Fig. 7.42.

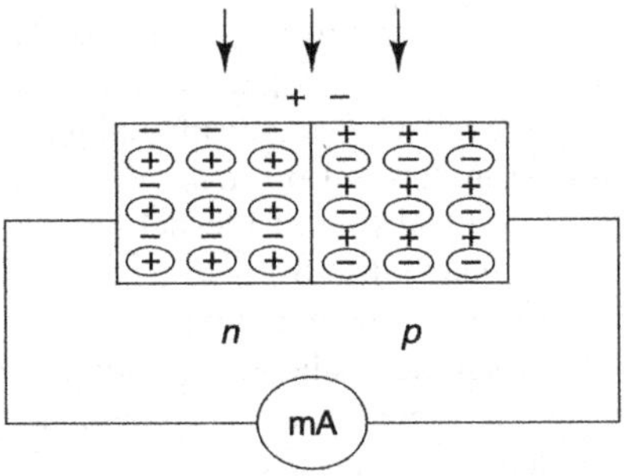

Fig. 7.42 *An* n-p *junction photo-diode*

If the junction is maintained at constant temperature in a dark background, no current in shown by the milliammetre. The reason for this is that an internal emf is set up at the junction which does not allow the electrons to cross it. If light is shone on the *n-p* junction, it is absorbed freeing more electrons and producing 'holes'. Due to the potential difference in the junction, the electrons move to the right and the 'holes' to the left and a current is shown by the milliammetre. This is the principle of the so-called solar battery. Maximum current response is obtained when light falls in the immediate neighbourhood of the junction.

Another form of photo-transistor employs an *n-p-n* type of construction in which only the base is photo-sensitive.

Since the energy of light quantum varies inversely with the wavelength of light, it follows that the photo-conductivity response of photo-transistor will depend upon the wavelength of light used. In Fig. 7.43, this dependence is shown in the case of a certain transistor. The graph shows that as the wavelength becomes smaller than two microns, the response rises sharply and then slowly starts to fall off.

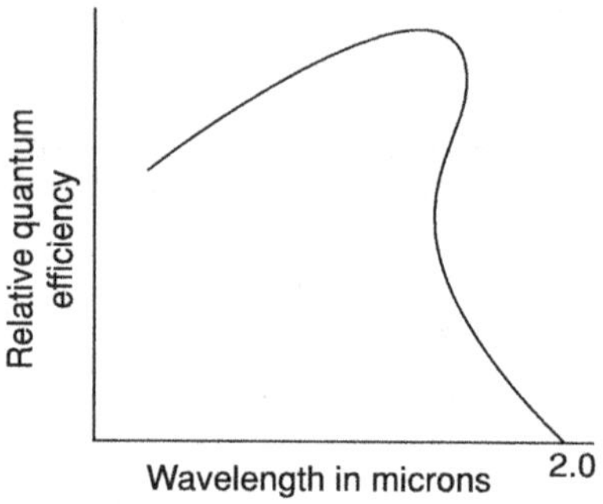

Fig. 7.43 *Relative quantum efficiency vs. wavelength of incident radiation*

The photo-transistor is a more efficient device than the photo-diode because the collector current in this is very high, and therefore, it can detect very weak electromagnetic radiations. Compound semiconductors GaAs and AlGaAs are used for the photo-transistor.

7.39 SOLAR CELL

It is a semi-conducting device which absorbs and converts the radiant energy of sunlight into electrical energy. Solar cells may be employed as light detectors, e.g. in cameras. A number of solar cells connected in series and parallel may be employed to obtain the required values of current and voltage for electric power generation.

Mostly, solar cells are made from single-crystal silicon. The same is shown in Fig. 7.44.

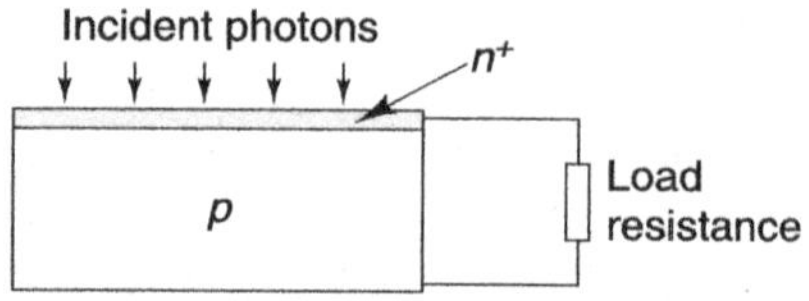

Fig. 7.44 *Schematic diagram of silicon solar cell*

Solar cells have been very expensive as a source of power generation. They have found use in space satellites and remote areas where conventional power sources are not available. The cost of solar cells may be minimised by making their area smaller and focusing sunlight on them with the help of mirrors of Fresnel lenses. Another way to reduce the cost is to replace the single-cell crystal with thin films of amorphous as polycrystalline silicon, gallium arsenide, cadmium sulphide and other compounds. GaAs or silicon is a preferred material for solar cell. The reason for this can be understood by the following graph (Fig. 7.45) which has been plotted between solar intensity and energy of incident radiation.

The maxima of the curve is near 1.5 eV or photo excitation, $h\nu > E_g$. Hence, semiconductor with band gap of 1.5 eV or lower is likely to give better solar conversion efficiency. Silicon has band gap energy of 1.1 eV; GaAs has 1.53 eV.

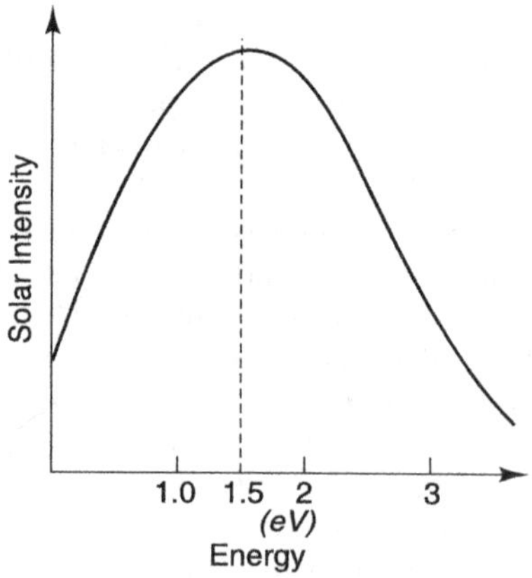

Fig. 7.45 *Solar intensity vs. energy*

Therefore, GaAs will be better option—although its band gap is higher than the band gap of silicon—due of its higher absorption coefficient. If we choose other material like CdS or CdSe, which have a band gap $Eg \sim 2.4$ eV, we can use only the high energy component of the solar energy for photo-conversion, and a major portion of energy will be of no use.

The question that arises at this stage is a why do we not use a material like PbS ($E_g \sim 0.4$ eV), which satisfies the condition $h\nu > E_g$ for ν maxima corresponding to the solar radiation spectra? If we do so, most of the solar radiation will be absorbed on the top-layer of solar cell and will not reach in or near the depletion region. For effective electron-hole separation—due to the junction field—we want the photo-generation to occur in the junction region only.

(i) Principle of Operation

The conversion of sunlight into electricity in a solar cell involves three major processes namely, absorption of light by the semiconductor material, production and separation of free electrons, and holes to different regions of the solar cell, creating a voltage in solar cell and transfer of these charges through the terminals to the load as electric current.

The absorption of sunlight by solar cell depends on the intensity and quality of light, the amount of light reflected from the front surface of solar cell, the band gap in the semiconductor—which is the minimum light energy to be absorbed by the material—and the thickness of the layer. Silicon and some other materials require tens of micrometres to absorb most of the sunlight, while others as gallium arsenide, cadmium telluride and copper sulphide require only a few micrometres.

The sunlight absorbed by the semiconductor creates electrons and holes. The electrical junction, which is the heart of the solar cell, separates these electrons and holes from one another. An electrical junction may be formed by the contact of a metal to a semi-conductor (this junction is referred to as Schotty barrier), a liquid to a semiconductor forming a photo-electrochemical cell, or two semiconductor regions known as a *p-n* junction.

As an illustration of electrical junction, consider the silicon *p-n* junction. The fixed charges at the junction due to the bound boron and phosphorus atoms create a permanent dipole charged layer with a weak electric field. When photons of sunlight fall on the crystal and create electron-hole pairs near the junction, the internal electric field at the junction forces the holes to the *p*-side and electrons to the *n*-side. This separation of charges results in a potential difference between the two regions of the crystal, with *p*-region as plus and *n*-region as minus. When a load is connected at the terminals, a current flows and useful electric power is available to the load.

(ii) Characteristics

The maximum power output of a solar cell, at full sunlight, is about 11 MW/cm^2 at an output voltage of 0.45 V. The overall conversion efficiency of the cell from solar to electric energy under these conditions operation is 11%. The output power and current are proportional to the irradiated surface area, whereas the output voltage can be increased by connecting the cells in series. The efficiency of commercial solar cells is around 10 to 12% under normal operating condition, though experimental silicon cells having efficiency up to 18% have been produced.

(iii) Arrays

Individual solar cells (size about 40 cm^2), at a 15% conversion efficiency, can deliver about 0.6 watt at 0.5 volt when in full sunlight. Higher power and voltage can be obtained by connecting a number of solar cells in arrays. The output voltage can be multiplied by connecting the cells in series and the output current by connecting the cells in parallel.

Practical solar cells made of silicon crystals were first developed only in 1955, and since 1958 they have become most exclusive power source for satellites. The Skylab, launched in 1973, had a 20 kilowatt solar cell array—the most powerful to be used so far.

7.40 PHOTO DIODE

The photo diode is a device which provides a useful means of illumination measurement level by converting time-varying optical signals into electrical signals. In most detectors, the speed of response of the device is critical. Therefore, it is desirable that the width of depletion region be large enough so that most of the photons are absorbed with in this width, rather than in the neutral *n-p* regions.

A photo diode has a depleted semiconductor region with a high electric field that serves to separate photo-generated electron-hole pairs. Such reverse biased *p-n* junction, if exposed to light, the current varies with the intensity of exposed light. This simple principle (effect) is utilised in a photo diode. It is a very sensitive device for detecting the optical signals.

A general photo-detector basically has three processes:

- Carrier generation by incident light
- Carrier transport
- Interaction of current with external circuit to provide output signal

Figure 7.46(a) shows a *p-n* junction (reverse biased) embedded in a transparent plastic. When photons (Em radiations) are allowed to incident upon this *p-n* junction, the current through the junction increases. The corresponding circuit and *I–V* characteristic are shown in Fig. 7.46(b).

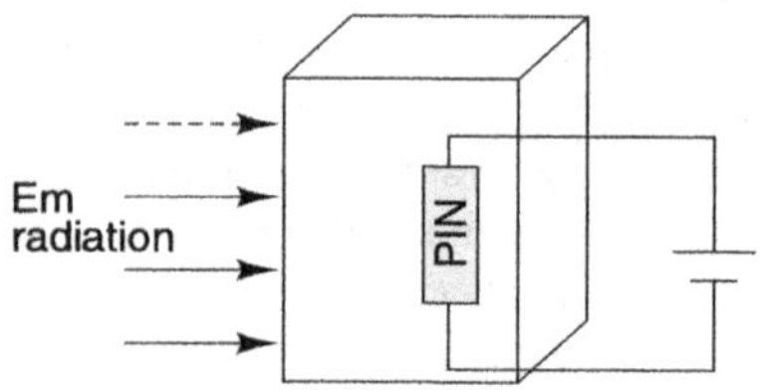

Fig. 7.46(a) *A reverse biased p-n junction*

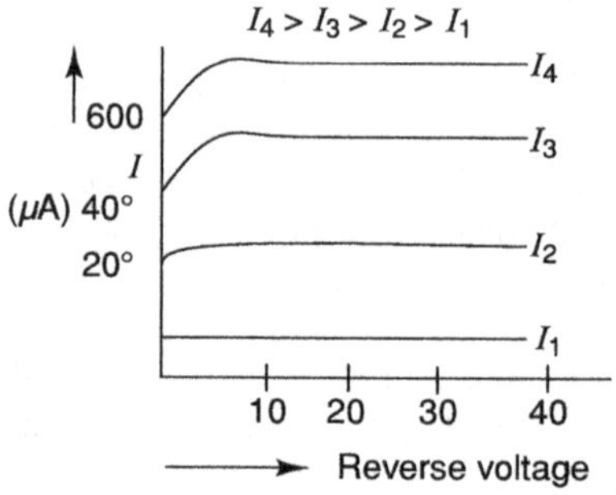

Fig. 7.46(b) *I-V characteristics at different intensities*

The entire unit is very small—nearly about 1/10th of an inch. The photons incident on the junction excite electrons from valence band to conduction band, which then generate new electron hole pairs. The electron and holes are swept by the applied field at the *p-n* junction. They move towards *n*-type region and holes toward *p*-type region. This flow of charge carriers (minority carriers) contribute to the junction current. Hence, the total current under large reverse bias is given by

$$I = I_0 + I_s$$

where I_0 is the current (reverse saturation current) known as dark current and is in the short-circuited current which is due to the illumination of light upon *p-n* junction. It is clear from the I–V characteristics that current is never zero (i.e. the curve does not pass through origin).

(i) Sensitivity of Photodiode with Position of Illumination

The current in a reverse-biased photodiode depends upon the diffusion of minority carriers to the junction. If the radiation is focussed on a small spot quite far from the junction, the injected minority carriers can recombine before diffusing to the junction, giving therefore a very small current and vice versa. The photocurrent as a function of distance from the junction is shown in Fig. 7.47.

Every photon of suitable energy (more than band gap energy) can generate an electron-hole pair. The minimum energy of a photon required for electron-hole pair generation is given by

$$E_{min} = h\nu_{min} = \frac{hc}{\lambda_{max}} = E_g$$

$$\boxed{\lambda_{max} = \frac{hc}{E_g}}$$

where λ_{max} is the maximum permitted wavelength limit known as threshold wavelength, which is characteristics of the given diode.

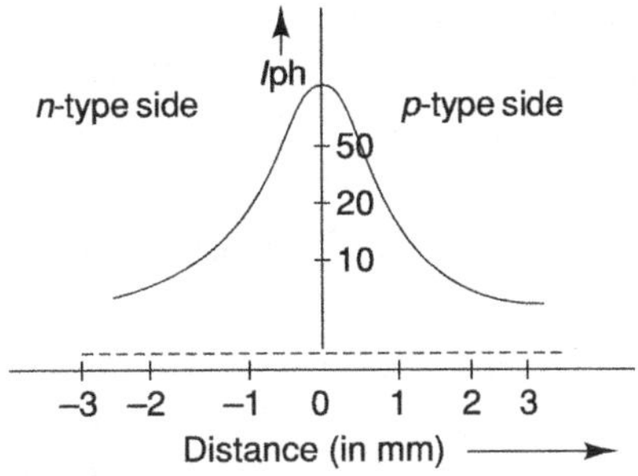

Fig. 7.47 *Photocurrent as a function of distance from* p-n *junction*

(ii) Detecting the Light Signal

Initially, the p-n junction is reverse biased and the current is noted. This reverse saturation current in the absence of light is known as dark current. The diode is then illuminated and the new current is noted. The difference between the new current and the dark current gives the photocurrent I_{ph} which then can be amplified and detected. Since the change in current is directly proportional to the intensity of incident radiation, a radiation even with small intensity can be detected. The smaller the reverse saturation current, the greater will be the sensitivity of the device.

(iii) Importance of Photodetector

Photodetectors are important in optical fibre communication systems operated near an infrared region (0.8 to 1.6 μm). They demodulate optical signals that are they convert the optical signal into electrical signal that are subsequently amplified and further processed. The photodetectors must exhibit high sensitivity at operating wavelengths, high response speed and the minimum noise. In addition to these, they should be compact in size, use low biasing voltages or current, and be reliable under operating conditions.

- For high speed operations, the depletion region is kept thin to reduce transit time.
- To increase the quantum efficiency, the depletion layer must be sufficiently thick to allow a large fraction of incident light to be absorbed.

7.41 LIGHT EMITTING DIODE

When carriers are injected across a forward-biased junction, the current is usually accounted for recombination in transition region and in the neutral region near the junction. In a semiconductor with an indirect band gap, such as silicon are germanium, the recombination releases heat to the lattice. On the other hand, in a material characterised by direct recombination, considerable light is emitted from the junction under forward bias. This effect is known as injection electro luminescence. It provides an important application of diodes as generators of light. Another important device making use of radiative recombination under forward bias is a *p-n* junction in semiconductor laser.

LED is a heavily doped *p-n* junction which under forward bias emits spontaneous electromagnetic radiation. The emitted radiation may be ultraviolet, visible or infrared depending on the type of impurity used in the semiconductor. The diode is encapsulated with a transparent cover so that emitted light can come out. The structure of LED is shown in Fig. 7.48.

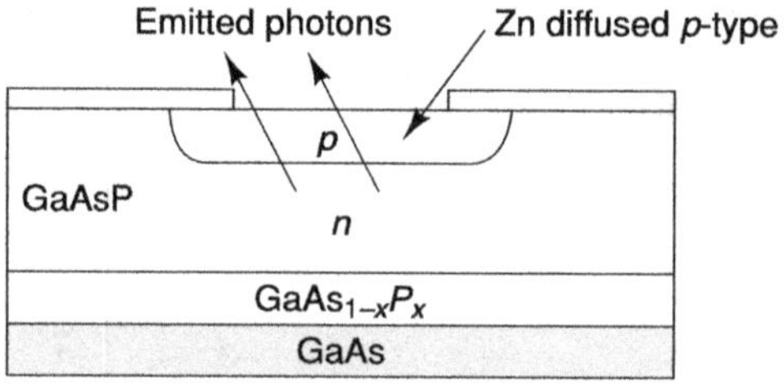

Fig. 7.48 *Schematic diagram of LED*

When the diode is forward biased, electrons are ejected into *p*-region and holes are sent into *n*-region. At the junction boundary, the concentration of minority carriers increases compared to the equilibrium concentration (i.e. when there is no bias). Thus, at the junction boundary on either side of the junctions, excess minority carriers are there that recombine with majority carriers near the junction. On recombination, the energy may be released in the form of photons. Photons with energy equal to or slightly less than the intensity of light emitted is small. This process is called **injection electro luminescence.** The recombination of electrons and holes may not always necessarily emit photons. There are other processes in which the photon is not released. Such a recombination is known as nonradiative recombination. The energy released in such recombinations is absorbed by the lattice itself. As the forward current increases, it results in a decrease in light intensity. LEDs are biased such that the light emitting efficiency is maximum.

The *V–I* characteristics of LED is similar to that of a Si junction diode. But the threshold voltages are much higher and slightly different for each colour. The reverse breakdown voltages of LEDs are very low—typically around 5 V. So, care should be taken that high reverse voltages do not appear across them.

LEDs that can emit red, yellow, orange, green and blue light are commercially available. The semiconductor used for fabrication of visible LEDs must at least have a band gap of 1.8 eV (spectral range of visible light is from about 0.4 μm to 0.7 μm, i.e. from about 3 eV to 1.8 eV). The compound semiconductor Gallium Arsenide-Phosphide ($GaAs_{1-x}P_x$) are used for making LEDs of different colours – $GaAs_{0.6}P_{0.4}$ ($E_g \sim 1.9$ eV) is used for red LED, GaAs ($E_g \sim 1.4$ eV) is used for making infrared LED, etc. These LEDs find extensive use in remote controls, burglar alarm systems, optical communication, etc. Extensive research is being done for developing white LEDs that can replace incandescent lamps in near future.

LEDs have the following advantages over conventional incandescent low-power lamps:

- Low operational voltage and less power consumption.
- Fast action and no warm-up time required.

- The bandwidth of emitted light is 100 Å to 500 Å. In other words, it is nearly monochromatic.
- Long life and ruggedness.
- Fast on-off switching capability.

Summary

- A *p-n* junction is the key to all semiconductor devices. When such a junction is made, a space charged region and depletion layer is formed consisting of immobile ion-cores devoid of their electrons or holes. This creates a junction potential barrier.
- Diodes can be used to rectify an ac voltage. With the help of a capacitor or a suitable filter, a dc voltage can be obtained.
- Zener diode is a special purpose diode. In reverse bias, after a certain voltage, the current suddenly increases (breakdown voltage) in a zener diode. This property has been used in voltage regulator circuits.
- *p-n* junctions have also been used to obtain many photonic or optoelectronic devices where photon excitation results in a change of reverse saturation current, which helps us to measure light intensity.

  Solar cells which convert photon energy into electricity, light emitting diode and diode Laser in which electron excitation by a bias voltage results in the generation of light.
- A transistor is an *n-p-n* or *p-n-p* junction device. The central block (thin and lightly doped) is called 'base', while the other electrodes are Emitters and collectors. The emitter-base junction is forward biased, while collector-base junction is reverse biased.
- The transistors can be connected in such a manner so that C or E or B is common to both the input and output. This gives the three configurations in which a transistor is used: Common Emitter (CE), Common Collector (CC) and Common Base (CB). The plot between I_C and V_{CE} for fixed I_B is called output characteristics, while the plot between I_B and V_{BE} with fixed V_{CE} is called input characteristics. The important transistor parameters for CE-configuration are

  Input resistance, $r_1 = \left(\dfrac{\Delta V_{BE}}{\Delta I_B}\right)_{V_{CE}}$

  Output resistance, $r_0 = \left(\dfrac{\Delta V_{CE}}{\Delta I_C}\right)_{I_B}$

  Current amplification factor, $\beta = \left(\dfrac{\Delta I_C}{\Delta I_B}\right)_{V_{CE}}$
- Transistor can be used as an amplifier and oscillator. In fact, an oscillator can also be considered as a self-sustained amplifier in which a part of output is fed back to the input in the same phase (positive feedback). The voltage gain of a transistor amplifier in common emitter configuration is

  $A_v = \left(\dfrac{v_0}{v_i}\right) = \beta \dfrac{R_C}{R_B}$, where R_C and R_B, respectively are the resistances in collector and base sides of the circuit.

- When the transistor is used in the cut-off or saturation state, it acts as a switch.
- A heavily doped *p-n* junction, called the zener diode, is used for voltage regulation. A zener diode is operated in reverse bias, near the breakdown voltage V_Z.
- Tunnel diode is a degenerate heavily doped *p-n* junction. Under forward bias, the diode has negative resistance in a certain voltage range and this characteristic of the diode is made use of in its applications in high frequency oscillator, pulse and digital circuits.
- A junction transistor, (*p-n-p* or *n-p-n*), by virtue of its construction and nature of biasing, is an inherent power amplifier. In the transistor, the emitter is more heavily doped than the base and the base region is made very thin, the width being less than the diffusion length of the charge carriers.
- A solar cell is a specially constructed *p-n* junction which converts light energy into electrical energy. It is also known as a photo-voltaic cell. The *p-n* junction is so designed as to collect maximum incident radiation, at the junction. The efficiency of a solar cell is about 18.7%.
- A photo diode is a *p-n* junction diode which is very sensitive to electromagnetic radiation. It is used in reverse bias. When electromagnetic radiation falls on the junction, the reverse current increases proportional to the intensity of the incident radiation. The diode is used to detect weak light signals.
- A photo transistor is also a device that detects electromagnetic radiation. It is more sensitive than a photo diode, as the current generated due to the incident radiation gets inherently amplified due to the transistor action. In a photo-transistor, it is the base-collector junction that generates the photocurrent. It can detect extremely weak light signals.
- LED is a *p-n* junction device that emits electromagnetic radiation when forward biased. LEDs are made of compound semiconductors like GaAsP and InP. The wavelength range of the emitted radiation depends on the type of semiconductor and the impurities. LEDs that emit red, yellow and green light are commonly used for displays. Infrared LEDs are used in remote control devices.
- There are two types of field effect transistors (FET)—junction FET (JFET) and metal oxide semiconductor FET (MOSFET). MOSFETs are more widely used in all modern electronic devices (ICs). FETs are uni-polar devices, i.e. the charge carriers are of only one type—either electrons or holes.
- In a JFET, the current through the channel from source to drain is controlled by the reverse gate voltage (i.e. the reverse bias applied between the channel and the gate).
- MOSFET is a four terminal device. The MOSFET is different from JFET. In the MOSFET, there is a thin surface oxide layer between the source and the drain. The conduction between the source and the drain is controlled sensitively by the positive voltage applied between the gate and the substrate.
- IC chips are devices in which a large number of electronic components (millions) are integrated on a single Si crystal of a very small size. All modern electronic appliances use IC chips.

Exercise

1. Using drift and diffusion current in a semiconductor, find an expression of continuity equation. Establish Einstein's relationship, as well.
2. What is a *p-n* diode? Explain the common technique used for the fabrication of *p-n* diode.
3. Write volt-ampere equation of a *p-n* diode. Write the expression if $V >> V_T$.
4. Define reverse saturation current. Explain its temperature dependence.
5. Draw *V–I* characteristic of a diode showing temperature dependence.
6. Explain transition capacitance and diffusion capacitance for *p-n* diode.
7. Write short notes on:
   (i) Mobility and conductivity of semiconductors
   (ii) Continuity equation
   (iii) Space charge capacitance
8. What are semiconductor devices? Give some examples.
9. Explain the constructional details of a field effect transistor (FET). Draw the schematic symbol of an *n*-channel FET.
10. With the help of neat sketches, explain the working and characteristics of a JFET.
11. Define the pinch-off voltage V_p. Sketch the depletion region before and after pinch-off.
12. What is the different between FET and BJT?
13. What are the advantages and disadvantages of FET over bi-polar transistor and vacuum tubes?
14. Explain the FET parameters, transconductance g_m, drain resistance rd and amplification factor μ. Establish the relation between them.
15. Describe the transfer characteristics of JFET. Show that

$$I_D = I_{DSS}\left[1 - \frac{V}{V_{GS(off)}}\right]$$

    where $V_{GS(\text{off})}$ is the gate-source cut-off voltage.
16. Explain how the mutual conductance of a FET varies with drain current and gate voltage.
17. What are the relative merits of an *n*-channel and a *p*-channel FET?
18. Explain why a low power FET is called as a square law device.
19. Sketch the cross-sectional view of an enhancement mode MOSFET. Explain its operation and characteristics.
20. Write short notes on the following:
    (i) Characteristics of JFET
    (ii) JFET's merits and demerits
    (iii) MOSFET
    (iv) Characteristics of MOSFET
    (v) Comparison between BJT, JFET and MOSFET
21. For semiconductors, the hole mobility is usually less than that of electron mobility. Explain this phenomenon on physical basis. How does charge carrier mobility change with doping level and temperature?
22. What is a zener diode? Discuss its formation, volt-ampere characteristics and applications.
23. Explain various charge carriers in semiconductors and their method of operation. Discuss the volt-ampere characteristics and the application of the following semiconductors: (a) Zener diode; (b) Tunnel diode.

Numerical Problems

1. A *p-n* photodiode is fabricated from a semiconductor with band gap of 2.8 eV. Can it detect a wavelength of 6000 nm? **[Ans. 4466 Å]**
2. When gate-source voltage is –12 V and gate current is 1×10^{-6} mA, determine the resistance between the gate and source of given JEFT. **[Ans. 12×10^9 ohms]**
3. In case of JFET, the variation in drain current is of 100 μA when drain-source voltage is varied by 1.25 V, keeping gate-source voltage VGS constant. Determine the dynamic resistance of the JFET. **[Ans. 1.25×10^4]**
4. In a JEFT, the drain current is changed by 0.2 mA, the gate-source voltage is changed by 0.125 V, keeping drain-source voltage constant. Calculate the transconductance of the given JEFT. **[Ans. $1.6 \times 10^{-3} \mu$ mho]**
5. A visible radiation of 5 W falls on a solar panel. It generates a current of 125 mA at 6 V. What is the efficiency of the solar cell? **[Ans. 10.4%]**
6. The energy gap in an LED is 1.5 eV. Find the wavelength of electromagnetic radiation it would emit? **[Ans. 8.33×10^{-7} m]**
7. The threshold wavelength of a photodiode is 750 nm. Calculate the energy gap in the photo diode in electron volts. **[Ans. 1.66 eV]**
8. When a reverse voltage of 10 V is applied between gate and source of JEFT, the gate current is 0.001 μA. Determine resistance between gate and source. **[Ans. 10^{10} ohms]**
9. When drain-source voltage is changed by 1.6 volts, the change in drain current is of 118 μA. The gate-source voltage remaining unchanged. Determine the ac drain resistance of the JEFT. **[Ans. 13.5 kΩ]**
10. A JEFT has V_p = –4.5 V, I_{DSS} = 10 mA and I_{DS} = 2.5 mA. Determine the transconductance. **[Ans. 2.22 m A/V]**
11. An *n*-channel JEFT has a pinch-off voltage of – 4.5 V and I_{DSS} = 9 mA. What value of V_{GS} will I_{DS} equal to 3 mA? What is its g_m at this I_{DS}? **[Ans. –19.02 V, 2.31 mA/V or 2.31 mS]**

Self-quiz Questions

1. What do you understand by reverse saturation current of a diode?
2. What types of carriers are present in space-charge region?
3. What is the effect of reverse bias on the width of a *p-n* junction?
4. What properties of a semiconductor are determined from a hall effect?
5. What is the effect of temperature on the conductivity of intrinsic semiconductors?
6. Define diffusion current in a semiconductor.
7. In which bands do the movement of electrons and holes take place?
8. Define mean lifetime of a carrier.
9. What is the main factor for controlling the thermal generation and recombination?
10. Why does a pure semiconductor behave like an insulator at absolute zero?
11. Give the names of at least two semiconductors. What is the order of conductivity of semiconductors at absolute zero?
12. Why are silicon and germanium the two widely used semiconductor materials?
13. Which of the two semiconductor materials Si or Ge has larger conductivity at room temperature? Why?
14. What is intrinsic semiconductor?
15. What is hole current?

16. Define a hole in a semiconductor.
17. What is the energy gap?
18. Classify the following as conductors or semiconductors:
    (i) Gold (ii) Germanium (iii) Silver (iv) Silicon

Objective type Questions

1. An FET has three terminals, namely, source, drain and
   (a) gird (b) substrate
   (c) ground (d) gate
2. FET is a solid state device equivalent to a
   (a) diode valve
   (b) tetrode valve
   (c) pentode valve
   (d) none of these
3. An FET is essentially a
   (a) current-driven device
   (b) voltage-driven device
   (c) power-driven device
   (d) none of these
4. An FET is a
   (a) uni-polar transistor
   (b) bi-polar transistor
   (c) tri-polar transistor
   (d) none of these
5. The input impedance of a FET is of the order or
   (a) 102° ohms
   (b) hundreds of mega ohms
   (c) hundred ohms
   (d) a few ohms
6. An FET is characterised by
   (a) current gain
   (b) voltage gain
   (c) power gain
   (d) none of these
7. The noise level in an FET is
   (a) more than BJT
   (b) negligibly small
   (c) slightly less than BJT
   (d) none of these
8. The properties of JEFT resemble those of
   (a) thermionic valve
   (b) *npn* transistors
   (c) *pnp* transistors
   (d) UJT
9. In a semiconductor, the resistivity decreases .... with temperature.
   (a) linearly
   (b) non-linearly
   (c) exponentially
   (d) any of these
10. A thermistor has .... temperature co-efficient of resistance.
    (a) zero (b) a negative
    (c) a positive (d) any of these
11. When radiation is incident on a semiconductor, its conductivity increases. This effect is called
    (a) Hall effect
    (b) Seebeck effect
    (c) photo-conductive effect
    (d) none of these
12. In a *p-n-p* transistor, majority charge carriers are
    (a) electrons
    (b) holes
    (c) both (a) and (b)
    (d) either (a) or (b)
13. In a *p-n-p* transistor, the collector current is always
    (a) less than the emitter current
    (b) equal to the emitter current
    (c) greater than the emitter current
    (d) none of these
14. Emitter arrow in a transistor shows the direction of
    (a) flow of conventional current
    (b) electron flow
    (c) either of these
    (d) none of these
15. As compared to thermionic tubes, transistors claim which of the following advantages?
    (a) Compact size, light in weight
    (b) Instantaneous operation
    (c) Long life, if operated with in the permissible limits of temperature

(d) Operating voltage quite low
(e) All of these

16. Conduction of heat in a semiconductor takes place in which of the following ways?
(a) By the thermal vibrations of the atoms
(b) By the electrons
(c) Both (a) and (b)
(d) None of these

17. Non-linear resistors are also called
(a) thermistors
(b) varistors
(c) either of these
(d) none of these

18. Photo-volatic cells find applications in which of the following?
(a) Automatic control systems
(b) Television circuits
(c) Sound motion picture recording and reproducing equipment
(d) All of these

19. The change of resistance in a magnetic field is called the
(a) magneto-resistance effect
(b) Seeback effect
(c) photoelectric effect
(d) none of these

20. In an *n*-type silicon, which of the following statements is true?
(a) Electrons are majority carriers and trivalent atoms are the dopants.
(b) Electrons are minority carriers and pentavalent atoms are the dopants.
(c) Holes are minority carriers and pentavalent atoms are the dopants.
(d) Holes are majority carriers and trivalent atoms are the dopants.

21. In an unbiased *p-n* junction, holes diffuse from the *p*-region to *n*-region because
(a) free electrons in the n-region attract them
(b) they move across the junction by the potential difference
(c) hole concentration in p-region is more as compared to *n*-region.
(d) all of these

Answers

1. (d)	2. (d)	3. (b)	4. (a)	5. (b)
6. (c)	7. (b)	8. (d)	9. (c)	10. (c)
11. (c)	12. (b)	13. (c)	14. (a)	15. (e)
16. (a)	17. (b)	18. (d)	19. (a)	20. (c)
21. (c)				

Fill in the Blanks

1. The tunnel diode is used for .... applications.
2. FET amplifier has .... gain than a bi-polar amplifier.
3. The device in which charge carriers are of single kind know as .... device.
4. The conduction in semiconductor devices takes place due to ....
5. .... is a uni-polar device.
6. Width of depletion region varies .... as concentration of impurity.
7. The order of magnitude of potential barrier in *p-n* junction is ....
8. The .... remains constant with applied reverse voltage in a *p-n* junction.
9. Einstein's relation gives a relationship between diffusion co-efficient and .... of charge carrier.
10. The reverse saturation current in *p-n* junction is due to .... charge carriers.
11. Zener diode is used as ....

12. The continuity equation states the .... of charge carries.
13. Hall effect is used to determine the .... of charge carriers.
14. Tunnel diode is a .... heavily doped *p-n* junction.
15. Solar cell utilise *I–V* characteristics in .... quadrant.
16. In phototransistor, electrical conductivity .... on illumination.

Answers

1. high frequency
2. much less
3. uni-polar
4. drift and diffusion
5. field effect transistor
6. inversely
7. 0.1 to 0.3 volts
8. reverse currents
9. mobility
10. minority
11. voltage regulator
12. conservation
13. sign
14. non-degenerate
15. fourth
16. increase

Reading List

1. Millman, Jacob; Halkias, Christos C *Integrated Electronics: Analogue and Digital Circuits and Systems*. McGraw-Hill, Kogakusha Ltd.
2. Ryder, J D *Electronic Fundamentals and Applications.* New Delhi: PHI Learning Pvt Ltd.
3. Dekker, A J (1998) *Solid State Physics.* New Delhi:Macmillan India Ltd.
4. Vijaya, M S, Rangarajan G *Material Science.* New Delhi: McGraw-Hill Education Pvt Ltd.
5. Seth, S P *Electrical Engineering Materials*. New Delhi: Dhanpat Rai & Sons.
6. Dekker, A J (2000) *Electrical Engineering Materials.* New Delhi: PHI Learning Pvt Ltd.
7. Streetman, Ben G *Solid State Electronics Devices*. 2ed. New Delhi: PHI Learning Pvt Ltd.
8. Rajaraman, V; Vishwanathan, T R; Mehta, G K *Electronics for Scientists and Engineers (Devices, Circuits and Systems)*. New Delhi: PHI Learning Pvt Ltd.
9. Mottershead, Allen *Electronic devices and circuits: An Introduction*. New Delhi: PHI Learning Pvt Ltd.
10. Van Der Ziel, A (1957) *Solid-State Electronics*. Prentice-Hall, Inc., Englewood Cliffs, N.J.
11. Gray, P E (1967) *Introduction to Electronics, Chap. 4. New York: John Wiley & Sons. Inc.*

8

Mechanism of Conduction in Semiconductors

Chapter Objectives

- To provide a knowledge of semiconductor classification.
- To understand how photons interact in direct and indirect bandgap semiconductors.
- To understand the remarkable characteristics and applications of compound semiconductors.
- To know the growth and refining techniques (modern) of semiconductors.
- To understand the transport mechanism in semiconductor materials and various kinds of currents like drift and diffusion current.
- To know Fermi level and its position in intrinsic and extrinsic semiconductors.
- To get an idea of the effective mass of charge carriers.
- To learn the measurements of various parameters like photoconductivity, dark conductivity and band gap in semiconductor materials.

8.1 INTRODUCTION TO SEMICONDUCTING MATERIALS

Semiconductors are—technologically—the newest class of material. Metals and insulators have been known since long, but semiconductors came into being only around 1950s. During this period, a revolution took place in the electronics industry with the advent of transistor, followed by a more advanced technology known as microelectronic circuitry. The beauty of these devices is their low cost, compactness and high efficiency.

The key reason behind this sudden change was very simple—the preparation of extremely pure semiconductor, and thereafter the possibility of controlling impurity in them. In the beginning, extremely pure semiconducting materials were prepared and then a desired amount of impurity was deliberately incorporated. This was a development around the 1940-50s, when by using various methods, namely crystal pulling, zone refining and epitaxial method an extremely pure semiconductor containing impurity of only 1 part in 101° was obtained. Such high purity could not be obtained even in traditionally known engineering materials.

This chapter will deal with various methods of preparation of high-purity materials. It will also discuss why the semiconducting materials begin to conduct although they are absolute insulators at very low temperature. What are the various parameters that affect the conduction process and how their conductivity can be controlled—a nice flexibility is not available with their counterpart metals.

8.2 POSITION IN THE PERIODIC TABLE

There are many materials that exhibit the semiconducting property, but the Ge and Si are the most popular. These are also known as elemental semiconductors as they exhibit only one kind of atoms.

Table 8.1 *A portion of the periodic table*

IIIA	*IVA*	*VA*	*VIA*	*VIIA*
B	C	N	O	F
Al	Si	P	S	Cl
Ga	Ge	As	Se	Br
In	Sn	Sb	Te	I
Tl	Pb	Bi	Po	At

The semiconductors shown in Table 8.1 belong to IV-A subgroup of the periodic table. There are a large number of compound semiconductors, as well, for example, metallic-oxides and sulphides. These are extremely promising as they exhibit a wide range of applications. For instance, PbS is used in photoconductive devices, BaO in oxide coated cathodes, CsI in photomultiplier tubes, etc. The detailed description about compound semiconductors will be given in the forthcoming sections.

8.3 BASIC STRUCTURE OF Ge AND Si

Silicon and germanium belong to group IV of the periodic table along with carbon. Carbon is a very interesting element. It is one of the typical examples of a material that exists in two different crystal forms, exhibiting entirely different properties. Diamond is a perfect electrical insulator and is one of the hardest known substance, whereas graphite is a good conductor and soft material. Diamond is used as a cutting tool due to its hardness, while graphite is used as lubricant due to its softness. In solid state, they crystallise into diamond structure represented in Fig. 8.1(a). In this structure, a given atom is surrounded by four other atoms occupying their position at corners of a regular tetrahedron.

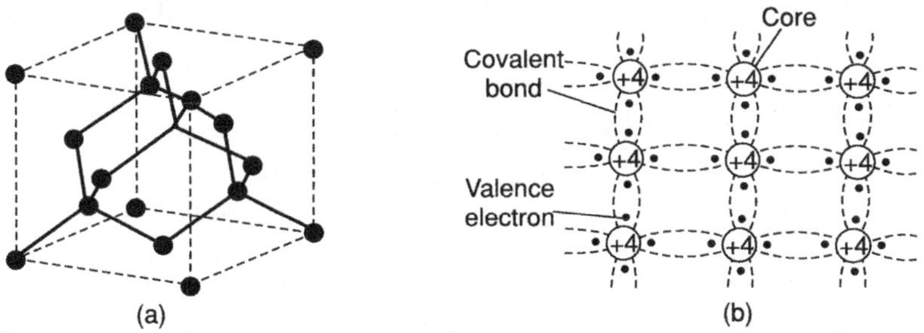

Fig. 8.1 *(a) Diamond lattice (b) Bonding arrangement in one-dimensional lattice*

The bonding arrangement between the atoms in this structure and their neighbours is known as electron-pair bonds or covalent bonds because bonds of this kind are accomplished by a pair of electrons—each partner atom contributes one electron to the bond. Since each atom has four valence electrons, it is just enough to provide for electron-pair bonds with four other atoms. A two-dimensional picture of bonding arrangement is also shown in Fig. 8.1(b) The crystal structure and parameters lead to the energy-band structure with a wide bandgap of 5.4 eV in diamond and nearly 1.0 eV in Ge or Si. Such high value of bandgap is responsible for the insulating property of diamond.

In an actual three-dimensional lattice, the angle between the bonds is nearly 109° in such a way as it is found in CH_4 molecule. Such type of bonds, in which the atoms share electrons with each other, are known as homopolar.

At absolute zero—in both semiconductors—all the electrons participate in bond formation. Therefore, no free electron is available in these and this is the reason why these elements behave like insulators. Thus, at absolute zero, silicon and germanium are insulators as there are no mobile charge carriers. Also, these materials exhibit negative temperature coefficient of resistance. Therefore, if the temperature is raised some of the valence electrons may then absorb sufficient energy and become free from their bonds. As a result, they can move through the crystal on applying external electric field. At this stage, the material can show a substantial conductivity.

The variation (increase) in conductivity of these materials follows the following equation:

$$\sigma = \sigma_0 e^{-E_g/kT}$$

where E_g is the bandgap energy and T is the temperature of semiconductor material.

This feature separates semiconducting materials from insulators. Thus on supplying thermal energy, a number of bonds may be broken and conduction may be observed as a result of the movement of electrons and holes under the influence of an electric field. Therefore at higher temperature, a vacant state in the conduction band is occupied by a mobile electron leaving behind another vacant state in some other bond, which is again occupied by the other electron, and so on. This process of generation and recombination of charge carriers may be written in the form of a chemical reaction as follows:

$$\text{Bound electron} \rightleftharpoons \text{Free electron} + \text{Hole}$$

The extent to which this reaction proceeds to the right at a given temperature depends on the chemical bond. If the electrons are bound strongly, the number of charge carriers will be relatively small.

Comparison between Certain Parameters of Ge and Si

Some important parameters related with Ge and Silicon are described in Table 8.2.

Table 8.2 *Certain parameters of Ge and Si*

Parameter	*Silicon*	*Germanium*
Atomic number	14	32
Structure	Diamond cubic	Diamond cubic
Valency	4	4
Energy gap (at RT)	1.12 eV	0.66 eV
Mobility of electrons at RT	0.14 m^2/Vs	0.38 m^2/Vs
Mobility of holes at RT	0.05 m^2/Vs	0.18 m^2/Vs
Dielectric constant	12	16
Melting point	1410°C	937°C

(Contd)

Parameter	*Silicon*	*Germanium*
Specific gravity	2.33	5.32
Operating temperature	200°C	80 – 85°C
Economy	costly	cheaper

8.4 CLASSIFICATION OF SEMICONDUCTORS

The band theory provides a clear-cut classification of the materials; according to which there are three categories namely, metals, insulators and semiconductors. Most of the metals exhibit zero bandgap, insulators 5 eV or more, while semiconductors exhibit around 1.0 eV bandgap. These are materials that lie intermediate between metal and insulators from the resistivity point of view. Table 8.3 gives the electrical resistivities of various materials at 200°C.

Table 8.3 *Electrical resistivity of various materials at 20°C (in Ohm-m)*

Metals		*Semiconductors*		*Insulators*	
Material	*Resistivity*	*Material*	*Resistivity*	*Material*	*Resistivity*
Ag	1.6×10^{-8}	Germanium	0.47	Glass	$10^{10} - 10^{11}$
Cu	1.7×10^{-8}				
Al	2.8×10^{-8}	Silicon	3000	Mica	3.0×10^{14}
Fe	10×10^{-8}	In Sb	2.0×10^{4}	Diamond	10^{14}
Constantan	49×10^{-8}				

With the advance of technical knowledge other varieties of semiconducting materials have also been developed. They have been proved to be extremely promising, efficient and useful. These newly developed materials have replaced their traditional counterparts namely, germanium and silicon. In fact, they are now being used in countless devices. The classification or kinds of semiconductor materials and some parameters related with them are shown in the flowchart in Fig. 8.2 and Table 8.4.

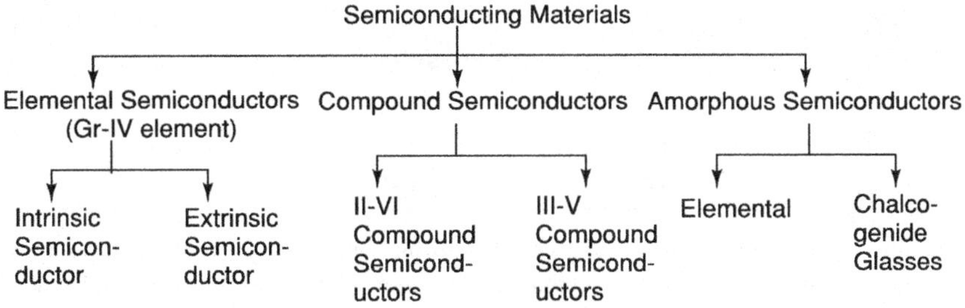

Fig. 8.2 *Classification of semiconductors*

Table 8.4 *Bandgap of some semiconductors*

Group	*Material*	*Bandgap (in eV)*
IV	Carbon	5.4
	Silicon	1.1
	Germanium	0.67
	Silicon Carbide	5.4

(Contd)

Group	*Material*	*Band-gap (in eV)*
III-V		
	AlAs	2.2
	AlSb	1.6
	GaN	3.42
	GaP	2.42
	GaAs	1.42
	In N	1.89
	In As	0.36
	In Sb	0.17
	ZnS	3.67
II-VI	ZnSe	2.58
	ZnTe	2.26
	CdS	2.42
	CdSe	1.74
	CdTe	1.50

8.5 ELEMENTAL SEMICONDUCTORS

Elements belonging to the IVth group of periodic table are known as elemental semiconductors. These are well-known traditional semiconductors; their atomic number, outermost shell configuration and lattice spacing is given in Table 8.5.

Table 8.5 *Electronic configuration of Group IV materials*

Element	*Atomic Number*	*Outermost shell Configuration*	*Lattice spacing (Å)*
C	6	$2s22p^2$	3.56
Si	14	$3s23p^2$	5.43
Ge	32	$4s^24p^2$	5.66

The above elements in Table 8.5 exhibit diamond-like lattice. In all these elements—Ge, Si or C—diamond lattice is surrounded by four nearest neighbours, each with four electrons in the outermost orbit. In these crystals each atom shares its valence electrons with its four neighbours.

Bonding between the nearest neighbour atoms in the diamond lattice diagram is shown in the preceding section. The bonding forces arise from a quantum mechanical interaction between the shared electrons. This is known as covalent bonding; each electron pair constitutes a covalent bond. In the sharing process it is irrelevant to think which electron belongs to a particular atom—both belong to a bond. The two electrons are indistinguishable except that they are characterised by opposite spins to fulfil the requirements of Pauli's exclusion principle.

Similar to ionic crystals, these elements also do not have any free electrons in the diamond lattice. so, Ge and Si must be insulators. Of course this is true but at 0 K. However, at moderate temperatures an electron can be thermally or optically excited out of a covalent bond, and therefore can show conduction. This is one important feature of the above semiconducting materials.

8.6 CLASSIFICATION OF ELEMENTAL SEMICONDUCTORS

Elemental semiconductors are interesting materials, but not too much as their electrical properties cannot be altered to a large extent. These materials are classified as

- Intrinsic semiconductor
- Extrinsic semiconductor

Intrinsic Semiconductor

The semiconducting material in its pure (no impurity added) form is generally known as intrinsic semiconductor. In such materials there are no charge carriers at 0 K as the valence band is completely full of electrons and the conduction band is empty. At higher temperatures, electron-hole pairs are generated as the electron of valence band can cross the energy gap by incident thermal energy. These electron hole pairs are the only charge carriers in these materials denoted by EHP.

Since electrons and holes are created in pairs, the conduction band electron concentration n is equal to the concentration of holes in valence band p. Each of these intrinsic carrier concentrations is commonly denoted by n_i. Therefore, for intrinsic materials

$$n = p = n_i$$

At a given temperature, there is a certain concentration of electron-hole pairs n_i. Obviously, if a steady state carrier concentration is maintained, there must be recombination of EHPs at the same rate at which they are generated. Recombination occurs when an electron in the conduction band makes a transition (direct or indirect) to some empty state (hole) in the valence band; thus, annihilating the pair. Denoting the generation rate by g_i and recombination rate by r_i, in equilibrium:

$$r_i = g_i$$

Each of the above rates is temperature dependent. For example, g_i increases with the rise in temperature. As a consequence a new carrier concentration is established such that the higher recombination rate r_i balances the generation. At any temperature, we can predict that the rate of recombination of electrons and holes r_i is proportional to the equilibrium concentration of electrons n_0 and the concentration of holes, p_0.

$$r_i \propto p_0 n_0 \qquad \propto n_i^2 = g_i$$

where α is a constant which depends on the particular mechanism by which recombination takes place.

8.7 IMPURITY ADDITION IN SOLID MATERIALS—DOPING

The impurities—whether they are substitutional or interstitial—act as lattice defects in the solid material. Some impurities such as oxygen in silicon, tend to be definitely inert. However, other impurities, e.g. gold or phosphorus in silicon can drastically change the electrical properties of the material.

Deliberate addition (in controlled amounts) of impurities in a semiconductor material alters the conductivity favourably. This technique of impurity addition to change the electrical characteristics is commonly known as doping. There are two general methods of doping:

- Impurity diffusion
- Ion implantation

Impurity diffusion occurs when a semiconductor crystal is placed in a high temperature nearly 1000°C in gaseous atmosphere containing the desired impurity atom. At such a high temperature, many of the crystal atoms can randomly move in and out of their single-crystal lattice sites. Vacancies may be created by this random motion so that impurity atoms can move through the lattice by hopping from one vacancy to the other. Impurity diffusion is the process by which impurity atoms move from a high concentration region to lower one within the crystal. When temperature goes down, these impurity atoms become permanently frozen into the substitutional lattice sites.

Ion implantation generally occurs at low temperature. A beam of impurity ions is accelerated to high energies of order of 50 keV or more, which is then directed to hit the semiconductor surface. The high energy impurity ions enter the crystal and come to rest somewhere within the lattice. This is the method that carries an advantage—one can control the number of impurity atoms into the semiconductor lattice. A disadvantage of this technique is that the incident impurity atoms collide with the crystal atoms causing lattice displacement damage. This lattice damage can be removed by thermal annealing.

8.8 EXTRINSIC SEMICONDUCTOR

Extrinsic semiconductors can be obtained by doping the pure semiconductor by IIIrd or Vth group elements of the periodic table. The real power of semiconductor materials is realised when these are doped by group III or V materials. Even a very little impurity addition (nearly 1 : 10^{12} atom) leads to a drastic change in the electrical properties of these materials. This is the reason why these materials have gained extreme importance compared to their counterparts. The impurity addition to elemental intrinsic semiconductor—in controlled amounts—makes them extrinsic. When impurities or lattice defects are introduced in a perfect semiconducting crystal, additional levels are created in the energy gap as shown in Fig. 8.3.

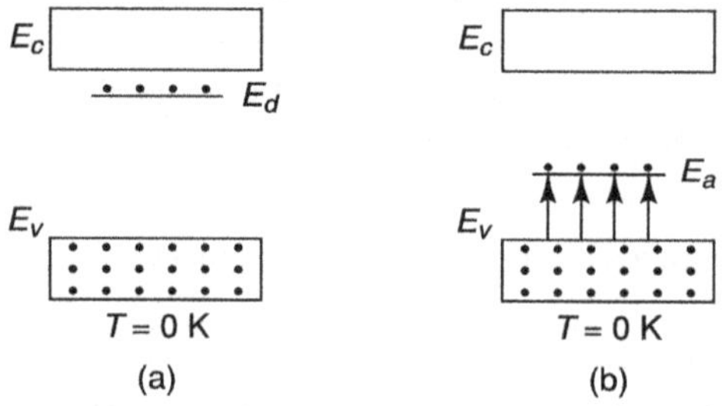

Fig. 8.3 *(a) Donor levels at various temperatures*
(b) Accepter levels at various temperatures

Introducing a group V element in a pure semiconductor creates a level close to the conduction band, and that of group III element creates a level near to the valence band. Actual position of these impurity levels will be estimated in the forthcoming sections. The impurity levels created by group V impurity atoms is known as donor level, and those created by group III element is called acceptor level. Thus, addition of group III impurity leads to increase in number of holes, and that of group V element leads to increase in the number of electrons. As these materials are highly sensitive to impurity addition, one can use these to fabricate the various semiconducting devices. Depending on the type of impurity (group III or V) addition, these materials can be categorised into either *p*- or *n*-type semiconductors. The majority carriers in *p*-type semiconductors are 'holes', while in *n*-type are 'electrons'. Many more important characteristics of these materials and their conduction mechanism will be discussed the in forthcoming sections. The bonding arrangement in *p*- and *n*-type semiconductors is shown in Fig. 8.4.

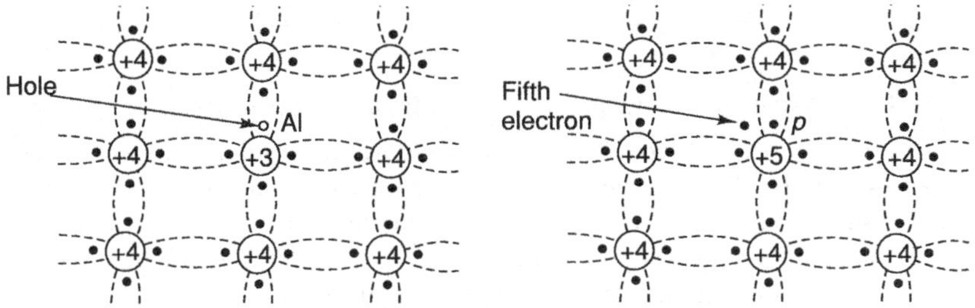

Fig. 8.4 *(a) p-type semiconductor (b) n-type semiconductor*

8.9 COMPOUND SEMICONDUCTORS

Besides germanium and silicon—that were earlier mastered—there are other semiconductors that are the compounds of groups IIIB-VB and group IIB-VIB, respectively. Figure 8.5 gives the classification of compound semiconductors.

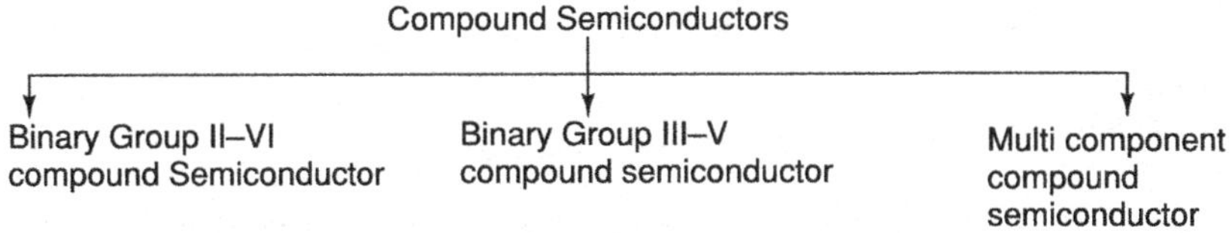

Fig. 8.5 *Classification of compound semiconductors*

In this section, these semiconductors and their excellent features will be discussed in detail.

(i) Binary III-V Compound Semiconductors

Let us talk first about binary III-V compound semiconductors. A binary compound is the one which is composed of two elements. Table 8.6 illustrates some III-V compounds (binary semiconductors).

Table 8.6 *Some III-V compound semiconductors*

Group III elements	Group V elements	Group III-V Compound Semiconductor
B, Al, Ga, In	N, P, As, Sb	BN, AlN, AlP, AlAs, GaN, GaP, GaAs, GaSb, InP, InAs. InSb

Most of the III-V compound semiconductors crystallise in the Zinc Blende structure. This is very similar to the diamond structure with C atoms replaced alternately with a III and V or II and VI elements. Therefore, all bonds are between unlike atoms. The same is shown in Fig. 8.6 and Fig. 8.7.

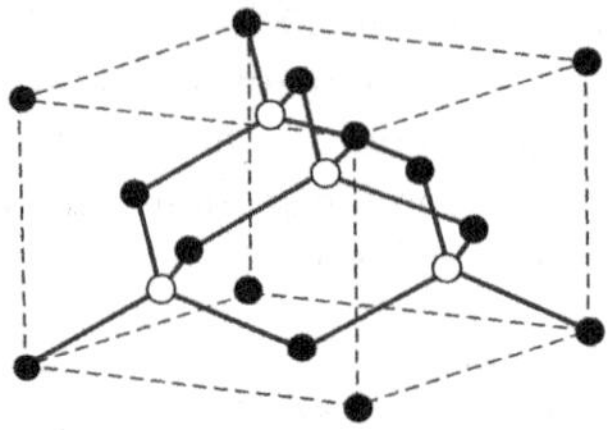

Fig. 8.6 *Zinc Blende Structure*

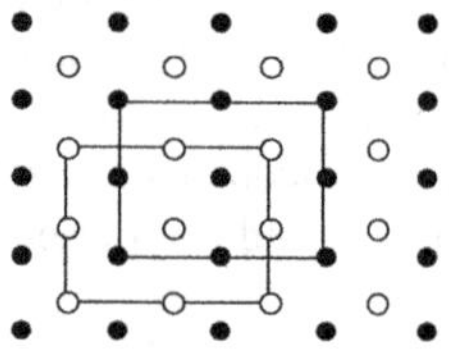

Fig. 8.7 *Top view of Zinc Blande*

Similar to germanium and silicon, these compounds are also insulator at low temperature. This is because all the electrons participate in bond formation and none are available free which could show conduction. Not only that, at higher temperatures, breaking of covalent bond takes place thereby making the free electrons available into the conduction band. The only feature that distinguishes these compounds from elemental Ge or Si is that the bonding is ionic in these, while in Ge or Si it is covalent. As an example, in the case of GaAs the three outer electrons of Ga make a tie-up with five electrons of As in order to complete the equilibrium (stable) configuration.

GaAs is one of the oldest and last developed (technologically) compound semiconductors that got extreme attention around 1950, and has since been a preferred material in numerous devices. The reason was simple—arsenic remained untouched, as it was dangerous. One striking feature that was noticed in GaAs was its high value of mobility as compared to Ge or Si. Fig. 8.8 illustrates this fact.

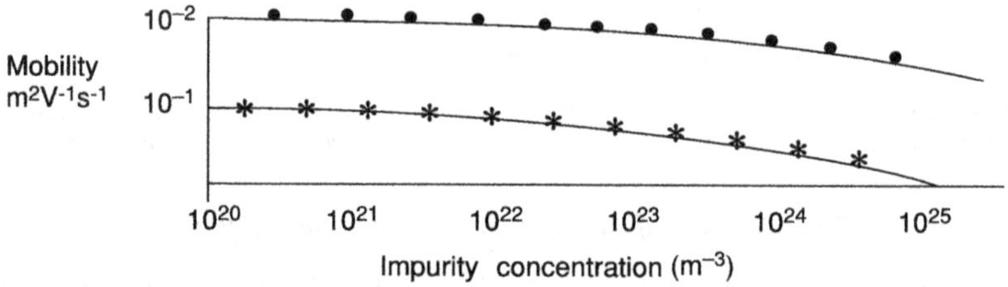

Fig. 8.8 *Electron and hole mobilites in GaAs with impurity concentration*

Band-gap of this material is also quite large—1.42 eV comparid to Ge or Si. This shows that it is not easy to break a covalent bond in GaAs. Again, the reason to this is ionic bonding. An interesting feature of GaAs is that one can tailor the bandgap by suitably selecting a proper combining material. This flexibility made these materials quite useful material for device applications. It is clear from Table 8.7, how the bandgap increases when we move upward in the table and decreases as we move downwards.

Table 8.7 *Variation of bandgap in GaAs*

Parent Element	*Combining Element*	*Atomic No.*	*Group in Periodic Table*	*Compound*	*Band-gap (in eV)*
	N ↑	7		GaN	3.42
	P	15		GaP	2.24
Ga	As ↓		VB	GaAs	1.42
	Sb	51		GaSb	0.67

The same is true with compounds of Al and In, as shown in Table 8.8.

Table 8.8 *Variation of bandgap in various compounds of Al and In*

Parent Element	*Combining Element*	*Atomic No.*	*Group in Periodic table*	*Compound*	*Band-gap (in eV)*
	P ↑	15		Al P	3.34
Al	As	33	VB	Al As	2.2
	Sb ↓	51		Al Sb	1.6
	N ↑	7		In N	1.89
	P	15		In P	1.27
In	As	33	VB	In As	0.36
	Sb ↓	51		In Sb	0.17

The general rule that can be observed is that the downside elements lead to a decrease in bandgap, and vice versa. This is quite justified as downside elements exhibit a high atomic number as against a low atomic number of the upside elements. Hence, valence electrons are quite farther in downside and nearer in upside combining elements. Electrons are tightly bound in upside compounds and loosely bound in the downside case. Thus, bonding force is weaker for downside compounds and vice versa.

(ii) Binary II-VI Compound Semiconductors

Table 8.9(a) shows some II-VI compound semiconductors. These semiconductors exhibit an ionic character less than III-V compounds. The variation of bandgap in these compounds follows exactly the same rule as in III-V compounds, which are described in Table 8.9(b).

Table 8.9(a) *Some II-VI compound semiconductors*

Group II elements	*Group VI elements*	*Group II-VI Compound Semiconductors*
Cd, Zn	S, Te, Se	CdS, CdTe, CdSe, ZnS, ZnTe, ZnSe

Table 8.9(b) *Variation of bandgap in II-VI compounds*

Parent Element	*Combining Element*	*Atomic Number*	*Group in Compound*	*Band-gap*	*Periodic table (in eV)*
	Se	34	VI B	Zn Se	2.58
Zn					
	Te	52		Zn Te	2.26
	Se	34		Cd Se	1.74
Cd			VIB		
	Te	52		Cd Te	1.50

The main problem that arises with II-VI materials is that some compounds could be made *p*-type, some others *n*-type, but no single compound could be made both *p*- and *n*-type.

The possibility to make III-V compounds either *p*- or *n*-type appears easy in principle, but practically of course it is not so easy. If there is excess Ga, it will be *p*-type; and if

As is in excess, it will be *n*-type. Group IV elements, such as silicon and germanium, can also be impurity atoms of gallium arsenide. If a silicon atom replaces a gallium atom, the silicon impurities will act as a donor; but, if the silicon atom replaces an arsenic atom, the silicon impurity will act as an acceptor. The same is true for germanium as an impurity atom. Such impurities are known as amphoteric. Experimentally, it is found in gallium arsenide that germanium is predominantly an acceptor and silicon is predominantly a donor. Table 8.10 lists the ionisation energies for the various impurity atoms in gallium arsenide.

Table 8.10 *Impurity ionisation energies in gallium arsenide*

Impurity (Donors)	*Ionisation energy (eV)*	*Impurity (Acceptors)*	*Ionisation energy (eV)*
Selenium	0.0059	Beryllium	0.028
Tellurium	0.0058	Zinc	0.0307
Silicon	0.0058	Cadmium	0.0347
Germanium	0.0061	Silicon	0.0345
		Germanium	0.0404

Previous discussion was confined to binary compounds. In principle, there is no difficulty in preparing a ternary compound containing three elements.

(iii) Ternary and Multi-component Compound Semiconductors

The excellent opportunity to tailor the bandgap of compound semiconductors—by adding one or more element—makes these extremely useful for device applications. For different devices, different bandgap material is required. This is only possible when a third material is added in compound semiconductor lattice. We know that the more Al is added, the larger will be the bandgap of GaAs. The sources for optical communications, of course, require a material that can only be obtained by suitably incorporating a third element to GaAs lattice. As an example, the preferred material for blue-green semiconductor laser is GaN, but its energy gap puts it in the UV region. This energy gap can be adjusted by replacing some of the Ga atoms with In (heavier the element lower will be the gap). The usefulness of III-V compounds is directly linked with the ability to vary the mixture of elements on each of the two interpenetrating FCC sub-lattices of the zincblende crystal. For example, in the ternary compound AlGaAs, it is possible to vary the composition of the ternary alloy by choosing the fraction of Al or Ga atoms on the column III sub-lattice. It is common to represent to composition by assigning subscripts to the various elements. For example, $Al_xGa_{1-x}As$ refers to a ternary alloy in which the column III sub-lattice in the zincblende structure contains a fraction x of Al atoms and $1-x$ of Ga atoms. The composition $Al_{0.3}\,Ga_{0.7}As$ has 30% Al and 70% Ga on the column III sites, with the interpenetrating column V sub-lattice occupied entirely by As atoms. It is extremely useful to be able to grow ternary alloy crystals such as this with a given composition. For the Al_xGa_{1-x} As example, we can grow crystals over the entire composition range from $x = 0$ to $x = 1$; thus, varying electric and optical properties of the material from that of GaAs ($x = 0$) to that of AlAs ($x = 1$). To vary the properties even further, it is possible to grow four-element (quaternary) compounds such as $In_x\,Ga_{1-x}\,As_y\,P_{1-y}$ having a very wide range of properties.

8.10 DIRECT AND INDIRECT BAND-GAP SEMICONDUCTORS

A moving electron within the material experiences a periodic potential that gives rise to certain allowed values of its energy which can be plotted *E* vs. wave vector *k*. The plot between *E* vs. *k* is very interesting and useful in understanding the characteristic property of a material. Since the periodicity of most lattices is different in different directions, the *E-k* diagram must be plotted for different directions. Such a diagram provides a full relationship between *E* and *k* that is normally a complex surface which should be visualised in three dimensions.

Let us examine the band structure of GaAs. It has a minima of conduction band and maxima of valence band at same *k*-value ($k = 0$). A semiconductor with this property is said to be a direct bandgap semiconductor. In such materials, the transition between two allowed bands can take place with no change in crystal momentum. This direct nature has a significant effect on the optical properties of the material.

GaAs and other direct bandgap materials are ideally suited for semiconductor laser and other photonic devices. Figure. 8.9 presents the *E-k* diagram for direct and indirect bandgap semiconductors.

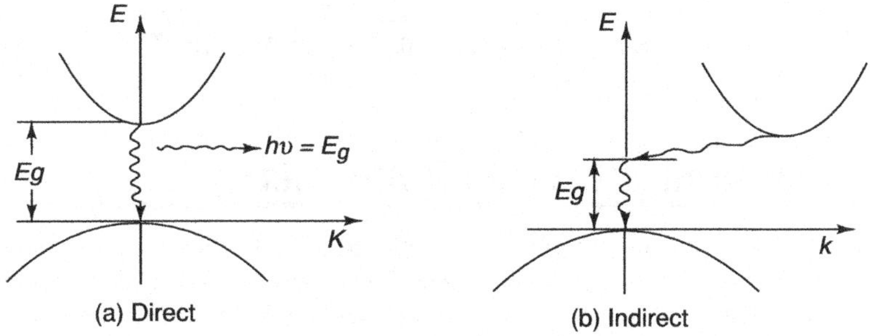

Fig. 8.9 *Direct and indirect bandgap semiconductor*

The *E-k* diagram for silicon does not have the conduction band minima and valence band maxima at the same *k*-value. The difference between the minimum conduction-band energy and maximum valence-band energy is still defined in terms of the bandgap energy, E_g. A semiconductor material with this characteristic is known as indirect bandgap semiconductor. When electrons make transition from conduction to valence band, there must be a change of momentum of the electron involved. GaP and AlAs are some materials of this kind.

In a direct bandgap semiconductor (e.g. GaAs), an electron from a conduction band jumps to an empty state in the valence band, giving off the energy difference in the form of visible light photon. On the other hand, in an indirect material, there is no such visible photon. The electron of conduction band in such materials does not fall directly to the valence band, but it may go through some defect state. In an indirect transition, which involves a change in *k*, the energy is generally given up as heat to the lattice rather than as emitted photon.

The above difference between direct and indirect band structures is very important in deciding which semiconductor material can be suitable for devices requiring light output. For example, semiconductor light emitters and laser are generally made from direct bandgap materials or from indirect bandgap materials with vertical transitions between the defect states.

Example 8.1

GaAs, a well-known and direct bandgap semiconductor—having energy gap of 1.4 eV—is irradiated by photons of energy 1.5 eV. If the effective mass of electrons and holes in GaAs are 0.07 and 0.068 times the free electron mass, respectively, calculate the kinetic energy and momenta of the carriers.

Solution

$$E_{photon} = E_g + E_{KE}$$

$$E_{KE} = E_{photon} - E_g = 1.5 - 1.4 = 0.1 \text{ eV}$$

Momentum of carriers

$$p = \sqrt{2mE}$$

Momentum of electrons $pe = \sqrt{2m_e E_{KE}}$

$$= \sqrt{2 \times 0.07 \times 9.1 \times 10^{-31} \times 0.1 \times 1.6 \times 10^{-19}}$$

$$= 4, 5 \times 10^{-26} \text{ kg m/s}$$

Momentum of holes

$$p_h = \sqrt{2m_e E_{KE}}$$

$$= \sqrt{2 \times 0.068 \times 9.1 \times 10^{-31} \times 0.1 \times 1.6 \times 10^{-19}}$$

$$= 4.4 \times 10^{-26} \text{ kg m/s}$$

Ans.

8.11 AMORPHOUS SEMICONDUCTORS

Amorphous semiconductors are important from the point of view of the economy and whenever we require large surface area. These can be prepared in thin-film form by thermal evaporation on a substrate, or by rapid cooling of the melt. The material—in bulk—is sealed in a quartz ampoule under high vacuum, and then kept in a furnace at high temperature (nearly 1000°C or more). Rocking of melt (stirring of the melt) is done to maintain homogeneity. This ampoule is suddenly quenched into the ice-water or LNT. The material, so obtained, exhibits a glassy structure in which the crystalline periodicity or long-range order is found missing. Therefore, Block theorem is not applicable in the case of such materials, and hence, the electron states cannot be described by well-defined k-values. The momentum selection rule for optical transition is relaxed, and therefore, all Infrared and Raman modes contribute to the absorption spectra. The optical absorption edge is also featureless. Allowed bands and energy gaps still exist because of the presence of short range order.

Both electrons and holes can carry current in amorphous semiconductors. The carriers may be scattered strongly by disordered structure, so that the mean free path may sometimes be of order of the scale of disorder. Amorphous materials also appear to behave similar to intrinsic semiconductors. The position of Fermi-level lies nearly in the middle of the gap. The pinning of Fermi level is found in these materials, which is due to the presence of defects such as dangling bonds or other misfit in the structure. These defects produce localised states in the energy gap. The glassy semiconductors were found almost insensitive to impurity addition in the past, but only around 1967 it was shown (experimentally) that these materials can also be doped to n- or p-type, similar to their crystalline counter parts. With this possibility, these materials then also become attractive for a use in device applications.

Two distinct classes of amorphous semiconductors are widely-studied:

- Tetrahedrally bonded amorphous solids like Ge or Si
- Chalcogenide glasses

The tetrahedrally bonded materials have similar properties to those of crystalline counterparts, provided the defects due to dangling bonds are compensated with hydrogen. These materials can be doped and their conductivity can be sharply modified by injection of free-charge carrier from a metallic contact.

Amorphous hydrogenated silicon is a material that is, nowadays, being used in solar cells. Amorphous silicon is economical as compared to single-crystal silicon. In early days, attempts were made to use pure amorphous silicon, but it could not be translated into reality because of the presence of structural defects commonly known as dangling bonds. Introduction of hydrogen was found quite successful in removing the above-stated defects. Normally, large amounts of hydrogen—of the order of 10% or more—are incorporated.

Compound semiconductors prepared in the glassy form belong to a new class of semiconductors known as glassy semiconductors. These materials have acquired a great scientific interest due to their easy preparation in thin film or bulk form. Besides, these materials also exhibit some special characteristics that can be utilised in many high-performing optical and electronic devices. Examples of some semiconducting glassy alloys are As-se, As-S, As-S-Se, Se-Te, Ge-Se, etc. These materials are also called chalcogenide glasses and they contain at least one chalcogen element (S, Se, or Te). These are the elements of VI group of periodic table.

8.12 REMARKABLE CHARACTERISTICS OF CHALCOGENIDE GLASSES

- Amorphous materials have same short-range structure like its crystalline counterpart, but have distributed bond-length and bond-angle.
- Chalcogenide glasses are solids with covalent bonds—maximum ionic conductivity is 9%.
- The properties of chalcogenide glasses are significantly different from oxide glasses. They are, therefore, also called non-oxide glasses.
- Chalcogenide glasses shows electronic conduction, while oxide glasses, e.g. phosphate silicate and borate glasses do not show such a type of conduction.
- Chalcogens are usually combined with Sb, As or P, and often with Ge, Si or B, and form binary and multi-component system.
- These glasses are extremely promising materials for various optical and photonic applications.
- They are also known as tailor-made materials, as they can be prepared as per the requirements by changing composition in a particular glassy system.
- Chalcogenide glasses—pure or doped—can be used for fibres production.
- Chalcogenide fibres are transparent in the IR region, and can be used for sensing and remote sensing of many chemical substances in chemistry, medicine, environmental applications, optical-signal processing, non-bleeding surgery, dental surgery, etc.

8.13 SELECTION OF SEMICONDUCTOR MATERIALS FOR VARIOUS DEVICES

The flowchart in Fig. 8.10 lists the names of suitable materials that could be used for various given devices.

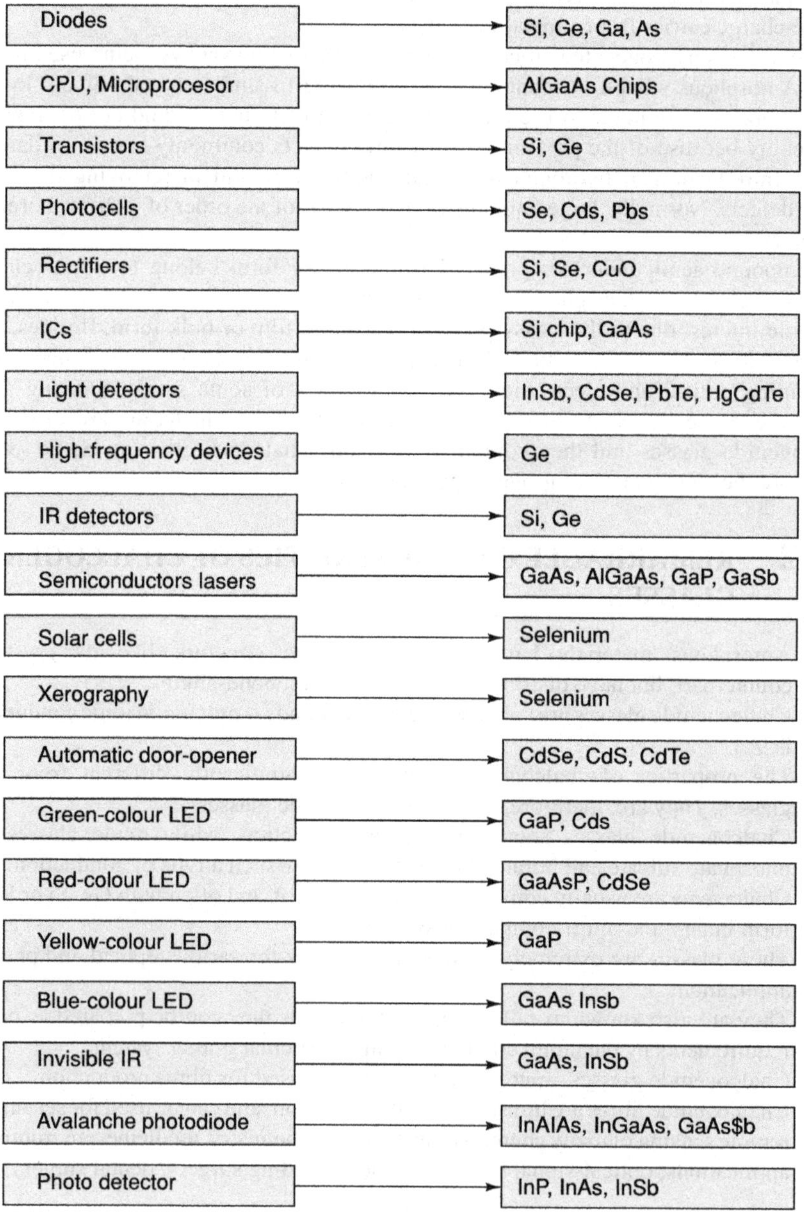

Fig. 8.10 *Semiconductor materials for various devices*

8.14 GROWTH AND REFINING OF SEMICONDUCTOR MATERIALS

The progress of solid state device technology, since the invention of transistor in 1948, has depended not only on the development of device concept, but also on improvement of the materials. The fact that ICs, LSI and VLSI can be made today, is the result of a considerable breakthrough in the growth of pure, single-crystal of Si around 1950s.

8.15 PREPARATION OF ELECTRONIC GRADE SILICON (EGS)

SiO_2 is the raw material for silicon crystal. It is reacted with carbon in an arc furnace at a temperature nearly 1000°C to reduce SiO_2, as given in the below reaction:

$$SiO_2 + 2C \longrightarrow Si + 2CO$$

Silicon obtained in this way is of metallurgical grade (MGS); it contains impurities like Fe, Al and heavy metals. The concentration of these impurities range from several hundred to several thousand Parts Per Million (PPM). The silicon so obtained is not suitable to be used for electronic applications. Besides, it is also not in a single-crystal form.

The MGS, therefore, is further refined and is made of EGS grade which involve the following reaction:

$$Si + 3HCl \longrightarrow SiHCl_3 + H_2$$

$SiHCl_3$ is a liquid known as trichloro silicon, having a boiling point 32°C. The impurities present in $SiHCl_3$ are removed by fractional distillation. Now, it is converted to highly pure EGS by a reaction with H_2:

$$2SiHCl_3 + 2H_2 \longrightarrow 2Si + 6HCl$$

8.16 GROWTH OF SEMICONDUCTOR CRYSTALS FROM THE MELT—CZOCHRALSKI METHOD

This is the simplest method for preparing semiconducting crystals. It involves the growth of crystal by a gradual layer-to-layer condensation of the melt. The highly purified materials (of the order of few parts per billion) are melted in a crucible, as given in the Fig. 8.11. The seed crystal is immersed in the melt, in such a way that only a small portion of the crystal is inside. Cooling is done at a slow rate and the crystal is pulled gradually upwards, maintaining a critical pulling rate. Since the pointed end cools faster the crystal seeds at the bottom, it grows through the melt. Under well-controlled conditions, a growth of single crystal takes place.

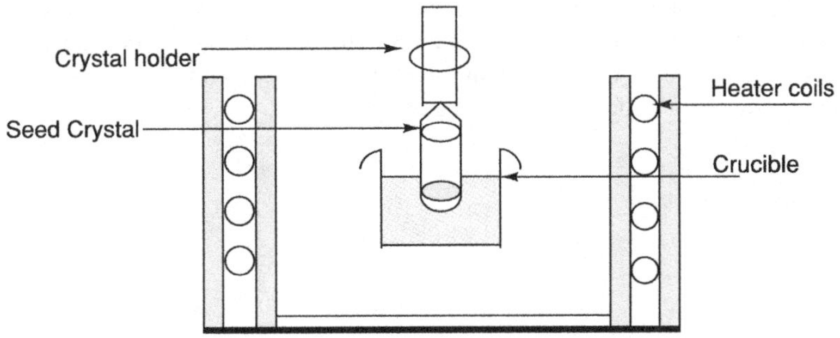

Fig. 8.11 *Crystal growth from the melt*

Controlled amounts of specific impurities, e.g. boron or phosphorous, may be added to the melt so that the grown semiconductor crystal—deliberately doped with impurity atom—exhibits different electronic properties. This technique is widely used in growing Si, Ge and some of the compound semiconductors. The atmosphere of the crystal growth is controlled by enclosing the unit in an evacuated chamber. This prevents contamination and oxidation. If necessary, the chamber may be filled with hydrogen or some inert gas at suitable pressure to maintain equilibrium vapour pressure. For GaAs, arsenic atmosphere is used.

8.17 EPITAXIAL GROWTH—VAPOUR-PHASE EPITAXY

The word 'epitaxial' is derived from a Greek word meaning 'arranged upon'. The technique of growing an oriented crystal layer on a substrate wafer is called epitaxial growth or epitaxy. There are several ways to carry out such a growth. This is one of the most versatile methods to deposit a thin crystal layer on a wafer of compatible crystal. The substrate crystal may be a wafer of the same material as the grown layer, or a different material with a similar lattice structure. In this process, the substrate serves as the seed crystal onto which the new crystalline material grows. The growing crystal maintains the crystal structure and orientation of the substrate.

Wafers of single-crystal silicon are contained in a tubular furnace, which is maintained at a temperature of 1250°C. Silicon tetrachloride vapour in a stream of hydrogen is passed through the furnace by which the below given reaction takes place.

$$SiCl_4 + 2H_2 \rightleftharpoons Si + 4HCl$$

Silicon is deposited on the silicon wafer as a single-crystal layer following the crystal arrangement of the substrate. Sometimes the silane reaction is present to get non-corrosive products.

$$SiH_4 \rightleftharpoons Si + 2H_2$$

In addition to the above, there are several methods that could be used to provide the appropriate atoms to the surface of the growing layer. These are

- Chemical vapour deposition
- Growth from the melt (liquid phase epitaxy)
- Molecular beam epitaxy

(i) Chemical Vapour Deposition

One epitaxial growth technique that has been used extensively is called Chemical Vapour-phase Deposition (CVD). Silicon epitaxial layers, for example, are grown on silicon substrates by the controlled deposition of silicon atoms onto the surface from a chemical vapour containing silicon. In one method, the chemical reaction take, place as shown above. A sharp demarcation between the impurity doping in the substrate and in the epitaxial layer can be achieved using the CVD process. This technique allows great flexibility in the fabrication of semiconductor devices.

(ii) Liquid Phase Epitaxy

This technique is mainly used with compound semiconductors. The substrate crystal is held above the melt on a quartz plate and immersed into the molten semiconductor as shown in Fig. 8.12.

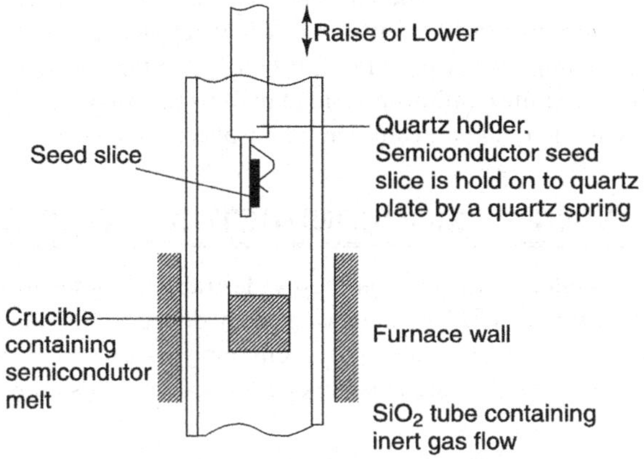

Fig. 8.12 *Liquid phase epitaxy*

By controlling the cooling rate better, a single-crystal layer can be grown epitaxially on the crystal. In recent years, this technique is widely used in growing semiconductors for lasers. It is simple and quite fast and is capable of growing a material of a desired bandgap.

(iii) Molecular Beam Epitaxy (MBE)

This is one of the best and most versatile techniques for growing epitaxial layers. In this method, the substrate is kept under a high vacuum. Each material component to be deposited on the substrate sits in a cylindrical cell which is heated to a temperature usually above the melting point. The molecular or atomic beams coming from different cells impinge upon the surface of the substrate. The schematic diagram of this method is shown in Fig. 8.13.

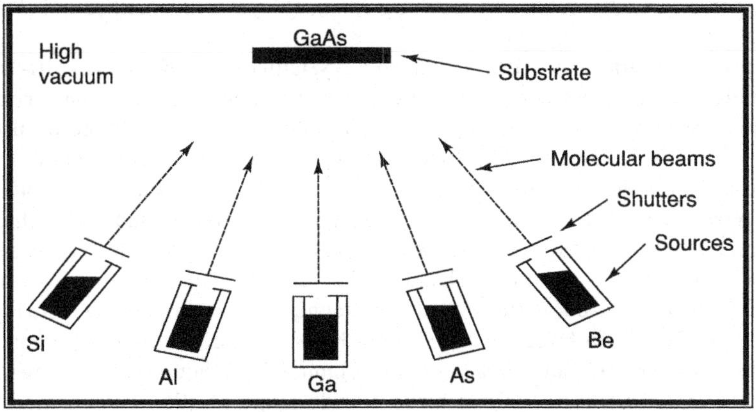

Fig. 8.13 *Molecular beam epitaxy set-up*

The amount of depositing materials on the substrate can be guarded by controlling the opening of various cylindrical cells. Because of high vacuum and close controls involved, MBE requires a sophisticated set-up.

This is a very precise method capable of excellent composition control and was used to develop ternary and quaternary compounds such as GaAlAs. This method is exclusively successful for II-VI compounds like ZnSe and CdS, where it seems difficult to dope both *p*- and *n*-type. These compounds can be made *n*-type, but *p*-type was the stumbling block. MBE avoids thermodynamic equilibrium and maintains the *p*-type centres active in these materials, provided temperature and ambient atmosphere are controlled.

8.18 REFINING OF SEMICONDUCTOR MATERIAL

Refining of a semiconductor material is highly desirable in order to use it for device application. After extreme refining, the material in doped deliberately to get desired electrical properties. Zone refining technique is commonly used to purify the semiconductor materials. The details of this technique is beyond the scope of this book.

8.19 LATTICE MATCHING IN EPITAXIAL GROWTH

On depositing Si epitaxial layers on Si substrates, there is a natural matching of the crystal lattice, resulting in high-quality single-crystal layers. Sometimes, it is desirable to obtain epitaxial that differ somewhat from the substrate, which is known as hetero-epitaxy. This can be achieved easily if the lattice structure and the lattice constant match for the two materials. For example, GaAs and AlAs both have the zincblende structure, with a lattice constant of about 5.65 Å. As a result, epitaxial layers of alloy AlAs can be grown on GaAs substrates with little lattice mismatch. Similarly, GaAs can be grown on Ge substrates.

Since AlAs and GaAs have similar lattice constants, it is also quite natural that the ternary alloy AlGaAs has necessarily the same lattice constant over the whole range of compositions—from AlAs to GaAs. As a result, one can choose the composition x of the ternary compound $Al_xGa_{1-x}As$ to fit the particular device requirement, and grow this composition on a GaAs wafer. The resulting epitaxial layer will be lattice-matched to the GaAs substrate.

Figure 8.14 illustrates the energy bandgap E_g vs. lattice constant for many III-V ternary compounds over different composition ranges. For example, as the ternary compound InGaAs is varied by choice of composition on the column III sub-lattice from InAs to GaAs, the bandgap changes from 0.36 to 1.43 eV, while the lattice constant of the crystal varies from 6.06 Å for InAs to 5.65 Å for GaAs. Clearly, this ternary compound cannot be grown over the entire composition range on a particular binary substrate, which has a fixed lattice constant. Figure 8.14 illustrates, however, that it is possible to grow a specific composition of InGaAs on an InP substrate. The vertical (invariant lattice constant) line from InP to the InGaAs curve shows that a midrange ternary composition (actually, $In_{0.53}Ga_{0.47}As$) can be grown lattice-matched to an InP substrate. Similarly, a ternary InGaP alloy with about 50% Ga and 50% In on the column III sub-lattice can be grown lattice-matched to a GaAs substrate. To achieve a wide range of alloy compositions grown lattice-matched on particular substrates, it is helpful to use quaternary alloys such as InGaAsP. The variation of compositions on both column III and V sub-lattices provides additional flexibility in choosing a particular bandgap providing lattice-matching to convenient binary substrates such as GaAs or InP.

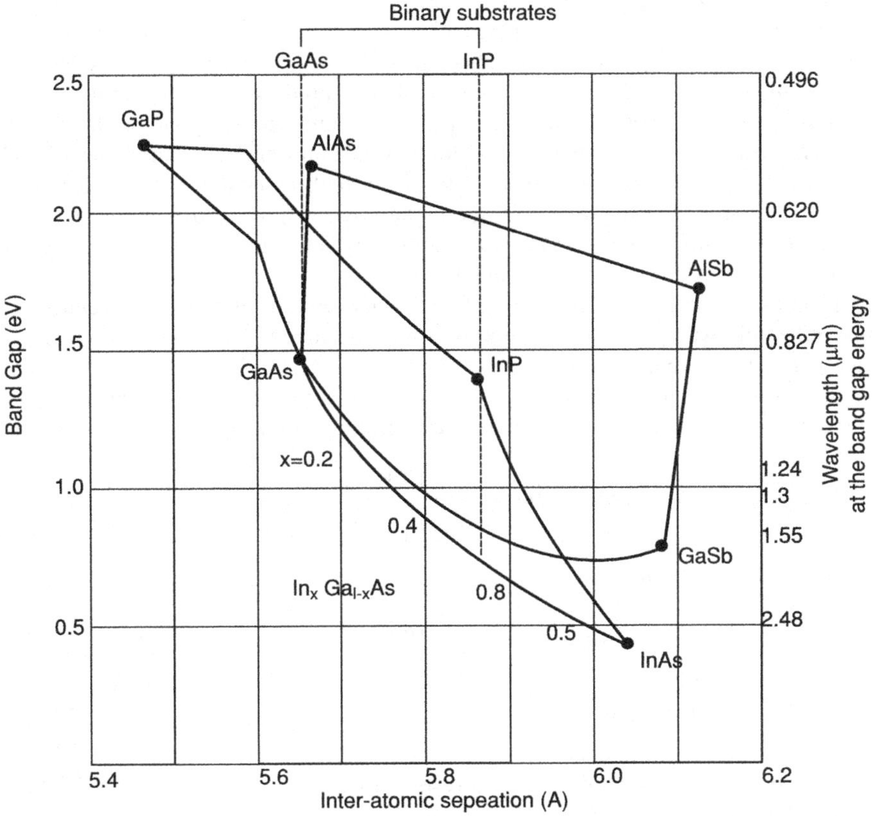

Fig. 8.14 *Energy bandgap vs. lattice constant*

In the case of GaAsP, the lattice constant is intermediate between that of GaAs and GaP, depending upon the composition. For example, GaAsP crystals, used in red LEDs, have 40% phosphorus and 60% arsenic on the column V sub-lattice. Since such a crystal cannot be grown directly on either a GaAs or a GaP substrate, it is necessary to gradually change the lattice constant as the crystal is grown. Using a GaAs or Ge wafer substrate, the growth is possible at a composition near GaAs. A region ~ 25 μm thick is grown while gradually introducing phosphorus until the desired As/P ratio is achieved. The desired epitaxial layer (e.g. 100 μm thick) is then grown on this graded layer. By this method, epitaxial growth always occurs on a crystal of similar lattice constant. Despite some crystal dislocations occur due to lattice strain in the graded region, such crystals are of high quality and can be used in LEDs.

8.20 MOTION OF ELECTRON IN SOLID MATERIAL (CONCEPT OF EFFECTIVE MASS)

The movement of an electron in a solid material is entirely different to that of its motion in free space. The electron moves under the influence of periodic potential of the positive ions. Hence, in addition to external applied force due to electric field, the internal forces also play their role. So, the net force on the electron is the sum total of the two forces mentioned above. It is very difficult to account for the internal forces due to the presence of positive ions. Hence, one has to replace the mass in the equation of motion by effective mass, represented by m_e or m_h—depending upon the kind of charge carrier (electron or hole). The m_e or m_h include the particles mass along with the effect of internal forces.

The motion of electron in a solid material can be understood by taking the example of a stone dropped in a container containing water and another containing oil. The stone moves faster in water as compared to oil. In both cases the external force is the gravitational pull, while the internal force is the viscous force of different liquids. The same stone takes different time to reach the bottom, which makes one think as if the mass of stone is different in the above two situations.

The plot between E and k in quite useful to understand the effective mass concept. In semiconducting materials, there are not only fully occupied bands, but empty bands as well separated by forbidden gap.

Let us consider a free electron whose E vs. k plot is given below. The energy values and momentum of such an electron, which are permitted, may be given by the relation

$$E = \frac{P^2}{2m} = \frac{\hbar^2 k^2}{2m} \tag{8.1}$$

$$P = \hbar k \tag{8.2}$$

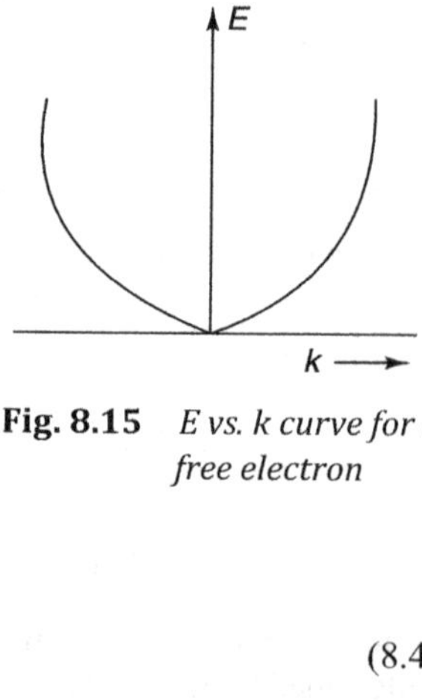

Fig. 8.15 *E vs. k curve for a free electron*

From the above equation,

$$\frac{dE}{dk} = \frac{\hbar^2 2k}{2m} = \frac{\hbar P}{m} \tag{8.3}$$

or
$$\frac{1}{\hbar}\frac{dE}{dk} = \frac{P}{m} = v$$

where v is the velocity of the particle.

Hence, the first differential of E with respect to k is linked with the velocity of the particle.

On further differentiating, we get

$$\frac{d^2E}{dk^2} = \frac{\hbar^2}{m} \tag{8.4}$$

which may be rearranged as,

$$\frac{1}{\hbar^2}\frac{d^2E}{dk^2} = \frac{1}{m} \tag{8.5}$$

Therefore, the above equation

or
$$m_{eff} = \frac{\hbar^2}{d^2E/dk^2} \tag{8.6}$$

This shows that the second derivative of E with respect to k is inversely proportional to the mass of the particle. For electron, the mass is constant which means $\frac{d^2E}{dk^2}$ is also constant. The figure (E vs. k) also indicates that $\frac{d^2E}{dk^2}$ is a positive quantity, which simply means that mass of electron is also a positive quantity.

8.21 CONCENTRATION OF CHARGE CARRIERS IN INTRINSIC SEMICONDUCTOR

The charge carriers namely, electrons and holes in semiconducting materials are fermi particles; therefore, their concentration in a given energy range E and $E + dE$ can be obtained by applying the principles of Fermi–Dirac statistics.

The concentration of electron in energy range E and $E + dE$ is $n_c(E)dE$ = density of states in the above energy range × Fermi function.

Density of states in energy range E and $E + dE$ are

$$g_e\,(E)\,dE = 4\pi\left[\frac{2m_e}{\hbar^2}\right]^{3/2}(E - E_C)^{1/2} \tag{8.7}$$

where energy E in measured with respect to the bottom of conduction bond, E_C.

$$\text{Fermi-function (probability of occupation) } f(E) = \frac{1}{e^{(E-E_F)/kT}+1} \tag{8.8}$$

where E_F is the Fermi energy. For electrons in the conduction band $(E - E_F) >> kT$; 1 may be neglected in the denominator.

Therefore,

$$n_e(E)dE = 4\pi\left[\frac{2m_e}{\hbar^2}\right]^{3/2}.(E - E_C)^{1/2}\,dE.e^{-(E-E_F)/kT} \tag{8.9}$$

On integrating the above from E_C to ∞, we get the concentration of electrons in the conduction band,

i.e.
$$n_e = \int_{E_C}^{\infty} n_c(E)dE = 4\pi\left[\frac{2m_e}{\hbar^2}\right]^{3/2}\int_{Ec}^{\infty}(E - Ec)^{1/2}e^{(E-E_F)/kT}\,dE$$

On substituting the value of integral in the above equation, we get

$$n_e = 2\left[\frac{2\pi m_e kT}{\hbar^2}\right]^{3/2}.e^{(E_F-E_C)/kT} \tag{8.10}$$

where m_e stands for effective mass of electron.

Similarly, one can get the concentration of holes in the valence band.

For holes in valence band,

$$g_h(E)dE = 4\pi\left[\frac{2m_h}{\hbar^2}\right]^{3/2}(E_V - E)^{1/2} \tag{8.11}$$

Fermi function for holes is given by

$$f_h = 1 - f_e = 1 - \frac{1}{1+e^{(E-E_F)kT}}$$

Neglecting 1 in the denominator

$$f_h = e^{-(E_F-E)/kT}$$

Therefore, concentration of holes in the valance band in the energy range E and $E + dE$

$$n_h(E)dE = 4\pi\left[\frac{2m_h}{\hbar^2}\right]^{3/2}\int_{-\infty}^{E_v}(E_V - E)^{1/2}e^{-(E_F-E)/kT}\,dE \tag{8.12}$$

On integrating $-\infty$ and E_v, we get

$$n_h = 2\left[\frac{2\pi m_h kT}{\hbar^2}\right]^{3/2} e^{(Ev-E_f)/kT} \tag{8.13}$$

For intrinsic semiconductor, since the number of holes is equal to number of electrons, we can write

$$n_e = n_h = n_i$$

where n_i is known as the intrinsic carrier concentration.

$$n_e.n_h = n_i^2 = 4\left[\frac{2\pi kT}{\hbar^2}\right]^{3/2} (m_e m_h)^{3/2} e^{-Eg/2kT} \tag{8.14}$$

where $E_g = E_c - E_v$, commonly known as energy of the forbidden gap.

8.22 POSITION OF FERMI LEVEL IN INTRINSIC SEMICONDUCTOR

We know that the number of electrons in a conduction band and the number of holes in a valence band are equal to each other. Therefore, on equating these

$$n_e = n_h$$

$$2\left[\frac{2\pi m_e kT}{\hbar^2}\right] e^{(E_F-E_c)/kT} = 2\left[\frac{2\pi m_h kT}{\hbar^2}\right]^{3/2} e^{(E_v-E_F)/kT}$$

i.e. $$\left[m_e\right]^{3/2} e^{(E_r-E_c)/kT} = \left[m_h\right]^{3/2} e^{(E_v-E_F)/kT}$$

On further simplifying,

$$e^{2E_F/kT} = \left[\frac{m_h}{m_e}\right]^{3/2} e^{(E_c-E_v)/kT}$$

On taking log,

$$\frac{2E_F}{kT} = \frac{3}{2}\log\frac{m_h}{m_e} + \frac{(E_c - E_v)}{kT} \tag{8.15}$$

$$\frac{2E_F}{kT} = 0 + \frac{E_g}{kT}$$

Therefore, $E_F = \dfrac{E_g}{2}$ since $(m_e = m_h)$

Hence, the Fermi level of an intrinsic semiconductor lies in the middle of energy gap. However, since m_h is slightly more than m_e, E_F lies slightly above the middle of the forbidden band. Since in fact $m_h \neq m_e$; therefore, on increasing the temperature, E_F also increases.

As $$n_i = 2\left[\frac{2\pi kT}{\hbar^2}\right]^{3/2} (m_e m_h)^{3/4} e^{-E_g/2kT}$$

$$n_i = CT^{3/2} e^{-E_g/2kT} \tag{8.16}$$

where $$C = 2\left[\frac{2\pi k}{\hbar^2}\right]^{3/2} (m_e m_n)^{3/4}$$ is a constant for a given semiconductor.

In the above equation, it is clear that the temperature occurs in the pre-exponential factor ($T^{3/2}$), as well as in the exponential factor. The temperature dependence of intrinsic carrier concentration is decided by the factor $e^{-Eg/2kT}$.

8.23 TEMPERATURE DEPENDENCE OF CARRIER CONCENTRATION

It is very clear that energy gap is the function of temperature for the reason that it depends on the lattice constant, which does vary with temperature. Since the position of fermi-level in intrinsic semiconductor is given by

$$E_F = \frac{E_g}{2} + \frac{3}{4} kT \log_e \frac{m_h}{m_e} \tag{8.17}$$

The concentration of electrons is given by

$$n_e = n_c \exp\left[\frac{(E_F - E_C)}{kT}\right]^{3/2}, \text{ where } n_c = 2\left[\frac{2\pi m_e kT}{h^2}\right]^{3/2} \tag{8.18}$$

and the concentration of holes is

$$n_h = n_v \exp\left[\frac{E_V - E_F}{kT}\right] \tag{8.19}$$

Therefore, the intrinsic carrier concentration is

$$n_i = 2.\left[\frac{2\pi kT}{\hbar^2}\right]^{3/2} (m_e m_h)^{3/4} e^{-E_g/2kT} \tag{8.20}$$

The above relation shows that exponential temperature dependence dominates and we find that n_e and n_h are proportional to $\exp\left(-\frac{E_g}{2kT}\right)$. Hence, log n_i vs. $\frac{100}{T}$ will be linear, as shown in Fig. 8.16.

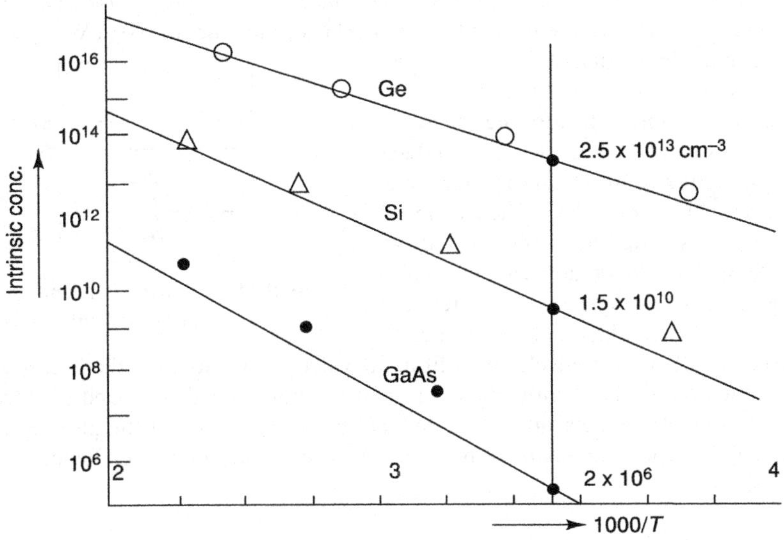

Fig. 8.16 *Intrinsic carrier concentration vs. temperature*

The value of n_i—at any temperature—is a definite number for a given semiconductor and is known for most materials. The temperature dependence of electron concentration in a doped semiconductor can be seen in Fig. 8.17.

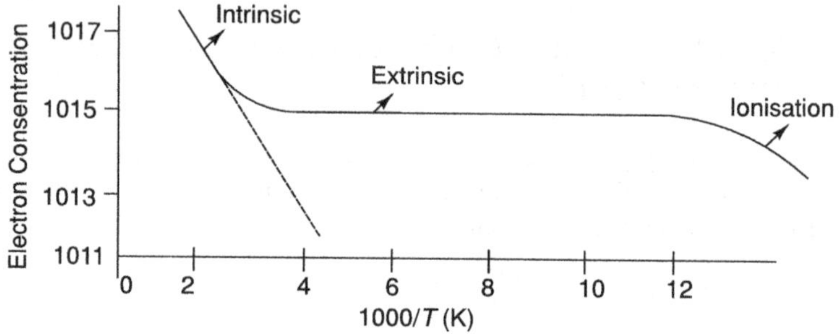

Fig. 8.17 *Temperature dependence of electron concentration in an extrinsic semiconductor*

8.24 POSITION OF FERMI LEVEL IN EXTRINSIC SEMICONDUCTORS

(i) n-type Extrinsic Semiconductor

When a small amount of pentavalent impurity is added to the pure semiconducting crystal, it creates extra electrons without adding any new holes. This impurity introduces new energy levels into the forbidden gap. The location of these new donor levels is slightly below the bottom of the conduction band. The width of the gap *Eg* for germanium 0.72 eV. The energy required to move an electron from a donor impurity into the conduction band conduction band is of the order of 0.01 eV. Since at normal ambient temperature, the thermal energy is considered to be E_F about 0.02 eV, it is therefore clear that almost all the electrons E_D are detached from donor atoms and have conduction band energies. In the case of silicon doped with donor impurities, the energy required to move an electron from donor impurity level to the conduction band is of the order of 0.05 eV.

The energy band diagram for a *n*-type semiconductor is shown in Fig. 8.18.

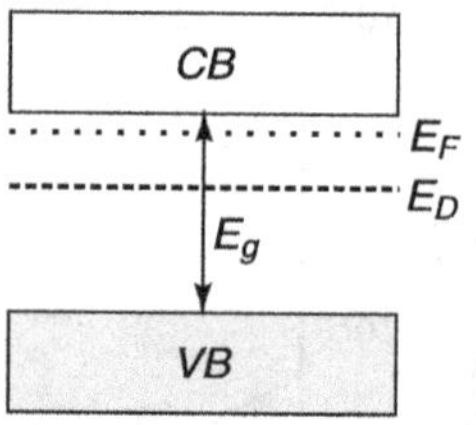

Fig. 8.18 *Position of Fermi level in extrinsic semiconductors*

In intrinsic semiconductors, Fermi Level lies very close to the middle of forbidden energy gap E_g, indicating equal concentrations of free electrons and holes. When a donor type impurity is added to the crystal—if we assume that all the donor atoms are ionised—the donor electrons will occupy the states near the bottom of the conduction band. Hence, it will be more difficult for the electrons from the valence band to cross the energy gap by thermal agitation. Consequently, the number of holes in the valence band is decreased. Since Fermi level is a measure of the probability of occupancy of the allowed energy states, E_F for *n*-type semiconductor must move closer to the conduction band, as shown in Fig. 8.18.

Mathematically, if the density of ionised donor atoms is N_d, then at usual temperatures all the donor levels will be fully activated and ionised. It means that the density of electrons in the conduction band will be approximately equal to the donor atoms density, i.e. $n = N_d$.

Then
$$n = N_d = 2\left(\frac{2\pi m_e kT}{h^2}\right)^{3/2} e^{(E_F - E_c)/kT} \tag{8.21}$$

$$= N_c e^{(E_F - E_c)/kT}$$

where
$$N_C = 2\left(\frac{2\pi m_h kT}{h^2}\right)^{3/2} \text{ is a constant.}$$

Therefore,
$$\frac{N_C}{N_d} = e^{-(E_F - E_C)/kT} \tag{8.22}$$

Taking the logarithm of both the sides,

$$\log e\frac{N_C}{N_d} = -\frac{E_F - E_C}{kT}$$

or
$$E_F = E_C - kT \log e\frac{N_C}{N_d} \tag{8.23}$$

It shows that the Fermi level lies below the bottom of the conduction band.

(ii) P-type Extrinsic Semiconductor

When a Group III impurity is added, it also modifies the energy-level diagram of the semiconductor, again making it easier to bring out conduction. The presence of impurity creates new energy levels that are near to the top of the valence band, as shown in Fig. 8.19. Ambient temperatures result in ionisation of most acceptor atoms, and thus, an apparent movement of holes from the acceptor levels to the valence band starts. The energies for holes are highest near the valence band and decrease vertically upward in the energy-level diagram. Alternatively, one might say that electrons are accepted by the acceptors; these electron are supplied from the valence band, leaving a preponderance of holes in the it.

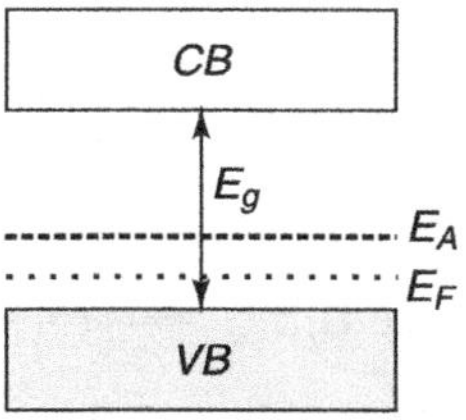

Fig. 8.19 *Position of Fermi level in a p-type semiconductor*

The energy-band diagram for a *p*-type semiconductor is shown in Fig. 8.19, where E_A represents the energy levels corresponding to the acceptor impurities.

When an intrinsic semiconductor is doped with acceptor-type impurity, the concentration of holes in the valence band is more than the concentration of electrons in the conduction band; Fermi level shifts towards the valence band, as shown in Fig. 8.19. The acceptor level lies immediately above the Fermi level.

If we assume that there are only acceptor atoms present and that these are all ionised, we have $p = N_A$. Then

$$p = N_A = 2\left(\frac{2\pi m_h kT}{h^2}\right)^{3/2} e^{(E_V - E_F)/kT}$$

$$= N_V e^{(E_V - E_F)/hT}$$

where $$N_V = 2\left(\frac{2\pi m_h kT}{h^2}\right)^{3/2} \rightarrow \text{is a constant}$$

$$\therefore \quad \frac{N_V}{N_A} = -\frac{(E_v - E_F)}{kT}$$

or $$E_F = E_V + kT \log e \frac{N_V}{N_A}$$

It shows that the Fermi level lies above the top of the valence band.

Example 8.2

The position of Fermi energy level is 0.25 eV above the valance band energy. The value of N_v for silicon at $T = 300$ K is 1.04×10^{19} cm^{-3}. Calculate the thermal equilibrium hole concentration in silicon at $T = 400$ K.

Solution The parameter values at $T = 400$ K are found as:

$$N_v = (1.04 \times 10^{19})\left(\frac{400}{300}\right)^{3/2} = 1.60 \times 10^{19} \text{ cm}^{-3}$$

and

$$kT = (0.0259)\left(\frac{400}{300}\right) = 0.03453 \text{ eV}$$

The hole concentration is then

$$p_0 = N_v \exp\left[\frac{-(E_F - E_v)}{kT}\right] = (1.60 \times 10^{19}) \exp\left(\frac{-0.25}{0.03453}\right)$$

or $$p_0 = 7.17 \times 10^{15} \text{ cm}^{-3}$$ **Ans.**

Example 8.3

Find the intrinsic carrier concentration in GaAs at 300 K and at 450 K. The values of N_c and N_v are 3.8×10^{17} cm^{-3}. and 6.5×10^{18} cm^{-3}, respectively. Both N_c and N_v vary as $T^{3/2}$. Assume the bandgap energy of GaAs to be 1.42 eV and it does not vary with the temperature. The value of kT at 450 K is equal to 0.03885 eV.

Solution The intrinsic carrier constriction is given by

$$n_i^2 = N_c N_v e^{-E_g/kT}$$

at $$n_i^2 = (3.8 \times 10^{17})(6.5 \times 10^{18}) \exp\left(\frac{-1.42}{0.0259}\right) = 3.81 \times 10^{12}$$

so that $$n_i = 1.94 \times 10^6 \text{ cm}^{-3}$$

At $T = 450$ K, we find

$$n_i^2 = (3.8 \times 10^{17})(6.5 \times 10^{18})\left(\frac{450}{300}\right)^3 \exp\left(\frac{-1.42}{0.03885}\right) = 1.48 \times 10^{21}$$

so that $$n_i = 1.10 \times 10^{10^\circ} \text{ cm}^{-3}$$ **Ans.**

Example 8.4

Find the position of intrinsic Fermi level with respect to the middle of the bandgap in silicon at 300 K. The density of states effective carrier masses in silicon are m_e = 1.08 m_0. And m_h = 0.56 m_0.

Solution The position of Fermi level with respect to middle of the bandgap is begin by

$$E_F - E_g = \frac{3}{4} kT \log\left(\frac{m_h}{m_e}\right)$$

$$= \frac{3}{4}(0.0259)\log\left(\frac{0.56}{1.08}\right)$$

$$= -12.8 \text{ meV}$$ **Ans.**

8.25 INTRINSIC CONDUCTIVITY OF A SEMICONDUCTOR

Let n_h be the number of holes and n_e be the number of electrons in a pure semiconductor. Then, the total conductivity will be given by

$$\sigma = e[n_e\mu_e + n_h\mu_h]$$

where μ_e and μ_h are corresponding drift mobilites.

Since $n_e = n_h = n_i$

$$\sigma_i = n_i e(\mu_e + \mu_h)i$$

$$= 2e\left[\frac{2\pi kT}{\hbar^2}\right]^{3/2} (m_e m_h)^{3/4} e^{-Eg/2kT}(\mu_e + \mu_h)$$

As the mobilities are dependent upon temperature by simple power law, the temperature dependence of conductivity is decided by $e^{-E_g/2kT}$

i.e. intrinsic conductivity. $\sigma \,\alpha\, e^{-E_g/kT}$

or resistivity $\rho = Ae^{E_g/2kT}$

i.e. $\log \rho = \log A + \frac{E_g}{2k}$

This shows that for a pure semiconductor log ρ varies inversely as temperature as shown in the Fig. 8.20. The figure shows the plot for two semiconducting materials, namely Si and Ge.

By taking the slope of the above curve, we can get bandgap energy of the semiconductor.

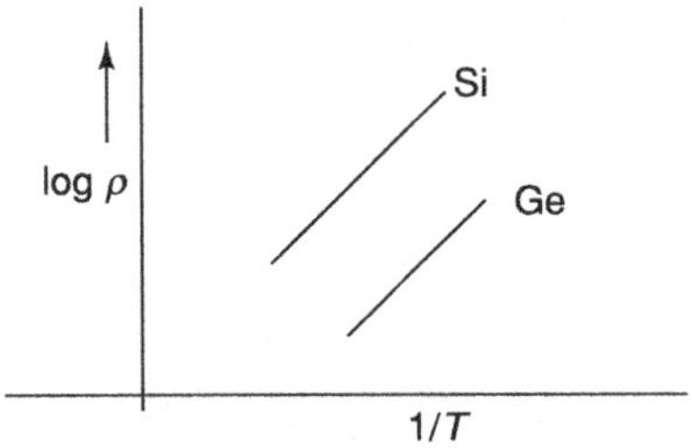

Fig. 8.20 *Plot for two semiconducting materials, namely Si and Ge*

8.26 LOCATION OF DONOR ENERGY LEVEL

When the element of Group V is added to a pure semiconductor, e.g. silicon, four electrons make bond with silicon atom and one excess electron remains unbonded. Since impurity element has donated its four electrons for bond formation, it becomes positively charged. The extra un bonded electron (loosely bound) can be thought of moving around the field of positive ion. The dielectric medium between the electron and the positive ion is silicon. This extra free electron definitely has a high energy, which is able to start conduction by picking a very little amount of energy. In band picture, its position will be close to the conduction band and can be calculated as follows:

According to the quantum theory of H-like atom, the ground state energy is given by

$$E = -\frac{me^4}{2[4\pi\varepsilon_0\hbar]^2} = -13.6 \text{ V} \quad (8.24)$$

where ε_0 is the permittivity of free space.

The radius of revolving electron is

$$r = \frac{4\pi\varepsilon_0\hbar^2}{me^2} = 0.53 \text{ Å} \quad (8.25)$$

(Using the same equation, the ionisation energy of donor atom in *n*-type semiconductor)

On dividing the above

$$\frac{E_d}{E} = -\frac{m_{eff}.e^4}{2[4\pi\varepsilon_0\varepsilon_r\hbar]^2} \times \frac{2[4\pi\varepsilon_0\hbar]^2}{me^4}$$

$$\frac{E_d}{E} = \frac{m_{eff}}{m\varepsilon_r^2} \quad \text{or} \quad E_d = \left[\frac{m_{eff}}{m}\right]\frac{13.6}{\varepsilon_v^2} \text{ eV}$$

For silicon $\quad \frac{m_{eff}}{m} = 0.2, \varepsilon_r = 11.7$

$$E_d = 20 \text{ meV}$$

This shows that the ionisation energy of donor electrons is very low, i.e. it will lie very close to the conduction band. Therefore, the energy required to liberate such electrons is very low for starting the conduction process.

On calculating the radius of this loosely bound electron, it comes out quite large and is nearly 31 Å.

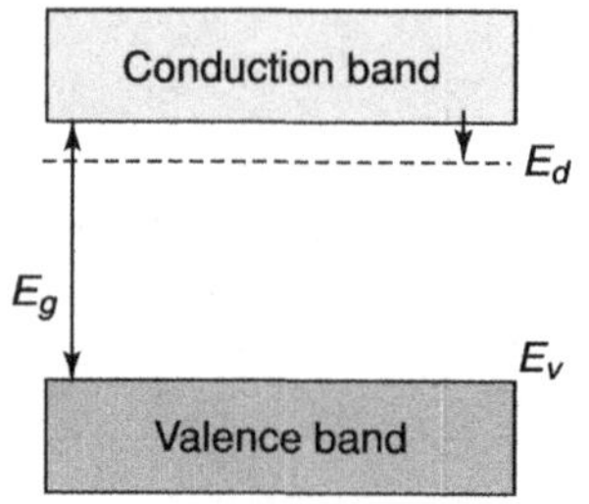

Fig. 8.21 *Donor levels in an extrinsic semiconductor*

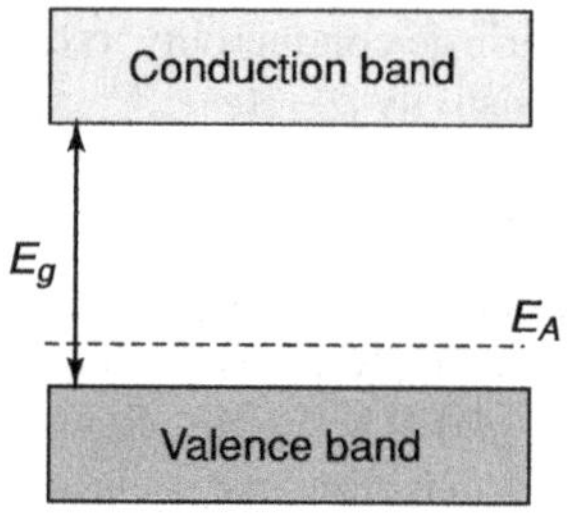

Fig. 8.22 *Acceptor levels in an intrinsic semiconductor*

8.27 LOCATION OF ACCEPTOR ENERGY LEVELS IN A *p*-TYPE SEMICONDUCTOR

Similarly, a Group III impurity material is used to get a *p*-type semiconductor whose position will be close to the valence band. Its position can be estimated in the same way as in an *n*-type semiconducting material.

Table 8.11 illustrates the position of impurity level (donor or acceptor) in Ge and Si.

Table 8.11 *Position of donor and acceptor levels in Ge and Si (in eV)*

	Impurity	*Ge*	*Si*
Donors	Antimony (Sb)	0.0096	0.039
	Phosphorus (P)	0.0120	0.045
	Arsenic (As)	0.0127	0.049
Acceptors	Indium (In)	0.0112	0.160
	Gallium (Ga)	0.0108	0.065
	Boron (B)	0.0104	0.045
	Aluminium (Al)	0.0102	0.057

Example 8.5

Calculate the approximate donor binding energy for GaAs.
(e_r = 13.2, m_e = 0.067 m_0)

Solution The donor binding energy in given by

$$E = \frac{m_n e^4}{8(\varepsilon_0 \varepsilon_r)^2 h^2}$$

$$= \frac{0.067(9.11\times10^{-31})(1.6\times10^{-19})^4}{8(8.85\times10^{-12}\times13.2)^2(6.63\times10^{-34})^2}$$

$$= 8.34 \times 10^{-22}\, J = 0.0052 \text{ eV}$$ **Ans.**

Example 8.6

A Si sample is doped with 10^{17} As atoms/cm^3. What is the equilibrium hole concentration p_0 at 300 K? Find the location of E_F relative to E_i, assuming intrinsic concentration 1.5×10^{10} cm^{-3}.

Solution Since $N_d >> n_i$, we can approximate $n_0 = N_d$ and

$$p_0 = \frac{n_i^2}{n_0} = \frac{(1.5\times10^{10})^2}{10^{17}} = 2.25\times10^3 \text{ cm}^{-3}$$

Position of Fermi level is given by

$$E_F - E_i = kT \log\frac{n_0}{n_i} = 0.0259 \log\frac{10^{17}}{1.5\times10^{10}} = 0.176 \text{ eV}$$

The energy-band diagram is

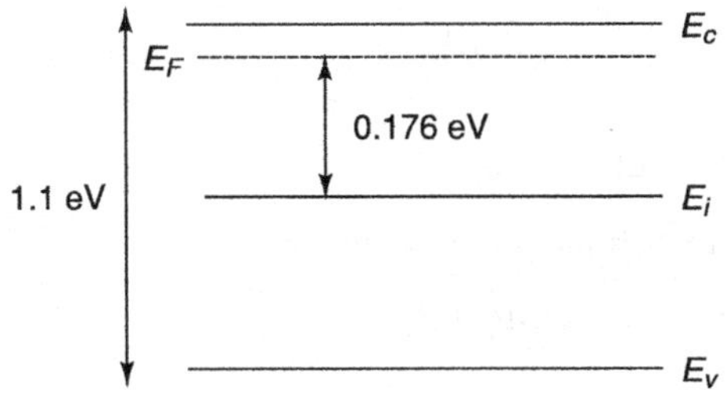

Fig. 8.23

8.28 TRANSPORT PHENOMENON IS SEMICONDUCTORS

The transport of charge carriers in semiconductors takes place in two ways

- Carrier drift
- Carrier diffusion

Carrier Drift and Drift Mobility

When electric field is applied to a semiconductor material, it exerts a force on electrons and holes. The motion of electrons and holes which was random in the absence of electric field, tends to acquire an average motion (net movement) and gives rise to drift current. Average motion or net movement of electrons or holes is a consequence of external electric field. The average velocity gained by the charge carrier per unit field is known as mobility, which is sometimes also named as drift mobility μ,

i.e.
$$\mu = \frac{v_d}{E}$$
$$= \frac{\text{Average drift velocity}}{\text{Applied electric field}}$$

For electron,
$$\mu_e = \frac{v_e}{E} \quad (8.26)$$

For hole,
$$\mu_h = \frac{v_h}{E} \quad (8.27)$$

Mobility is an important parameter, since it describes how well a charge carrier will move under the infhence of an electric field.

8.29 CONDUCTIVITY OF SEMICONDUCTOR

Since both types of charge carriers (electrons and holes) are available in a semiconductor and both are influenced by external applied electric field, therefore the net conductivity of a semiconductor will be a sum of conductivity of holes and electrons. If n be the number of electrons and p be the number of holes in a semiconducting specimen, then

the current density due to electrons $J_e = n.e.v_e$

and the current density due to holes $J_h = P.ev_h$

The net current density of a semiconductor
$$J = J_e + J_h$$
$$= n.e.v_e + p.ev_h$$
$$= eE\left[n\frac{v_e}{E} + P\frac{v_h}{E}\right]$$
$$\boxed{J = eE[n\mu_e + p\mu_h]} \quad (8.28)$$

And conductivity due to drift motion of charge carriers
$$\sigma = \frac{J}{E} = e[n\mu_e + p\mu_h]$$

For an intrinsic semiconductor,
$$n = p = n_i$$

The conductivity will, therefore, be

$$\boxed{\sigma = n_i e(\mu_e + \mu_h)} \tag{8.29}$$

The reciprocal of conductivity is resistivity, therefore

$$\rho = \frac{1}{\sigma} = \frac{1}{e(n\mu_e + p\mu_h)}$$

If we have a bar of semiconductor material with a voltage, V applied over it and carrying a current I, and if the length and area of the specimen be l and A then

Current density, $J = \dfrac{I}{A}$

Electricity field, $E = \dfrac{V}{l}$

Since $J = \sigma E$

$$\frac{I}{A} = \sigma \frac{V}{l}$$

$$V = \left(\frac{l}{\sigma A}\right) I = \left(\rho \frac{l}{A}\right) I = IR \tag{8.30}$$

The above equation is ohm's law for a semiconductor. The resistance is a function of resistivity or conductivity, as well as the geometry of the semiconductor.

Comparison of v_d versus E plot of GaAs, Ge, Si

The drift velocity vs. electric field E characteristic of GaAs is more complicated than Ge or Si, as shown in Fig. 8.24.

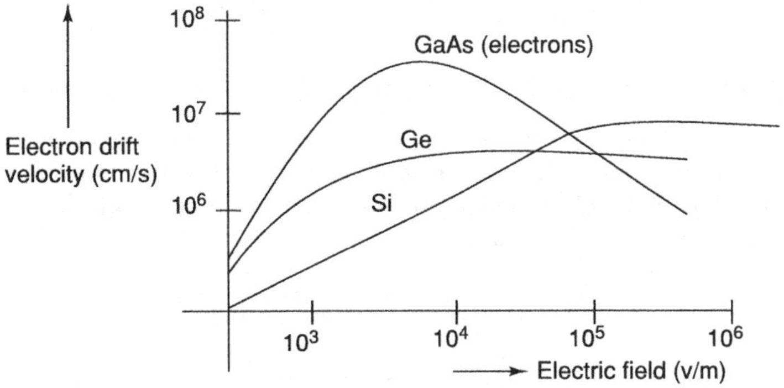

Fig. 8.24 *Drift velocity vs electric field of GaAs, Ge and Si*

At low fields, the slope of v_d vs. E curve is constant, which is known as low-field electron mobility. This is nearly 8500 cm^2/v–s for GaAs. This is much larger for GaAs compared to silicon. As the field increases, the electron drift velocity attains a peak value and then decreases. A differential mobility is the slope of v_d vs. E curve at a particular point on the curve, and the negative slope therefore represent a negative differential mobility. The negative differential mobility produces a negative resistance. This characteristic of GaAs makes it useful for the design of oscillators.

Example 8.7

Calculate electrical conductivity of pure silicon at 200°C. Given that the electrical resistivity of pure silicon is 2.3 × 10³ Ω-m at room temperature 27°C, E_g (Si) = 1.10 eV and k = 8.62 × 10⁻⁵ eV/K.

Solution

$$\sigma = \sigma_0 \exp\left(-\frac{E_g}{2kT}\right) \quad (1)$$

$$\sigma(473\text{ K}) = \sigma_0 \exp\left(-\frac{E_g}{2k \times 473}\right) \quad (2)$$

$$\sigma(300\text{ K}) = \sigma_0 \exp\left(-\frac{E_g}{2k \times 300}\right)$$

Dividing Eq. (1) by Eq. (2) gives

$$\frac{\sigma(473)}{\sigma(300)} = \exp\left(-\frac{E_g}{2k \times 473} + \frac{E_g}{2k \times 300}\right)$$

$$= \exp\left[-\frac{-1.1\text{eV}}{2(8.62 \times 10^{-5}\text{eV/K})}\left(\frac{1}{473\text{K}} - \frac{1}{300\text{K}}\right)\right]$$

In $$\frac{\sigma(473)}{\sigma(300)} = 7.777$$

$$\sigma(473) = \sigma_{300}(2385) = \frac{1 \times 285}{2.3 \times 10^3} = 10.36 \times 10^{-5}\text{ cm}^{-1}\text{ m}^{-1}$$ **Ans.**

Example 8.8

The mobilities of electrons and holes in a sample of intrinsic germanium at room temperature are 0.35 and 0.15 m²/V, respectively. If the electron and hole densities are each equal to 2.5 × 10¹⁹/m³, calculate the resistivity of germanium.

Solution For intrinsic semiconductor,

$$\sigma = n_i q(\mu_n + \mu_p)$$

$$= (2.5 \times 10^{19}/\text{m}^3)\,(1.6 \times 10^{-19}\text{C})\,(0.35 + 0.15)\text{ m}^2/\text{V}^{-s}$$

$$= 2.0\,\frac{\text{C/m}}{\text{V}-\text{s}} = 2.0/\text{ohm-m}$$

Resistivity $$\rho = \frac{1}{\sigma} = \frac{1}{2.0} = 0.50\ \Omega\text{-m}$$ **Ans.**

Example 8.9

The resistivity of Si at room temperature is 3.16 × 10³ ohm-m. Calculate the intrinsic carrier density. Mobility of electrons and holes in Si are 0.14 m²/V-s and 0.05 m²/V-s, respectively.

Solution The conductivity is given by

$$\sigma = ne(\mu_e + \mu_h)$$

$$n = \sigma/e(\mu_e + \mu_h)$$
$$= 1/\rho e(\mu_e + \mu_h)$$
$$= 1/3.16 \times 10^3 \times 1.6 \times 10^{-19}\ (0.14 + 0.05)$$
$$= 1.06 \times 10^{16}/\text{m}^3$$

Ans.

Example 8.10

The carrier density in pure germanium at 27°C is 1.5×10^{19} m^{-3}. It is doped with an n-type impurity of concentration 1 ppm. Assuming that all the impurity atoms are ionised, calculate (i) the factor by which the majority concentration is more than the intrinsic carrier concentration, (ii) hole concentration, (iii) conductivity. Density of germanium = 5.32×10^3 kg/m^3 and its atomic weight = 72.59.; assuming the electron and hole mobilities 0.36 and 0.18 m^2/volt-s.

Solution Number of Ge atoms per unit volume

$$N = \frac{\rho N_{\text{Avogadro}}}{\text{Atomic weight}}$$

$$= \frac{5.32 \times 10^3 \times 6.023 \times 10^{26}}{72.59}$$

$$= 4.4 \times 10^{28}/\text{m}^3$$

Number of pentavalent impurity atoms/m^3

$$N_d = N \times 10^{-6} = 4.4 \times 10^{28} \times 10^{-6} = 4.4 \times 10^{22}/\text{m}^3$$

Since all the impurity atoms are ionised,

Number of majority carriers (electrons)/m^3, $n_e = N_d = 4.4 \times 10^{22}/\text{m}^3$

(i) Factor by which the majority carrier concentration is more than the intrinsic carrier concentration

$$f = \frac{N_d}{n_i} = \frac{4.4 \times 10^{22}}{1.5 \times 10^{19}} = 2933$$

(ii) Hole concentration in the n-type Ge,

$$n_h = \frac{n_e^2}{n_e} = \frac{(1.5 \times 10^{19})^2}{4.4 \times 10^{22}} = 5.1 \times 10^{15}/\text{m}^3$$

(ii) Conductivity

$$\sigma = n_e|e|m_e + n_h|e|\mu_h = 1.6 \times 10^{-19}(4.4 \times 10^{22} \times 0.36 + 5.1 \times 10^{15} \times 0.18)$$
$$= 2534\ \text{ohm}^{-1}\ \text{m}^{-1}$$

Ans.

Example 8.11

The resistivity of Ga-Sb at 300 K is found to be 5×10^{-3} ohm-m. The electron and hole mobilities are 0.3 and 0.1 m^2/volt-s, respectively. Calculate the carrier density.

Given $e = 1.6 \times 10^{-19}$ **Coulomb.**

Solution Conductivity $= \dfrac{1}{\text{Resistivity } \rho} = \dfrac{1}{5 \times 10^{-3}} = 200/\text{ohm-m}$

As $\sigma = n_e e \mu_e + n_h e \mu_h$

μ_e and μ_h = 0.3 and 0.1 m^2/V. sec respectively, and talking $n_e = n_h = n$, we get

$$200 = n.e(\mu_e + \mu_h)$$

or $$n \times 1.6 \times 10^{-19}(0.3 + 0.1) = 200$$
$$n = 3.1 \times 10^{21}/\text{m}^3$$ **Ans.**

Example 8.12

A current of density 500 A/m² flows through an *n*-type germanium crystal. This crystal has a resistivity of 0.05 ohm-m and the electron mobility of 0.4 m²/Vs. Calculate the drift velocity and time taken by the electrons to travel 100 m in the crystal.

Solution The given data are

$$J_d = 500\ \text{A/m}^2, \quad \rho = 0.05\ \text{ohm-m},$$

travel length, $$l = 100\ \mu\text{m} = 100 \times 10^{-6}\ \text{m}$$

and $$\mu_e = 0.4\ \text{m}^2/\text{Vs}.$$

Let t be the time taken by electrons to travel a distance of l. For a n-type semiconductor, we know that $n_e << n_h$. Therefore, neglecting the term $n_h\, e\, \mu_h$.

$$\sigma = n_e . e . \mu_e$$

$$\therefore \quad \frac{1}{0.05} = n_e \times 1.602 \times 10^{-19} \times 0.4$$

or $$n_e = \frac{1}{0.06 \times 1.602 \times 10^{-19} \times 0.4} = 3.13 \times 10^{20}/\text{m}^3$$

We also know that

$$J_d = n_e v_d$$

$$\therefore \quad v_d = \frac{J_d}{n_e} = \frac{500}{3.13 \times 10^{20} \times 1.602 \times 10^{-19}} = 10\ \text{m/s}$$

The distance l travelled in time t at drift velocity v_d, by an electron may be obtained from

$$l = t v_d$$

$$t = \frac{l}{v_d} = \frac{100 \times 10^{-6}}{10}$$

$$= 10 \times 10^{-6}\ \text{s}$$ **Ans.**

Example 8.13

If the conductivity of an intrinsic semiconductor is to be increased by 30%, to what temperature should it be heated above the room temperature?

Solution For an intrinsic semiconductor, the conductivity is given by

$$\sigma = \sigma_0 \exp\left(\frac{-E_g}{2kT}\right)$$

For increase in conductivity by 30%, let it be heated to a temperature of $T + \Delta T$, where T is room temperature = 300 K

$\therefore$ from the above equation,

$$1.3 = 1 \times \frac{\exp(-E_g/2k(T + \Delta T)}{\exp(-E_g/2kT)}$$

or
$$1.3 = \exp\left[\frac{-E_g}{2k}\left(\frac{1}{T+\Delta T} - \frac{1}{T}\right)\right]$$

$$T + \Delta T = \frac{1}{\dfrac{1}{T} - \dfrac{2k}{E_g}\log_e 1.3}$$

$$T + \Delta T = 305.9 \text{ K}$$ **Ans.**

Therefore a temperature rise of $\Delta T = 305.9 - 300 = 5.9$ K will lead to a rise of 30% in conductivity.

8.30 CARRIER DIFFUSION

In addition to the drift, there is also a second mechanism that can induce a current in semiconductor. This is known as 'diffusion'. We can understand the mechanism by taking an example from classical physics. Let us take a container, as shown in Fig. 8.25 divided by a membrane. The left side contains gas molecules at a particular temperature and right side is initially empty. The gas molecules are in a continuously random motion, so that, when the membrane is broken, the gas molecules flow into right side of the container. This is 'diffusion' by which gas molecules flow from a region of high concentration to another region of low concentration.

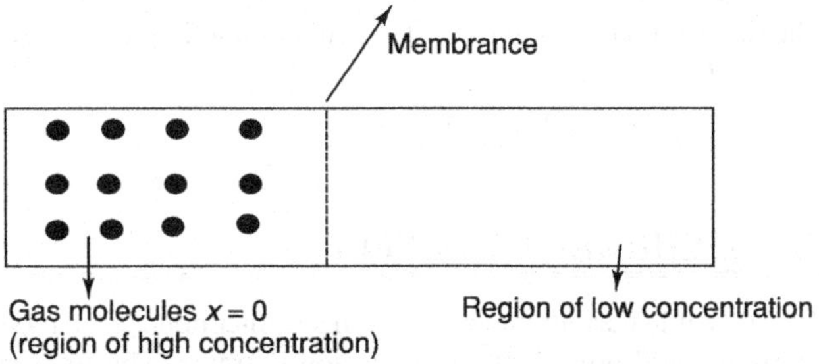

Fig. 8.25 *Demonstration of diffusion mechanism*

If the gas molecules are replaced by charge carriers, which happens in the case of semiconductor, the flow of these will constitute the diffusion current.

8.31 DIFFUSION CURRENT—CARRIER-DIFFUSION MECHANISM

During the formation of junction, the holes and electrons enter on either sides, and therefore, the electron and hole density changes by several orders of magnitude on either side of junctions $\frac{dP}{dx}$ and $\frac{dn}{dx}$, in the junction region as shown the Fig. 8.26. This sets up a diffusion process from the region of higher concentration to the region of lower concentration.

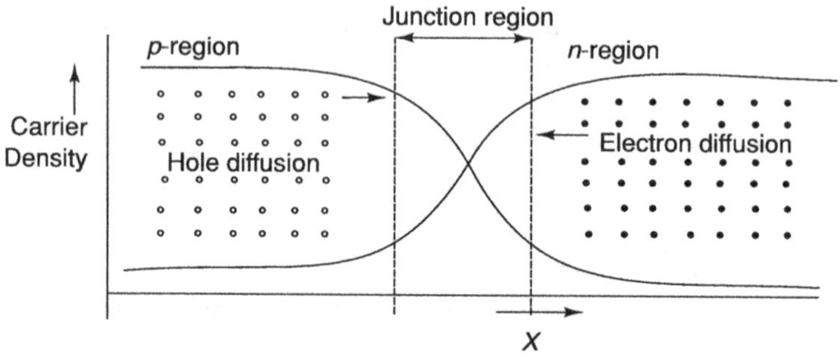

Fig. 8.26 *Diffusion of charge carriers across the p-n junction*

Due to this diffusion process, diffusion currents are set-up across the junction. The rate of diffusion is proportional to the concentration gradient. This rate, when multiplied by charge of charge carrier, gives the diffusion current density.
Therefore,

$$J_h = -eD_h \frac{dP}{dx} \text{ for holes}$$

$$J_e = eD_e \frac{dn}{dx} \text{ for electrons}$$

where D_h and D_e are called hole diffusion constant and electron diffusion constant, respectively.

Therefore, the diffusion current density in a semiconducting material would be

$$J = J_h + J_e$$

$$J = e\left[D_e \frac{dn}{dx} - D_h \frac{dP}{dx}\right] \tag{8.31}$$

8.32 TOTAL-CURRENT DENSITY

We now have four possible current mechanisms in a semiconductor, namely electron drift and diffusion currents, and hole drift and diffusion currents. The total current density, therefore, will be the sum of all these.

In a one-dimensional case

$$J = ne\mu_e E_x + ep\mu_h E_x - eD_p \frac{dp}{dx} + e\,D_n \frac{dn}{dx} \tag{8.32}$$

which may be generalised in three dimensions

$$J = ne\mu_e E + p\mu_h E + eD_n \nabla n - eD_p \nabla P \tag{8.33}$$

The electron mobility gives an indication of how will an electron move in a semiconducting material under the influence of electric field. In a similar way, electron diffusion co-efficient gives an indication of how well an electron moves in a semiconductor in the presence of density gradient.

8.33 EQUATION OF CONTINUITY

The continuity equation states a condition of dynamic equilibrium for the concentration of mobile carriers in any elementary volume of the semiconductor. The above equation indicates that the rate of thermal generation of holes equals the rate of recombination under equilibrium conditions.

Whenever the equilibrium concentrations of carriers are disturbed in a semiconductor, the concentrations of holes or electrons vary with time, approaching the equilibrium value exponentially. In general, however, the carrier concentration in the body of semiconductor is a function of both time and distance. The differential equation governing this functional relationship, called the continuity equation, is based on the conservation of charge.

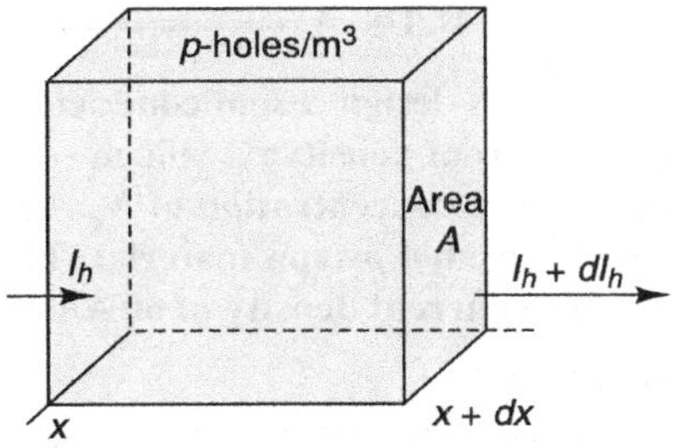

Fig. 8.27 *Small volume element under consiteration*

Consider a small element of volume of area A and length dx, as shown in Fig. 8.27, with the average hole concentration as p. If τ_h is the mean lifetime of holes, then p/τ_h equals the holes per second lost by recombination per unit volume. Assuming e to be the electronic charge, the per second decrease of charge within the volume will be

$$= eAdx\,\frac{p}{\tau_h}$$

If g is the thermal rate of generation of electron-hole pairs per unit volume, the per second increase of change within the volume will be

$$= e\,A\,dx\,g$$

In general, the current varies with distance within the semiconductor. If the current entering the volume at x is I_h and leaving at $x + dx$ is $I_h + d\,I_h$, the per second decreased of change within the volume will be

$$= d\,I_h$$

Because of the three effects enumerated above, the hole concentration must change with time, and the total number of coulombs per second increase within the volume will be

$$= eA\,dx\,\frac{dp}{dt}$$

Since the charge must be conserved,

$$eA\,dx\,\frac{dp}{dt} = -eAdx\frac{p}{\tau_h} + eAdxg - dI_h \tag{8.34}$$

Also, since the hole current Ih is the sum of drift current and diffusion current so

$$I_h = AEpe\mu_h - AeD_h\frac{dp}{dx} \tag{8.35}$$

If the semiconductor is in thermal equilibrium with its surroundings and is subjected to no applied fields, the hole density will attain a constant value p_0. Under these conditions $I_h = 0$ and $\frac{dp}{dt} = 0$.

So we get

$$g = \frac{p_0}{\tau_h} \tag{8.36}$$

Combining Eqs (8.34), (8.35) and (8.36) we have the equation of conservation of charge, called the continuity equation

$$\frac{dp}{dt} = -\frac{p - p_0}{\tau_h} + D_h\frac{d^2p}{dx^2} - \mu_h\frac{d(pE)}{dx}$$

Example 8.14

In order to design a semiconductor resistor with a specified resistance to handle a given current densitya a silicon semiconductor at T = 300 K is initially doped with donors at a concentration of $N_d = 5 \times 10^{15}$ cm^{-3}. Acceptors are to be added to form a compensated p-type material. The resistor is to have a resistance of 10 kΩ and handle a current density of 60 A/cm^2, when 5 V is applied.

Solution

For 5 V applied to a 10^{-k} Ω resistor, the total current is

$$I = \frac{V}{R} = \frac{5}{10} = 0.5 \text{ mA}$$

If the current density is limited to 50 A/cm^2, then the cross-sectional area is

$$A = \frac{I}{J} = \frac{0.5\times10^{-3}}{60} = 0.83\times10^{-5} \text{ cm}^2$$

If we, somewhat arbitrarily at this point, limit the electric field to E = 100 V.cm, then the length of the resistor is

$$L = \frac{V}{E} = \frac{5}{100} = 5\times10^{-2} \text{ cm}$$

he conductivity of the semiconductor is

$$\sigma = \frac{L}{RA} = \frac{5\times10^{-2}}{(10^4)(0.83\times10^{-5})} = 0.60\ (\Omega\text{–cm})^{-1}$$

The conductivity of a compensated p-type semiconductor is

$$\sigma \approx e\mu_p p = e\mu_p(N_a - N_d)$$

where the mobility is a function of the total ionised impurity concentration $N_a + N_d$.

Using trial and error, if $N_a = 1.25 \times 10^{16}$ cm^{-3}, then $N_a + N_d = 1.75 \times 10^{16}$ cm^{-3}, and the hole mobility, from approximately μ_p = 410 cm^2/V-s. The conductivity is then

$$\sigma = e\mu_p(N_a - N_d) = (1.6 \times 10^{-19})(410)(1.25 \times 10^{16} - 5 \times 10^{15}) = 0.492$$

which is very close to the value we need. **Ans.**

Example 8.15

Assuming that in an n-type gallium arsenide semiconductor at T = 300 K, the electron concentration varies linearly from 1×10^{18} to 7×10^{17} cm^{-3} over a distance of 0.10 cm. Calculate the diffusion current density, if the electron diffusion co-efficient is D_n = 250 cm^2/s.

Solution The diffusion current density is given by

$$J_{n/dif} = eD_n\frac{dn}{dx} \approx eD_n\frac{\Delta n}{\Delta x}$$

$$= (1.6\times10^{-19})(225)\left(\frac{1\times10^{18} - 7\times10^{17}}{0.10}\right) = 120 \text{ A/cm}^2$$ **Ans.**

8.34 MEASUREMENT OF SEMICONDUCTOR PARAMETERS

In dealing with the semiconductor materials, the main properties which are looked upon are

- Mobility of charge carrier

- Effective mass of charge carrier
- Band-gap
- Carrier lifetime of charge carrier
- Hall co-efficient

(i) Mobility of Charge Carrier

The charge carriers within the semiconductor material are always in random motion in the absence of electric field which tends to convert into a net resultant motion when electric field is applied. This is known as carrier drift. The ratio between drift velocity v_d to the applied electric field is known as mobility of charge carrier, which is given as

$$\mu = \frac{v_d}{E}$$

The most direct way to measure the mobility is to measure the drift velocity caused by a known dc field. The circuit diagram is given in Fig. 8.28.

The drift velocity can be measured by measuring the time needed to a charge carrier to reach from point A to point C. A less direct way to measure the mobility is to measure the conductivity and to use the relation

$$\mu = \frac{\sigma}{Ne}$$

The method normally in practice is called 'four-probe method', which is shown in Fig. 8.29.

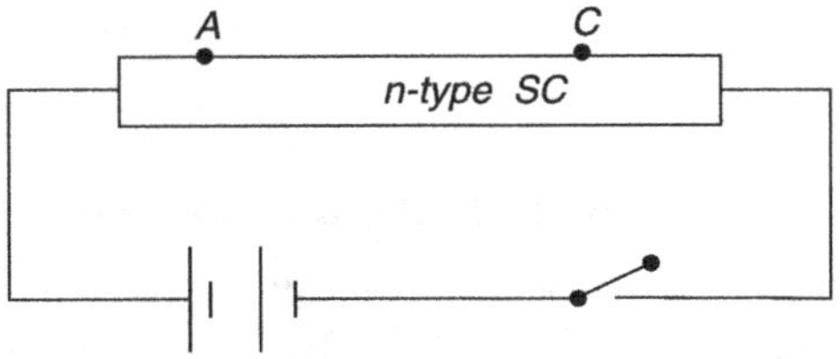

Fig. 8.28 *Measurement of drift velocity in a semiconductor*

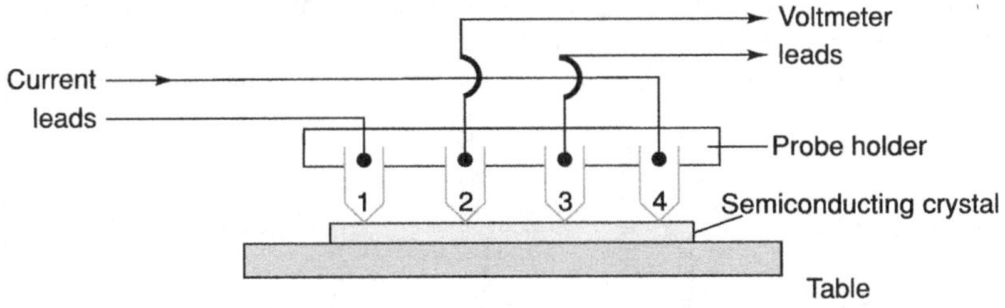

Fig. 8.29 *Four-probe method for determination of conductivity*

The current is passed from contact 1 to 4 and the voltage drop is measured using a voltmeter of very high impedance between points 2 and 3. Using these two parameters, one can measure conductivity from which mobility can be measured.

It is important to realise that sometimes mobility is a function of field. Since Ohm's law is valid for low fields, we may define the low-field mobility as a constant. For high fields the differential mobility is usually the important parameter in device applications, which can be written as

$$\mu_{diff} = \frac{dv_d}{dE}$$

(ii) Effective Mass

Effective mass of charge carriers can be measured by using the phenomenon of cyclotron resonance absorption. It is essentially an interaction of EM wave with charge carriers, which leads to vibration at the same frequency as that of applied electric field. For the resonant absorption to be observable, the electron must travel an appreciable part of the period without collisions. This is achieved by using low temperature, high frequency electric field and a high-intensity magnetic field. Figure 8.30 shows the set-up for cyclotron resonance experiment.

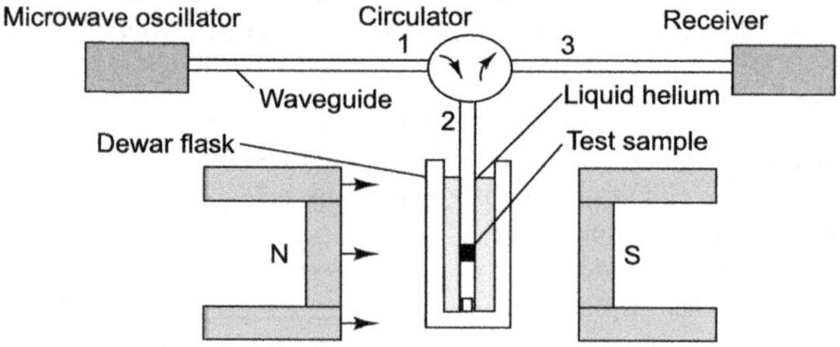

Fig. 8.30 *Schematic diagram of cyclotron resonance experiment*

The measurements are generally made in the microwave region (10^{10} Hz) at liquid He temperature, or in IR about (10^{13} Hz) at LNT. Employing a wave of known frequency and variable magnetic field, the effective mass can be measured by using the formula

$$m^* = \frac{eB}{w_c}$$

where B is the magnetic field corresponding to an absorption as shown in Fig. 8.31.
On plotting absorption vs. magnetic field, there are observed number of absorption peaks corresponding to the various holes and electrons present, as shown below in the figure.

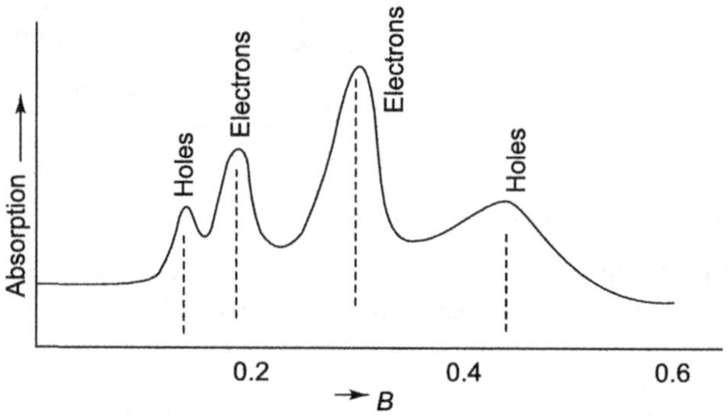

Fig. 8.31 *Absorption peaks in semiconductor*

From the shape of these peaks, information about the density and collision times of various carriers can be obtained.

(iii) Bandgap Measurement

It can be achieved in two ways, namely: by study of variation of log $\sigma v_s \frac{1}{T}$ or by optical transmission study.

• **Conductivity Method** The variation of conductivity with temperature is the simplest way to measure the bandgap of semiconductor. The total conductivity of a semiconductor material is given by

$$\sigma = (n_e\mu_e + n_h\mu_h)e$$

For intrinsic material

$$n_e = n_h = n_i \quad \text{(intrinsic-carrier concentration)}$$

Therefore,

$$\sigma = n_i e(\mu_e + \mu_h)$$

$$= 2e\left[\frac{2\pi kT}{\hbar^2}\right]^{3/2}(m_e m_n)^{3/4}e^{-E_g/2kT}(\mu_e + \mu_h)$$

$$\boxed{\sigma = \sigma_0 e^{-E_g/2kT}} \text{ where } \sigma_0 \text{ us a constant.} \quad (8.37)$$

As we know that at high temperatures, semiconductors behave as intrinsic and at low temperature they look like pseudo-intrinsic with an energy-gap equal to the gap between the impurity level and the band edge. Thus, we would expect two definite straight line regions with entirely different slopes on the plot of $\log_e\sigma$ vs. $\frac{1}{T}$ as shown in Fig. 8.32.

In the region between these slopes, the temperature is high enough to ionise the donors fully, but not so high to ionise an appreciable number of electrons from the host lattice. Hence, in this temperature region the carrier density will not be greatly influenced by temperature and the variations in the mobility and $T^{3/2}$ factor that we neglected will determine the shape of the curve.

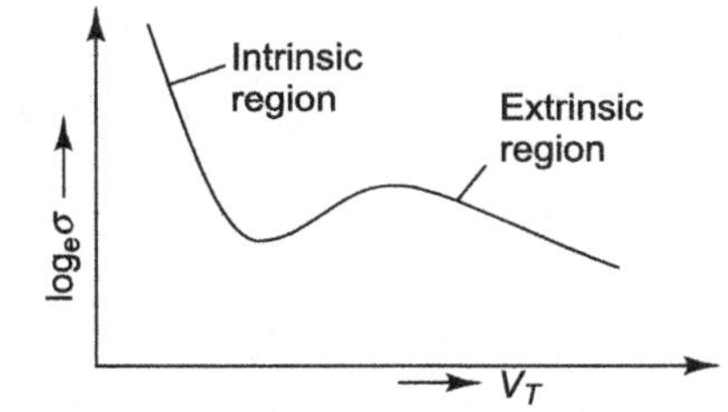

Fig. 8.32 *$\log_e\sigma$ vs. $1/T$ in semiconductor material*

• **Optical-transmission Method** The study of optical transmission through a slice of semiconductor material is another tool to measure the bandgap. The light is shone upon a semiconductor material and the amount of transmission is plotted with wavelength. The schematic diagram is shown in Fig. 8.33.

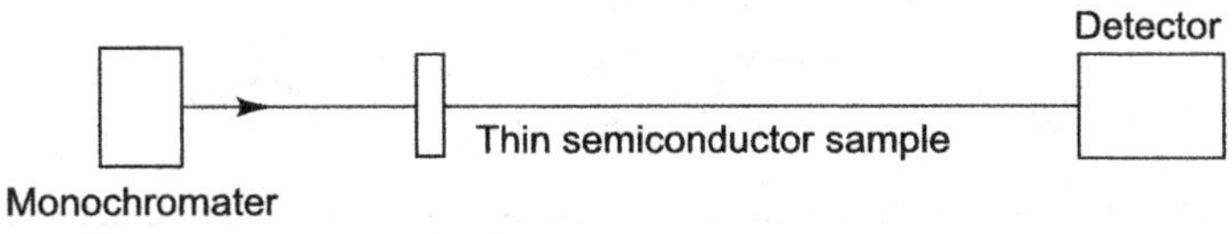

Fig. 8.33 *Schematic diagram for measurement transmission*

If the incident wavelength is sufficiently small, the incident photons have enough energy to promote electrons from the valence band to conduction band. Most of the photons are then absorbed, and the transmission is close to zero. On increasing the wavelength, there will be a particular value $\left(\lambda = \frac{c}{v} = \frac{h_C}{E_g}\right)$ when band-to-band transitions are no longer possible. The absorption then suddenly drops and corresponding transmission sharply increases as shown in Fig. 8.34.

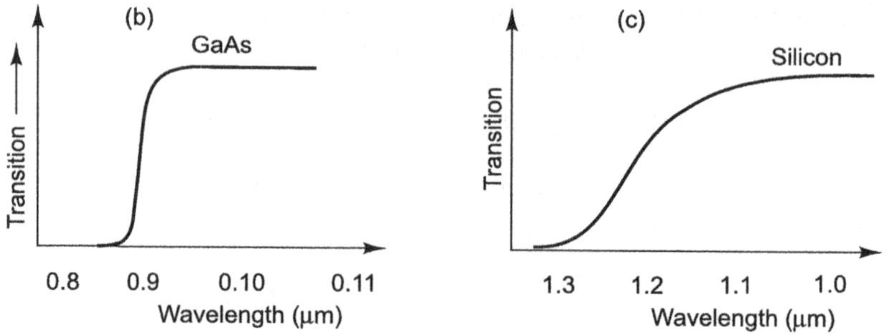

Fig. 8.34 *Transmission vs. wavelength in silicon and GaAs*

The point where sudden rise starts may be estimated—found 880 nm for GaAs—which gives the value of bandgap equal to 1.14 eV for this material.

Example 8.16

Energy band gap in a Ge crystal is 0.75 eV. Determine the wavelength at which Ge starts to absorb light.

Solution Given are $E_g = 0.75 \text{ eV} = 0.75 \times 1.6 \times 10^{-19}$ J

$$\because \qquad E_g = h\nu = \frac{h \cdot c}{\lambda}$$

$$\lambda = \frac{hc}{E_g}$$

$$= \frac{(6.62 \times 10^{-34}) \times (3 \times 10^8)}{0.75 \times 1.6 \times 10^{-19}} = 16560 \text{ Å}$$ **Ans.**

Example 8.17

Find the cut-off wavelength for In P having bandgap energy 1.35 eV. Take appropriate values suitably.

Solution Given, $E_g = 1.35 \text{ eV} = 1.35 \times 1.6 \times 10^{-19}$ eV

$$\lambda = \frac{hc}{E_g} = \frac{(6.63 \times 10^{-34}) \times (3 \times 10^8)}{1.35 \times 1.6 \times 10^{-19}} = 0.902 \ \mu\text{m}$$ **Ans.**

Example 8.18

If the wavelength of the absorption edge of a semiconductor material is $\lambda = 1771$ nm, then estimate the bandgap energy.

Solution Since $E_g = \frac{h \cdot c}{\lambda}$

Therefore, on putting the appropriate values, we get

$$E_g = \frac{(2.99 \times 10^8) \times (6.62 \times 10^{-34})}{1.771 \times 10^{-6}}$$

$$= 1.12 \times 10^{-19}$$

$$= 0.698 \text{ eV}$$ **Ans.**

(iv) Carrier Lifetime

Whenever we talk about carrier lifetime, it means a minority carrier lifetime. This is simply because on injection or by optical generation, the minority carrier density will be more compared to its value at thermal equilibrium, whereas the change in majority carriers is generally insignificant.

For example, consider silicon with a 10^{22} fully ionised impurity atoms per m^3. Then, since N_i for silicon is about 10^{16} per m^3 at room temperature, N_h will be about 10^{10} per m^3. Now, suppose that on incidence of light 10^{15} EHP per m^3 are created. The hole density in silicon will then increase by a large factor (105), but the change in electron density will be an imperceptible 10–5%. Thus, to observe hole current is much simpler. Thus, the current flowing in the circuit given below consists mainly of holes created by the input light.

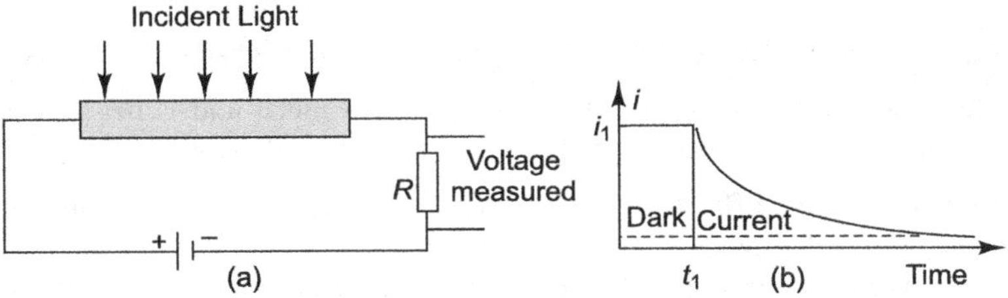

Fig. 8.35 *Photoconduction experiment in semiconductor material*

If the light is switched off at $t = t_1$, the current and voltage across resistance R decrease exponentially as $\left(-\dfrac{t}{\tau_p}\right)$. By measuring the current decay, the carrier lifetime τ_p can be obtained.

(v) Hall Effect and Hall Coefficient

When a magnetic field is applied at right angles to a current-carrying conductor, a voltage in the direction perpendicular to the current and magnetic field set-up. This phenomenon is named as Hall effect, and the voltage so developed is called Hall voltage. The Hall effect is a consequence of the forces that are exerted on the moving charges by electric and magnetic fields.

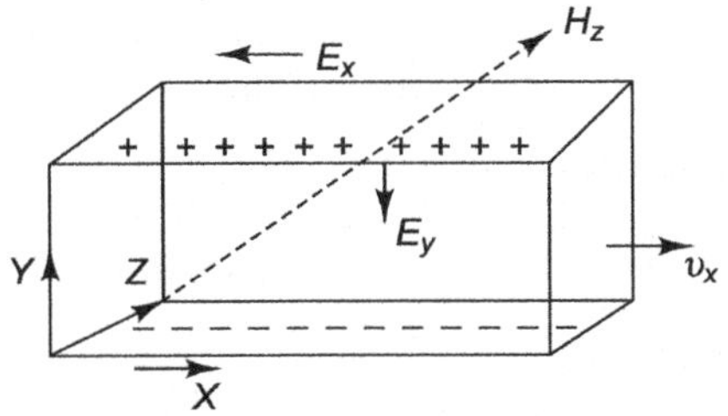

Fig. 8.36 *Accumulation of charge-carriers in Hall effect experiment*

❖ **Expression for Hall Voltage** Let us consider a slab of semiconducting material in rectangular shape. The electric field is directed along x-axis and the magnetic field is directed along z-axis, as shown in Fig. 8.36. If the conduction in this material is assumed to be by electrons having charge e, then the external magnetic field H_z will exert a force on electrons—towards the bottom of the specimen—thereby making a lower surface negatively charged and the upper surface positively charged, as shown in Fig. 8.36. This accumulation of negative charge on the bottom and corresponding positive charge on the upper side gives rise to a transverse electric field, represented by E_y, which is known as Hall field E_y. The Hall field E_y, so developed, opposes the transverse drifting of electrons. Ultimately, an equilibrium is established and the force due to accumulation of electrons becomes equal to the magnetic force, and so, the flow of electrons stops. Thus, the net force on electrons become zero, $F_y = 0$.

Hence, $$F_y = -eE_y + e(V_x H_z) = 0$$

or $$E_y = \upsilon_x H_z \tag{8.38}$$

Current density, $$J_x = -nev_x$$

where $n \rightarrow$ number of conduction electrons per unit volume or carrier concentration.

Therefore,

$$\upsilon_x = -\frac{J_x}{ne}$$

and $$E_y = -\frac{J_x H_z}{ne} = R_H . J_x H_z \tag{8.39}$$

where $R_H = \frac{-1}{ne}$ and is known as Hall coefficient. For the substance of the specimen under consideration, Hall co-efficient is constant for a metal and is negative for free electrons. If the majority carriers are holes, then the Hall coefficient is positive.

• **Significance of Hall Effect** The Hall effect measurements are very important from the following points if view:

- The sign (electron or hole) of charge carries can be known.
- The concentration of majority charge carriers can be determined.
- The mobility of majority charge carriers can be measured directly.
- It can be used to determine whether the material is a metal, insulator or semiconductor.
- From the knowledge of Hall voltage, the unknown magnetic field can be measured.

Example 8.19

An *n*-type germanium sample has a donor density of 1021/m³. It is arranged in a Hall experiment having magnetic field of 0.6 *T* and the current density is 500 A/m². Find the Hall voltage if the sample is 3 mm wide.

Solution Given that

$$n = N_D = 10^{21}/\text{m}^3, \quad H = 0.6\ T$$

$$J = 500\ \text{A/m}^2, d = 3\ \text{mm} = 3 \times 10^{-3}\ \text{m}$$

Field
$$E_y = -\frac{J_x H_z}{ne}$$

$$= -\frac{500 \times 0.6}{10^{21} \times 1.6 \times 10^{-19}}$$

$$= \frac{30}{16}\ \text{V/m}$$

Hall voltage
$$V_H = E_y \cdot d$$

$$= \frac{30}{16} \times 3.0 \times 10^{-3}$$

$$= 5.6\ \text{mV}$$ **Ans.**

Example 8.20

An electric field of 80 V/m is applied to a sample of *n*-type semiconductor whose Hall coefficient is –0.0125 m³/C. Determine the current density in the sample assuming $\mu_e = 0.36\ m^2V^{-1}\ s^{-1}$

Solution As Hall coefficient is given by

$$R_H = -\frac{1}{ne}$$

$$n = -\frac{1}{R_H e} = -\frac{1}{(-0.0125)(1.6\times 10^{19})}$$

$$n = 5\times 10^{20}/m^3$$

Again,

$$\sigma = ne\mu_e$$

$$\sigma = \frac{J}{E}$$

$$J = \sigma E = ne\mu_e E$$

$$= (5\times 10^{20})(1.6\times 10^{-19})(0.36)(80)$$

$$= 2304\ \text{Ampere/m}^2$$ **Ans.**

Example 8.21

The resistivity of a semiconductor was known to be 0.00893 ohm-m at room temperature. The field Hz in the Hall model was 0.5 weber/m². Calculate the Hall angle for a Hall coefficient of $3.66 \times 10^{-4}\ m^3/C$.

Solution

$$\rho = 0.00893\ \text{ohm-m}$$

$$H_Z = 0.5\ \text{weber/m}^2$$

$$\theta_h = ?$$

$$R_H = 3.66\times 10^{-4}\ m^3\ C$$

As Hall angle, $$\theta_H = \tan^{-1}(\mu H_z)$$

where $$\mu = R_H\sigma$$

$$= \frac{R_H}{\rho} = \frac{3.66\times 10^{-4}}{0.00893}$$

$$= 0.0410\ m^2V^{-1}\ s^{-1}$$

$$\theta_H = \tan^{-1}(0.0401\times 0.5)$$

$$= \tan^{-1}(0.0205)$$

$$\theta_H = 1.1744°$$ **Ans.**

8.35 MERITS OF SEMICONDUCTOR MATERIALS

The following are certain merits of semiconductor materials which make them attractive for device application:

- Devices made from semiconductor material are smaller in size.
- They are also very light in weight, for example, weight of a transistor is only fraction of a gram.

- They consume negligible power.
- Their operating voltage is low.
- Such devices have a long life.
- These devices do not show ageing effect.
- They are least affected by shock loads and are almost shock-proof.
- They start operating instantly.

Summary

- Silicon and germanium crystallise in diamond-cubic structure. GaAs is one of the most widely used compound semiconductors; it crystallises in zinc blende structure.
- Pure semiconcucters are called intrinsic semiconductors. Semiconductors doped with impurity atoms are called extrinsic semiconductors.
- In intrinsic semiconductors, the number of electrons in the conduction band is equal to the number of holes in the valence band.
- Semiconductors doped with Group V elements are called *n*-type semiconductors, and the dopants are called donors. In *n*-type semiconductors, electrons are the majority carriers. The donor energy levels lie nearly 0.01 eV below the bottom of the conduction band.
- Semiconductors doped with Group III elements are called *p*-type semiconductors, and the dopants are called acceptors. In *p*-type semiconductors, holes are the majority charge carriers. The acceptor energy levels lie nearly 0.01 eV above the top of the valence band.
- For device applications, single crystals of semiconductors are required. The standard method used for growing single crystals of semiconductors is the Czochralski technique.
- Zone-refining technique is used for purifying semiconductors. Zone-levelling is used for getting uniform distribution of dopant concentration. Zone-melting technique is used for growing single crystals of semiconductors.
- The thermal equilibrium concentration of electrons in the conduction band is given by

$$n_0 = N_e \exp\left[\frac{-(E_c - E_F)}{kT}\right]$$

where N_e is the effective density of states in the conduction band.

- The thermal equilibrium concentration of holes in the valence band is given by

$$p_0 = N_e \exp\left[\frac{-(E_F - E_v)}{kT}\right]$$

where N_v is the effective density of states in the valence band.

- The intrinsic carrier concentration can be obtained from the relation

$$n_i^2 = N_c N_v \exp\left[\frac{-E_g}{kT}\right]$$

- MBE is a highly precise technique capable of excellent composition control and is used to get epitaxial layer of ternary, quaternary and multi-component semiconductor materials.
- *p*-type II-VI compound semiconductor can be prepared using MBE technique.
- Chalcogenide glasses are new class of materials which exhibit short-range order and exhibit semiconducting properties. These glasses have attained a surge of interest because of low-cost and easy preparation.
- Growing a layer on a substrate of such a material, having close matching of lattice constant, is known as epitaxy.
- Zone-refining method provides extremely pure semiconductor materials to be used in the preparation of various devices.
- The two fundamental transport mechanisms are drift—due to an applied electric field, and diffusion—due to a density gradient.
- Carriers attain an average drift velocity in the presence of an applied electric field, due to scattering events. Two scattering processes within a semiconductor are lattice scattering and impurity scattering.
- The average drift velocity is a linear function of the applied electric field for small values of electric field, but the drift velocity reaches a saturation limit that is in the order of 107 cm/s at high electric fields.
- Carrier mobility is the ratio of the average drift velocity and applied electric field. The electron and hole mobilites are functions of temperature and of the ionised impurity concentration.
- The drift current density is the product of conductivity and electric field (a form of Ohm's law). Conductivity is a function of the carrier concentrations and mobilities. Resistivity is the inverse of conductivity.
- The diffusion current density is proportional to the carrier diffusion co-efficient and the carrier density gradient.
- The diffusion co-efficient and mobility are related through the Einstein's relation.
- The Hall effect is a consequence of a charged carrier moving in the presence of perpendicular electric and magnetic fields. The charged carrier is deflected, inducting a Hall voltage. The polarity of the Hall voltage is a function of the semiconductor conductivity type. The majority carrier concentration and mobility can be determined from the Hall voltage.

Exercise

1. What are different types of semiconductor materials? Discuss their applications. Describe the merits and characteristic properties of semiconducting materials?
2. Enumerate different types of semiconductors. Intrinsic semiconductors are not suitable for applications in electronic devices. Why? Show that the Fermi level for intrinsic germanium lies in the middle of its forbidden gap.
3. Differentiate between intrinsic and extrinsic semiconductors. Discuss, in brief, the mechanism of conduction in *n*- and *p*-type semiconductors.
4. What are semiconductors? Discuss their classification. Differentiate between *n* and *p*-type semiconductors and how are they produced?

5. Explain how electrons and holes both conduct in a pure-silicon crystal? How is conductivity influenced by mobility of electrons and holes?
6. Differentiate between *n*- and *p*-type semiconductors. Name various impurities used to prepare them from intrinsic elements. Draw energy diagrams for both the above types, show salient levels on them, and explain the mechanism of conduction in them.
7. Describe 'effective mass' and 'density of states', and discuss their importance in the designing of solid state devices.
8. Describe the effects of temperature on (a) carrier concentration, and (b) mobility of carriers. Also discuss the effects of doping on mobility.
9. Explain direct and indirect semiconductors. Discuss their characteristic features. Enlist their examples, as well.
10. What do you mean by degenerate semiconductors? What are their different types? Also discuss the effects of heavy doping.
11. Compare the following:
    (i) Direct and indirect bandgap semiconductor
    (ii) Degenerate and non-degenerate semiconductor
12. What is Hall effect? Explain its origin and derive the relation between Hall co-efficient and carrier density.
13. What are the basic semiconductor parameters which are measured to characterise the semiconductor material?
14. Write short notes on
    (i) mobility and conductivity of semiconductors
    (ii) continuity equation
    (iii) zone refining
    (iv) Czochralski method
    (v) Space-charge capacitance
15. Explain the Czochralski method for growing single crystal used in the fabrication of 1 Cs.

Numerical Problems

1. Calculate the current produced in a small germanium plate of 1 cm^2 area and of 0.3 mm thickness, when a potential difference of 2 V is applied across the faces. Given that the concentration of free electron in germanium is 2×10^{19}/m^3 and mobilities of electrons and holes are 0.36 m^2/Vs and 0.17 m^2/V-s, respectively.

   [Ans. 1.13 A]
2. Mobilities of electrons and holes in a sample of intrinsic germanium at room temperature are 3600 cm^2/ V-s and 1700 cm^2 / V-s, respectively. If the electron and hole densities are each equal to 2.2×10^{13}/cm^3, calculate the conductivity.

   [Ans. 1.865 mho/m]
3. The resistivity of pure germanium at room temperature is 0.47 Ω m. Find out the carrier density of germanium at room temperature for the electron mobility of 0.32 m^2/ volt-s and hole mobility = 0.20 m^2/ volt-s. **[Ans. 2.55 × 10^{19}/m^2]**
4. The mobilites of silicon are μ_e = 0.17 m^2/V-s and μ_h = 0.035 m^2 V-s, respectively, at room temperature. If the carrier density in the material is known to be 1.1×10^6, calculate the resistivity of silicon. **[Ans. 2771.6 ohm–m]**

5. Calculate the drift velocities of holes and electrons in (i) silicon, and (ii) germanium at 300 K when applied electric field is 60 volt/cm. Take $\mu_p = 500$ cm²/volt-s and $\mu_n = 1500$ cm²/volt-s for silicon and $\mu_h = 3700$ cm²/volt^{-s} and $\mu_n = 3600$ cm²/volt^{-s} for germanium. **[Ans. (i) 300 m/s, 900 m/s, (ii) 1020 m/s, 2160 m/s]**
6. In germanium, the energy-gap is about 0.75 eV. What is the wavelength at which germanium starts to absorb light? **[Ans. 16562Å]**
7. A current of 1000 A/m² density flows through an n-type germanium crystal. This crystal has a resistivity of 0.05 ohm -m and the electron mobility of 0.4 m²/Vs. Calculate the drift velocity and time taken by the electrons to travel 100 m in the crystal. **[Ans. 5 ms]**
8. Calculate the conduction electron and hole densities at room temperature in Si-doped with 10^{19} donors/m³, if the intrinsic carrier concentration is 1.5×10^{16}/m³. **[Ans. 2.25 × 10^{13}/m³]**
9. If the conductivity of an intrinsic semiconductor is to be increased by 30%, to what temperature should it be heated above the room temperature? **[Ans. 5.9 K]**

Self-quiz Questions

1. What is a semiconductor?
2. What are the characteristics of semiconductors?
3. What is an intrinsic semiconductor? How does it differ from an extrinsic semiconductor?
4. Explain the following:
   (i) Thermistors and sensistors
   (ii) Photoconductors
5. List the applications of semiconductor materials.
6. State the advantages and disadvantages of semiconductor devices.
7. Differentiate between semiconductors and insulators.
8. What is Hall effect?
9. What are the applications of Hall effect?
10. How is mobility determined experimentally?
11. What are the applications of semiconductor materials in electrical engineering?
12. What is meant by the term 'intrinsic donor' and 'acceptor' when applied to a semiconductor? Differentiate between a p- and n-type semiconductor.

Objective-type Questions

1. Hall effect is utilised to find
   (a) type of semiconductor
   (b) energy gap
   (c) Fermi energy level
   (d) endoscope investigations
2. Density of state indicates the
   (a) number of electrons across the energy band
   (b) donar energy level
   (c) mobility of compound semiconductors
   (d) effective mass
3. Hall effect is most significant in
   (a) silver
   (b) sodium
   (c) lithium
   (d) gold
4. The energy gap in Si and Ge are, respectively
   (a) 0.66 and 1.12 eV
   (b) 0.56 and 0.7 eV
   (c) 1.12 and 0.66 eV
   (d) 0.7 and 0.56 eV

5. The type of bonding in GaAs is
   (a) covalent
   (b) ionic
   (c) partially covalent and partially ionic
   (d) van der Waals
6. *n*-type silicon is obtained by doping Si with
   (a) boron
   (b) arsneic
   (c) antimony
   (d) indium
7. Which of the following statements is true?
   (a) Mobility of holes is greater than that of electrons
   (b) Mobility of electrons is equal to the mobility of holes
   (c) Mobility of electrons is greater than that of holes
   (d) Mobility of electrons and holes are nearly equal
8. The number of silicon atoms per unit cell of Si crystal is
   (a) 2 (b) 4
   (c) 6 (d) 8
9. Crystal structure of GaAs is
   (a) diamond cubic
   (b) hexagonal
   (c) face centred
   (d) zinc blende
10. The position of a donor level in an extrinsic semiconductor is
    (a) 0.01 eV just below the condition band
    (b) exactly in middle of the forbidden band
    (c) 0.01 eV above the valence band
    (d) within the valence band
11. Semiconductor materials belong to
    (a) sixth group of periodic table
    (b) fourth group of periodic table
    (c) seventh gap of periodic table
    (d) zero gap of periodic table
12. Bonding in Ge and silicon is
    (a) Ionic
    (b) covalent
    (c) van der Waals
    (d) electrovalent
13. Chalcogen elements are
    (a) S, Se, Te
    (b) As, Sb, Bi
    (c) N, P, Cl
    (d) none of these
14. Atomic spacing in silicon is
    (a) 2.50 Å
    (b) 5.43 Å
    (c) 7.52 Å
    (d) 1.0 Å
15. In a *p*-type semiconductor, the position of Fermi level is
    (a) between valence band and acceptor level
    (b) near the middle of the forbidden gap
    (c) near the conduction band
    (d) with the conduction band
16. The temperature co-efficient of resistance for semiconductor material is
    (a) positive
    (b) negative
    (c) neither positive nor negative
    (d) none of these
17. By measuring the absorption, one can get
    (a) bandgap of semiconductor
    (b) carrier life time of semiconductor
    (c) kinds of semiconductor
    (d) kinds of impurity
18. MBE technique is used to get
    (a) *n*-type II-VI semiconductor
    (b) *p*-type II-VI semiconductor
    (c) both *p*-type and *n*-type III-V semiconductor
    (d) none of these
19. The position of donor-level in extrinsic semiconductor is
    (a) 20 meV below the conduction band
    (b) 10 meV above the conduction band
    (c) in the middle of conduction band
    (d) 10 meV above the valence band

20. The dielectric constant of silicon is
    (a) 20 (b) 100
    (c) 16 (d) 12

21. For high-frequency devices, the suitable semiconductor material is
    (a) GaAs (b) Ge
    (c) Se (d) Si

Answers

1. (a)	2. (a)	3. (b)	4. (c)	5. (c)
6. (b & c)	7. (c)	8. (d)	9. (d)	10. (a)
11. (b)	12. (b)	13. (a)	14. (b)	15. (a)
16. (b)	17. (a)	18. (a)	19. (a)	20. (d)
21. (b)				

Fill in the Blanks

1. The electronic configuration of silicon .... in its outermost orbital.
2. On doping with aluminium, germanium becomes ....type of semiconductor.
3. The number of nearest neighbours is germanium crystal is ....
4. The structure of gallium arsenide is ....
5. In an intrinsic semiconductor, the Fermi level lies in the .... of valence and conduction band.
6. The conductivity of an extrinsic semiconductor in terms of the mobility of the charge carriers is given by ....
7. The standard method used for growing single crystals of semiconductors is ....
8. The Hall coefficient is related to the density of charge carriers by the equation ....

Answers

1. $3s^2 3p^2$	2. p	3. Four	4. Zinc Blende
5. Middle	6. $\sigma = e[n\mu_n + p\mu_n]$		7. Czochrilaski method

8. $R_H = -\dfrac{1}{ne}$

Reading List

1. Solymar *Electrical Properties of Materials*. New Delhi: Oxford University Press.
2. Basak, T K *Electrical Engineering Materials*. New Delhi: New Age Press International.
3. Streetmann, Ben G; Banerjee, Sanjay Kumar *Solid State Electronic Devices*. 2ed. New Delhi: PHI Learning Pvt Ltd
4. Neamen, Donald A (1992) *Semiconductor physics and Devices: Basic Principles*. Homewood, IL: Irwin.
5. Rajput, R K *Electrical Engineering Materials*. New Delhi: Laxmi Publications Pvt Ltd.
6. Gupta, Navneet; Shahdev S K (2001) *A Textbook of Basic Electronics*. 1ed. New Delhi: Dhanpat Rai & Co. Ltd.
7. Dekker, A J (1998) *Solid State Physics*. New Delhi: Macmillan India Ltd.

8. Dekker, A J (2000) *Electrical Engineering Materials.* New Delhi: PHI Learning Pvt Ltd.
9. Kittle, Charles (1996) *Introduction to Solid State Physics* 7ed. New York: John Willey & Sons.
10. Indulkar, C S; Triruvagdan *An Introduction to Electrical Engineering Materials* New Delhi: S.Chand Publications.
11. Allison, John *Electrical Engineering Materials and Devices.* New Delhi: Tata Mc-Graw-Hill Education Pvt Ltd.

9

Mechanical Properties of Materials

Chapter Objectives

- To discuss the important mechanical properties of materials from the point-of-view of their application in electrical devices and machines.
- To test their hardness, compression, etc.
- To understand the desirable mechanical characteristics of material meant to be used for electrical purposes.

9.1 INTRODUCTION

Materials that are used for various purposes—e.g. in spacecrafts, marine automobiles, aircrafts, or in any other kind of machine—are required to exhibit certain special mechanical characteristics for best results. The chief mechanical properties of materials that are looked upon are given below:

- Elasticity
- Plasticity
- Ductility

The mechanical behaviour of materials is measured by testing some of the parameters given below:

- Fatigue
- Creep
- Resilience
- Modulus of elasticity (stiffness constant)
- Ultimate tensile strength
- Yield strength
- Hardness
- Toughness

Besides these, there are many other parameters too that characterise the mechanical behaviour of materials. The knowledge of mechanical properties of materials is very essential in order to construct a mechanically sound structure needed for any machine or project. The response of materials to the applied forces depends on the type of bonding, and the structural arrangement of atoms and molecules of the materials.

9.2 ISOTROPIC, ANISOTROPIC AND ORTHOTROPIC MATERIALS

(i) Isotropic Materials

In an isotropic material, all the planes are planes of symmetry, and thus their properties are independent of direction. The number of independent elastic constants is only two, in this case. Isotropic materials obey Hooke's law. All metals are isotropic.

(ii) Antisotropic Materials

In an anisotropic material no plane is a plane of symmetry, and thus the properties are direction dependent. The number of elastic constants in most general anisotropic case is 36, out of which 21 are independent. Anisotropic materials, such as composites, obey generalised Hooke's law which is different from Hooke's law.

(iii) Orthotropic Materials

Orthotropic material is a special kind of anisotropic material. The number of elastic constant is 12, out of which 9 are independent. An unidirectional (U/D) fibre composite is an example of such a material. It has three orthogonal planes of symmetry.

(iv) Necessity of Different Properties for Different Applications

- Elasticity is desirable in precision instruments and machines.
- Stiffness is desired in beams and columns.
- Resilience is desired in springs.
- Ductility is desired in cold wire drawing.
- Hardness and toughness in required for cutting tools.

9.3 TESTING OF MECHANICAL BEHAVIOUR

The mechanical behaviour of materials is tested by studying the response of the material to the applied load. Therefore, the following parameters become essential:

- Stress
- Strain
- Elastic coefficient

The applied load is measured in terms of internal force per unit area, which is known as 'stress' represented by σ. The deformation is measured in terms of strain, which is the ratio between the change in dimensions of the material to the initial dimensions. The strain, being the ratio of two similar quantities, is therefore unitless.

Elastic Coefficient

In most of the materials, at low stress values, the stress is always proportional to the strain,

i.e. stress α strain

or $$\sigma = E\in$$

where $\in$ is the strain.

Therefore, for low stress-value ratio of strain and stress, is equal to a constant E which is known as Hooke's law, and E is known as elastic modulus or stiffness constant. The unit of elastic modulus is N/m^2 or Pascal. The graph of stress vs. strain is an important characteristic to understand the mechanical behaviour of materials, which is explained in Fig. 9.1.

The stiffness constant E is a measure of the ability to withstand the applied load, without much deformation or resistance of the material to deformation.

Keeping in view that the stress and strain are tensors, the above relation can be written in a more generalised way:

$$\sigma_{ij} = E_{ijkl}\,\varepsilon_{kl}$$

where E_{ijkl} is a fourth rank tensor representing 81 components of stiffness constant. As we discuss only the isotropic material, therefore tensor treatment has not been taken into consideration.

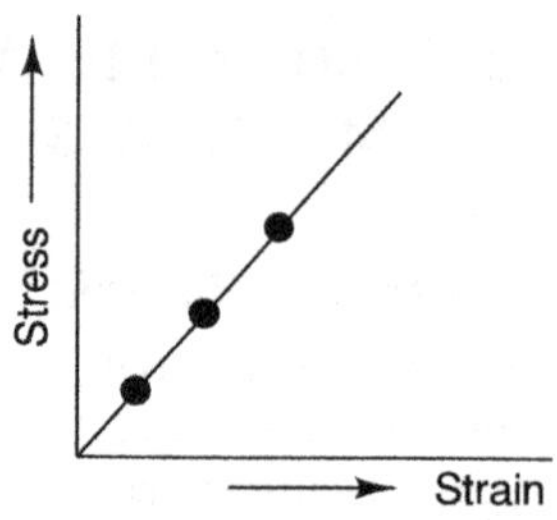

Fig. 9.1 *Stress-strain plot of electric material*

9.4 ELASTIC LIMIT

The above relationship between strain and stress is only valid for low values of stress. This limit is known as elastic limit. Above this limit the material acquires permanent deformation. Most metals are elastic at low values of stress.

Example 9.1

A steel wire of 30 cm length and 3 mm radius is subjected to tensile force of 8482 newtons. Find the elongation and the lateral contraction. (Young's modulus of steel is 207 GPa and its Poisson's ratio is 0.30).

Solution

$$Y = \sigma/\varepsilon = \frac{F/A}{dl/l_0} = \frac{Fl_0}{Adl}$$

The elongation

$$dl = \frac{Fl_0}{YA} = \frac{8482\times 0.30}{\pi\times(3\times 10^{-3})^2\times 207\times 10^9} = 4.35\times 10^{-4}\ \text{m}$$

$$= 0.435\ \text{mm}$$

$$\varepsilon_1 = \frac{dl}{l_0} = 0.435\times 10^{-3}/0.30 = 1.45\times 10^{-34}$$

The Poisson's ratio $\nu = \dfrac{-\varepsilon_2}{\varepsilon_1}$

Lateral strain $\varepsilon_2 = -\nu\varepsilon_1 = 0.30\times 1.45\times 10^{-3} = -4.35\times 10^{-4}$

$\varepsilon_2 = dr/r$

Lateral contraction $dr = \varepsilon_2 r = -4.35\times 10^{-4}\times 3\times 10^{-3}$

$= 1.30\times 10^{-6}\ \text{m} = 1.30\ \mu\text{m}$ **Ans.**

Example 9.2

A perfectly elastic material is subjected to a tensile load. Show that the work done per unit volume is (1/2) stress × strain.

Solution Let the load on the material of length l and area of cross-section A be F.

$$\sigma = F/A$$

If dl is the elongation at any stage, the corresponding strain = $d\varepsilon = dl/l$

$$\text{Work done} = \int F dl = \int \sigma A dl = \int \sigma A l d\varepsilon$$

$$= Al \int \sigma d\varepsilon$$

$$= Al \int E \varepsilon d\varepsilon$$

where E is the Young's modulus or stiffness constant.

$$\text{Work done} = VE \int F dl = \int \sigma A dl = \int \sigma A l d\varepsilon$$

$$= Al \int \sigma d\varepsilon$$

$$= Al \int E \varepsilon d\varepsilon$$

where E is the Young's modulus or stiffness constant.

$$\text{Work done} = VE \int \varepsilon d\varepsilon = VE\varepsilon^2/2$$

where V is the volume of the specimen.

Work done per unit volume = $E\varepsilon^2/2$

$$= \frac{\sigma}{\varepsilon} \times (\varepsilon^2/2)$$

$$= (1/2)\ \sigma \times \varepsilon$$

$$= (1/2)\ \text{stress} \times \text{strain}$$

Ans.

9.5 PLASTIC BEHAVIOUR

In some materials, the stress–strain relationship is not linear as shown in Fig. 9.2. For low value of stress, the material is elastic, but when it is plastic the original dimension of the material can never be recovered. The stiffness constant (elastic modulus), i.e stress strain behaviour is variable. In order to get the elastic modulus, the slope of the tangent drawn at that point gives its value. Figure 9.2 represents the plastic behaviour of the material.

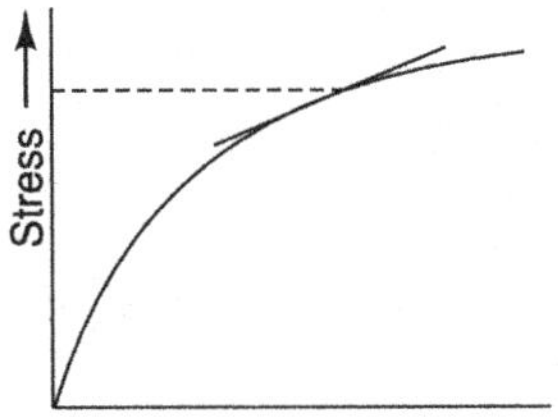

Fig. 9.2 *Stress–strain plot of plastic material*

The stiffness constant (elastic modulus) of ceramic material is more than as compared to the metals. However, polymers exhibit a quite low value of stiffness constant. The values of stiffness constant of some selected materials are given in Table 9.1.

Table 9.1 *Stiffness constants of some materials*

Material	*Stiffness constant (GPa)*	*Material*	*Stiffness constant (GPa)*
Diamond	786	Nickel	207
Aluminium oxide	345	Nylon	2.75
Copper	110	Natural rubber	0.007—0.06

9.6 STRESS AND ITS KINDS

Depending upon the type of deformation (compression or elongation, etc.), stress is of four kinds:

- Tensile or longitudinal stress
- Compressive stress
- Shear or tangential stress
- Bulk stress

The elastic coefficient of these four types is therefore:

- Young's modulus
- Bulk modulus
- Modulus of rigidity
- Poisson's ratio

Tensile or compressive stress results in either increase or decrease in length, while shear stress or tangential stress results in the change of shape. Young's modulus or stiffness constant is accompanied by linear changes in the material, when subjected to stress. Bulk stress is applicable is fluids, i.e. in liquids or gases, that undergoes a change of volume. The elastic coefficient related with this is known as bulk modulus. The reciprocal of bulk modulus is known as compressibility.

9.7 POISSON'S RATIO

When a material is subjected to tensile stress the length increases in the direction of stress, which is always accompanied by a decrease (contraction) in the cross-section. In other words, a linear or tangential strain produced in the wire is accompanied by a transverse or lateral strain of an opposite kind in a direction at right angles to the direction of applied force. Therefore, Poisson's ratio is defined as the ratio of lateral strain β to the tangential strain (α) within the elastic limit. The lateral strain (β) is proportional to the linear or tangential strain (α) for the material of a given body. It is denoted by the symbol υ, and is a unitless quantity.

$$\upsilon = \frac{\beta}{\alpha}$$

The normal value of Poisson's ratio for metals is in the range 0.25–0.35.

The stiffness constant E, the Poisson' ratio and shear modulus G are related as given below:

$$E = 2G(1 + \upsilon)$$

9.8 THE STRESS AND ENGINEERING STRESS

On subjecting a material to tensile stress, the cross-sectional area of the material decreases at a particular position. This is known as necking. The area of cross-section of the neck regularly changes as the load increases. Therefore, the force divided by instantaneous area of cross-section is commonly known as **true stress** at that instant, which is denoted by υ_T.

Therefore, true stress $$\sigma_T = \frac{\text{Force (tensile stress)}}{\text{Instantaneons area a of cross section}} = \frac{F}{A_i}$$

On the other hand, the force divided by original cross-sectional area is known as engineering stress.

Engineering stress $\sigma = \frac{F}{A_0}$

True Strain and Engineering Strain

True strain is defined as the ratio of change in the instantaneous length of test specimen to the original length of the specimen.

True strain, $\varepsilon_T = Ln\left[\frac{l_i}{l_0}\right]$ where l_i is the instantaneous length and l_0 is the original length.

The engineering strain is defined as the ratio of $(l_i - l_0)$ to l_0,

i.e. engineering strain $\varepsilon = \frac{l_i - l_0}{l_0} = \frac{l_i}{l_0} - 1$

Therefore, true strain ε_T and engineering strain E are related as

$$\text{In}\left[\frac{l_i}{l_0}\right] = \ln(E + 1)$$

or $\varepsilon_T = \ln(E + 1)$

Considering that there is no change of volume in the test specimen during deformation

$$A_0 l_0 = A_i l_i$$

or $A_i = A_0\left[\frac{l_0}{l_i}\right]$

Therefore, true stress $\sigma_T = \frac{F}{A_i} = \frac{F}{A_0} \cdot \frac{l_i}{l_0}$

or $\sigma_T = \sigma(E + 1)$

The curve in Fig. 9.3 gives a comparison between true stress σ_T and engineering stress σ.

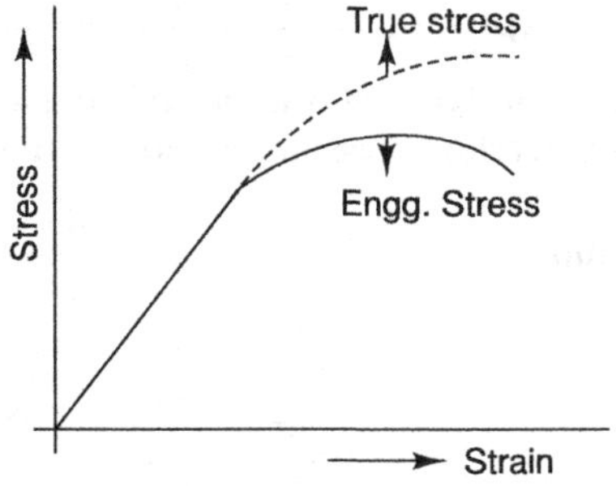

Fig. 9.3 *Graphical representation of true and engineering stress*

9.9 BRITTLE AND DUCTILE MATERIALS

The stress-strain behaviour of two materials (1 and 2) as given in Fig. 9.4.

The stress–strain behaviour of Material 1 is exactly linear from 0 → *F*. Such a material is known as brittle material. Point *F* is known as the breaking point, beyond which it breaks. Examples of this category are concrete, cast iron, glass, etc. Silicate is the most brittle material.

On the other hand, the stress–strain behaviour of Material 2 is entirely different. Up to *Y* it is perfectly linear

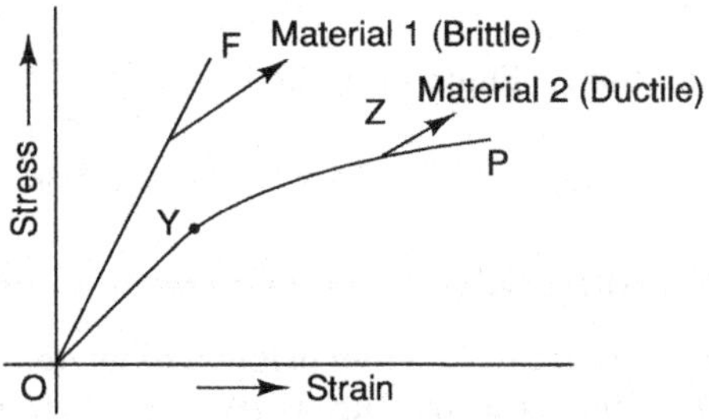

Fig. 9.4 *Stress–strain behaviour comparison of two materials*

(elastic). Beyond *Y*, it is non-linear and deforms completely, hence does not regain its original condition, i.e. it undergoes a change from elastic to plastic. This limiting point *Y* is known as the yield point. The stress of the material at this point is known as the yield strength. Beyond *P*, the material breaks. Examples of ductile materials are copper, silver, aluminium, gold, etc.

Measurement of Ductility

Ductility of a material is measured in terms of percentage elongation. If the length at which the material fractures (breaks) is l_F and l_0 is the original length then

$$\text{Ductility} = \frac{l_F - l_0}{l_0} \times 100$$

It is also measured in terms of percentage reduction in the area of cross-section (normal to the tensile stress), as given below:

$$\text{Ductility} = \frac{A_0 - A_F}{A_0} \times 100$$

where A_0 is the original (initial area) and A_F is the final area of cross-section, when the material undergoes a fracture. Ductility is a measure of the degree to which a material can be deformed during designing. It also indicates the ability to which a material can be stretched into wires.

Example 9.3

A 6 mm steel rod is under the action of a tensile force of 400 newtons. Calculate the engineering stress in the rod in MPa.

Solution

$$\sigma = \frac{P}{A_0} = \frac{400N}{\pi r^2} = \frac{400}{\frac{\pi}{4}(6 \times 10^{-3})^2 m^2}$$

$$\sigma = \frac{1600 \times 10^6}{3.14 \times 36} \text{N/m} = 14.15 \times 10^6 \text{ N/m}^2$$

$$= 14.15 \text{ MPa}$$

Ans.

Example 9.4

The original gauge length marked on an aluminium test piece is 40 mm. The test piece is strained in tension so that the guage length becomes 43.3 mm. Calculate the engineering strain and the per cent elongation which the sample undergoes.

Solution: Engineering strain $e = \frac{L_f - L_0}{L_0} = \frac{(42.3 - 40)\text{mm}}{40 \text{ mm}} = 0.0575$

% elongation—0.0575 × 100% = 5.75%

Ans.

Example 9.5

A steel wire of 12.8 mm original diameter is subjected to a tensile load up to fracture. Its diameter at fracture is 10.7 mm. Find its ductility.

Solution Ductility is measured as % elongation or % reduction in area.

$$\text{Percent reduction in area} = \frac{A_0 - A_f}{A_0} \times 100\%$$

$$= \frac{\pi r^2 - \pi r^2}{\pi r_0^2} \times 100$$

$$= \frac{(6.4 \times 10^{-3})^2 - (5.35 \times 10^{-3})^2}{(6.4 \times 10^{-3})^2} \times 100$$

$$= 0.302 \times 100 = 30.2\ \%$$

Ans.

9.10 TENSILE STRENGTH

The stress–strain behaviour graph is very useful to understand the tensile strength of the material. Figure 9.5 represents the stress–strain behaviour of a metal.

The total behaviour of a material can be understood in three parts:

(1) $OY \rightarrow$ Linear behaviour (elastic)

(2) $YP \rightarrow$ Non-linear behaviour (plastic); this is the maximum stress point where formation of neck takes place

(3) $PF \rightarrow$ Necking increases, T_s decreases and at F fracture occurs.

Table 9.2 gives yield strength, tensile strength and ductility of some materials.

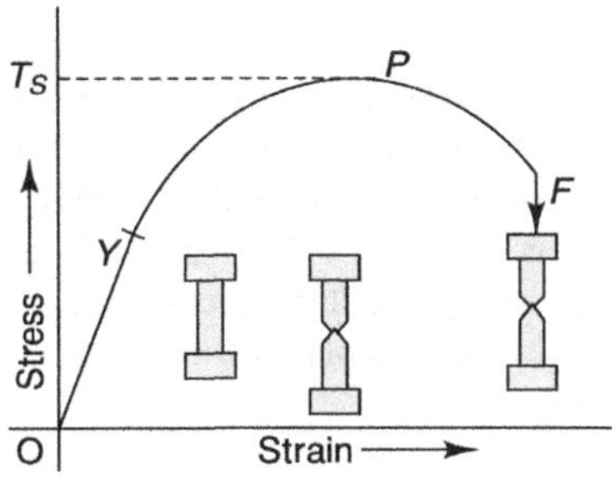

Fig. 9.5 *Typical engineering stress-strain curve for a ductile material up to fracture*

Table 9.2 *Yield strength, tensile strength and ductility of some materials*

Metals	*Yield strength in mPa*	*Tensile strength mpa*	*ductility in % Elongation*
Aluminium	35	90	40
Copper	69	200	45
Steel	180	380	25
Brass	75	300	68

9.11 SUITABILITY OF MATERIALS

The suitability of a martial is decided on the basis of its strength and ductility. The material should not break if it is loaded beyond yield stress, but it should undergo plastic deformation without breaking. To develop an alloy and composite is a constant effort of material scientists.

Stress–Strain Behaviour of Single Crystals

The behaviour of a single crystal depends on its orientation with respect to the stress direction. Usually, in single crystals yield occurs at a much lower stress than in polycrystalline material. Figure 9.6 compares the stress-strain curve for a polycrystalline material and a single-crystal form of the same.

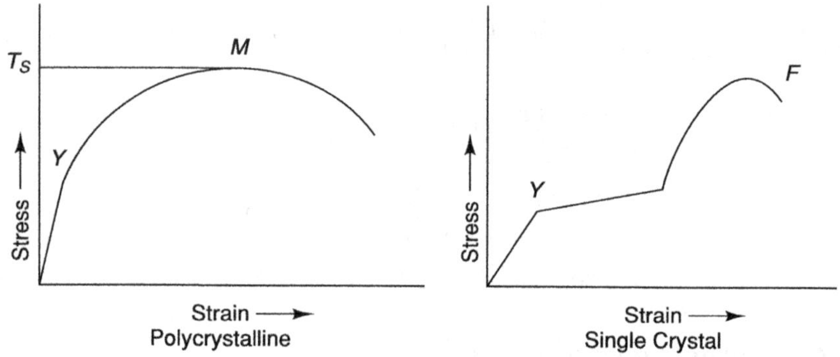

Fig. 9.6 *Stress-strain curves for a polycrystalline material and its single-crystal form*

In a single crystal, after the yield point, deformation takes place at almost constant stress. Eventually work hardening (i.e. continued deformation under ever-increasing stress for a given deformation rate) commences, finally resulting in rupture. The deformation takes place by shear, with blocks of crystals sliding over one another. For a given metal, there is a preferred slip plane and slip direction. A critical shear-stress resolved on the slip plane and in the slip direction is needed to produce the gliding. Slip occurs more readily on close-packed planes (ones with largest interplanar spacing) and in directions of closest packing.

9.12 TOUGHNESS OF MATERIALS

Toughness is defined as the ability of a material to absorb energy up to fracture. It is measured by the area under the stress–strain curve. Its unit is energy per unit volume. Ductile materials have a higher toughness than brittle ones.

9.13 ANELASTIC MATERIALS

In certain materials, elastic behaviour is time dependent. This is known as **anelasticity.** Such materials may not respond immediately to the applied stress. When a fixed stress is applied, the elastic deformation will slowly continue for some time, before it reaches a constant strain value. If stress is removed, it again takes some time to recover back its original form.

9.14 VISCO-ELASTIC MATERIALS

Those materials which deform anelastically, but do not completely come back to their original form—after any length of time after removal of deforming stress—are known as viscoelastic materials. For example, a thermoplastic material exhibits this property.

9.15 HARDNESS

The capability of a material to resist to the localised plastic deformation, like dent or scratch, is known as hardness. It is a surface property. It plays a crucial role during designing the tools and some special machine components. The various hardness tests depend upon following factors:

- Nature of scratch
- Nature of indentation
- Nature of rebound

It is measured by making a small dent in the material by forcing a small indenter on its surface under standard controlled loading conditions. The depth of the dent (or size) is taken to be a measure of hardness of the material. There cannot be an absolute value for hardness, as the size and depth of dent depends upon the external loading conditions. A standard method of loading is defined and the hardness of different materials is compared. There are several hardness scales (test) which are in use—Rockwell hardness scale, Brinell hardness scale, Moh's hardness scale and Vicker's hardness scale. The hardness test is carried out in different ways in the above scales, mainly giving a variation in the size of indentor and the method of applying load. The hardness scales method mentioned above is indicated in Table 9.3.

Table 9.3 *Hardness measurement scales*

Hardness scale	*Maximum*	*Minimum*
Moh's hardness scale	10	1
Brinell hardness scale	10000	5
Rockwell hardness scale	110	0
Vicker's hardness scale	1100	100

The relative hardness of materials is assessed by using Moh's scale, as given in Table 9.4. This is the oldest method used by mineralogists. Table 9.4 consists of a list of materials arranged in order that any mineral in the list will scratch one below it. Thus diamond, the hardest known material, is on the top of the list with a hardness index of 10,while talc is at the bottom of the list with a hardness index of 1.

Table 9.4 *Moh's scale*

Mineral	*Hardness index*
Diamond	10
Corundum	9
Topaz	8
Quartz	7
Feldspar	6
Apatite	5
Fluorite	4
Calcite	3
Gypsum	2
Talc	1

The details of hardness measurement scales in beyond the scope of this book.

9.16 ELASTIC AFTER EFFECT

On applying the deforming force some bodies return to the original position immediately, while some bodies return after some time on the removal of deforming force. For example, quartz fibre immediately comes to original position, but a glass fibre takes hours to do so. This delay in returning to the original state by an elastic body after the removal of deforming forces is known as elastic after effect.

9.17 ELASTIC FATIGUE

When a body undergoes tension repeatedly by deforming forces, it becomes less elastic. A disc suspended by a thin wire executing torsional oscillation is the best example in which the wire undergoes tension alternately to the twisting couples. The same wire if again undergoes torsional vibrations, then its vibrations die away quickly than before—as if the wire becomes tired or fatigued. Thus, the elastic fatigue is the property of an elastic body by virtue of which its behaviour becomes less elastic under the action of repeated alternating deforming forces.

This is exactly similar to physical fatigue in human beings. As a tired person becomes fresh when takes rest, similarly the elastic body regains its full elasticity when it is allowed to rest for some time.

The main effects of fatigue on the properties of materials are following:

- Loss of ductility
- Loss of strength
- Enhanced uncertainty in strength and the service life of materials

9.18 ELASTIC HYSTERESIS

A rubber wire does not follow the well-known Hooke's law even for small stresses, and has no permanent strain even for quite high stress values. It has a large elastic region in its stress–strain curve, but it does not retrace the curve during unloading. The stress-strain behaviour of rubber is given below in Fig. 9.7. The complete closed curve during loading and unloading is known as the hysteresis loop.

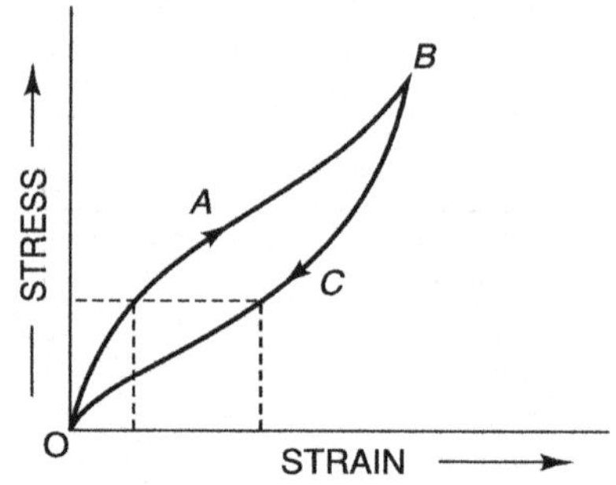

Fig. 9.7 *Elastic hysteresis curve*

The area of the loop is proportional to the energy dissipated by the material as heat, during stress loading and unloading.

9.19 RESILIENCE

This is the property of a material which enables it to store energy, and resist shock and impact. Therefore, it can be defined as the ability of the material to store elastic energy without a permanent deformation into it. The maximum amount of elastic energy that may be so stored in the material during its first loading cycle is called proof resilience. Numerically, it is equal to the area under the stress–strain curve within the elastic limit.

Materials that exhibit a high value of resilience are used for springs. Annealed copper is not used as it exhibits low elastic limit, while cold-worked copper is best suitable for it. Figure 9.8 illustrates the above-mentioned proof resilience.

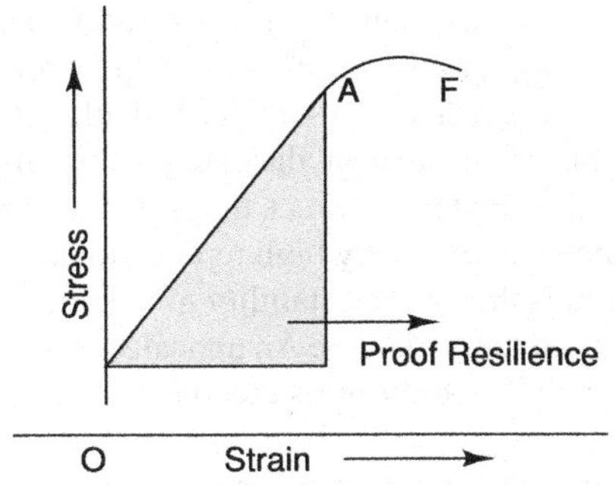

Fig. 9.8 *Demonstration of proof resilience*

9.20 CREEP

Creep is the continuous deformation of a material with a constant load. The deformation is permanent and non-recoverable. In general, creep occurs in the material at high temperatures. Figure 9.19 shows a creep curve. The creep test of a material is done at constant load and constant temperature.

Creep can be classified in three categories:

- Primary creep
- Secondary creep
- Tertiary creep

Curves of all the above kinds have been shown in Fig. 9.9.

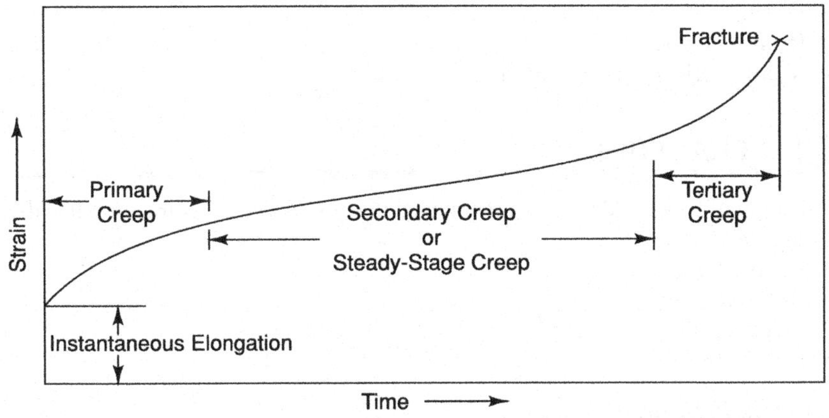

Fig. 9.9 *Different kinds of creep*

The slope of the curve measures the creep rate. During primary creep, creep-rate decreases continuously, while in the secondary creep state, this rate remains constant. The material becomes softer at this stage. During third stage (tertiary creep), creep rate increase,. At this time, the material is said to have ruptured.

9.21 FACTORS AFFECTING CREEP

Creep is highly affected by grain size, microstructure and pre-strain history.

Grain size is a major factor in creep. Coarse-grained materials exhibit better creep resistance than fine-grained ones, as the former have a greater amount of grain boundary materials which behave as a quasi-viscous material with a high tendency to flow at elevated temperatures. This is the reason that single crystals exhibit a higher creep resistance than poly-crystalline materials. Since tungsten filaments of electrical lamps are made of single crystals, they can bear very high temperatures.

Another important factor is the thermal stability of the microstructure of alloys and its resistance to oxidation at high temperatures. An annealed specimen, for instance, having greater thermal stabilities is far superior in its creep resistance to a quenched steel for its poor thermal stability.

Cold working has also a major effect on creep which increases at a high rate in some metals during re-crystallisation. This accelerated creep is followed by a period of relatively slow creep in the re-crystallised metal.

Pure metals, which have high melting points and compact atomic structure, usually show good creep resistance at high temperatures. The creep resistance of pure metals can be considerably increased by alloying them with suitable elements.

9.22 IMPORTANCE OF CREEP CONSIDERATION

Consideration of creep is extremely important in the following applications:

- In industrial belts
- In blades of steam turbine
- In rockets and missiles
- In tubes of heat exchanges
- In the pistons of IC engines
- In nuclear reactors
- In polymeric and electrometric constrictions

9.23 FRACTURE

The breaking of a material when it is subjected to stress is known as fracture. It is divided into four kinds:

- Ductile fracture
- Brittle fracture
- Creep fracture
- Fatigue fracture

(i) Ductile Fracture

In ductile fracture, the material undergoes substantial plastic deformation with high energy absorption before fracture and so the toughness value of ductile materials is high. Fracture occurs in these materials due to necking. Examples are metals and alloys.

(ii) Brittle Fracture

In brittle fracture, the material remains perfectly elastic up to the fracture point. The energy of absorption accompanying brittle fracture is low, i.e. toughness values of brittle materials is low. The fracture in brittle materials occurs mainly due to pre-existence of tiny cracks on the surface of the material, which get introduced during fabrication of the material. Brittle fracture may be prevented by suitable surface treatment like coating, etching, tempering, etc. (Tempering is a kind of heat treatment by which compressive stresses are created on the surface of the material).

(iii) Creep Fracture

Creep fracture is normally referred to as rupture. In the third state of the creep curve, where the creep rate increases steeply, the material ultimately ruptures. Rupture occurs in metals at high temperatures and at ordinary temperatures in low melting materials

(iv) Fatigue Fracture

A material, when subjected to cyclic loading, ultimately fractures at a stress level considerably lower than the tensile or yield strength under static load. Fatigue failure is normally brittle-like nature even in ductile materials, in that there will not be much plastic flow before failure. Presence of crack on the surface of the material initiates this type of fatigue failure. All types of materials (metals, polymers and ceramics) are susceptible to fatigue failure. Glasses normally do not undergo a fatigue failure.

The fracture that occurs in all above-mentioned four type of behaviour is distinct in the way the materials respond to the external load before fracture occurs.

9.24 FACTORS AFFECTING THE MECHANICAL PROPERTIES

The following are the various factors that influence mechanical properties of materials:

- Grain size
- High temperature
- Low temperature
- Radiation exposure
- Heat treatment
- Rate of loading
- Moisture absorption

9.25 METALS AND ALLOYS OF IMPROVED MECHANICAL PROPERTIES

The well-known popular alloy metals which are very common in use are alloys of Al, Cu, Mg. They exhibit unique mechanical properties. Titanium and its alloys have also becomes very popular due to their superior mechanical properties and are now widely used. Titanium exhibits low density (4500 kg/m^3), high melting temperature (1670°C), high elastic constant (107 GPa) and very high tensile strength (1400 MPa). Alloys of titanium are highly ductile and easily machineable. These properties make them suitable for air plane structure and space vehicles.

(i) Refractory Metals

Niobium, molybdenum, tungsten and tantlum are another group of metals which exhibit very high melting temperatures. The range of MP of these metals vary from 2468°C – 3410°C. They also exhibit high stiffness constant, high tensile strength and hardness even at elevated temperatures. This is the reason why these are used as refractory metals. These metals are alloyed with stainless steel to improve their corrosion resistance. Alloys of these metals are used in space vehicles, X-ray and welding electrodes.

(ii) Super Alloys

Super alloys are alloys that have unique combination of properties. They are alloys of cobalt, nickel and iron with refractory metals, or chromium, titanium, etc. These alloys have superior mechanical properties and are highly corrosion-resistant, even at very high temperatures. They can withstand extreme environmental conditions, such as a severely oxidising environment and extremely high temperatures, for considerable length of time. They are specially suited for aircraft turbine blades, components in nuclear reactors and equipment in petro chemical industries.

Summary

1. The chief mechanical properties of interest are elasticity, ductility and plasticity.
2. The parameters that are associated with the mechanical behaviour of materials are elastic modulus or stiffness constant, tensile strength, yield strength, hardness and toughness.
3. Stress vs. strain curve is very useful in the study of mechanical properties of materials.
4. Perfect elastic materials obey Hooke's law. The slope of stress–strain curve within elastic limit measures the stiffness constant of materials.
5. Most materials undergo plastic deformation beyond a certain stress called the yield strength. Beyond the yield strength, the stress vs. strain curve is non-linear.
6. Materials which undergo fracture after considerable plastic deformation are called ductile materials, and materials which remain perfectly elastic up to fracture are called brittle materials.
7. Tensile strength is the stress corresponding to saturation in the stress–strain curve.
8. A ductile material undergoes necking near the stress corresponding to the tensile strength, and the cross-sectional area subsequently decreases.
9. In a crystal, the mechanical behaviour depends on the orientation of the crystal with respect to the stress direction.
10. Toughness is defined as the ability of a material to absorb energy up to fracture. It is the area under the stress-strain curve and is expressed in energy per unit volume.
11. After stress is applied in an elastic material, it continue to deform for some time at constant load. When the load is removed, the original form is gradually recovered.
12. In a viscoelastic material, deformation continues for some time at a constant load. But, when the load is removed, the original form is not recovered after any length of time.
13. Hardness is a measure of the resistance of material to localised deformation, like a dent or a scratch.
14. Creep is deformation of the material as a function of time under constant load. Creep deformation is non-recoverable. It normally occurs in materials at high temperatures. Creep curves are drawn between strain and time.
15. The four types of fracture are ductile, brittle, creep (rupture) and fatigue.
16. Fatigue is the behaviour of materials under cyclic loading.
17. Titanium and its alloys, refractory metals and super alloys are special materials with superior mechanical properties. They are also highly corrosion resistant. They find applications in spacecraft, aircraft, nuclear reactors, petrochemical industries, etc.

Exercise

1. Explain briefly the following mechanical properties of metals:
   Elasticity, Plasticity, Ductility, Malleability.
2. Define and explain the following mechanical properties:
   (i) Strength (ii) Strain energy
   (iii) Stiffness (iv) Resilience
   (v) Toughness (vi) Hardness
3. Define and explain the following mechanical properties
   (i) Isotropy, anisotropy and orthotropy
   (ii) Homogeneity and heterogeneity
   (iii) Fatigue
   (iv) Creep
   (v) Impact
4. Describe briefly the following mechanical properties
   (i) Elasticity (ii) Plasticity
   (iii) Toughness (iv) Fatigue
   (v) Creep

Numerical Problems

1. A steel wire of 11.8 mm diameter is subjected to a tensile load up to fracture. Its diameter at fracture is 10.5 mm. Find the ductility. **[Ans. 20.82%]**
2. A steel wire having 40 cm length and 6 mm diameter is subjected to a tensile load of 8472. Find the elongation and lateral contraction. Given y = 207 GPa and Poisson's ratio is 0.25. **[Ans. 1.08 μm]**
3. A perfectly elastic material is subjected to a temperature load. Find the work done per unit volume if Young's modulus is 207 GPa and stain is 1.38×10^{-3}.
   [Ans. 1.42×10^5 J/m³]
4. A wire is subjected to a tensile load. Its length increases by a factor of 1.5. Calculate the engineering strain and true strain. **[Ans. 0.5]**
5. A 4.0 m long and 0.3 mm in diameter wire is stretched by a load of 800 g wt. If the extension in the length is 1.5 mm, calculate the energy stored in the wire.
   [Ans. 5.88×10^{-3} J]
6. Find the amount of work done in testing a steel wire of 1 mm radius and 25 cm length through an angle of 45°. Given: the modulus of rigidity 8.0×10^{11} dy/cm².
   [Ans. 0.1547 J]

Self-quiz Questions

1. What is Hooke's law?
2. Distinguish between elastic and plastic materials.
3. Define stiffness constant of a material.
4. What is Poisson's ratio? What is its lower and upper limit?
5. What is meant by resistance of a material?
6. What is creep and its various kinds?
7. Distinguish between true states and engineering states. How are these related to each other?
8. Name various hardness scales that are used to test the hardness of a material.
9. What is ductility of a material and how is it measured?

10. Define toughness of a material.
11. What is elastic fatigue and elastic hysteresis?
12. What is fracture and how it is measured?
13. What is the importance of creep consideration?
14. What are refractory metals and super alloys?

Objective-type Questions

1. Time-dependent deformation in which the material does not recover its original dimension is called
   (a) elasticity
   (b) plasticity
   (c) visco-elasticity
   (d) anealasticity
2. Brittle fracture occurs in materials due to
   (a) the presence of impurities
   (b) presence of cracks
   (c) rise in temperature
   (d) low bond strength
3. Among the following materials, which one has the highest hardness?
   (a) Steel
   (b) Copper
   (c) Cast iron
   (d) Silicon carbide
4. Creep curve is a plot of
   (a) strain vs. time
   (b) strain vs. temperature
   (c) stress vs. strain
   (d) elastic modulus vs. temperature
5. Materials that normally undergo creep are
   (a) metals at room temperature
   (b) metals at high temperature
   (c) low melting point materials
   (d) high melting point materials
6. Yield strength of the material is the stress at which the material
   (a) fractures
   (b) develops cracks
   (c) becomes plastic
   (d) ruptures
7. Among the following types of materials, which one has the highest modulus of elasticity?
   (a) Metals
   (b) Ceramics
   (c) Polymers
   (d) Semiconductors
8. On increasing the temperature of a metal, its ductility
   (a) decreases
   (b) increases
   (c) remains the same
   (d) partially increases
9. A material is said to be elastic if its stress–strain curve is
   (a) linear
   (b) non-linear
   (c) closed loop
   (d) none of these
10. Materials exhibiting non-linear stress–strain curve have a—fatigue resistance.
    (a) good (b) poor
    (c) excellent (d) none
11. Creep occurs
    (a) in the materials which have low melting point
    (b) in the materials which have high melting point
    (c) in the materials which have moderate melting point
    (d) in all materials
12. The failure of a material due to creep is known as
    (a) rupture
    (b) brittleness
    (c) hardness
    (d) none of these
13. The area under the stress–strain curve gives
    (a) hardness of the material
    (b) toughness of the material
    (c) strength of the material
    (d) fatigue of the material

14. The converse of hardness is known as
    (a) malleability
    (b) toughness
    (c) softness
    (d) none of these
15. Uniaxial anisotropy can be induced in bulk materials by which of the following methods?
    (a) Cold working
    (b) Magnetic annealing
    (c) Magnetic quenching
    (d) All of these
16. In carbon steels, if the percentage of carbon is increased, it has which of the following effects?
    (a) Resistivity is increased
    (b) Saturation temperature is lowered
    (c) Permeability is decreased
    (d) Coercive force and retentivity is increased
17. Among the following types of materials, which one has the highest modulus of elasticity?
    (a) Metals
    (b) Ceramics
    (c) Polymers
    (d) Semiconductors
18. Yield strength of the material is the stress at which the material
    (a) fractures
    (b) develops cracks
    (c) becomes plastic
    (d) ruptures
19. Materials that normally undergo creep are
    (a) metals at room temperature
    (b) metals at high temperature
    (c) low melting point materials
    (d) high melting point materials
20. Creep curve is a plot of
    (a) strain vs. time
    (b) strain vs. temperature
    (c) stress vs. strain
    (d) elastic modulus vs. temperature
21. Among the following materials, which one has the highest hardness?
    (a) Steel
    (b) Copper
    (c) Cast iron
    (d) Silicon carbide
22. Brittle fracture occurs in materials due to
    (a) the presence of impurities
    (b) presence of cracks
    (c) rise in temperature
    (d) low bond strength
23. Time-dependent deformation in which the material does not recover its original dimensions is called
    (a) elasticity
    (b) plasticity
    (c) visco-elasticity
    (d) anelasticity

Answers

1.	2.	3.	4.	5.
6.	7.	8.	9.	10.
11.	12.	13.	14.	15.
16.	17. (b)	18. (c)	19. (b) (c)	20. (a)
21. (d)	22. (b)	23. (c)		

Fill in the Blanks

1. In an .... material, properties are identical in all the direction.
2. All .... are isotopic.
3. Orthotropic material is special kind of .... material.
4. Modulus of elasticity is also known as .... constants.

5. Hooke's law states that within .... stress is proportional to the strain.
6. The material in which stress–strain behaviour is nor linear are known as .... material.
7. .... is applicable to the fluids.
8. .... is reciprocal of bulk modulus.
9. Poisson's ratio is a .... quantity.
10. The ability of materials to store elastic energy without permanent deformation is known as ....
11. The metals which are suitable for high temperature applications are called .... metals.
12. Ductility of a metal .... with decreasing temperature.
13. A material is said to be elastic if stress–strain curve is ....
14. For a good fatigue resistance material, the slope of the S-N curve is ....
15. Materials that exhibit non-linear stress-strain curve have .... fatigue resistance.
16. Creep is more likely to occur in .... melting point materials.
17. .... type of fracture is accompanied by high energy absorption.
18. The failure of a material due to creep is normally called ....
19. The area under the stress-strain curve gives .... of the material.

Answers

1. isotropic	2. metals	3. anisotropic	4. stiffness
5. elastic limit	6. plastic	7. bulk stress	8. compressibility
9. dimensionless	10. resilience	11. refractory	12. decreases
13. low	14. linear	15. poor	16. low
17. brittle	18. rupture	19. toughness	

Reading List

1. Dekker, A J *Electrical Engineering Materials.* New Delhi: PHI Learning Pvt Ltd
2. Rajput, R K *Electrical Engineering Materials.* New Delhi: Laxmi Publications Pvt Ltd.
3. Indulkar, C S; Triruvagdan *An Introduction to Electrical Engineering Materials* New Delhi: S. Chand Publications.
4. Hones, Lan P *Material Science for Electrical and Electronic Engineering.* New Delhi: Oxford University Press.
5. Seth, S P *A Course in Electrical Engineering Materials.* New Delhi: Dhanpat Rai and Sons.
6. Raghvan V *Material Science and Engineering.* New Delhi: PHI Learning Pvt Ltd
7. Vijaya, M S; Rangaranjan, G *Materials Science.* New Delhi: Tata McGraw-Hill Education Pvt Ltd.
8. Shackelford, James F, Madanapalli, K; Muralidhar A *Introduction to Material Science for Engineers.* 6ed. New Delhi: Pearson Education.
9. Rajendra, V; Marikani A *Materials Science.* New Delhi: PHI Learning Pvt Ltd
10. Callister, W D *Materials Science and Engineering* New Delhi: Wiley and Sons Inc.

10

Optical Properties of Materials

Chapter Objectives

- To discuss the interaction of electromagnetic radiations (EM) radiations with the matter
- To highlight various other phenomena related with EM radiation
- To know the theories related with optical phenomena
- To understand laser and the techniques to obtain a laser beam
- To know the materials suitable for producing a laser beam
- To discuss the applications of laser in various areas including the fibre optic Communication
- To understand fibre optics and its need in the present age
- To know about some significant and interesting properties of optical materials

10.1 INTRODUCTION

In modern times, optical materials and devices play a crucial role. In the present age, it is next to impossible to imagine a smooth existence without the use of materials. As these materials are getting an overwhelming response a new area which in a hybrid of optics and electrons has emerged. This area is better known as optoelectronics or photonics. The major areas covered by these materials and devices are optical communication, laser, imaging displays for medical applications, optical coatings, devices for energy conservations, lenses, filters, optically active materials for detecting, displaying and recording the images in microscopy, photography, media and information applications. It is surprising to note that a specially purified glass used in optical fibres is able to transmit nearly 99.99% of light through it, which has proved to be a boon for electronic communication and data transfer over long distances. Not only this, many other optical and electronic devices used by us today including cathode-ray tubes, image intensifiers, photomultiplier tube, photodiodes, lasers utilise optoelectronic material. High speed applications in electronic industry have become a reality only with the advent of a large variety of optoelectronic materials. Today's electronic market has been captured by 80% of optical materials. The beauty of a fibre is that it can transmit a frequency of 200,000 GHz through a fibre cable of the thickness of human hair.

The optical materials as a whole are classified in two categories:

- Active optical materials
- Passive optical materials

The materials which exhibit some special characteristics in response to electrical, magnetic, optical, thermal and other stimulus interactions are called active optical materials. Examples are lasers, LEDs, photodiodes, image to optical recording media, luminescent materials, polymer photo-resists and LCDs. The passive ones include from the ones missing from the above discussion. This chapter is mainly devoted to some basics of em radiations, the response of em radiation to the material and the properties of optical materials.

10.2 ELECTROMAGNETIC RADIATION SPECTRUM

Electromagnetic radiations exist throughout the universe; a part of them is visible to the human eye, known as visible region. This extends nearly from 4000 Å to 7800 Å. The radiation on left side is more intense, while the one on right side is extremely energetic. Figure. 10.1 shows the complete electromagnetic radiation spectrum.

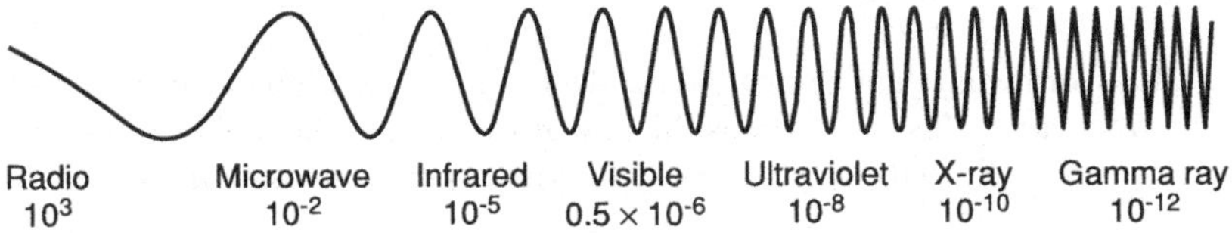

Fig 10.1 *EM radiation spectrum*

10.3 BASIC FEATURES OF ELECTROMAGNETIC RADIATION

Certain key features related with EM radiations are listed below:

- Electromagnetic radiation travels with speed of light in vacuum and less than c in material medium.
- Electromagnetic waves are transverse in nature, i.e. E-vector and H-vector comprising EM wave are mutually perpendicular and also perpendicular to the direction of propagation.
- Ratio of electric field vector to magnetic field vector is equal to c.
- The electromagnetic waves carry energy with them and transfer it to the particles on their way during propagation. This rate of transfer of energy can be expressed in terms of E and H vectors.
- The total energy of an electromagnetic wave is half electric and half magnetic.
- The electromagnetic wave does not penetrate into a conducting material as a result they appear shiny.

10.4 RESPONSE OF THE MATERIAL TO EM RADIATION

When EM radiation incidents on a material, the material responds to it in several ways. This response is due to an interaction between the radiation and the matter. Optical properties of the materials are determined by the interaction of the electrons in the matter with the EM radiation. Electromagnetic radiation is a combination of time-varying electrical and magnetic fields, and therefore, the electrons of matter come under the influence of these time-varying fields.

Here are some basic phenomena related with the materials when a radiation is incident upon them:

- Reflection
- Photoconductivity
- Refraction
- Total internal reflection
- Scattering
- Transmission
- Absorption
- Luminescence

All the above phenomena can be understood on the basis of interaction of EM radiation with the electrons present in the material.

Before going in details, let us briefly discuss these phenomena.

(i) Reflection

The electromagnetic waves, similar to light waves, undergo the phenomenon of reflection when they fall at the interface between two dissimilar media. The angle of incidence (θ_i) is equal to the angle of reflection θ_r in the same media,

i.e. $\theta_i = \theta_r$

It is also worthwhile to note that the incident, reflected and refracted waves all have the same frequency. This is known as the law of frequency. The incident, reflected and refracted waves all lie in the same plane, but normal to the boundary surface.

(ii) Photoconductivity

The interaction of EM radiation with a semiconducting material is also interesting. In the absence of light, an intrinsic semiconductor or a pure single crystal may exhibit negligible electrical conductivity. It is because the energy of the incident photon is very small as compared to the energy gap between the top of the valence band and bottom of the conduction band. However, when an electromagnetic radiation including white light is incident, considerable increase in the electrical conductivity may be observed. Thus, photoconductivity is the increase in the electrical conductivity of certain crystalline solids, as a result of exposure to electromagnetic radiation. An increase in the electrical conductivity is due to the generation of electron-hole pairs by the absorbed photons. This phenomenon is used in light meters and solid state infrared detectors. A simple experimental setup for measuring the photoconductivity is illustrated in Fig. 10.2.

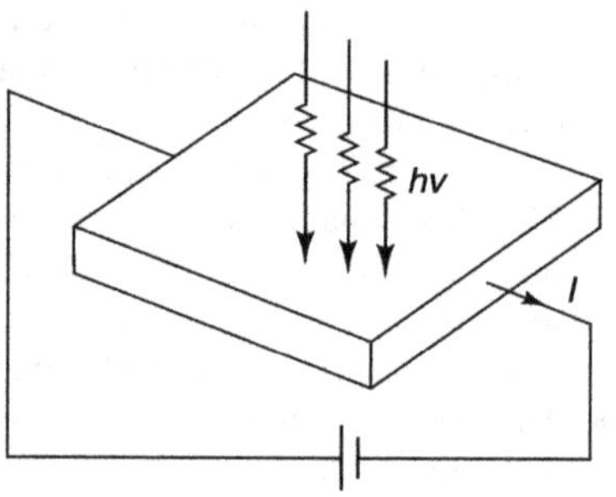

Fig. 10.2 *The generation of photocarriers in a photoconductor*

Let us consider a semiconducting slab placed in the dark. Since negligibly small current flows in dark, the dark conductivity is written as

$$\sigma_d = ne\mu_e + pe\mu_h = 0 \tag{10.1}$$

where n and p are free electron, hole concentrations, μ_e and μ_h are the electron and hole mobilities and e is the electronic charge. When the electromagnetic radiation of an appropriate frequency is allowed to fall on the semiconductor slab, the electrical conductivity increases. The change in the conductivity (assuming $\Delta n = \Delta p$) is given by

$$\Delta\sigma = \sigma - \sigma_d = \Delta ne(\mu_e + \mu_h) \tag{10.2}$$

and
$$\frac{\Delta\sigma}{\sigma_d} = \frac{\Delta ne(\mu_e + \mu_h)}{\sigma_d} \tag{10.3}$$

If τ is the carrier lifetime and f is the rate of their generation (i.e. number of electron-hole pairs produced per second by the absorbed photons) then the average photo induced concentration is given by

$$\Delta n = f\tau \tag{10.4}$$

where $\Delta n \approx 10^{20}$ m^{-3} for germanium.

In a practical device, the sensitivity is expressed as gain G, which is the ratio of the carrier photocurrent I to the rate of generation of carriers. Thus,

$$G = \frac{I}{ef} \tag{10.5}$$

If t is the transit time of the carriers between the electrodes then

$$G = \frac{\tau}{t} \tag{10.6}$$

Thus, photoconductivity is large when the lifetime is large. A pure material is generally found to have low τ (~ 10^{-6} s) due to the greater probability of electron-hole recombination, and hence is relatively insensitive. However, a careful doping can suppress recombination and hence, increase τ and sensitivity. For example, in CdS careful doping of iodine increases τ to 10^{-3} s. This process is represented in Fig. 10.3—iodine, substituting for sulphur, acts as donor. To adjust into the divalent ionic lattice, the iodine atom should accept two electrons. However, it shows a strong affinity to one while the second is only weakly bound and can be easily removed thermally. This gives rise to a charge imbalance. In order to compensate this, Cd^{2+} vacancies are generated at the rate of one per two iodine donors. These divalent positive ion vacancies act as efficient hole traps. During photo excitation, holes are readily trapped and electrons have much less chance of recombination and hence, average lifetime is increased.

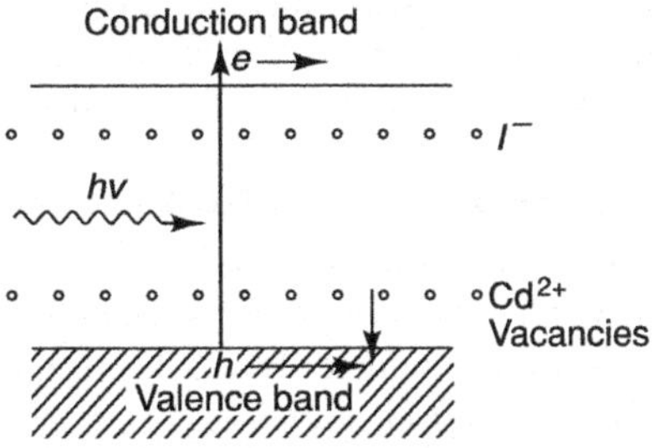

Fig. 10.3 *Model for increasing the electron lifetime in CdS as a result of doping with iodine*

CdS and CdTe single crystals are widely used as photoconductors because of their high sensitivity and response to visible light wavelengths.

(iii) Scattering

The phenomenon of scattering is a consequence of interaction between the time-varying electric field of EM radiation with the electrons present in the material. The oscillating electric field of the EM wave compels the electrons of the material to oscillate with the frequency of incident radiation, and therefore, these electrons become the source of EM radiation. As the electrons oscillate they give rise to the radiation of the same frequency as that of incident one. This radiation produced from the oscillating electrons travels in all directions and is known as scattering. Nearly every material object scatters radiation due to which they are visible to us.

(iv) Refraction

The radiation produced by oscillating field of the EM wave interferes with the incident radiation and propagates into the material with a phase retardation. The EM wave slows down when it penetrates into the material, and therefore, bends towards or away from the normal. The general law governing the refraction is known as Snell's law, which is given as

$$\frac{\sin\theta_1}{\sin\theta_r} = \frac{n_2}{n_1} = n_{21}$$

where θ_1 and θ_r are angles of incidence and angle of refraction, respectively; n_1 and n_2 are the refractive indices of the media under consideration.

A characteristic parameter known as index of refraction can be defined in terms of velocity of em wave in vacuum and in material:

$$n = \frac{\text{Velocity of EM wave in vacuum}}{\text{Velocity of EM wave in material}} = \frac{c}{\upsilon} \quad (10.7)$$

The index of refraction defined above is termed as absolute refractive index as it reveals with respect to vacuum. If the EM radiation travels from the medium 1 of absolute refractive index n to the medium 2 of absolute refractive index n_2 then

$$\frac{n_2}{n_1} = \frac{v_1}{v_2} = n_{21} \quad (10.8)$$

n_{21} is known as the refractive index of the medium 2 with respect to the median 1. n_{21} will be greater than 1 when the medium 2 is optically denser than the medium 1 and will be less than 1, if the medium 2 is optically rarer than the medium 1. During propagation, the incident EM wave is partially absorbed by the material and partially reflected into the same medium, and rest is transmitted into the second medium.

1. Refraction in Non-metals

Refraction is vital for other applications such as,

- Optical fibres—keeps the light in
- Semiconductor laser—keeps the light in the amplifying cavity of the laser

In the Eq. (10.7),

$$v = \frac{1}{\sqrt{\varepsilon\mu}} \text{ and } c = \frac{1}{\sqrt{\varepsilon_0\mu_0}} \begin{cases} \mu \to \text{permeability of medium} \\ \mu_r \to \text{relative permeability} \\ \mu_0 \to \text{permeability of free space} \end{cases}$$

From above, we have

$$n = \sqrt{\mu_r\varepsilon_r} = \sqrt{\varepsilon_r}; \in_r \to \text{relative permittivity or dielectric}$$

or $$n^2 = \varepsilon_r$$

Since most substances are only slightly magnetic, $\mu_r = 1$.

Since light is an electromagnetic wave, the connection between the dielectric permittivity (ε_r) and the magnetic permeability (μ) is not surprising. The index of refraction is therefore a consequence of electrical polarisation, especially the electronic the size of the constituent atoms or ions has a considerable influence on the magnitude of this effect. This means the larger an atom or ion, greater will be the electronic polarisation, slower the velocity, greater the index of refraction.

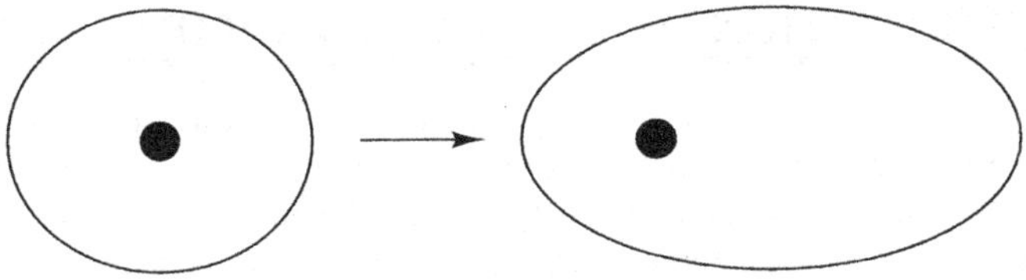

Fig. 10.4 *Electronic polarisation*

Electronic polarisation tends to be easier for a larger atom, and so n is higher in those materials, e.g. glass: $n \sim 1.5$ and lead crystal: $n \sim 2.1$. n can be anisotropic for crystals which have non-cubic lattices.

2. Reflection and Transmission

When EM wave (light wave) is incident upon a material, it may penetrate only a very short distance inside. This happens due to the damping of EM wave inside the material (the decreases in amplitude being exponential with increasing depth). In such situations, only a fraction of incident energy of the wave is dissipated as heat and the major part is reflected in the same media.

The reflectivity R of a material is therefore defined as the ratio of reflected intensity to the intensity of incident light.

$$R = \frac{I_R}{I_0} \tag{10.9}$$

Assuming the normal incidence of light on the material surface,

$$R = \left(\frac{n_2 - n_1}{n_2 + n_1}\right)^2 \tag{10.10}$$

Total Internal Reflection

Total internal reflection is a special class of refraction and is the basics of light propagation through optical fibre. Figure 10.5 shows the refraction occurring at the boundary of two media.

When light passes from denser to rarer medium, the ray bends away from the normal. According to Snell's law

$$n_1 \sin \phi_1 = n_2 \sin \phi_2 \tag{10.11}$$

On increasing angle of incidence ϕ_1 in denser media, angle of refraction in rarer media ϕ_2 also increases. At a particular situation, i.e. $(\phi_1 = \phi_c)$; $\phi_2 = \frac{\pi}{2}$ when ϕ_1 is further increased beyond ϕ_c, the refracted ray gets completely reflected in the same media. This phenomenon is known as total internal reflection. This is the technique which provides a reflected wave without any loss of intensity.

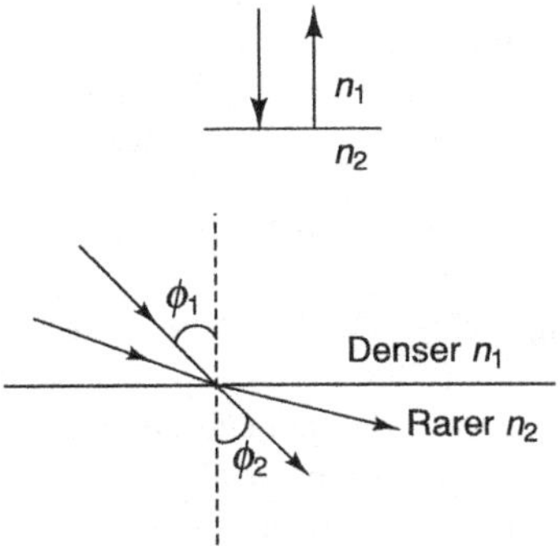

Fig. 10.5 *Refraction across the boundary of two media*

10.5 INTERACTION OF LIGHT WITH SOLID

When light travels from air and passes through a solid material, some of its radiations get transmitted through the medium, some get absorbed, and some get reflected at the interface between the two media. If intensity I_0 of the beam incident to the surface of the solid medium then—because of conservation of energy—we can say that

$$I_0 = I_T + I_A + I_R \tag{10.12}$$

where I_0 is the intensity of incident light and subscripts refer to transmitted, absorbed or reflected.

An alternate form of the above equation is

$$T + A + R = 1 \tag{10.13}$$

where T, A and R are fractions of the amount of incident light, given as T (transmissivity) $= \frac{I_T}{I_0}$, A (absorptivity) $= \frac{I_A}{I_0}$, R (reflectivity) $= \frac{I_T}{I_0}$, etc.

So, on this basis materials can be broadly classified as:

- *transparent*: relatively little absorption and reflection
- *translucent*: light is scattered within the material
- *opaque*: relatively little transmission

If the material is not perfectly transparent, the intensity decreases exponentially with distance inside the material. To understand this, consider a small thickness of material δx in which the fall of intensity is δI. So,

$$\delta I = -\alpha.\delta x.I$$

where α is the absorption co-efficient

In the limit of x tends to 0, we get

$$\frac{dI}{dx} = -\alpha I$$

The solution of which is

$$I = I_0 \exp(-\alpha x)$$

Taking natural log of both sides, we have,

$$\alpha x = In\left(\frac{I}{I_0}\right) \tag{10.14}$$

which is known as Lambert's law. Thus, on plotting in I against x, we can find α from the gradient.

Depending on the material and the wavelength, light can be absorbed by

- *nuclei*—all materials
- *electrons*—metals and small band-gap materials

Luminescence

When electromagnetic radiation is incident on a material, emission of photons from the materials is observed. This happens because the electrons get excited to high energy levels, where they cannot stay for long time and therefore, come back to the ground state. During this transition, the photon of energy $hv = E_2 - E_1$ is emitted. This type of emission is caused by spontaneous falling of electrons to the lower energy state. Basically, there are two kinds of spontaneous emission:

- Florescence
- Phosphorescence

These two are different in the time for which the excited electrons stay in higher energy level before falling spontaneously to the lower energy level.

If the excited electrons have a very short lifetime—less than 10^{-8} seconds—in the higher energy level, the emission is almost instantaneous after excitation and the emission takes place only as long as the exiting radiation is present. Such an emission is called fluorescence and materials which exhibit this property are called *fluorescent materials*.

If the excited electrons stay in the higher energy level for greater than several seconds, the emission continues even after the exciting radiation is switched off. Such emission is called phosphorescence and materials that exhibit this property are called phosphorescent materials.

Depending on the processes of excitation, luminescence may be classified as

- Photoluminescence
- Cathedoluminescence
- Electroluminescence

In **photoluminescence**, excitation is achieved by incident electromagnetic radiation. Examples are X-ray detectors and fluorescent lamps. In fluorescent lamps, the exciting radiation is the ultraviolet mercury radiation. The UV radiations are allowed to incident on a coating of calcium phosphate doped with antimony and manganese. The spontaneous emission in the visible range from the fluorescent coating causes the lamp to glow.

Cathedoluminescence is a process in which the excitation is produced by bombardment of the material by high energy electrons, e.g. cathode ray tube and TV tubes. The screens are coated with zinc sulphide. High-energy electrons, that strike the screen, produce spontaneous emission due to which bright spots are observed on the screen.

Electroluminescence is a process in which the excitation is done by applying electric field. Examples are light emitting diodes in which the electric field causes the electrons and poles to recombine at the junction, thereby emitting light photons.

10.6 ELECTROMAGNETIC THEORY OF ABSORPTION AND REFRACTION IN MATERIALS

The absorption and refraction of electromagnetic radiations can be understood using the Maxwell's equation. Let us consider that electromagnetic wave is propagating in a dielectric material. The Maxwell's equation for E and H vectors are given by

$$\overline{\nabla} \times \overline{E} = -\frac{\partial B}{\partial t} = -\mu_0 \frac{\partial \overline{H}}{\partial t} \tag{10.15}$$

$$\overline{\nabla} \times H = J + \frac{\partial D}{\partial t} \tag{10.16}$$

$$= \sigma E + \frac{\partial}{\partial t}.\{\epsilon_0 E + P\}$$

$$= \sigma E + \epsilon_0 \frac{\partial E}{\partial t} + \frac{\partial P}{\partial t}$$

We know that for dielectric material $\sigma = 0$, therefore

$$\overline{\nabla} \times \overline{H} = \epsilon_0 \frac{\partial E}{\partial t} + \frac{\partial P}{\partial t} \tag{10.17}$$

Taking curl of Eq. (10.15), we get

$$\overline{\nabla} \times \overline{\nabla} \times \overline{E} = -\mu_0 \frac{\partial}{\partial t}\left[\overline{\nabla} \times \overline{H}\right]$$

$$\text{grad. div } E - \nabla^2 E = -\mu_0 \frac{\partial}{\partial t}\left[\epsilon_0 \frac{\partial E}{\partial t} + \frac{\partial P}{\partial t}\right]$$

$$-\nabla^2 E = -\mu_0 \epsilon_0 \frac{\partial^2 E}{\partial t^2} - \mu_0 \frac{\partial^2 P}{\partial t^2} \qquad \text{[Since grad. div } E = 0]$$

Therefore,

$$\nabla^2 E - \mu_0 \epsilon_0 \frac{\partial^2 E}{\partial t^2} - \mu_0 \frac{\partial^2 P}{\partial t^2} = 0 \tag{10.18}$$

The above equation can also be written as (for wave travelling along z-axis and polarised along x-axis)

$$\frac{\partial^2 E_x}{\partial z^2} - \frac{1}{c^2}\frac{\partial^2 E_x}{\partial t^2} = \frac{1}{\epsilon_0 c^2}\frac{\partial^2 P_x}{\partial t^2} \tag{10.19}$$

Since $\quad \dfrac{1}{c^2} = \mu_0 \epsilon_0$

The solution of the above equation can be given as

$$E_x = E_{x_0} e^{J\omega\left(t - \frac{z}{v}\right)} = E_{x_0} e^{J\omega\left(t - \frac{n}{c}z\right)} \tag{10.20}$$

where n is the index of refraction of dielectric material. The above equation represents the equation of oscillating electric field whose frequency of the polarisation P_x is in the x-direction, and it is assumed that it varies as $e^{J\omega t}$.

i.e. $\quad P_x = P_0 e^{J\omega t} \qquad (10.21)$

Using Eqs (10.20) and (10.21) in Eq. (10.19), we get

$$-\frac{\omega^2 n^2}{c^2} E_x + \frac{\omega^2}{c^2} E_x = -\frac{\omega^2}{\epsilon_0 c^2} P_x \tag{10.22}$$

On simplifying, we get

$$n^2 = 1 + \frac{P_x}{\epsilon_0 E_x} \tag{10.23}$$

We know that index of refraction n and electric susceptibility χ_E are related, as

$$x^2 = 1 + \chi_E = \epsilon_r \tag{10.24}$$

where ϵ_r is relative permittivity.

The above equation is important in the sense that it relates the optical parameters n with electrical parameter χ_E or ϵ_r. The polarisation discussed above will be electronic simply because at optical frequencies it is only the electrons which undergo oscillations with the applied electromagnetic field.

The dielectric constant also will be complex, which can be expressed as

$$n^* = n - jk \tag{10.25}$$

where n is the real part and k, the imaginary part of refractive index.

Using the above, the equation of electromagnetic wave can be given as

$$E_x = \left[E_{x_0} e^{-\frac{\omega}{c}kz}\right] . e^{J\omega\left[t - \frac{nz}{c}\right]} \quad (10.26)$$

The above equation contains two terms. The term outside the parenthesis contains the real part of the refractive index. The imaginary part of the refractive index appears inside the parenthesis, which is the amplitude part of the electromagnetic wave. This shows that the amplitude of electromagnetic wave is not constant but decreases exponentially inside the material. Due to this, the wave gets attenuated as the wave propagates into the dielectric material. The imaginary refractive index gives an indication that how much the wave gets attenuated. The reason of this attenuation is energy lesser in atomic oscillators.

The intensity of the electromagnetic wave can therefore be given (using the term in the parenthesis

$$I = I_0 e^{-\frac{2\omega}{e}kZ} = I_0 e^{-\beta Z} \quad (10.27)$$

where $$\beta = \frac{2\omega}{c}k \quad (10.28)$$

This is known as **absorption coefficient.** Sometimes it is also known as **absorption constant.** The corresponding attenuation of the electromagnetic wave is shown below in Fig. 10.6.

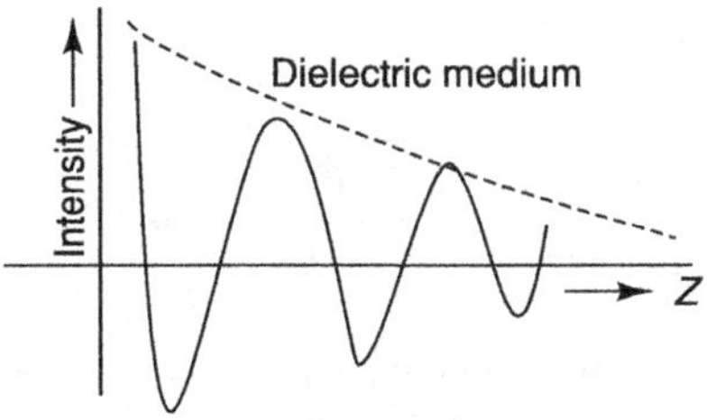

Fig. 10.6 *Attennation of EM wave in dielectric*

Therefore, the two terms of equation (1) are equally significant as one represents the attenuation (decrease in amplitude) and the other represent the decrease in velocity of electromagnetic wave when it progresses into the dielectric material.

1. Penetration Depth (β)

Since a decrease in the amplitude of electromagnetic wave is exponential, one can define a characteristic depth known as **penetration depth (β).** It is the distance travelled by the electromagnetic wave inside the dielectric when its intensity reduces by $\frac{1}{e}$ to its parent value.

Therefore, if $z = \frac{1}{\beta}$ is substituted in Eq. (10.27), we get

$$I = \frac{I_0}{e}$$

i.e. $$w = \frac{1}{\beta} = \frac{c}{2\omega k} \quad (10.29)$$

Therefore, penetration depth is the reciprocal of absorption coefficient.

As we can write

$$\in_r = n^{*2} = (n - jk)^2$$
$$= n^2 - k^2 - 2jnk$$

or $$\in'_r - j\in''_r = n^2 - k^2 - 2jnK$$

Therefore,

$$\epsilon'_r = n^2 - k^2 \tag{10.30}$$

$$\epsilon''_r = 2nk \tag{10.31}$$

Since ϵ''_r is related to ac conductivity, we get

$$\epsilon''_r = \frac{\sigma_{ac}}{\epsilon_0\, \omega} = 2nk \tag{10.32}$$

For a perfect insulator $\sigma_{ac} = 0$, and therefore, $k = 0$

Hence, $\epsilon'_r = n^2$

i.e. dielectric constant is equal to the square of the real part of index of refraction for perfect insulator and there will be no attenuation of electromagnetic fields in these materials.

For those materials whose sac $\sigma_{ac} \neq 0$, the attenuation will follow the pattern given in the above figure.

2. Relation between Dielectric Constant and *k*

We know that $\epsilon_r = n^2 - k^2$ (10.33)

Since, $\dfrac{\sigma_{ac}}{\epsilon_0\, \omega} = 2nk$

Therefore, $k = \dfrac{\sigma_{ac}}{2n\, \epsilon_0\, \omega}$

On substituting the value of *k* in Eq. (10.33), we get

$$\epsilon_r = n^2 - \left[\frac{\sigma_{ac}}{2n\, \epsilon_0\, \omega}\right]^2$$

On simplifying, we get

$$(2\, \epsilon_0\, \omega)^2 n^4 - (2\, \epsilon_0\, \omega)^2\, \epsilon'_r\, n^2 - \sigma_{ac}^2 = 0$$

or

$$n^4 - \epsilon'_r\, n^2 - \left[\frac{\sigma_{ac}}{2\, \epsilon_0\, \omega}\right]^2 = 0$$

The solution of the above equation is given by

$$n^2 = \frac{1}{2}\left[\epsilon'_r + \sqrt{\epsilon'^2_r + \left(\frac{\sigma_{ac}}{\epsilon_0\, \omega}\right)^2}\,\right] \tag{10.34}$$

or

$$n^2 = \frac{1}{2}\left[\epsilon'_r + \sqrt{\epsilon'_r + \epsilon''^2_r}\right] \tag{10.35}$$

From Eq. (10.33),

$$n^2 = \epsilon'_r + k^2$$

$$\epsilon'_r + k^2 = \frac{1}{2}\left[\epsilon'_r + \sqrt{\epsilon'^2_r + \epsilon''^2_r}\right]$$

i.e.

$$k^2 = \frac{1}{2}\left[\epsilon'_r + \sqrt{\epsilon'^2_r + \epsilon''^2_r}\right] - \epsilon'_r$$

or

$$k^2 = \frac{1}{2}\left[\sqrt{\epsilon'^2_r + \epsilon''^2_r} - \epsilon'_r\right]$$

3. Relation between Reflectivity and Conductivity

It is well known that metals exhibit a high value of conductivity. Therefore, the ratio $\frac{\sigma}{\epsilon_0 \omega}$ will be large. Hence, $\epsilon_{r'}$ can be neglected compared to $\frac{\sigma}{\epsilon_0 \omega}$ in Eq. (10.34) and in Eq (10.30), i.e.

$$n^2 - k^2 = \epsilon_r'^2 = 0$$

$$n^2 = k^2$$

and

$$n^2 = \frac{1}{2}\left[\frac{\sigma_{ac}}{\epsilon_0 \omega}\right] \quad (10.36)$$

i.e.

$$n^2 = k^2 = \frac{\sigma_{ac}}{2\epsilon_0 \omega}$$

Therefore,

$$R = 1 - \frac{4n}{2n^2 + 2n + 1}$$

Neglecting $(2n + 1)$ compared to $2n^2$ (for $n >> 1$), we have

$$R = 1 - \frac{2}{n} = 1 - 2\sqrt{\frac{2\epsilon_0 \omega}{\sigma_{ac}}} \quad (10.37)$$

This equation implies that at low frequencies for metals,

$$R = 1$$

which suggests that good electrical conductors are good reflectors also. This was verified experimentally also.

4. Transmission Coefficient

When the electromagnetic radiation is incident upon a material, the part which is not reflected is transmitted or absorbed into the material. Transmission coefficient is defined as the ratio of the intensity of transmitted light through the material to the incident one,

i.e.

$$T = \frac{I_T}{I_0} \quad (10.38)$$

In case of no absorption

$$R + T = 1 \quad (10.39)$$

10.7 ATOMIC THEORY OF ABSORPTION

(i) Conducting Materials

Since the number of electrons in metals are very large, therefore these electrons readily absorb the incident radiation and undergo collision with the lattice atoms. The extra energy is dissipated in the material as heat and we call it absorption. Now, if the probability of collision is small, the electron will emit a photon when it comes back to the ground level. This is the origin of strong reflection in metals in the visible and the infrared regions.

Most metallic elements reflect the complete portions of visible spectrum and thus appear white. Gold and copper preferably absorb in the green region, and these metals respectively assume the colour of reflected yellow and red light.

Semiconductors have strong variation in the absorption coefficient with wavelength. The several orders of magnitude change in α at the wavelength corresponding to the energy band-gap has important implications in semiconductor optoelectronic devices.

(ii) Insulating Materials

Insulating materials completely devoid of the free electrons. The electromagnetic radiation interacts in a different way than in conductors. Both electrical and optical phenomena in dielectrics, as well as in conductors, originate from applied ac electric fields. When high-frequency electromagnetic radiation incident upon a dielectric, it is then subjected to a high-frequency electric field. The charges and electric dipoles in the dielectric respond to the oscillating electric field of electromagnetic radiation. The electrons and ions therefore oscillate to and fro, while molecular dipoles try to reorient in an attempt to respond the rapidly alternating electric field. The dielectric constant ε_r reflects the extent of polarisation when the frequency is low. When the frequency is high, only electronic polarisability contributes then

$$\varepsilon_r = n^2$$

where n is the index of refraction. n increases with the number of charges or dipoles per unit volume that interact with the electric field. High value of n is for a material that has high mass density, containing atoms with high atomic number, large ionic polarisability and large degree of covalent bonding.

10.8 MATERIALS FOR PHOTONIC DEVICES (LIGHT-EMITTING MATERIALS)

The elemental semiconductors Ge and Si have been quite popular since their invention to fabricate many devices. Yet interest in these materials has been limited, owing to the fixed bandgap. This was the biggest hurdle due to which these could not attain a wide application in the electronic industry. To get a material of a desired bandgap was a long-lasting desire of semiconductor engineers. This desire translated into reality with the invention of compound semiconductors (binary, ternary and quaternary). The development of these materials brought a revolution in the modern electronic industry.

It has been discussed previously that the band structures of III–V ternary and quaternary alloys can be varied by varying the composition. Figure 10.7 illustrates how the band structure changes in a ternary compound $Al_x Ga_{1-x} As$ (with $x = 0$, i.e. GaAs to $x = 1$, i.e. AlAs). The binary compound GaAs is a direct ($k = 0$) material with bandgap 1.43 eV, represented by Γ in the figure. Besides this, there are two indirect minima far above Γ. These have been represented by L (lowest lying minima) and the other by X. On the other hand, in AlAs the direct minima Γ is quite apart than the indirect minima X, and hence, this material is indirect with a bandgap of 2.16 eV at room temperature.

In the tennary alloys $Al_x Ga_{1-x} As_x$, all of these conduction band minima move up, as the composition is varied from $x = 0$ to $x = 1$. However, the indirect minimum x moves up much less than the other, and for composition greater than nearly 38% Al. This indirect minimum become lowest lying conduction band. Therefore, this ternary compound is direct material for aluminium composition up to 38% and indirect for more than 38%. The band picture of GaAs binary, AlAs binary and $Al_x Ga_{1-x} As$ has been shown below in Fig. 10.8.

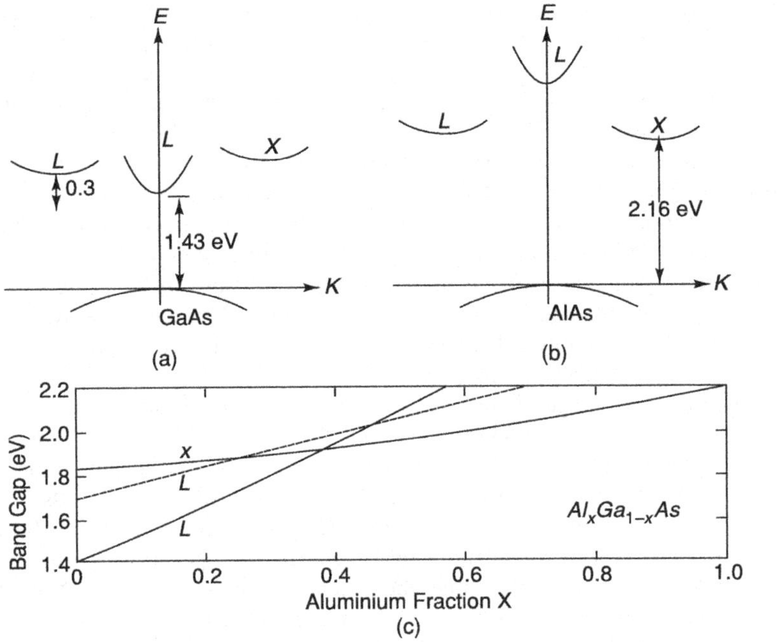

Fig. 10.7 *(a) Band structure of binary GaAs, (b) Band structure of binary AlAs (c) Variation of bandgap with x*

The above trick has been utilised to develop the compound semiconductors of desired band-gap. Due to the success in getting a material of known band-gap, it has become now possible to get a LED emitting a desired colour (green, yellow or red).

Figure 10.8 illustrates the bandgaps of many compound semiconductors along with their band-gaps from (GaN; band-gap 3.4 eV to InSb bandgap 0.18 eV).

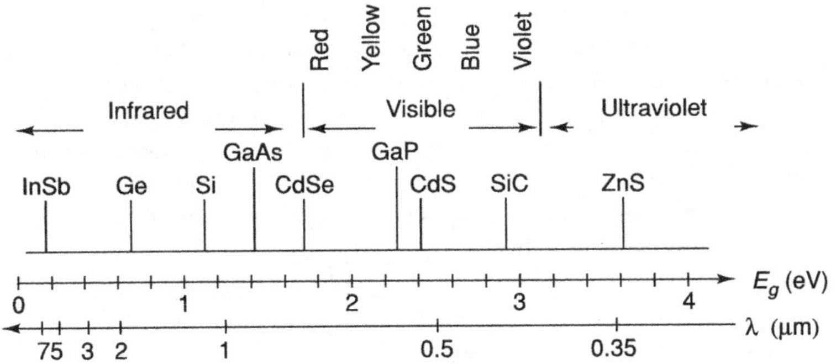

Fig. 10.8 *Band-gaps of some popular binary semiconductors*

In the above list, many more promising materials can be added (over a wide range of composition) using ternary and quarterly compounds. Of course, by doing so one can feel free to develop a particular material required for some specific application. The interest in developing the materials emitting the primary colours (blue, green and red) was at the top priority, simply because by mixing these colours one can get intense white light source which might be extremely efficient and economic. It may become a replacement for an incandescent bulb that consumes much more power. Such a white light source will have a life four times (at least) as compared to the ordinary one, along with high conversion efficiency. Such a source may fulfill the increasing global demand for low power-consuming source. Arrays of red, green and blue emitters are being used in outdoor displays, TV screens, as well as in tail-light of vehicles and traffic signals.

Below is an example of a ternary compound of $GaAs_{1-x}P_x$ that has flexibility of being a direct band-gap material at some value of x, and indirect at some other composition. The light emitting from compound semiconductors may not always be visible. Some materials, such as GaAs and InP and mixed alloys of these compounds, emit a radiation in the infrared region, which are equally good and may be used in TV remote control and in optical fibre communication systems. The optical fibre communication system requires a laser light which is highly coherent and intense and carries the information, while the ordinary LED emits incoherent light.

In view of a wide range of applications, the III–V materials are therefore of extreme interest. Besides AlGaAs and GaAsP systems, the AlGaInN is a strong emitter in blue and green and InAlGaP is useful for red, yellow and orange wavelengths. More recently, in around 1990s, much higher brightness and efficiency (~ 30 lumens/ watts) red-orange yellow LED have been developed using InAlGaP system. AlGaN system believes direct band-gap over the entire composition range, and hence, appears very attractive since it emits highly efficient light in blue and green part of the spectrum. Now, the only parameter that needs attention is the cost which, if could be reduced substantially (nearly 100 times), will fulfill the goal of opto-electronic community.

10.9 LASERS

The emission of EM radiation by the materials has been exploited skillfully to develop devices that amplify the radiation to produce an extremely intense, coherent, monochromatic and highly directional beam. The beam exhibiting the above remarkable characteristics is known as LASER. Laser action is based upon the principle of stimulated emission. The necessary requirements for a laser are given below:

(i) There must be an active medium. It may be a collection of atoms, molecules, ions, radicals or atomic nuclei (in liquid or solid form) that can absorb photons of a particular energy and emit radiations in the optical regions of the spectrum.

(ii) There must be a population inversion of atoms, molecules, ions, radicals or atomic nuclei. They will be treated in common under the general term as atom.

(iii) There must be some form of optical feedback present in the system for the continuous production of the laser beam.

In lasers, an active system is placed between a pair of mirrors that are facing each other. Out of these, one is fully silvered and the other is partial so that an intense beam can emerge out of this system which is also known as cavity resonator. It is shown in Fig. 10.9.

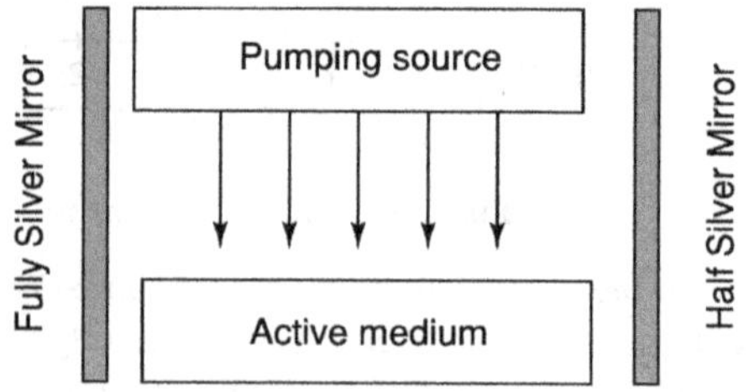

Fig. 10.9 *Main components of laser*

10.10 KINDS OF TRANSITION

To understand laser action, it is very important to appreciate the interaction of radiation with matter. A material medium is composed of identical atoms or molecules, each of which is characterised by a set of discretely allowed energy states. An atom can move from one energy state to another when it receives or releases an amount of energy equal to the energy difference between those two states. This process is known as transition. There are three kinds of transition processes.

- Stimulated absorption
- Spontaneous emission
- Stimulated emission

(i) Stimulated Absorption

When a photon is incident on a system with energy $h\nu = E_2 - E_1$, the atom in the ground state may absorb it to occupy a higher energy state as shown in Fig. 10.10. Absorption obviously requires external stimulation—that is why it is known as stimulated absorption. In this process, the photon disappears.

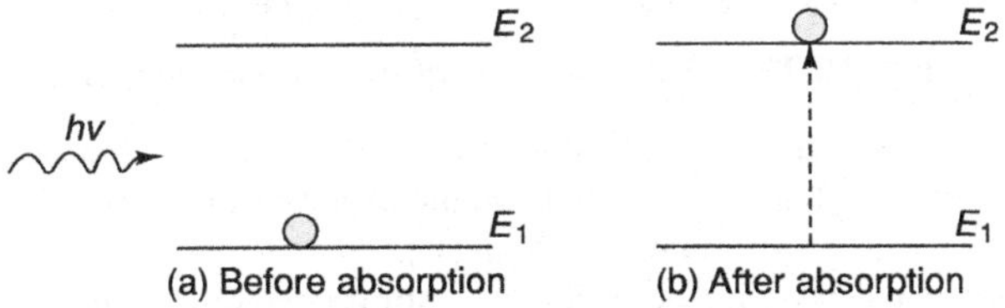

Fig. 10.10 *Absorption of energy by an atom*

In the absorption process, the probable rate of transition from $E_1 \rightarrow E_2$ depends upon the properties of state E_1 and state E_2, and proportional to the energy density $u(\upsilon)$ of the incident radiation of frequency υ.

i.e. $$P_{12} = B_{12}\, u(\upsilon)$$

where B_{12} is a constant that depends upon the property of state E_1 and E_2, which is called the Einstein's coefficient of stimulated absorption.

(ii) Spontaneous Emission

The atom in excited level E_2 may drop down to energy level E_1 spontaneously without the need of any external radiation giving out a radiation of energy $h\,\upsilon$ equal to $E_2 - E_1$. Ordinary sources of light emit the radiation by this mechanism. Naturally, the emitted light will be incoherent. The emission of this kind is shown in Fig. 10.11.

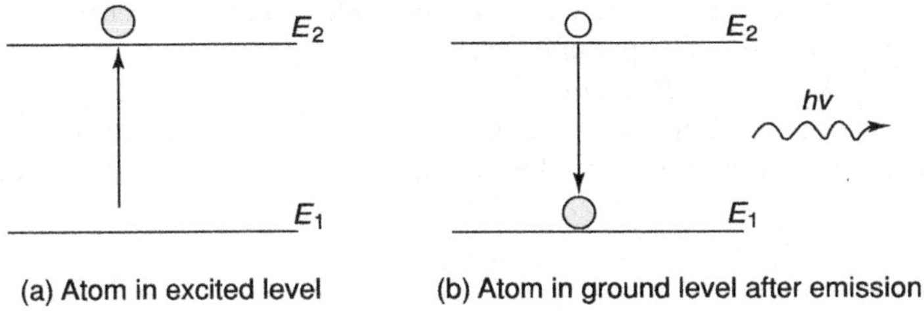

Fig. 10.11 *Demonstration of spontaneous emission*

The probability of spontaneous emission from state E_2 to E_1 is given by

$[P_{21}]$ spontaneous $= A_{21}$

where A_{21} is known as Einstein's co-efficient of spontaneous emission. Many transitions do not emit radiations. They are commonly known as non-radiative transitions.

(iii) Stimulated Emission

If a photon of energy $h\upsilon = E_2 - E_1$ is incident upon a system in which the atom is already in the excited level, as shown in Fig. 10.11, it will cause the atom to come down to the ground level emitting a photon of the same energy. These two photons will therefore be completely in-phase.

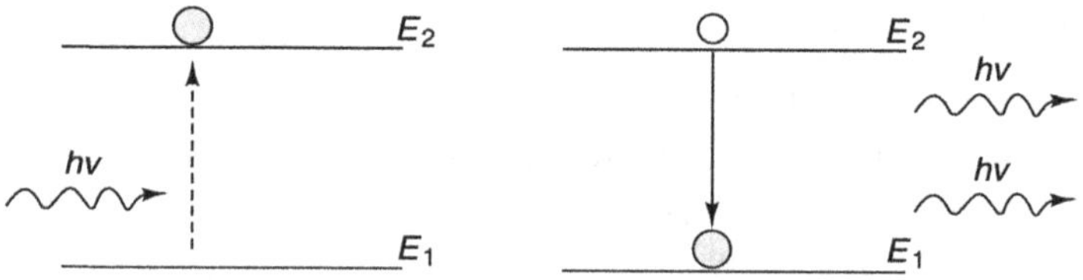

(a) System in excited state (b) System after stimulation

Fig. 10.12 *Demonstration of stimulated emission*

The probability of stimulated emission from state E_2 to E_1 depends on the energy density of the incident radiation, as well as on the properties of two energy states.

i.e. $$[P_{21}] \text{ stimulated} = B_{21}\, u(\upsilon) \quad (10.40)$$

where B_{21} is known as Einstein's co-efficient of stimulated emission.

Therefore, the total probability of emission of radiation from state E_2 to E_1 is the sum of spontaneous and stimulated emission probabilities.

i.e. $$[P_{21}] \text{ total} = A_{21} + B_{21}\, u(\upsilon) \quad (10.41)$$

Normally, the mean lifetime of spontaneous emission is 10^{-8} s, but in some materials there are states that have quite a large lifetime—nearly 10^{-3} s. Such states are known as meta-stable states. These states play a vital role in laser action.

10.11 EINSTEIN'S COEFFICIENT AND THE RELATION BETWEEN THEM

Let us assume an assembly of atoms in thermal equilibrium at temperature T with radiation of frequency υ and energy $u(\upsilon)$. Also, suppose N_1 and N_2 are the number of atoms in energy states 1 and 2 at any instant.

The probability of absorption that the number of atoms in the state 1 which can absorb a photon and rise to the state 2 per unit time:

$$N_1P_{12} = N_2B_{12}u(\upsilon) \quad (10.42)$$

The total probability of emission (spontaneous + stimulated) that the number of atoms in the higher state 2 can drop to the lower state 1 by emitting photon, per unit time is given by

$$N_2P_{21} = N_2[A_{21} + B_{21}\, u(\upsilon)] \quad (10.43)$$

At thermal equilibrium,

Absorption probability = Total emission probability

$$N_1P_{12} = N_2P_{21}$$

$$N_1B_{12}\, u(\upsilon) = N_2[A_{21} + B_{21}\, u(\upsilon)]$$

$$N_1 B_{12} u(\upsilon) = N_2 A_{21} + N_2 B_{21} u(\upsilon)$$
$$N_1 B_{12} u(\upsilon) - N_2 B_{21} u(\upsilon) = N_2 A_{21}$$
$$u(\upsilon)[N_1 B_{12} - N_2 B_{21}] = N_2 A_{21}$$
$$u(\upsilon) = \frac{N_2 A_{21}}{N_1 B_{12} - N_2 B_{21}}$$
$$= \frac{A_{21}}{\frac{N_1}{N_2} B_{12} - B_{21}} = \frac{A_{21}}{B_{21}\left[\frac{N_1}{N_2}\frac{B_{12}}{B_{21}} - 1\right]}$$
$$u(\upsilon) = \frac{A_{21}}{B_{21}} \frac{1}{\left[\frac{N_1}{N_2}\cdot\frac{B_{12}}{B_{21}} - 1\right]} \tag{10.44}$$

According to Einstein,
Probability of stimulated absorption = Probability of stimulated emission

$$B_{12} = B_{21}$$
$$u(\upsilon) = \frac{A_{21}}{B_{21}} \frac{1}{\left[\frac{N_1}{N_2} - 1\right]} \tag{10.45}$$

According to Boltzmann's distribution law, if the number of atoms N_1 and N_2 in energy state E_1 and E_2 in thermal equilibrium at temperature T, then

$$N_1 = N_0 e^{\frac{-E_1}{kT}}$$

and
$$N_2 = N_0 e^{\frac{-E_2}{kT}} \tag{10.46}$$

where N_0 is the total number of atoms in systems. (10.47)

$$\frac{N_2}{N_1} = \frac{e^{-E_2/kT}}{e^{-E_1/kT}}$$
$$= e^{-E_2/KT + E_1/KT} = e^{\frac{-(E_2 - E_1)}{kT}} = e^{\frac{-h\upsilon}{kT}}$$
$$\frac{N_1}{N_2} = \frac{1}{e^{\frac{-h\upsilon}{kT}}}$$

or
$$\frac{N_1}{N_2} = e^{h\upsilon/kT} \tag{10.48}$$

by Eqs (10.47) and (10.48), we get

$$u(\upsilon) = \frac{A_{21}}{B_{12}}\cdot\frac{1}{e^{\frac{h\upsilon}{kT}} - 1} \tag{10.49}$$

We know from Planck's radiation formula,

$$u(\upsilon) = \frac{8\pi h\upsilon^3}{c^3} \times \frac{1}{e^{h\upsilon/kT} - 1} \tag{10.50}$$

By Eqs (10.49) and (10.50), we get

$$\frac{A_{21}}{B_{21}} = \frac{8\pi h\upsilon^3}{c^3}$$

or

$$\boxed{\frac{A_{21}}{B_{21}} \alpha \upsilon^3}$$

10.12 POPULATION INVERSION

Under normal conditions, most of the atoms are in the ground state, so the chances of stimulated emission are negligible. If by some means the number of atoms is increased in the excited state compared to the ground state i.e., $N_2 > N_1$, stimulated emission is promoted. This is achieved by a special technique known as *pumping*.

Since a system in thermal equilibrium obeys Maxwell–Boltzmann distribution, therefore the number of atoms in various energy states at temperature T are given by

$$N_1 = N_0 e^{-E_1/kT} \tag{10.51}$$

$$N_2 = N_0 e^{-E_2/kT} \tag{10.52}$$

where N_0 is the total number of atoms in an assembly.

Thus, the relative population of atoms can be obtained on dividing Eqs (10.51) and (10.52),

i.e. $$\frac{N_2}{N_1} = e^{-(E_2-E_1)/kT} \tag{10.53}$$

where k is the Boltzmann constant. Graphically, the distribution of atoms at various energy levels has been shown in Fig. 10.13.

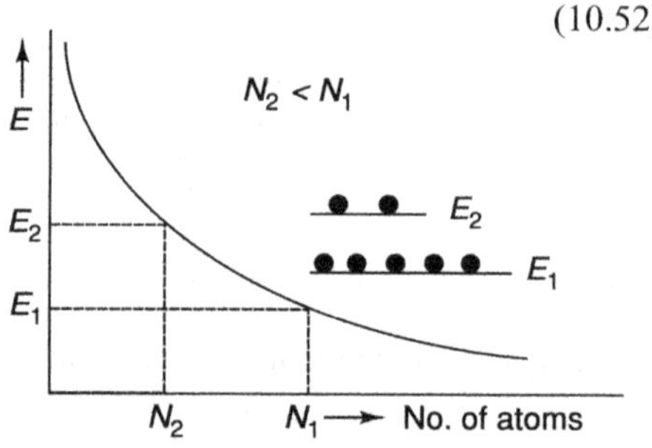

Fig. 10.13 *Distribution of atoms in normal state among various levels*

The negative exponent in Eq. (10.53) shows that $N_1 > N_2$. This means the number of atoms will be more in the ground energy level at normal temperature This is the natural tendency of any system.

Under such condition, the chances of stimulated emission is quite low rather than the probability of spontaneous emission. But since, in spontaneous emission, the photons emitted from various atoms have a random direction and a random phase, the light emitted in this way will be incoherent.

On the other hand, in stimulated emission, the radiation is completely coherent and amplified. Hence, the fundamental requirement for a laser is to have a predominately stimulated emission. To achieve this, two conditions are required:

(i) The higher energy state should have a longer lifetime, i.e. it should be a meta stable state.

(ii) The number of atoms in the higher energy state must be greater than in the ground state.

Condition (ii) is quite unnatural. If anyhow by some means we get it, the stimulated emission is promoted. This situation in which the number of atoms in high energy state is more than in the ground state is known as population inversion.

10.13 COMPARISON BETWEEN ORDINARY AND LASER RADIATION

The ordinary light differs a laser radiation in the following ways:

(i) The ordinary light is incoherent, but laser beam is highly coherent.

(ii) The coherence time for a laser beam is much larger as compared to ordinary light source.

(iii) The conventional light sources emit radiation in all directions, while a laser emits radiation only along one direction, i.e. laser is highly directional.

(iv) The light emitted by a laser is highly monochromatic than that of any conventional monochromatic source.

(v) Ordinary light spreads out more or less uniformly in all directions, whereas laser gives out light into a narrow beam and its energy is concentrated in very small regions.

(vi) Ordinary light is obtained as a result of spontaneous emission.

10.14 THREE-STAGE RUBY LASER

This is the first solid-state laser, developed in 1960. It consists of a pink ruby cylindrical rod whose ends are optically flat and parallel. One end is fully silvered and the other is only partially silvered. The rod passes through a coil-shaped flash lamp which is filled with xenon gas, as shown in Fig. 10.14.

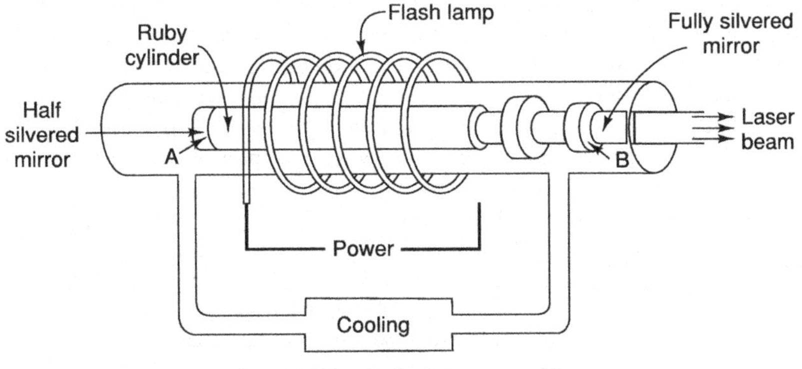

Fig. 10.14 *Ruby laser assembly*

Working Mechanism

The ruby rod is a crystal of aluminium oxide (Al_2O_3) doped with 0.05% chromium oxide (Cr_2O_3), so that some of the Al^{+++} ions are replaced by Cr^{+++} ions. These 'impurity' chromium ions give a pink colour to the ruby and also give rise to laser action.

Figure 10.14 shows a simplified energy-level diagram of chromium ion. It consists of an upper short-lived energy level (rather energy band) E_3, above its ground-state energy level E_1, the energy difference $E_3 - E_1$ correspond to a wavelength of about 5500 Å. There is an intermediate excited-state level E_2 which is metastable having a lifetime of 3×10^{-3} second.

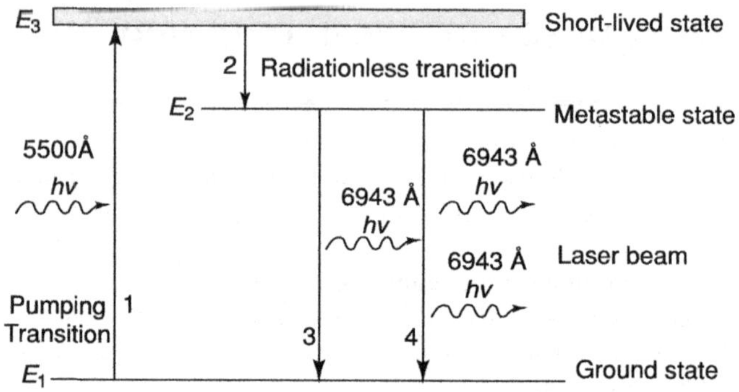

Fig. 10.15 *A three-level laser*

Normally, most of the chromium ions are in the ground state E_1. When a flash of light (which lasts only for about a millisecond) falls upon the ruby rod, the 5500 Å radiation photons are absorbed by the chromium ions which are "pumped" (raised) to the excited state E_3. The transition 1 is the (optical) pumping transition. Excited ions give up, by collision, part of their energy to the crystal lattice and decay to the 'metastable' state E_2. The corresponding transition 2 is thus a radiationless transition. Since, the state E_2 has a much longer lifetime, the number of ions in this state goes on increasing while, due to pumping, the number in the ground state E_1 goes on decreasing. Thus, population inversion is established between the metastable (excited state E_2) and the ground state E_1.

When an (excited) ion passes spontaneously from the metastable state to the ground state (transition 3), it emits a photon of 6943Å wavelength. This photon travels through the ruby rod and, if it is moving parallel to the axis of the crystal, is reflected back and forth by the silvered ends until it stimulates an excited ion and causes it to emit a fresh photon in phase with the stimulating photon. This 'stimulated' transition 4 is the laser transition. The process is repeated again and again because the photons thus multiply. When the photon beam becomes sufficiently intense, a part of it emerges through the partially silvered end of the crystal.

Below is given a brief summary of ruby laser.

Table 10.1 *Ruby laser in brief*

Active medium	Cr^3+ dopant in Al_2O_3 crystal (Cr^{3+} ions are responsible for laser action)
Pumping mechanism	Optical pumping (using xenon flash lamp)
Wavelength emitted	6943 Å
Nature of output radiation	Pulsed
Maximum output power	Nearly few kilowatts
Main applications	In maching, drilling of holes with ultra high precision, LIDAR application, measurement of distance using pulse echo technique,

10.15 DEMERITS OF THREE-STAGE LEVEL LASER

There is a drawback with the three-stage level laser such as ruby. The laser requires high pumping power because its transition is terminated at the ground state and more than one-half of the ground-state atoms must be pumped up to the higher state to achieve population inversion. Moreover, ions which happen to be in their ground state absorb the 6443 Å photons from the beam, as it builds up.

10.16 FOUR-STAGE LASER—He-Ne LASER

He–Ne is four-stage level laser in which the population inversion is achieved by electric discharge. A mixture of about 7 parts of helium and 1 part of neon is contained in a glass tube at a pressure of about 1 mm of mercury. At both ends of the tube optically plane and parallel mirrors are fitted, one of them being only partially silvered.

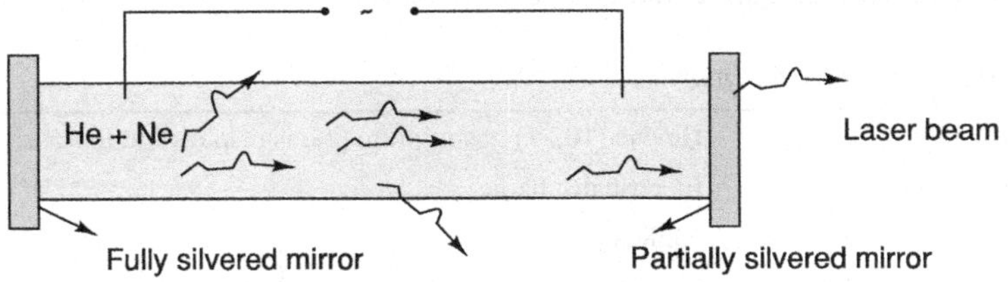

Fig. 10.16 *He-Ne laser assembly*

Spacing of the mirrors is equal to an integral number of half-wavelengths of the laser light. An electric discharge is produced in the gas-mixture by electrodes connected to a high-frequency electric source.

The electrons from the discharge collide with and 'pump' (excite) He and Ne atom to metastable states 20.61 eV and 20.66 eV, respectively above their ground states. Some of the excited He atoms transfer their energy to ground state Ne atoms by collision, with the 0.05 eV of additional energy being provided by the kinetic energy of atoms. Thus, He atoms help in achieving a population inversion in the Ne atoms.

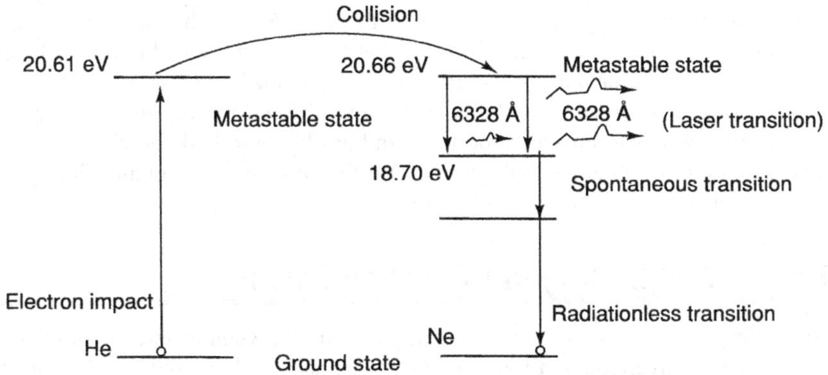

Fig. 10.17 *A four-level laser*

When an excited Ne atom passes spontaneously from the metastable state at 20.66 eV to 18.70 eV, it emits a 6328 Å photon. This photon travels through the gas mixture, and if it is moving parallel to the axis of the tube, is reflected back and forth by the mirror ends until it stimulates an excited Ne atom and causes it to emit a fresh 6328 Å photon in phase with stimulating photon. This stimulated transition from 20.66 eV level to 18.70 eV level is the laser transition. This process is continued and a beam of coherent radiation builds up in the tube. When this beam becomes sufficiently intense, a portion of it escapes through the partially silvered end.

From the 18.70 eV level, the Ne atom passes down spontaneously to a lower metastable state emitting incoherent light, and finally to the ground state through collision with the tube walls. The final transition is thus radiationless.

Obviously, the Ne atom in its ground state cannot absorb the 6328 Å photons from the laser beam, as happens in the three-level ruby laser. Also, because the electron impacts that excite the He and Ne atoms occur all the time, as in the ruby laser, the He–Ne laser operates continuously.

Further, since the laser transition does not terminate at the ground state, the power needed for excitation is less than that in a three-level laser.

Below is a brief summary about He-Ne laser.

Table 10.2 *He–Ne laser in brief*

Active medium	He–Ne (10 : 1) gas mixture (Ne is responsible for laser action)
Pumping mechanism	Electric discharge
Wavelength emitted	6328 Å
Nature of output radiation	Continuous
Maximum output power	100 mW
Main applications	In holograms, image processing bar-code reading, interferometric experiments, etc.

10.17 THE SEMICONDUCTING LASER

Laser action is also possible in semiconductors in *p-n* junctions. The incoherent light emitted by spontaneous emission from LEDs was converted into stimulated emission by achieving population inversion in the junction in around 1962 when the first *p-n* junction laser was built in infrared (GaAs) and visible (using GaAs P). The incoherent emission of radiation from LEDs was converted into highly coherent and directional beam of laser from semiconductor materials. Today, these junction lasers compete with He–Ne lasers from the point of view of the power output. Therefore, the main purpose of a semiconductor laser is to provide a portable and easily controlled source of low-power coherent radiation which is no doubt very useful for fibre optic communication systems, as such a beam works as carrier of information.

10.18 THE BASIC *p-n* JUNCTION LASER

Stimulated emission occurs in a direct band-gap material when electrons combine with holes. Population inversion and laser action can be achieved by applying forward bias. To get a *p-n* junction laser all that we need is a junction of direct band-gap material (GaAs) which is highly doped, and then to get a resonant cavity compatible (having proper geometric relationship) with the junction and also to make contact to the junction in a mounting which permit for efficient heat transfer. The first such laser was developed- in the beginning with a degenerate *n*-type material upon which a *p*-region was formed on one side (using Zn diffused *n*-type GaAs since Zn is a group II element of periodic table), and so it replaces Ga sites, and serves as an acceptor in GaAs. Therefore, the highly doped Zn diffused layers form the $p+$ region as shown in Fig. 10.18.

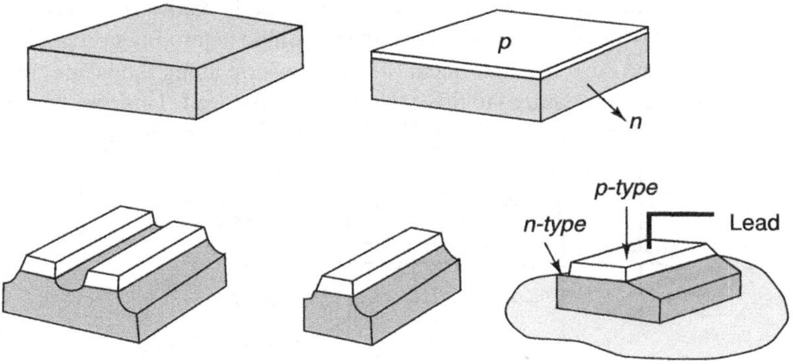

Fig. 10.18 *Fabrication of heterojunction laser*

At this stage, we have a large area planer *p-n* junction. After this the grooves are cut along the length of the sample, leaving a series of long *p*-regions isolated from each other. These *p-n* junctions can be cut or broken apart and then cleaved into devices of desired length.

It is essential to keep in mind that during the above fabrication process, front and back faces must be flat and parallel which can be accomplished by cleaving. The device is then mounted on suitable header and contact is made to the *p*-type alloys. Many techniques are used to provide adequate heat sinking of the device when it is operated at high values of forward current.

10.19 HETEROJUNCTION LASER : A BOON FOR OPTICAL COMMUNICATION

The device which was developed in the very beginning was simply a homojunction laser as it utilised only one type of material. To obtain a more efficient laser that can be operated at room temperature, it is necessary to use multiple layers in the laser structure, usually known as heterojunction lasers. This proved to be very useful for optical communication purposes, and was found suitable to be utilised at room temperature.

In the structure shown in Fig. 10.19, the injected carriers are confined to a narrow region in order that population inversion can be built up at lower current levels. This results in a lowering of the threshold current at which the laser action begins.

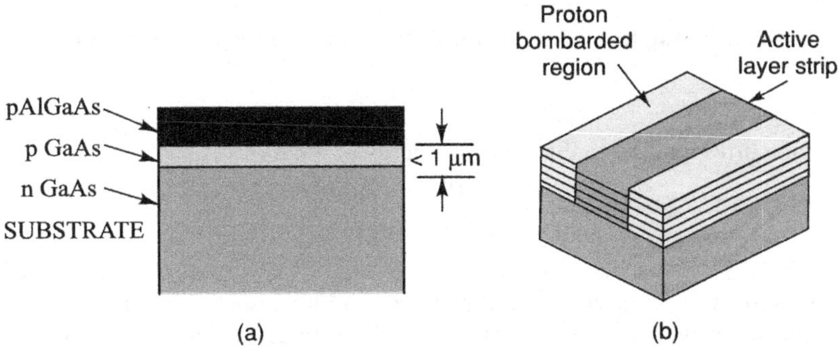

Fig. 10.19 *AlGaAs heterojunction grown on thin p-type GaAs layer*

A further improvement can be obtained by sandwiching the active GaAs layer between two AlGaAs layers. This double-heterojunction structure further confines injected carriers to the active region, and the change in refractive index at the GaAs–AlGaAs boundaries assists in confining the produced light waves. In the double heterojunction laser shown in Fig. 10.20, the injected current is restricted to a narrow strip along the lasing direction in order that the total current required to drive the device is reduced. This type of a laser was a major step forward in the development of lasers for fibre-optic communications.

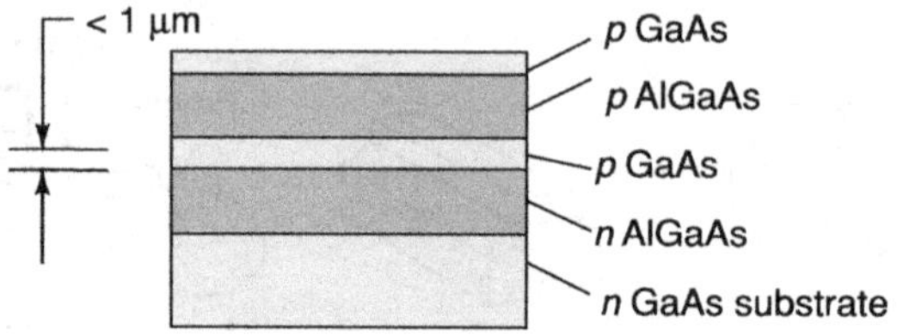

Fig. 10.20 *Double heterojunction laser structure*

One drawback of the double heterojunction laser is the fact that the carrier confinement and the optical wave guiding, both depend on the same heterojunction. It will be better to optimise these junctions by making use of a narrow confinement region in order to keep the carriers in a region of high recombination, and a somewhat wider optical waveguide region.

Example 10.1

An electron moves from an energy level of 5.56 × 10^{-19} J to an energy level of 2.36 × 10^{-9} J. Calculate the wavelength and colour of the photon.

Solution $\Delta E = 5.56 \times 10^{-19} - 2.36 \times 10^{-19}$ J

$$= 3.2 \times 10^{-19} \text{ J}$$

or
$$\Delta E = \frac{hc}{\lambda}$$

∴
$$\lambda = \frac{hc}{\Delta E} = \frac{6.626 \times 10^{-34} \times 3 \times 10^{8}}{3.2 \times 10^{-19}}$$

or
$$\lambda = 6.211 \times 10^{-7} \text{ m} = 6211 \text{ Å}$$

The colour of the photon is red.

Example 10.2

Diamond has a bandgap of 5.6 eV. Find the maximum wavelength for which it is opaque

Solution $E = hc/\lambda$

Wavelength corresponding to the bandgap energy of 5.6 eV

$$\lambda = hc/E = \frac{6.63 \times 10^{-34} \times 3 \times 10^{8}}{5.6 \times 1.6 \times 10^{-19}}$$

$$= 2.2 \times 10^{-7} \text{ m}$$

Radiation of wavelength below 2.2 × 10^{-7} m will be absorbed by diamond, and so it is opaque for all wavelengths below this value. Therefore, the maximum wavelength for which diamond is opaque.

$$I_{max} = 220 \text{ Å}$$ Ans.

Example 10.3

The band-gap of GaAs–GaP alloy increases approximately linearly with GaP additions (in mol %). To emit orange light of 0.60 μm, what should be the composition of GaAs–GaP semiconductor? (Band-gap energy of GaAs and GaP are 1.42 eV and 2.25 eV, respectively).

Solution Energy of radiation of the wavelength 0.60 μm

$$E = hc/\lambda = \frac{6.6\times10^{-34}\times3\times10^{8}}{0.60\times10^{-6}\times1.6\times10^{-19}} = 2.0625 \text{ eV}$$

Assuming linear variation of energy gap with mol % addition of GaP to GaAs, the rate at which the energy gap varies with addition of GaP.

$$\frac{E_g(\text{GaP}) - E_g(\text{GaAs})}{100} = \frac{0.83}{100}$$

$$= 0.0083 \text{ eV/ mol \% of GaP}$$

If x is the mol % of GaP to be added to get an energy gap of 2.0625 eV,

$$2.0625 = 0.0083 \times x + 1.42$$

$$x = \frac{2.0625 - 1.42}{0.0083} = 77.4 \text{ mol \%}$$ Ans

Example 10.4

In a material, transition occurs between a metasatble state of energy E_3 and an energy level $E_2 = 0.4 \times 10^{-19}$ J, just above the ground state. The wavelength of the photon emitted is 1.1 μm. Find the energy of the metastable state.

Solution $$E_3 - E_2 = \frac{hc}{1.1\times10^{-6}}$$

$$E_3 = 0.4\times10^{-19} + \frac{6.63\times10^{-34}\times3\times10^{8}}{1.1\times10^{-6}} = 2.2\times10^{-19} \text{ J}$$ Ans.

Example 10.5

Determine the frequency difference between the cavity modes of a laser that is 1.5 m long and contains a gas of refractive index 1.0204. If the bandwidth of the laser is 1.5×10^9 Hz then find the number of optical modes in the output radiation.

Solution $$\Delta v = \frac{c}{2Ln} = \frac{3\times10^{8}}{2\times1.5\times1.0240} = 9.8\times10^{7} \text{ Hz}$$

Number of optical modes:

$$N = \frac{1.5\times10^{9}}{9.8\times10^{7}} \cong 15$$

There are 1.5 optical modes in the output. Ans.

10.20 MATERIALS FOR SEMICONDUCTOR LASERS

Materials for semiconductor lasers must be efficient light emitters and also have the ability to be obtained as a junction or as a heterojunction barriers. Due to this requirement all the materials therefore are not practically useful. For example, indirect band-gap materials are not suitable since these materials are not efficient light emitters. The II-VI materials on the other hand are very good light emitters but to get a *p-n* junction or heterojunction from these materials is quite difficult. However by using the modern techniques like MBE and MOVPE one can get these in a *p-n* junction form that emit the radiation in green and blue-green part of em spectrum.

Recently there has been extreme surge of interest in developing a large band-gap semiconductor using GaN and its alloys with InN and AlN. The InAlGaN system has direct band-gaps over the entire alloy composition range and therefore is a very efficient light emitter. We find a band-gap beginning from 2 eV in InN, to 3.4 eV in GaN and even more 5.0 eV in AlN that cover a wide range of wavelength from 620 nm to 248 nm.

On the other hand In GaAs P/InP system which we have discussed earlier are well suitable for the lasers to be used for fibre-optic communication. The quaternary InGaAsP is particularly interesting for fabrication of Laser diode since it allows a considerable choice of wavelength range and flexibility in lattice matching. On varying the composition, the above said material can provide laser in infrared range from 1.3–1.55 μm for fibre optics. The pollution monitoring applications require wavelength in far infrared which can be obtained by using a ternary alloy PbSnTe. This provides the laser output that range from 7 μm to more than 30 μm at low temperature depending upon material composition. The short wavelength emitters are also important as these are required for storage application in DVDs.

10.21 INTRODUCTION TO OPTICAL FIBRE

Communication through optical fibre is one of the remarkable discoveries of the twenty-first century. It has brought a revolution in the modern times. The transfer of information over long distances was earlier performed through copper wires and coaxial cables. These devices could not fulfill the expectations and hence, were replaced by glass fibre. It was the effort of Alexander Graham Bell who, in 1880, used the light as a carrier of signal. Since the attenuation in the optical fibres was quite high, an attempt to minimise it was done to improve its efficiency. Today, its features are so extraordinary that the optical communication through glass fibre with low loss has become a reality. A large number of advantages of optical fibres over traditional wires and coaxial cables are no more hidden and have been accepted throughout the globe. Apart from the use in communication, optical fibres are widely used in other areas as well. In a nutshell, we can therefore say that fibre optics is the backbone of communication infrastructure.

10.22 STRUCTURE OF OPTICAL FIBRE

An optical fibre is a device that works upon principle of total internal reflection of light. It operates at optical frequencies to guide the information from one place to another, utilising a glass fibre which is as thin as human hair and is made of clear plastic or glass. A practical optical fibre has, in general, three coaxial regions as given below:

- Core-central region
- Cladding-outside the core
- Sheath-outer jacket

The innermost region is known as the core, through which the light wave transmits. The core is surrounded by a coaxial mid-region having lower refractive index called cladding, which provides the mechanical strength to the fibre, protects it from absorbing surface contaminants and reduces the losses due to scattering. The outermost region is called the sheath; it protects the core and cladding from external hazards like abrasions, contamination and the harmful influence of moisture. Figures 10.21 and 10.22 demonstrate an optical fibre and its cross-sectional view along with the dimensions.

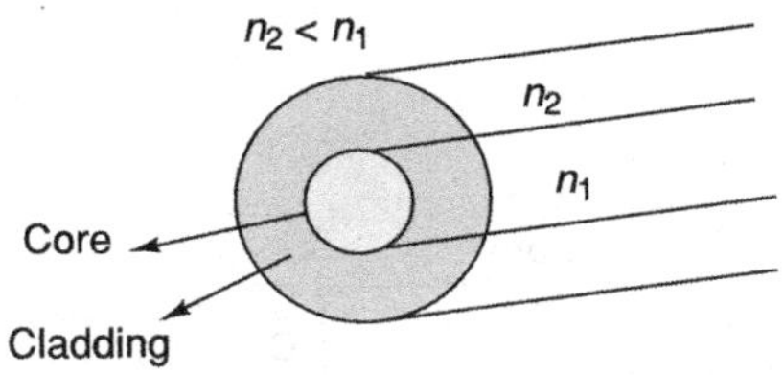

Fig. 10.21 *An optical fibre*

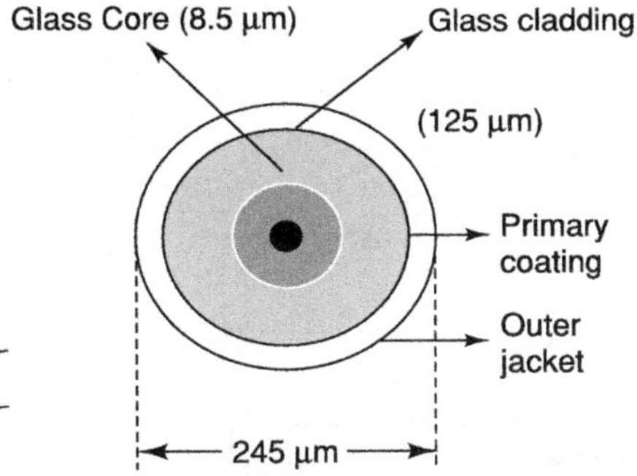

Fig. 10.22 *Cross-section view of an optical fibre*

In an optical fibre, the refractive index of the core region (n_1) is slightly more than cladded material (n_2), and therefore, the entering ray suffers TIR, is guided along its length and carries information from one end to the other.

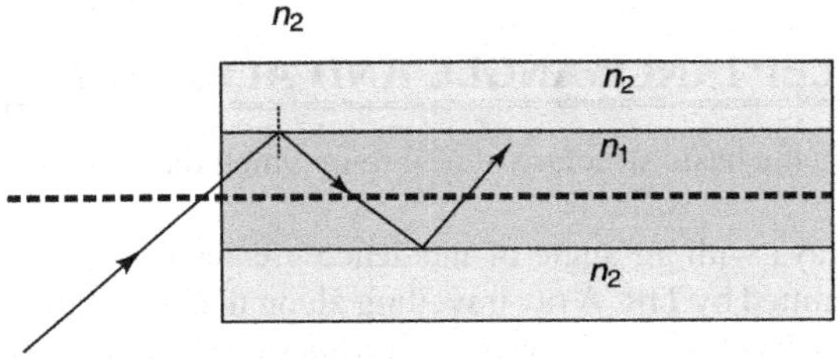

Fig. 10.23 *Transmission of optical ray through glass fibre*

The angle of incidence at core-cladding interface exceeds the critical angle $\left(\phi_c = \sin^{-1}\frac{n_2}{n_1}\right)$, and therefore the ray gets completely reflected in the same media and process along the length.

10.23 STEP INDEX FIBRE (SI)

In the SI fibre, the core is of constant refractive index n_1 and the cladding is of slightly lower refractive index. Thus, in this fibre there is a step change in refractive index at the core–cladding interface. The refractive index profile may be defined as

$$n(r) = n_1 \text{ for } r < a \quad \text{(Core)}$$
$$= n_2 \text{ for } r > a \quad \text{(Cladding)}$$

10.24 GRADED INDEX FIBRE (GI)

In GI fibres, the core index $n(r)$ varies (decreases from axis) with radial distance to a constant value n_2 beyond the core radius (a). The index variation may be given as

$$n(r) = n_1 \left\{1 - 2\Delta\left[\frac{r}{a}\right]^2\right\}^{1/2} \quad \text{for } r < a \text{ (Core)}$$

$$= n_2\,[1 - 2\Delta]^{1/2} \quad \text{for } r > a \text{ (Cladding)}$$

where
$$\Delta = \frac{n_1 - n_2}{n}$$

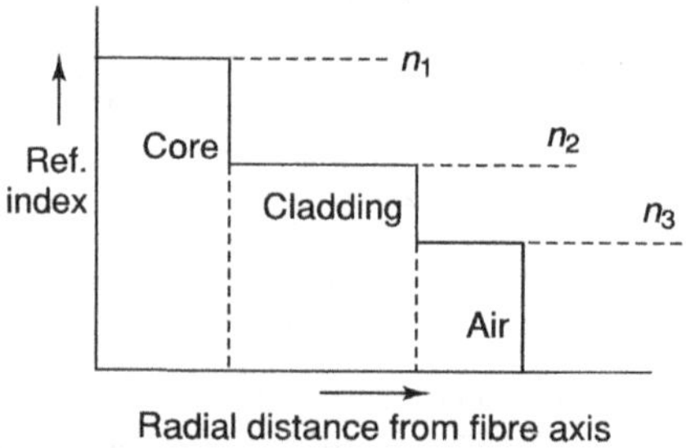

Fig. 10.24 *Step index fibre*

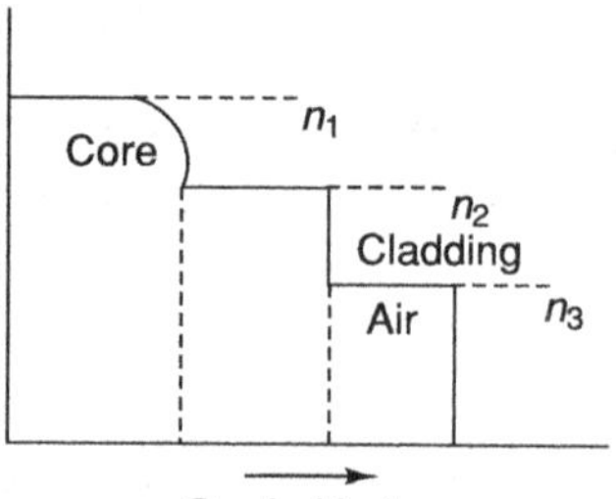

Fig. 10.25 *Graded index fibre*

10.25 ACCEPTANCE ANGLE AND ACCEPTANCE CONE

Figure 10.26 shows the basic structure of a fibre in which the optical ray is entering from the launching end.

Only incident rays with an angle of incidence greater than θ_c at the core–cladding interface are transmitted by TIR. A ray travelling along the axis of the fibre core is known as meridional ray. MP is the meridional ray. Considering the light rays entering the fibre, its angle of incidence with fibre axis at air-core interface in θ_a. The rays entering the fibre with the angle of incidence greater than θ_a are transmitted to the core–cladding interface at an angle less than θ_c. These rays don't undergo TIR and are lost by radiation, for example, the ray BP in Fig. 10.26. The rays entering the fibre-core within an acceptance cone, specified by conical half angle θ_a are transmitted.

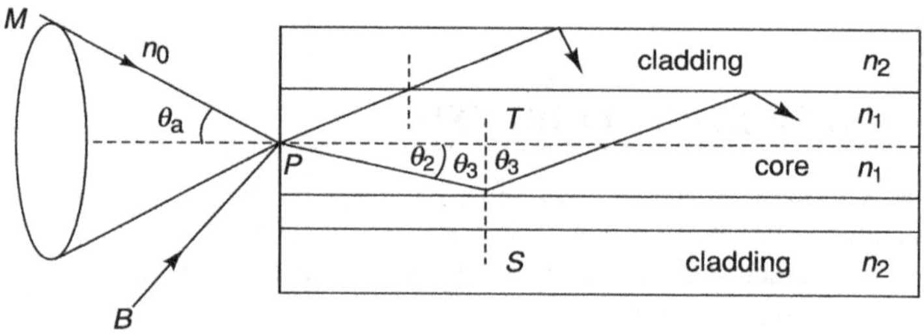

Fig. 10.26 *Total internal reflection inside an optical fibre*

Therefore, acceptance angle is the maximum angle of incidence, specified by the conical half angle for the light rays to be transmitted through the optical fibre." Acceptance cone and acceptance angle is shown in Fig. 10.26.

Numerical Aperture (NA)

NA is a more generally used term for ray transmission in optical fibre. It gives the relation between the acceptance angle and refractive indices of the three media involved–core, cladding and air. Consider the ray entering the fibre core at angle less than θ_a (the acceptance angle)

Applying Snell's law,

$$n_0 \sin\theta_1 = n_1 \sin\theta_2 \tag{10.54}$$

and $\theta_3 > \theta_C$

Therefore, $n_0 \sin\theta_1 = n_1 \sin\left(\frac{\pi}{2} - \theta_3\right) = n_1 \cos\theta_3$

$$n_0 \sin\theta_1 = n_1 \sqrt{1-\sin^2\theta_3} \tag{10.55}$$

In the limiting case,

$$\theta_1 \to \theta_a,\ \theta_3 \to \theta_c$$

$$\Rightarrow \quad n_0 \sin\theta_a = n_1 \sqrt{1-\sin^2\theta_c}$$

$$= n_1 \sqrt{1-\left(\frac{n_2}{n_1}\right)^2} = \sqrt{n_1^2 - n_2^2} \tag{10.56}$$

Numerical aperture is defined as

$$\text{NA} = n_0 \sin\theta_a = \sqrt{n_1^2 - n_2^2}$$

For air, $n_0 = 1$

Hence, $\text{NA} = \sin\theta_a$

$$= \sqrt{n_1^2 - n_2^2}$$

We know that $\sin\theta_1 = \sqrt{n_1^2 - n_2^2}$

Since $n_1^2 - n_2^2 = (n_1 + n_2)(n_1 - n_2)$

$$= \left(\frac{n_1+n_2}{2}\right)\left(\frac{n_1-n_2}{n_1}\right)2n_1$$

$$= \left(\frac{n_1+n_2}{2}\right)\Delta.2n_1$$

where $\Delta = \dfrac{n_1 - n_2}{n_1}$

is known as fractional change in refractive index. This parameter is always positive because n_1 is always greater than n_2, which is a prime requirement for total internal reflection. In order to guide light rays effectively through the fibre $\Delta \ll 1$, its value is normally nearly 10^{-1}.

Assuming $\frac{n_1 + n_2}{2} = n_1$, the equation (10.57) can be written as

$$n_1^2 - n_2^2 = 2n_1^2\Delta$$

Therefore, $$\text{NA} = \sqrt{2n_1^2\Delta}$$

Numerical aperture expresses the light–gathering ability of the optical fibre. It is a measurement of the amount of light that can be accepted by a fibre.

Example 10.6

In a step index fibre, the refractive index of the core is 1.55 and the refractive index of cladding is 1.53. Find the numerical aperture of the fibre.

Solution $$NA = \sqrt{n_1^2 - n_2^2} = \sqrt{(1.55)^2 - (1.53)^2}$$

$$= 0.248$$ Ans.

Example 10.7

Calculate the critical angle for light incident on water/air boundary from the water side, if the refractive index of water is 1.33.

Solution

$$\theta_c = \sin^{-1}\left(\frac{n_2}{n_1}\right) = \sin^{-1}\left(\frac{1}{1.66}\right) = 48.75°$$ Ans.

For angles above 48.75°, there will be total internal reflection in water.

10.26 MODAL CLASSIFICATION OF OPTICAL FIBRES

Fibres are classified as

- Single-mode fibre
- Multi-mode fibre

A single-mode fibre has a smaller core diameter and can support only one mode of propagation. On the other hand, a multimode fibre has a larger core diameter and supports a number of modes. Multimode fibres are further distinguished on the basis of index profile. Index profile is a plot of refractive index vs. the distance. The index profile of a multi-mode fibre can be either a step index type or a graded index type. The index profile of a single-mode fibre is usually a step index type.

There are three major configurations.

- Multimode step-index fibre
- Multimode graded-index fibre
- Single mode step-index fibre

10.27 ATTENUATION IN OPTICAL FIBRES

An optical signal propagating through a fibre will get progressively attenuated. The signal attenuation is defined as the ratio of the optical output power from a fibre of length L to the input optical power.

$$\alpha = \frac{10}{L}\log\frac{P_i}{P_0}$$

where P_i is the input power of optical signal launched at one end of the fibre and P_0 is the output power of the optical signal emerging from the other end of the fibre. In case of an ideal fibre, $P_0 = P_i$ and the attenuation would be zero. The unit of measurement of attenuation is decibel / kilometre (dB/km).

Kinds of Attenuation

❖ **Absorption by Material** This includes absorption due to light interacting with the molecular structure of the material, as well as loss because of material impurities. Even a highly pure glass absorbs light in specific wavelength regions. Strong electronic absorption occurs at UV wavelengths, while vibrational absorption occurs at IR wavelengths. These absorption losses are inherent property of the glass itself and are known as intrinsic absorption.

Impurities are a major source of losses in fibres. Hydroxyl radical ions (OH) and transition metals such as cooper, nickel, chromium, vanadium and manganese have electronic absorption in a near visible part of the spectrum. Their presence causes heavy losses. Losses due to impurities can be reduced by better manufacturing processes. Metal ions are practically negligible in improved fibres. The largest loss is caused by OH ions which cannot be sufficiently reduced.

❖ **Scattering** When light is scattered by an obstacle, it results in power loss. The local microscopic density variations in glass causes local variations in refractive index. These variations, are inherent and cannot be eliminated, act as obstructions and scatter light in all directions. This is known as **Rayleigh scattering.**

❖ **Waveguide and Microband Losses** The losses introduced during manufacturing or installation processes come in this category. Structural variations in the fibre or fibre deformation cause radiation of light away from the fibre. Microbands and very minute disturbances in core size also cause radiation of light.

10. 28 MERITS OF OPTICAL FIBRE

Optical fibres have many advantages that are not found in conducting wires. Some of these are listed below:

Optical fibres are made from silica (SiO_2), which is one of the most abundant materials on the earth and is also cheap. The overall cost of fibre optic communication is lower than that of equivalent cable communication system.

- Optical fibres are smaller in size, lighter in weight, flexible safe and yet strong. A wire communication link could be accidentally short circuit at high voltage.
- Optical fibres are immune to EMI and RFI. Light waves propagating along the optical fibre are completely trapped within the fibre and cannot leak out, and therefore no possibility of cross talk when optical fibre is used. Therefore, transmission is more secure and private.
- Because of a larger bandwidth, optical fibres have an ability to carry large amounts of information while a telephone cable composed of 900 pairs of wire can handle 10,000 calls, a 1 mm optical fibre can transmit 50,000 calls. They also exhibit low loss per unit length.

10.29 OPTICAL FIBRE COMMUNICATION

The main components of a general optical fibre communication system are shown in Fig. 10.27.

The transmission system has an information encoder or signal shaping circuit before the modulation which operates the optical source. Thus, the input signal is converted into optical signal. It is launched into an optical fibre from the launching end. The signal is received at the receiver end and is converted back into electrical signal by an optical detector. This electrical signal is then amplified and decoded or demodulated to obtain the information transmitted. The block diagram of a transmitting system is given in Fig. 10.27.

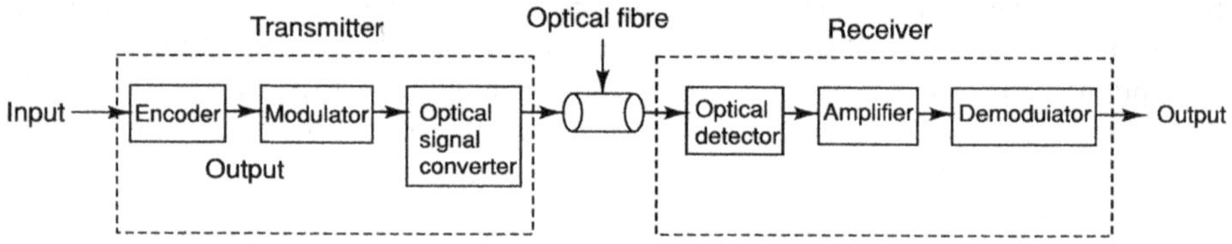

Fig. 10.27 *Various components of an optical-fibre communication system*

10.30 APPLICATIONS OF OPTICAL FIBRE

There are many advantages of using optical fibres over electrical cables. Optical fibres are widely used in the following systems:

1. In telecommunication networks, for example, telephone, cable TV, videophone, teleconferencing, etc.
2. Undersea cable systems are used in international telecommunication networks.
3. In digital transmission telecommunication networks.
4. In military applications—various factors are considered such as size, weight, deployability, survivability in conventional and nuclear attack, security, etc. In aircrafts, ships and tanks, optical fibres are used as they are small in size and less in weight. Optical fibre cables, being immune to electromagnetic interferences and noisy environment, are very useful in military mobiles.

10.31 LIGHT SOURCE MATERIALS FOR OPTICAL FIBRE COMMUNICATION

For optical fibre communication, laser is used as a carrier wave because it is highly monochromatic and allows a very large information bandwidth. Optical fibres are used as media for transmission of optical waves. In early optoelectronic system used for fibre optics, it was traditional to use well-known GaAs, Al–GaAs systems for making lasers. These light sources are highly efficient and good detectors can be made using it. However, they operate in a range of ~ 0.9 μm, where attenuation is greater than that for larger wavelengths. Today, the modern system therefore operates in the wavelength range of 1.3 and 1.55 μm minima. For such wavelengths, lasers can be made using InGaAs or InGaAsP grown on InP, and detectors can be made of the same materials. Such low-power coherent sources are particularly suitable for optical fibre communication.

10.32 SEMICONDUCTOR MATERIALS FOR SOLAR CELLS

The solar cell is considered to be a major source for getting the energy utilising the solar radiations. The first solar cell was developed by Chapin, Fuller and Pearson in 1954 using a diffused *p-n* junction. Afterwards, another candidate using II-VI compound CdS was developed. The main concern is to get maximum conversion efficiency, low cost and immunity from high energy radiation on being used in upper atmosphere, like in satellites.

In a conventional *p-n* junction, solar cell having a single band-gap (only those radiations contribute to the output whose energy exceed the band-gap), for example, any material exhibiting a band-gap between 1–2 eV, can be considered a solar cell materials. The maximum efficiency is found 31% for a material having band-gap 1.35 eV using III-V compound semiconductors. Many factors degrade the ideal efficiency, and therefore, practical obtainable efficiency is always lower. Research is going on to develop low cost, flat-panel solar cells. For terrestrial applications and in order to maintain at least 10% conversion efficiency, thin film CDS solar cells, as well as amorphous Si solar cells are important.

A typical *p-n* junction solar cell is shown in Fig. 10.28. It consists of a shallow *p-n* junction formed at the surface (by diffusion), a front ohmic contact stripe and fingers, a back ohmic contact that covers the entire back surface and an anti-reflection coating on the front surface.

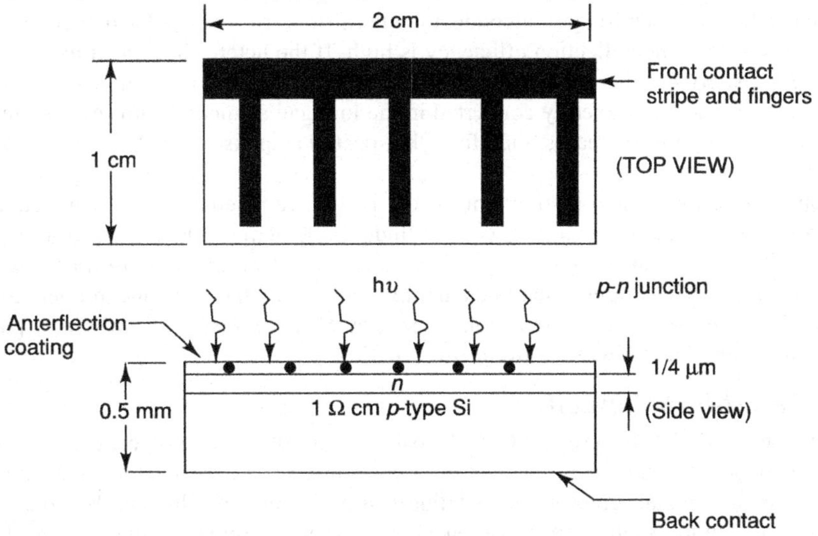

Fig. 10.28 *Schematic representation of a silicon p-n junction solar cell*

The single silicon solar cell of surface area 2 cm^2 provides an open-circuit voltage of 0.5 to 0.6 V and a current (open circuit) of 30–60 mA. Therefore, a need has been realised to construct such individual solar cells to make an array that could provide substantially high values of voltage and current. Such array delivers a power of about 10 and conversion efficiency of 11.5%. It has also been noticed that efficiency decreases linearly with temperature to about 200°C for silicon, and 300°C for GaAs. For sattelite applications, the high energy particle radiation in outer space produces defects in semiconductors that cause a reduction in output power. To improve the radiation tolerance, lithium has been incorporated which can diffuse and combine as the radiation induces point defects and neutralise the defects.

Further, the solar cell is covered with a cesium doped microsheet to minimise the number of high-energy particles striking its surface.

(i) Antireflection Coating

The addition of antireflection coating reduces the overall reflection to a few per cent. Reduced reflection enhances both short-circuit current and open circuit voltage, which in turn improves the conversion efficiency. Materials that serve the purpose of antireflection coating are described in the subsequent section.

(ii) Heterojunction, Interface and Thin film Solar Cells

The advantages of heterojunction solar cells over conventional *p-n* junction solar cells are as listed below:

- Enhanced short-wavelength spectral response
- Lower series resistance
- High-radiation tolerance, if first semiconductor is thick in addition to being high in bandgap

In a heterojunction solar cell a large band-gap material is added upon a *p-n* homojunction. It consists of *p*-type $Ga_{1-x}Al_x$ As/P-GaAs/*n*-GaAs. The wide-gap semiconductor acts as a window, allowing photons of energy less than Eg_1. Those photons which have energies in-between Eg_1 and Eg_2 will create carriers in homo junction. If the absorption coefficient is high in the lower band-gap semiconductor, the carriers are generated in depletion region or close to it. So, the collection efficiency is high. If the hetero-face material is indirect band-gap material with $x > 0.4$ and is thin enough, many photons with $h\,\nu > Eg_1$ will pass through and be efficiently converted in the low-gap semiconductor. As x increases, the bandgap Eg_1 also increases, therefore, the spectral response extends to higher photon energies.

One interesting heterojuction solar cell is the conducting glass semiconductor heterojunction. The conducting glasses include oxide semiconductors such as indium oxide with 3.5 eV bandgap, tin oxide with 3.5 eV bandgap and indium tin oxide with 3.7 eV bandgap. All these materials in thin film form have unique properties like good electrical conductivity and high transparency. They serve not only as a part of heterojunction, but also as an antireflection coating.

(iii) Thin-film Solar Cells

Solar cells can also be prepared by depositing amorphous or polycrystalline, layers of semiconductor materials, e.g., Cds, Si, GaAs, InP, etc. on a substrate such as glass, plastic ceramic metal, graphite or metallurgical grade silicon. This can be done using any one technique such as evaporation plasma, vapour growth or plating, etc. If the semiconductor's thickness is large than the inverse of the absorption co-efficient, most light will be absorbed; if the diffusion length is larger than the inverse of the absorption co-efficient, most light will be absorbed; if the diffusion length is larger than the film thickness, most photo-generated carriers can be collected.

The chief advantage of thin-film solar cell is its promise of low-cost by the use of relatively low-cost materials. The main disadvantages are low efficiency and long-term instability. The low efficiency is partly caused by the poor quality of semiconductor material grown on foreign substrates.

Another thin-film solar cell is a hetrojunction of $CuInSe_2$/Cds. The short circuit current and efficiency is dependent upon grain size. The max short circuit current (~ 2.6

mA / cm^2) is obtained for 2.5 μm grain radius and at this radius, efficiency is nearly 6.6%. Many ternary compounds are potential candidate for low-cost solar cell application as discussed earlier.

Amorphous silicon is also a suitable material to be used in solar cells. Layers of 1 to 3 μm thick are grown by RF glow discharge decomposition of Silane onto metal to ITO-coated glass substrates. The difference between crystalline and amorphous silicon is dramatic; crystalline Si has indirect band-gap of 1.1 eV, whereas hydrogenated a–Si has an optical absorption with a direct band-gap of 1.6 eV. Solar cells have been prepared in *p-n* junction as well as Schottky barrier form on thin film of hydrogenated amorphous silicon.

Summary

- The EM interaction with the valence electron is responsible for the optical properties of materials.
- The main optical properties of interest are scattering refraction, reflection, transmission absorption and luminescence.
- Scattering takes places due to oscillation of the valence electrons in the presence of the time changing electricity of the incident radiation. The electrons oscillate with frequency of the incident radiation and emit electromagnetic radiation of the same frequency in all directions.
- Refraction is the deviation in the path of the beam, when it travels from air into the material. The deviation arises due to decrease in velocity of the radiation in the material.
- Refractive index is defined as the ratio of the speed of the light beam in air to its speed in the material.
- Metals exhibit high reflectivity for light radiation in the range of infrared and visible range and low reflectivity in the ultraviolet and higher frequency range.
- The two types of transitions that may occur are: intra-band transition and inter-band transition.
- The transparency, opacity or colours of materials can be interpreted in terms of the absorption properties of the materials.
- The electrons that get excited to higher energy levels eventually drop down to lower energy levels spontaneously. The difference in the energy may be emitted in the form of a photon. This emission process is called luminescence.
- The two types of luminescence are fluorescence and phosphorescence. After excitation, if the emission is instantaneous it is called fluorescence. If the emission occurs even after incidence of radiation, it is called phosphorescence.
- Laser is a source of highly intense, monochromatic, coherent radiation. It is the acronym for Light Amplification by Stimulated Emission of Radiation.
- Laser works on the principle of population inversion.
- Population inversion is the process of making the upper energy level more highly populated than the lower energy level. Population inversion is a necessary condition, but not a sufficient one for laser action.
- A minimum of three energy levels system is required for laser action.

- High power lasers are used in materials processing in industries, such as cutting, welding, and drilling.
- Lasers are widely used in the medical field for eye surgery, skin treatment, cancer treatment, dental surgery, etc.
- Optical fibres are optical waveguides that carry light radiation. They are made of extremely thin wires of silica. They transmit light signals through long distances without much attenuation and distortion. They are used in optical communication for transmission of signals.

Exercise

1. Explain briefly the main optical properties of solids.
2. What is photo-electric emission? State and explain the laws of photo-electric emission.
3. What is the photo-electric emission? How is the electron emission affected if (i) the frequency, and (ii) the intensity of the incident radiations are increased? Name a material which is suitable for photo-emission.
4. What is luminescence? Explain the terms 'photoluminescence', 'cathode luminescence' and 'electroluminescence'.
5. What is electroluminescence? Explain the two basic mechanisms of electroluminescence.
6. What is LED? Give its principle of working, construction, circuit symbol, merits, demerits and applications.
7. Write short notes on the following:
   (i) Photoemissive materials and types of photocathode
   (ii) Electroluminescence
   (iii) LEDs
   (iv) Light emitting materials
   (v) Semiconductor lasers
   (vi) Materials for semiconductor lasers

Numerical Problems

1. A photo-emissive surface is irradiated by mercury light of wavelength 2,537 Å. If the emitted electrons have a maximum energy of 2.5 eV, calculate the work function of the cathode material. **[Ans. 2.388 eV]**

2. What is the minimum energy required to remove an electron from the surface of a photo material, threshold wavelength of which is 6,000 Å. **[Ans. 2.0677 eV]**
3. The bandgap of diamond is 5.6 eV. Find the maximum wavelength for which diamond is opaque. **[Ans. 2.2×10^{-7} m]**

4. The bandgap of GaAs–GaP alloy varies nearly linearly with Ga additions (in mol %). To emit orange light of 0.60 μm, what should be the composition of GaAs–GaP semiconductor? (Bandgap energy of GaAs and GaP, and 1.42 eV and 2.25 eV, respectively). **[Ans. 2.0625 eV]**
5. In a material, transition occurs between a metastable state of energy/energy levels $E_2 = 0.4 \times 10^{-19}$ J just above the ground state. The wavelength photon emitted is 1.1 μm. Find the energy of the metastable state. **Ans. 2.2**

6. Determine the frequency difference between the cavity modes of a 1.5 m long laser containing a gas of refractive index 1.0240. If the bandwidth laser beam is 1.5×10^9 Hz, find the number of optical modes in the radiation. **Ans. 15**
7. In a step index fibre, the refractive index of the core is 1.55 and the index of the cladding is 1.53. Find the numerical aperture of the fibre. **Ans. 0.248**
8. In an optical fibre, the core material has refractive index 1.6 and refractive index of cladding material is 1.3. What is the value of critical angle? Also calculate the value of angle of acceptance.

   Ans. 121°
9. A fibre cable has an acceptance angle of 30° and a core index of refraction of calculate the refractive index of the cladding. **Ans. 1.308**
10. Calculate the numerical aperture, acceptance angle, and the critical angle of the fibre from the following data: μ_1(core refractive index) = 1.50 and μ_2(cladding refractive index) = 1.45.

    Ans. 75.3°
11. If the fractional difference between the core and cladding refractive indices of a fibre is 0.0135 and numerical aperture NA is 0.2425, calculate the refractive indices of the core and cladding materials. **Ans. 1.456**
12. The man optical power launched into a fibre of 6 km length is 100 μW. The mean optical power at the fibre output is 5 μW. Calculate the signal attenuation per kilometre for the fibre.

    Ans. 2.16 dB/km
13. The sum of refractive indices of core and cladding is 2.89 and difference is 0.03. Determine the NA for the optical fibre. **Ans. 0.29**

Self-quiz Questions

1. What is laser?
2. What is the difference between mercury source and laser source?
3. In terms of quantum picture, when is a light beam said to be perfectly coherent?
4. What do you understand by a metastable state?
5. What is the wavelength of laser light from
   (i) ruby laser, and (ii) He–Ne laser
   How are the two lasers different?
6. Which is a better laser—ruby or He–Ne laser? Why?
7. What is population inversion? How is it achieved?
8. Why is population inversion necessary for laser action?
9. What is the active medium?
10. What is an optical resonator?
11. In He–Ne laser, lasing is through neon gas. What is the role of helium then?
12. Can we obtain light amplification in the absence of stimulated emission?
13. Why is helium used in an He–Ne laser?
14. Why must the refractive index of cladding always be lower than that of the core?
15. What are the advantages of using fibre–optic communication system?
16. Can light propagate through a fibre cable with the angle of incidence at the entrance end greater than acceptance angle?
17. Distinguish between acceptance angle and numerical aperture.
18. What is the purpose of cladding in optical fibre?
19. Define index profile.

20. What do you understand by normalised frequency?
21. How does light propagate in an optical fibre?

Objective-type Questions

1. In a dyne laser, the laser wavelength is tuned by
   (a) changing the rise time of the pumping source
   (b) changing the orientation of the grating in the cavity
   (c) changing the solvents
   (d) changing the intensity of pumping source
2. Cadmium sulphide has a band-gap of 2.4 eV. Which of the following wavelength radiation will it absorb?
   (a) 4000 Å
   (b) 5500 Å
   (c) 6500 Å
   (d) 7000 Å
3. Which of the following materials does not absorb visible radiation?
   (a) Sapphire (b) Ruby
   (c) Diamond (d) Silicon
4. Dielectric materials like enamels, porcelain, etc., are opaque (in the visible range) because
   (a) they have band-gap greater than 3 eV
   (b) they have band-gap les than 3 eV
   (c) they scatter light
   (d) they have a low refractive index
5. Which of the following relations between the velocity of light in the material v and its dielectric constant are true?
   (a) $v^2 = c^2/\varepsilon$ (b) $v^2 = c^2/\varepsilon^2$
   (c) $c = \upsilon\sqrt{\varepsilon}$ (d) $c = \sqrt{v\varepsilon}$
6. Among the following lasers, which one emits the largest wavelength radiation?
   (a) Ruby laser
   (b) CO_2 laser
   (c) Semiconductor laser
   (d) He–Ne laser
7. Which of the following are examples of direct band-gap semiconductors?
   (a) GaAs (b) Si
   (c) Ge (d) Se
8. In indirect inter-band transition, the electron undergoes
   (a) change in both energy and momentum
   (b) change in momentum, but no change in energy
   (c) no change in both energy and momentum
   (d) change in energy but no change in momentum
9. The ranges of electromagnetic radiation in which metals exhibit maximum reflectivity is
   (a) visible
   (b) beyond ultraviolet
   (c) ultraviolet
   (d) infrared

Answer

1. (b)	2. (a)	3. (c)	4. (c)	5. (a) and (c)
6. (b)	7. (a)	8. (a)	9. (a) and (b)	

Fill in the Blanks

1. The..................of electromagnetic spectrum extends from 4000Å-78000Å
2. Electromagenetic waves travel with..................in free space.
3. ....of material is equal to square of index of refraction.
4. Cds and CdTe are widely used as.................

5. Optical fibre communication is based upon.................phenomena.
6. Laser hight in.................while light emitted by LED is incoherent.
7. The emission of height in LED takes place due to.................of charge carrires.
8. Laser works on the principle of ....emission.
9. Stimulated emission occurs in.................semiconductors.
10. In p-n junction.................is achieved by applying forward bias.
11. Materials for.................must be effieicnet light emitters.
12. The material suitable for.................is In GaAsP. and InP.
13. ....takes place upto very high frequencies.
14. The innermost region through which light wave transmits in optical fibre is known as.................
15. The core region of optical fibre is surrounded by material of.................known as cladding.
16. ....expresses the light gathering ability of optical fibre.
17. The laser which emits largest wavelength radiation is.................
18. He-Ne laser is a.................laser.
19. The material which doses not absorb visible radiation is called.................
20. Metals exhibit................. in the UV and high frequency range.

Answers

1. visible region
2. Speed of light
3. Relative permittivity
4. photoconductors
5. Total internal reflection
6. highly coherent
7. recombination
8. stimulated
9. direct band gap
10. population inversion
11. semiconductors lasers
12. fibre optic communication
13. Electronic polarization
14. cenmtral core region
15. lower of index of refraction
16. Numerical aperture
17. Co_2 laser
18. four stage
19. diamond
20. low reflectivity

Reading List

1. Solymar *Electrical Properties of Materials.* New Delhi: Oxford University Press.
2. Streetmann, Ben G; Banerjee, Sanjay Kumar *Solid State Electronic Devices.* 2 ed. New Delhi: PHI Learning Pvt Ltd.
3. Neamen, Donald A (1992) *Semiconductor physics and Devices: Basic Principles.* Homewood, IL: Irwin.
4. Gupta, Navneet; Shahdev S K (2001) *A Textbook of Basic Electronics.* 1 ed. New Delhi: Dhanpat Rai & Co. Ltd.
5. Dekker, A J (1998) *Solid State Physics.* New Delhi: Macmillan India Ltd.
6. Dekker, A J (2000) *Electrical Engineering Materials.* New Delhi: PHI Learning Pvt Ltd
7. Indulkar, C S; Triruvagdan *An Introduction to Electrical Engineering Materials* New Delhi: S.Chand Publications
8. Bhattacharya, P (1944) *Semiconductor Optoelectronic Devices. NJ:* Prentice-Hall, Englewood Cliffs
9. Sze, S M *Physics of Semiconductor Devices.* 2 ed. New Delhi: Wiley Eastern Limited.

11

Special-Purpose Materials and Modern Techniques of Material Studies

Chapter Objectives

- To provide the reader a glimpse of recently developed modern materials that are useful for electrical engineering applications.
- To discuss the important characteristics of modern materials.
- To classify and characterise the materials on the basis of their properties are that useful in various electrical devices.
- To discuss various modern techniques for material studies like absorption spectroscopy, SEM,TEM and differential scanning calorimetery.

11.1 INTRODUCTION

The advent of new materials has become possible with the development of technology. It has added an unimaginable comfort and luxury in our lives. Polymers, composites, ceramics and nano materials, along with a wide range of semiconductors, have become part and parcel of our day-to-day life. In earlier times, the only available special purpose materials were metals and alloys. Now, alloys and metals have been completely replaced by plastics, Teflon, polymer, glass, composite, ceramics, etc. Many of the modern luxuries and comforts, be it a transportation, communication, entertainment or information technology, modern materials are being used everywhere. The use of modern materials has not only increased facility and comfort, but also reduced the cost many times. In general, the discovery of new and modern materials has completely changed the life of mankind. This chapter provides a glimpse of special-purpose materials.

11.2 OXIDE GLASSES

Glass is one of the important materials which is widely used in electrical engineering as it exhibits dielectric stability at elevated temperatures. This feature is useful under situations when organic insulation fails to operate. Glass can also withstand severe arcing conditions.

Glass is comprised of a complex system of oxides. Most commercial glasses contain silica as one of the essential structural constituent. Glasses are obtained by fusing silica with some other base like lime or lead oxide.

Pure silica exhibits best insulating properties. On addition with other elements its volume and surface resistivity decreases and loss tangent increases. The insulation properties of glass are found to be composition dependent. Some parameters related with glass are given in Table 11.1.

Table 11.1 *Some physical parameters of glass*

Parameter	*Value at normal temperature*
Volume resistivity	108–1018 ohm cm
Dielectric constant	3.7–10
Loss tangent	0.0003–0.01
Electric strength	25–50 kV/mm

Non-alkaline glasses are also used for electrical insulation. These normally contain high proportion of alumina and basic oxides like CaO, MgO. The fibre glass, which is also one of the important candidates, contains silica–53.5%, Al_2O_3–14.5%, B_2O_3–10%, CaO–17.5% and MgO–4.5%. Here are some characteristics that make fibre glass extremely useful for electrical engineering applications:

- Due to good insulation, it is used as envelope in bulbs and in many similar devices like electronic valves, switches, X-ray tubes, etc.
- Due to high dielectric constant, low temperature co-efficient and low power factor, fibre glass is used as a dielectric material in capacitors.
- Fibre glass can also be drawn in the form of thin sheets, and therefore, can be used as a substitute for mica.
- Potash, potassium-sodium pyrex and silica glasses are used for making various kinds of insulation support, antenna, line insulators and bushings. Glasses to be used for this purpose must have high mechanical strength and can sustain thermal shocks in the presence of chemicals.
- Glass wool is a well-known insulating material and is used in many electrical appliances.

11.3 ORGANIC POLYMERS

Polymers are amorphous (discovered) materials which are composed of big molecules. They may be broadly classified as natural and synthetic polymers. Those derived from plants and animals are known as natural polymers. Examples of such polymers are wood, rubber, cotton, wool, leather, starch, proteins, etc. Synthetic polymers are those that are made by using the modern technology. Examples are polyethylene, synthetic rubber, foam nylon, polyvinylchloride (PVC), polytetrtafluoroethylene (PTFE, better known as Teflon), etc. The advent of synthetic polymers has revolutionised technology and replaced metals in many modern equipments and electrical machines and several electrical appliances.

11.4 STRUCTURE OF POLYMERS

Polymers are made of big molecules called *macromolecules*. The bonding in them is covalent. Macromolecules are in the form of long chains. The basic building-block of polymer chain is a string of carbon atoms with a strong carbon–carbon bond, as shown in Fig. 11.1. Other two electrons of the carbon atom will form strong bonds with other atoms or radicals positioned adjacent to the carbon–carbon chain. The large molecular chain may be thought of as made up of identical structural units called *mer units* or *mers*. A single mer is called a *monomer* while several mers form a *polymer*.

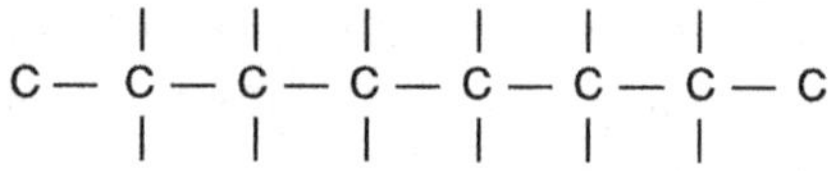

Fig. 11.1 *Backbone of a polymer chain*

The structure of polymers can be understood by considering the formation of polymer polyethylene from the mer unit ethylene. The molecular structure ethylene (C_2H_4) is shown in Fig. 11.2.

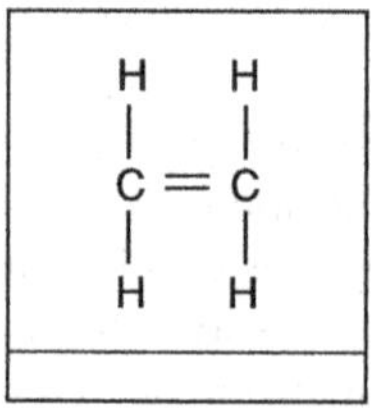

Fig 11.2 *Ethylene molecule*

In the above figure, two carbon atoms are linked through a double bond as shown below.

The formation of polymer polyethylene from the mer unit ethylene, i.e. the polymerisation process, takes place by the opening up of the double bond in two single bonds, which binds to the neighbouring ethylene molecules forming a chain of mer units as shown in Fig. 11.3.

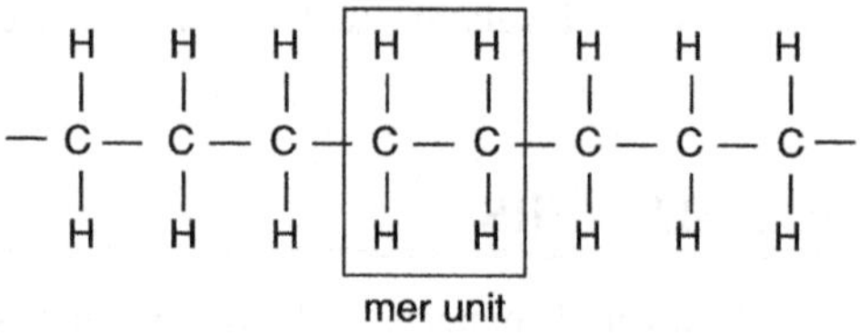

Fig. 11.3 *Schematic structure of polyethylene*

11.5 TEFLON AND PVC

If the hydrogen atoms in polyethylene are replaced by fluorine, the resulting polymer is called polytetrafluoroethylene, commercially known as Teflon. Teflon is one of the most widely used polymers for high temperature applications as it bears a high melting point. It can be made into thin tapes or sleeves for pressure seals and insulation. It is also used to make O-rings for ultra-high vacuum as it is chemically inert and does not outgas.

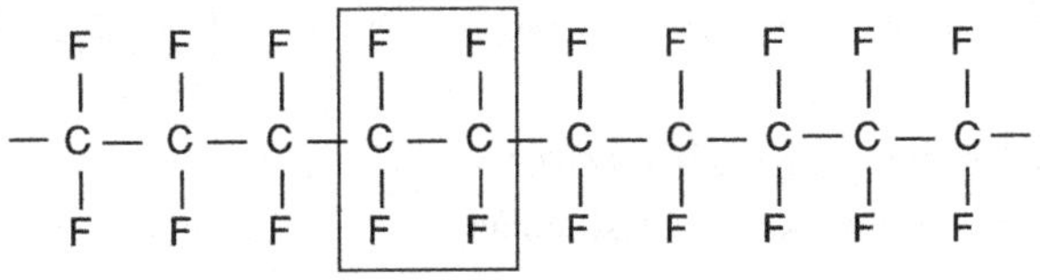

Fig. 11.4 *Schematic structure of polytetrafluoroethylene*

Polyvinylchloride (PVC) is a polymer in which one of the hydrogen atoms in the mer unit of polyethylene is replaced by chlorine atom as shown in Fig. 11.5.

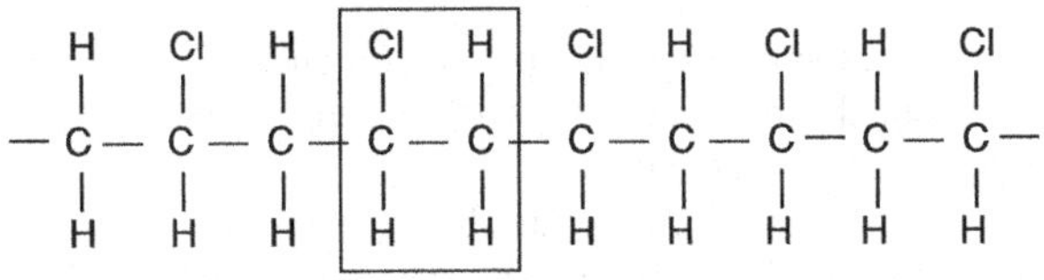

Fig. 11.5 *Schematic structure of polyvinylchloride*

Other ethylene derivative polymers are obtained by replacing chlorine in PVC by organic radicals such as CH_3, C_6H_5, CN, etc.

11.6 HOMOPOLYMERS AND COPOLYMERS

Polymers that contain identical mer units, as in the examples given above, are called *homopolymers*.

Copolymers are polymers in which more than one type of mer units get bonded in random sequence. Two or more polymers can be combined under specific conditions of temperature and pressure to form copolymers.

11.7 CLASSIFICATION OF POLYMERS—THERMOPLASTICS AND THERMOSETS

(i) Bonding Arrangement Based Classification

On the basis of bonding arrangement polymers are classified as:

- Thermoplastics
- Thermosets

In thermoplastics, the bonds between the polymer chains are weak secondary bonds. When they are heated, their plasticity increases and plastic flow occurs. The material softens and ultimately melts. The melting point of thermoplastics is of the order of few hundred °C. The main advantage of thermoplastics is that they can be re-melted and re-moulded, that is they can be recycled. Examples of thermoplastics are polyethylene, polypropylene, polyvinylchloride, nylon, etc.

Thermosets are polymers in which the mers join in more than one direction with several branches. The side chains form a three-dimensional network of primary covalent bonds. When thermosets are heated, further polymerisation reactions (cross-links) occur and on cooling to room temperature they become hard and brittle. On heating after hardening, they do not soften like thermoplastics. They decompose due to reaction with atmospheric oxygen. Examples of thermosets are vulcanised rubbers, epoxies, phenolic and polyester resins.

(ii) Structural Based Classification

On the basis of structure these are classified as:

- Linear polymers
- Branched polymers
- Cross-linked polymers

In linear polymers, mer units are joined together in single chains. Chains bond to each other by weak van der Waals forces. Example are polyethylene, polyvinylchloride, nylon, etc.

In branched polymers, the chains contain side branches. The side branch results from side reactions that occur during synthesis. The packing efficiency in these polymers is very small. So, they have a low density.

In cross-linked polymers, adjacent linear chains are joined to one another at various positions by covalent bonds. This kind of cross-linking is achieved by doing synthesis of the polymer or by chemical reactions at elevated temperature with additive atoms/molecules. One example of a polymer could be vulcanised rubber.

11.8 MECHANICAL PROPERTIES OF POLYMERS

Polymers are not as strong or as stiff as metals. Their mechanical properties—such as tensile strength, modulus of elasticity, fatigue strength—are sensitive to strain rate, temperature and environmental effects such as presence of water, oxygen, solvents, etc.

Polymers are elastic and obey Hooke's law for small deformations at relatively low temperatures. At very high temperatures (near melting point), they become viscous, i.e. they exhibit liquid-like behaviour. At moderate temperatures, they exhibit viscoelastic behaviour.

(i) Melting Point

Melting takes place over a range of temperature in polymers. Melting behaviour depends on the rate of heating. Higher the melting rate, greater will be the melting temperature. Normally, for polymers, a sharp melting point cannot be specified.

(ii) Glass Transition Temperature

On cooling, polymer crystallisation does not take place. There will be a gradual transformation from liquid state to a highly viscous state to a rigid solid state. The temperature at which it becomes a rigid amorphous solid is called the glass transition temperature. The glass transition temperature lies between 0.5 to 0.8 times the melting point. Melting point and the glass transition temperature define the upper and the lower temperature limits, respectively, for many applications. The fabrication processes used for a polymer depends on these temperature limits.

11.9 APPLICATIONS OF POLYMERS

Organic polymers have a wide range of application. Plastics have virtually replaced even metals and inorganic dielectrics (e.g. glass, etc.) in many industrial and household applications. They have the advantages of easy fabrication processes, easy mouldability and non-corrosive feature. They can be easily made into any forms like thin sheets, thin tapes, thin fibres, etc. They are usually good electrical and thermal insulators. They are used as packing materials, thermal insulators containers, coatings, adhesives, wires, etc. Figure 11.6 summarises some of the widely used polymers and their applications.

Material	Characteristic	Application
Polyethylene	Electrically insulating, easily machinable	Carry bags, packing sheets, etc.
Polyvinylchloride	Electrically insulating, non-corrosive	Insulating layer on copper wires, tubes for carrying electrical wires
Neoprene	Soft, recoverable strain	O-rings and gaskets for vacuum sealing, pressure cookers, etc.
Perspex (PMMA)	Easily machinable, electrically insulating, transparent	Electrical insulation, optical window, etc.
Thermocole (Polystyrene)	Thermally insulating, light weight, acoustic absorber	Ice-box or hot-pack, packing material.
Polyurethene foam	Thermally insulating	acoustic absorbers in auditorium,etc. Inner wall of refrigerator,etc
Teflon (PTFE)	Thermally and electrically insulating high melting point	Insulators at high temperatures, nonstick pans, non-reactive tapes, sheets and rods for use at high temperatures
Polyacetylene, Polyaniline	Electrically conducting	Conducting polymers, EMI shielding, electrodes, solid state batteries, polymer LEDs plastic transistors and integrated chips

Fig. 11.6 *Polymers and their applications*

11.10 CERAMICS

Ceramics are inorganic, non-metallic materials. Mostly oxides, nitrides, carbides and borides fall under the category of ceramics. The atomic bonding in ceramics is partially ionic and partially covalent, with a predominantly ionic character. During earlier times, materials which were normally classified as ceramics were glass, porcelain, bricks, tiles, etc. Presently, a long list comes under this classification. Ceramics are useful in a variety of applications such as high-strength tool materials, electronic and magnetic devices, aeroplane and marine engineering, communication engineering civil engineering, etc. Examples of some common ceramics are SiO_2, NaCl, MgO, Al_2O_3, Si_3N_4, Zn, SiC, BN, $BaTiO_3$, etc.

11.11 STRUCTURE OF CERAMICS

As the bonding in ceramics is mainly ionic, it consists of anions and cations. The crystal structure of ceramics is influenced by the radius ratios of the ions. The co-ordination number depends on the radius ratio of the bonding ions. Ceramics are classified on the basis of the structure as AX, AX_2, ABX_3, and AB_2X_4 types.

Examples under each type are given below:

AX: NaCl, CsCl, ZnS

AX_2: SiO_2, CaF_2, PuO_2, ThO_2

ABX_3: $BaTiO_3$, $SrZrO_3$, $SrSnO_3$

AB_2X_4: $MgAl_2O_4$, $FeAl_2O_4$

(i) Silicates

Silicates are ceramics that are most abundant in the earth's crust. Rock soil, clay and sand are different forms of silicates. They mainly consist of silicon and oxygen. The structure of various types of silicates can be understood in terms of the arrangements of SiO_4 tetrahedra. Silicon is tetravalent and oxygen is divalent. Thus, SiO_4 tetrahedron has a negative charge of 4 units. It is denoted by $(SiO_4)_4$ and its structure is shown in Fig. 11.7. Each atom of silicon is bonded to four oxygen atoms that are situated at the corners of the tetrahedron.

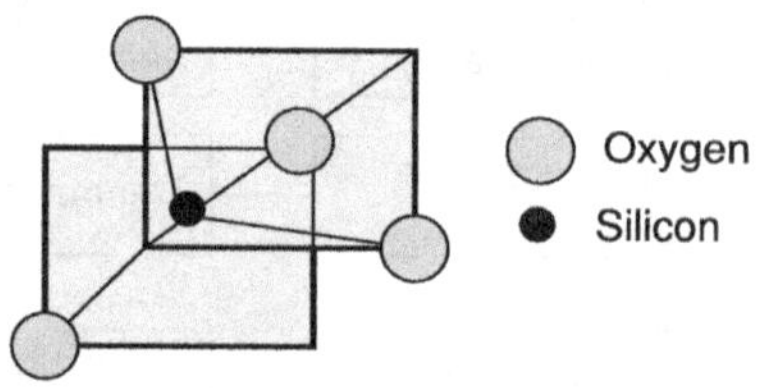

Fig. 11.7 *Silicate tetrahedron*

(ii) Silica

The simplest silicate is silicon dioxide. Its structure is a three-dimensional network generated by the tetrahedron, shown in Fig. 11.7, with each oxygen atom at the corner being shared by adjacent tetrahedra. The structure of silica is shown in Fig. 11.8. The material is electrically neutral and all the atoms have stable electronic structures. The chemical formula is SiO_2.

(iii) Quartz

This is one of the crystalline forms of silica. Silica can also exist in non-crystalline or glassy form, with high degree of atomic randomness. Such a material is called fused silica or vitreous silica. Soda lime glass (normally referred to as window glass) is a noncrystalline form of silicate doped with Na_2O and CaO. The alkali cations break up the network of silicate tetrahedra. Silicates also exist in other more complex structures.

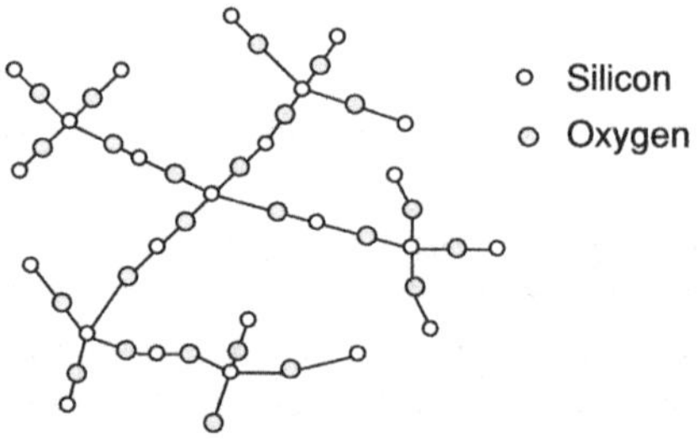

Fig. 11.8 *Structure of silica showing a three-dimensional network of SiO_4 tetrahedra*

11.12 MECHANICAL PROPERTIES OF CERAMICS

Most of the crystalline and non-crystalline ceramics are brittle. Normally, ceramics contain a large number of flaws like minute surface or interior cracks, internal pores, etc. So, when they are subjected to tensile load, stress gets magnified at the cracks and pores, and brittle fracture occurs. However, for compressive stresses, there will not be stress magnification at the flaws and so ceramics exhibit higher strength in compression than in tension. The stiffness constant of ceramics materials is much higher than that of metals lying in the range of 70–500 GPa.

They are the hardest known materials. Examples of ceramics of very high hardness values are silicon carbide, tungsten carbide, aluminium oxide, boron nitride, quartz, etc.

Creep occurs in ceramics at much higher temperatures than in metals.

Advanced ceramics with superior mechanical properties and some unique chemical, electrical and optical properties are being widely used in modern technology.

11.13 ELECTRICAL APPLICATIONS OF CERAMICS

Ceramics have high dielectric breakdown voltage. So, they are used as insulators in high-voltage applications. An example is porcelain.

They have high melting points and can withstand high temperatures. Therefore, they are used in high-temperature furnaces and crucibles for melting materials. Examples are alumina and silica for crucibles, and brick and asbestos for furnaces.

High T_C superconductors, ferrites, ferroelectrics are ceramics which have some special electrical and magnetic properties have been described in the earlier chapters.

11.14 COMPOSITE MATERIALS

Composites are materials obtained by a combination of two or more distinct type of materials. Such combination gives rise to superior properties not exhibited by the individual materials. In general, the mechanical properties are of interest.

There are many naturally occurring composite materials. For example, tissues in the body, wood, bamboo, muscles, etc. They exhibit high strength in addition to flexibility.

11.15 CLASSIFICATION OF COMPOSITE MATERIALS

In synthetic composites, the chief aim is to achieve optimum mechanical properties required for specific applications. Therefore, composites may be classified as given in Fig. 11.9.

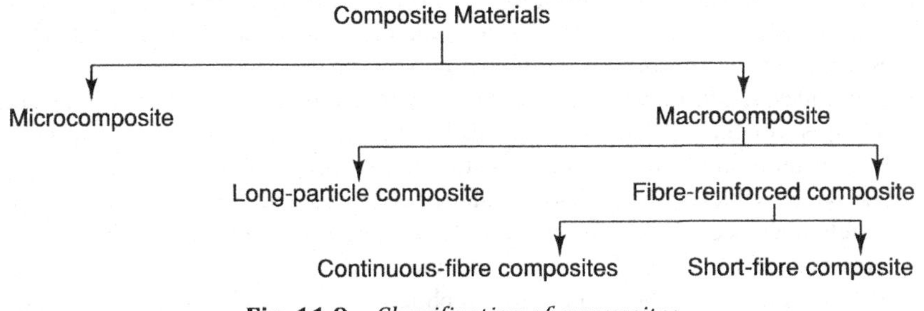

Fig. 11.9 *Classification of composites*

The detail description of composite materials is beyond the scope of this book.

11.16 REFRACTORY MATERIALS

The word 'refractory' implies resistance to melting or fusion. In technology, refractory refers to those materials which are used to withstand the effect of thermal, chemical and physical effects that are encountered in furnace procedures. Whenever very high temperatures are involved, for example, in furnaces and in other electrical heating apparatus—including resistances—the refractoriness provide the linings, supports and other filaments. Thus, refractories are those materials which are used for the construction of furnaces, ovens crucibles and retorts, etc. Refractories are available in the form of bricks, silica, magnesite, cromite, silicon carbide and zirconia.

These materials are classified in three categories:

- Acidic refractories
- Basic refractories
- Neutral refractories

The fusion temperatures of some refractories is given in Table 11.2.

Table 11.2 *Fusion temperatures of some refractories*

Material	*Temp. °C*	*Material*	*Temp. °C*
Fire clay bricks	1600–1750	Chrome bricks	1950–2200
Kaolite	1785	Alumina	2050
Silica bricks	1700	Silicon carbide	2700
Silica	1740	Magnesite bricks	2200
High alumina clay bricks	1800–1850	Zirconia bricks	2200–2700

11.17 SUPER REFRACTORIES

There are certain refractory materials which are capable of resisting the chemical changes brought about by slags, fluxes, molten salts, etc., very effectively and also the stresses and shocks due to sudden temperature fluctuations. This category includes basic types of compounds, for example,

(a) Carbides of various elements such as silicon carbide (carborundum), boron carbide, tantalum carbide, zirconium carbide, titanium carbide, vanadium carbide, etc.

(b) Borides of the metals of fourth, fifth and the sixth periodic table groups.

(c) Nitrides of B, Be, Si, Al, etc. and also the elements of lanthanides and actinides group. All these nitrides have high melting points.

(d) Sulphlides of cerium and thorium. For example, crucible made of cerium sulphide (CeS) is capable of melting the metals having melting points below 1800°C;

(e) Highly pure single oxides such as alumina (Al_2O_3), magnesia (MgO), zirconia (ZrO_2), beryllia (BeO) and thoria (ThO_2), etc., These pure oxide refractories have widely been used for the fabrication of tubes, muffles, and heavy refractory wares such as bricks. Oxides of beryllium and thorium have been used for the preparation of light refractory laboratory wares such as crucibles and tubes. Some important refractories are given below.

(i) Silicon Carbide Refractories

Silicon carbide is the most important refractory material which has a wide range of applications. Silicon carbide bricks are extremely refractory and exhibit high thermal conductivity, low expansion and high resistance to abrasion and spalling. They are mechanically strong and withstand loads in furnaces to temperatures up to 1650°C. Due to its high thermal conductivity it is used to a great extent in the manufacture of saggers, bais and other kiln furniture for supporting the ware in tunnel owens.

(ii) Carbon or Graphite Refractories

Carbon is not a ceramic materials but it is very important as a refractory lining under non-oxidising conditions. For example, it is used as a lining in the hearth of blast furnaces, smelting iron and electrical furnaces for melting various alloys. The coefficient of thermal expansion of graphite is very low, but thermal conductivity is very high. Carbon refractories are very resistant to sudden temperature changes. The carbon used in the manufacture of these refractories is either mineral graphite or retort graphite and the refractorieness depends upon the type of carbon used.

(iii) Pure Oxide Refractories

Refractory oxides of importance are alumina, magnesia, zirconaia, beryllia and thoria. Refractory oxides do not possess plastic properties. Clays or bentonites are used to make the plastic bodies, but they reduce the refractoriness of pure oxides to a great extent. Clays and bentonite are, therefore, added only in very small amount, but even then they decrease the melting point. The material that has the widest application as refractory is alumina.

(iv) Insulating Refractories

Refractory materials that are used to insulate the furnace walls to avoid excessive heat losses due to conduction are known as insulating refractories. Refractories of low thermal conductivities are used as heat insulators. These are of two types. They are

- **Bricks made from naturally occurring porous diatomaceous earth** (such as kieselguhr, asbestos, vermiculite, etc.): These heat insulators are generally employed for low-operating temperatures up to 900°. These heat insulators are not suitable for operations at high temperatures because kieselguhr undergoes large contraction. Vermiculite expands to a large extent on heating. Hence, porous refractories are very suitable under these circumstances.
- **Synthetic, light-weight bricks made by mixing saw dust, coke powder and clay, moulding and firing in a continuous type kiln**: Carbon bars during firing have a highly porous structure. These bricks have low conductivity and can be used for high temperature operations, because air inside the voids acts as poor conductor of heat.

(v) Ceramic Fibres

It is a new textile form for resistance to thermal shock conditions and useful in hypothermal environments. These fibres are used for thermal insulation and for high-temperature low-voltage wiring circuits of space and aircraft equipment. Ceramic-fibre reinforced metals, ceramic reinforced plastic and metal ceramic laminates have also been developed in order to withstand high temperatures.

(vi) Cermets

It is a special group of composite materials, consisting of an intimate mixture of ceramic and metallic components in the form of powder. In other words, cermets are hot-pressed or sintered materials consisting ceramics and metals. Refractory ceramics that have been used in cermets manufacture are high refractory carbides (such as zirconium carbide, molybdenum carbide tungsten carbide, chromic carbide and titanium carbide, etc), borides (of molybdenum, zirconium, titanium and tungsten, etc.), oxides (or aluminium and magnesium) and some silicides and sulphides. The metals that have commonly been used are Ni, Co, Al, Mo, Cr, Fe, etc. The composition of most cermets is 80% ceramic material, 20% metal and they are usually fabricated by making use of powder metallurgy technique. Cermets are mainly used as high refractories where high temperature, as well as high shock resistance is required. Thus, cermets are used in rocket and jet engine parts and spinning tools for hot metals.

(vii) Ferroelectric and Ferromagnetic Ceramics

Barium titanate ($BaTiO_3$) is the most common ceramic of this class. Titania and its compounds have been found to show unusual properties for electrical application. Ferromagnetic ceramic materials are used in television sets, computers, magnetic switches, transformers and recorders, etc. Procedures used in fabricating titania and titanate are ceramic in character.

(viii) Super Alloys

These are iron-based, nickel-based or cobalt-based alloys. They exhibit extremely low creep rates and are highly resistant to oxidation at high temperatures. They are mainly used for blades and motors of jet engines, which operate above 900°C continuously for long times.

Nimonic alloys are the super alloys that have recently been developed. Originally, the alloys were straight 80% Ni, 20% Cr alloys known as nicromes, but they are now modified by adding as Ti, Al, Co, Si, Mo, etc., to improve various properties. Some of them are given below:

- **Nimonic DS**: It contains Ni = 38%, Cr = 18%, C = 0.15%, Si = 2% and rest is iron. It can be used up to 950°C. It is used in furnace parts, gas turbine components, etc.
- **Nimonic 76**: It contains Cr = 20%, Fe = up to 5%, C = 0.1%, Ti = 0.4%, Si = 1% and rest is nickel. It can be used up to 750°C. It finds use in sheets for gas turbines, flame tubes, furnaces, combustion chamber, etc.
- **Nimonic 80**: It contains Cr = 20%, Fe up to 5%, Co = 2%, C = 0.1%, Al = 0.6%, Ti = 2%, Si = up to 1.0% and rest Ni. It can be used up to 700°C. It finds use in gas turbines, blades, drop forgings, etc.
- **Nimonic 90**: It contains Cr = 20%, Fe = up to 5%, Co = 20%, C = 0.1%, Al = 1.5%, Ti = 2.5% Si up to 1.0% and rest is Nickel. It can be used up to 850°C. It finds use in superior welding turbine, rotor blades, etc.
- **Nimonic 100**: It contains Cr = 10%, Fe = 2%, Co = 20%, C = 0.3%, Al 5%, Ti 1.5%, Si = 5% and rest is nickel. It can be used up to 900°C. It is used in forgings for turbine, rotor blades, etc.

Hastelloys are another series of super alloys, which originally contained nickel, molybdenum and iron only. In recent years, some iron in the alloys has been replaced by Cr, W and Co, etc. to get more desirable properties. Ti and Al have also been added in case of precipitation hardening alloys. The typical composition of hastelloy is Ni = 57–63%, Mo = 18–30% and Fe = 5–19%. They are also used in the cast form.

11.18 ORGANIC SEMICONDUCTOR

An organic material with semiconductor properties is known as organic semiconductor. Single molecules of short-chain (oligomers) and organic polymers can be semiconductive. Popular examples include the polycylic aromatic compounds pentacene, anthracene, and rubrene. Polymeric organic semiconductors include poly (3-hexylthiophene), poly (p-phenylene vinylene), as well as polyacetylene and its derivatives.

Organic semiconductors can be classified in two categories. These are organic charge-transfer complexes and various linear-backbone conducive polymers derived from polyacetylene. Linear backbone organic semiconductors include polyacetylene itself and its derivatives: polypurrole and polyaniline. The charge-transfer complexes often exhibit similar conduction mechanisms as that of organic semiconductors, owing to the presence of hole and electron conduction layers separated by a band-gap. Various processes like tunnelling, localised states, mobility gaps, and phonon-assisted hopping contribute to conduction, particularly in polyacetylenes. Organic semiconductors susceptible to doping, such as polyaniline (Ormecon) and PEDOT: PSS, are also known as organic metals.

The conduction in organic semiconductors is due to holes and electrons in π-bonds. Almost all organic solids are insulators. But when their constituent molecules have π-conjugate systems, electrons can move via π-electron cloud overlaps, especially by hopping, tunnelling and related mechanisms. Polycyclic aromatic hydrocarbons and phthalocyanine salt crystals are examples of this type of organic semiconductor.

Because of low mobility, even unpaired electrons may be stable in charge-transfer complexes. Such unpaired electrons can play a role of carriers. This type of semiconductor is also obtained by pairing an electron donor molecule with an electron acceptor molecule.

The first partially conductive organic material known as polyanilene was obtained in 1862 by Henry Letheby. In the 1950s, researchers discovered that polycyclic aromatic compounds formed semiconducting charge-transfer complex salts with halogens. In particular, high conductivity was reported in perylene-iodine complex in 1954. This finding indicated that organic compounds could carry current. In 1972, researchers found metallic conductivity in the charge-transfer complex TTF–TCNQ. Superconductivity in charge-transfer complexes was also reported first time in the Bechgaard salt (TMTSF)2 PF6 in 1980.

Melanin is a semiconducting polymer currently of extreme interest to researchers in the field of organic electronics in both natural and synthesised forms.

Shirakawa *et al.* (1977) reported high conductivity in oxidised and iodine-doped polyacetylene. They received the 2000 Nobel Prize in chemistry for "the discovery and development of conductive polymers".

Rigid-backbone organic semiconductors are now used as active elements in optoelectronic devices such as Organic Light-Emitting Diodes (OLED), organic solar cells, Organic Field-Effect Transistors (OFET), electrochemical transistors and recently in biosensing applications. Organic semiconductors have many advantages, such as easy fabrication, mechanical flexibility and low cost.

11.19 DIFFERENCE BETWEEN ORGANIC AND INORGANIC SEMICONDUCTORS

Organic semiconductors differ from inorganic counterparts in many ways. These include optical, electronic, chemical and structural properties. In order to design and model the organic semiconductors, the optical properties such as absorption and photoluminescence need to be characterised. Optical characterisation for this class of materials can be done using UV-visible absorption spectrophotometers and photoluminescence spectrometers. Semiconductor film appearance and morphology can be studied with Atomic Force Microscopy (AFM) and Scanning Electron Microscopy (SEM). Electronic properties such as ionisation potential can be characterised by probing the electronic band structure with ultraviolet photoelectron spectroscopy.

The transport properties of organic semiconductors are examined in a number of ways. For example, Time-Of-Flight (TOF) and Space Charge Limited Current (SCLC) techniques are used to characterise "bulk" conduction properties of organic films. Organic Field Effect Transistor (OFET) characterisation technique is probing "interfacial" properties of semiconductor films and alloys to study the charge carrier mobility, transistor threshold. Chemical composition and structure of organic semiconductors can be characterised by infrared spectroscopy, Secondary Ion Mass Spectrometry (SIMS) and X-ray Photoelectron Spectroscopy (XPS).

The conduction of charge carriers (electrons or holes) is described by quantum mechanical tunnelling. Charge transport depends on the ability of the charge carriers to pass from one molecule to another and occurs due to hopping.

Theoretically, it has been shown that in a low electric field the conductivity of organic semiconductor is proportional to $T^{-1/4}$ and in a high electric field is proportional to $e^{-(E/aT)}$, where a is a constant of the material. Another study shows that the AC conductivity of the organic semiconductor pentacene was frequency dependent and provides evidence that this behaviour was due to its polycrystalline structure and hopping conduction.

11.20 NANOMATERIALS—A BRIEF HISTORY

Nanotechnology deals with structures between 1nm to 100 nm. Roughly, the thickness of human hair is about 80,000 nm. This is one of the remarkable inventions of 21st century and it is still at the preliminary stage. The interesting feature which is worthwhile to note is the drastic change in the physical and chemical properties when a material is brought to nanoscale. Humans have been known to take advantage of the fascinating properties of nanoparticles in ancient times. Although scientists and researchers have been working on nanoparticles since long, they could only visualise the structure in the recent decade with the development of high-resolution microscopes. We know that atoms are extremely small having a size ranging from 0.1 nm to 0.5 nm, depending on the type of element. The carbon atom is one such element whose diameter is 0.15 nm.

A material as a whole is composed of grains, which in turn comprise many atoms. If we consider a material in bulk, its size can be expressed in three dimensions. For a planar material, the area of this is expressed in two dimensions and in a linear material, its length is expressed in one dimension. In general, a nanomaterial is defined as "materials whose constituent components have size around 100 nm". Figure. 11.10 describes the three-dimensional, two-dimensional and one-dimensional nanomaterials.

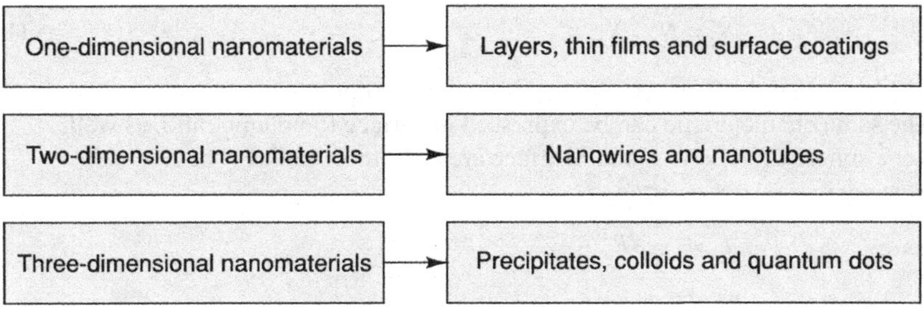

Fig. 11.10 *Types of nanomaterials*

11.21 REASON BEHIND THE DRASTIC CHANGE IN PROPERTIES AT NANOSCALE

Nanomaterials have attracted the attention of scientists and researchers worldwide because they exhibit magnificent properties. The same material having size of the order of micrometre seems entirely meaningless compared to when it is at nanometre size. Two major factors that are mainly responsible for the significant change in the properties are

- Increase in surface area to volume ratio
- Quantum confinement effects

(i) Increase in Surface-Area-to-Volume Ratio

The difference in the properties of material at nanoscale has its origin in a very fundamental physical property of the materials. In any material, an atom on the surface is different from an atom in the bulk. This difference is due to the different potentials that the two atoms experience. We will show how the surface area depends upon the number into which a material divided. For this, let us consider a large sphere of radius R, volume V and surface area S.

The volume of large sphere $V = \frac{4}{3}\pi R^3$

and the surface area of the same sphere $S = 4\pi R^2$

$$= 4\pi\left[\frac{3V}{4\pi}\right]^{2/3}$$

Let the big sphere be divided into N equal spheres of radius r.

The volume of small sphere $\quad v = \frac{V}{N} = \frac{4}{3}\pi r^3$

The surface area of small sphere $\quad S = 4\,\pi r^2$

i.e.

$$r = \left[\frac{3V}{4\pi N}\right]^{1/3}$$

Therefore, the new surface area when the material is divided (reduced in size)

$$S' = N.4\pi r^2 = N\left[\frac{3V}{4\pi N}\right]^{2/3}.4\pi$$

$$S' = N^{1/3}.\left[\frac{3V}{4\pi}\right]^{2/3}.4\pi$$

i.e. $$\boxed{S' = N^{1/3}.S} \quad (11.1)$$

which shows that the new surface area is dependent upon N.

The same characteristic can be expressed in surface to volume ratio, as well.

Let a sphere of volume V and Ssurface area S with its radius R be

Surface area $$S = 4\pi R^2$$

Volume $$V = \frac{4}{3}\pi R^3$$

$$\boxed{\frac{S}{V} = \frac{3}{R}} \quad (11.2)$$

which means that when radius decreases, surface to volume ratio goes on increasing. Figure. 11.12 shows how surface to volume ratio increases on dividing a cubical material into smaller pieces.

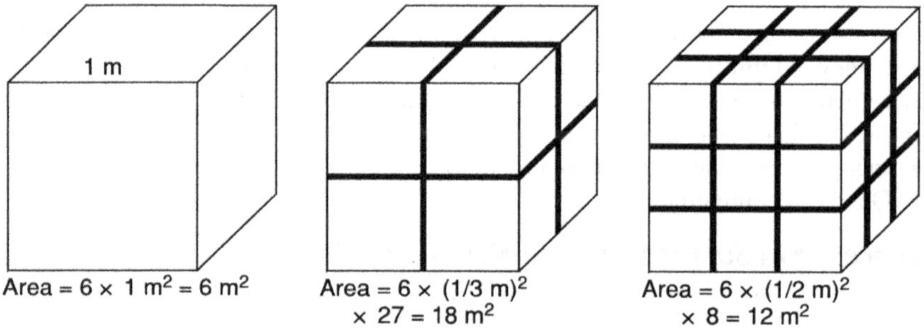

Fig. 11.11 *Increase in surface area on dividing a cube into smaller pieces*

Hence, as the particle size decreases, more and more atoms are found at the surface compared to those inside, causing a significant change in properties and thereby making the material more chemically active.

Quantum Confinement Effects

We know that an isolated atom exhibits sharp (discrete) energy levels. But, when a large number of atoms are brought together to form a solid, the energy levels split and overlap to form bands. Nanomaterials represent an intermediate stage.

It is well known that when a particle is confined in a potential box or well, and if the dimensions of the box are of the order of 'de Broglie wavelength' of electrons, the energy levels of electrons undergo a definite change. This is known as quantum confinement effect. Therefore, when the material size is sufficiently small (~ 10 nm or less), the arrangement of energy level also changes. It results is a significant change in the properties at nanoscale.

In the simplest buck ball or buckminsterfullerene which is the basic building block and contain, 60 carbon atoms with each carbon atom covalently bonded to three adjacent carbon atoms. The resultant sphere is around 1 nm is diameter. Other fullerenes exist with their name given by the symbol for carbon, C, followed by the number of carbon atoms present, e.g. C 60, C 70 and C 80.

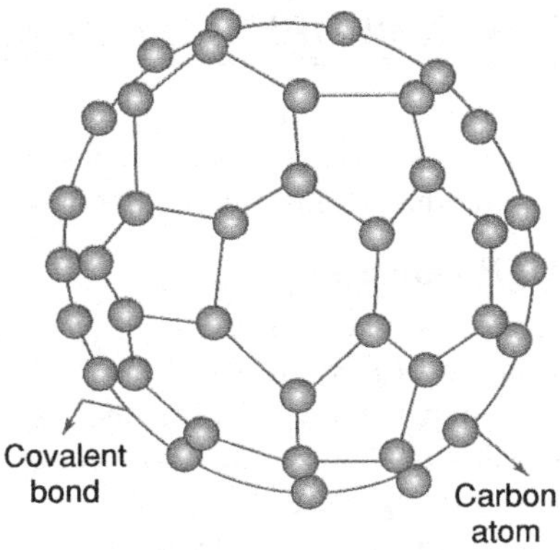

Fig. 11.12 *Variant of a buckyball*

11.22 CREATION OF BUCKYBALLS

Buckyballs spontaneously occur in nature from origins as simple as candle soot. The Japanese researcher Eiji Osawa of Toyohashi University of Technology proposed the existence of buckyballs in 1970.

In August 1985, Richard Smalley, Harry Kroto and Bob Curl performed an experiment at Rich University in the USA and produced buckyballs. They vaporised carbon with a laser and produced a very small quantity of buckyballs. For this work, they were awarded the 1996 Nobel Prize in Chemistry.

In order to produce larger quantities for study purposes, researchers from the Max Planck Institute and the University of Arizona combined forces to develop a method that used two carbon electrodes in a helium or neon atmosphere. When an arc was generated between the electrodes, buckyballs were generated along with carbon soot.

Several methods now exist for making buck balls in tonnage amounts, but they tend to be a closely guarded commercial secret. The first bulkyproduction method is known as *combustion synthesis*. Combustion synthesis burns a hydrocarbon and oxygen mixture at low pressure to produce 95% pure buckyballs.

11.23 USE OF BUCKYBALLS

A wide range of promising applications exist for buckyballs. They are:

- Buckyballs have hardness greater than that of diamond.
- They find use within armours.
- They are added to various polymers to make them stronger.
- Modified buckyballs are also being developed as antioxidants for use by humans.
- When a buckyball encounters a free radical, the unpaired electron in the free radical joins with a delocalised electron in the buckyball. These are modified in order to make it water soluble and suitable for medial use.
- Modifying a buckyball by adding or replacing an atom in order to change the properties of the buckyball is called functionalisation. Functionalised buckyballs are being developed for targeted drug delivery. The buckyball encases a minute dose of a particular drug.

11.24 VARIATION IN PROPERTIES OF NANOMATERIALS

Almost all the properties including the physical, electronic, magnetic and chemical properties of materials depend on size. Small particles display behaviours that are different from those of the individual atoms or bulk. Let us see a few of them in brief.

(i) Change in Physical Properties

Reducing the particle size creates more surface sites, and the surface to volume ratio also increases. This changes the surface pressure and results in a change in the inter-particle spacing. This effect is shown in Fig. 10.13 (in the case of Cu particles). The inter-atomic spacing decreases with size. This is due to the competition between long-range electrostatic forces and the short range core–core repulsion.

The change in inter-particle spacing and the large surface-to-volume ratio in particles have a combined effect on material properties. Variations in the surface free energy change the chemical potential. Material properties also change due to increase in the surface-to-volume ratio. Figure 11.14 shows the melting point of Au particles as a function of size. The melting point decreases with size and at very small sizes the decrease is faster.

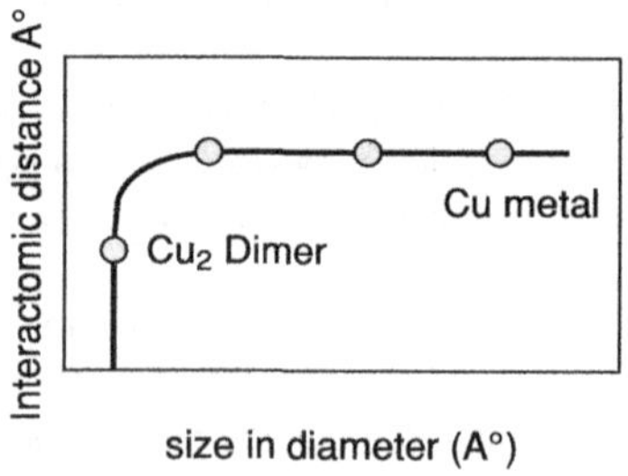

Fig. 11.13 *Inter-atomic distance in Cu as a function of size*

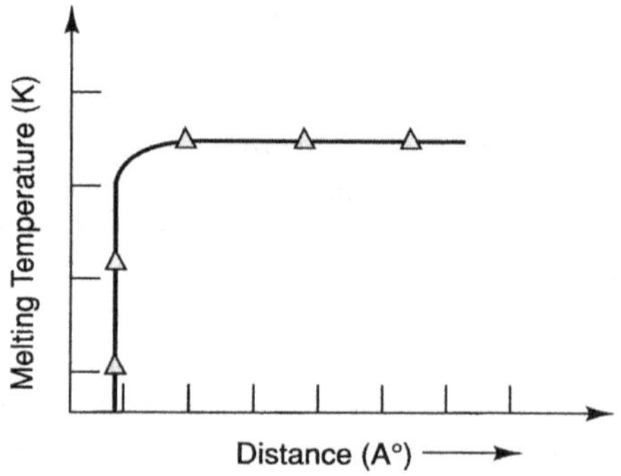

Fig. 11.14 *Melting point of small Au particles as a function of size*

(ii) Change in Chemical Properties

The large surface-to-volume ratio, the variations in geometry and the electronic structure have a strong effect on catalytic properties. As an example, the reactivity of small clusters has been found to vary by many orders of magnitude when the cluster size is changed. Figure. 11.15 shows this in the case of Fe clusters reacting with hydrogen.

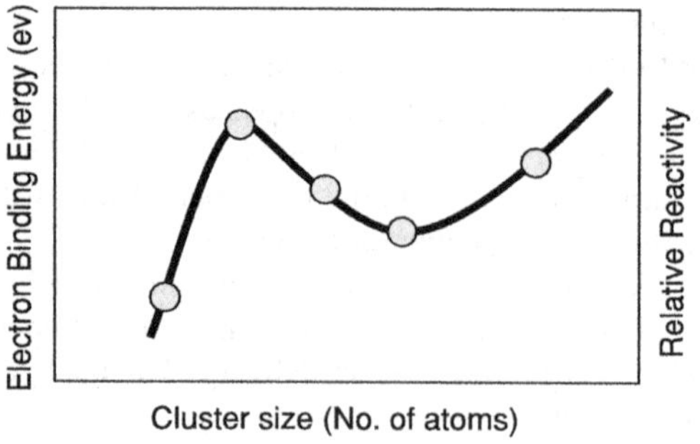

Fig. 11.15 *Ionisation potential and reactivity of Fe clusters as function of size*

(iii) Change in Electrical Properties

The ionisation potential at small sizes is higher than that for the bulk and show marked fluctuations as a function of size.

Due to quantum confinement, the electronic bands in metals become narrower. The delocalised electronic states are transformed to more localised molecular bands and these bands can be altered by the passage of current through these materials, or by the application of an electric field.

The change in electrical properties cannot be generalised. In nanoceramics and magnetic nanocomposites, the electrical conductivity increases with reduction in particle size, whereas in metals, electrical conductivity decreases.

(iv) Change in Optical Properties

Nanocrystalline materials have novel optical properties. Depending on the particle's size, different colours are seen in a suspension of nanoparticles. Gold nanospheres of 100 nm appear orange in colour, while 50 nm nanospheres appear green in colour. If semiconductor particles are made small enough, quantum effects come into play, which limit the energies at which electron and holes can exists in the particles. As energy is related to wavelength (or colour), this means that the optical properties of the particle can be finely tuned depending on its size. Thus, nanoparticles can be made to emit or absorb specific wavelengths (colours) of light, merely by controlling their size. The linear and nonlinear optical properties of such materials can be finely tailored by controlling the crystal dimensions and the chemistry of their surfaces.

(v) Change in Magnetic Properties

The strength of a magnet increases with a decrease in the grain size and an increase in the specific surface area of the grains. Therefore, nanomaterials present good properties in this field as well.

Table 11.3 gives a brief summary of the observed magnetic behaviour of very small particles, of various metals as well as their behaviour in bulk. In small particles, a large fraction of the atoms reside at the surface. These atoms have lower co-ordination numbers than the interior atoms. Experiments indicate that the local magnetic moment is primarily determined by the local co-ordination number. It is observed that as the co-ordination number decreases, the moment increases towards the atomic value, i.e. small particle, are more magnetic than the bulk material. The magnetic moment of Co particles of ~ 2 nm exhibited 20% higher value than that of the bulk.

Table 11.3 *Magnetism in bulk and in nanoparticles*

Metal	*Bulk*	*Cluster*
Na, K	Paramagnetic	Ferromagnetic
Fe, Co, Ni	Ferromagnetic	Super paramagnetic
Gd, Tb	Ferromagnetic	Rotors / Superparamagnetic
Cr	Antiferromagnetic	Frustrated paramagnetic
Rh	Paramagnetic	Ferromagnetic

Nanoparticles of even non-magnetic solids are found to be magnetic. It has been found–theoretically as well as experimentally–that the magnetism is special to small sizes and disappears in clusters containing more than 80 atoms. At small sizes, the clusters become spontaneously magnetic.

(vi) Change in Mechanical Properties

Most metals are made up of small crystalline grains; the boundaries between the grains slow down or arrest the propagation of defects when the material is stressed, thus increasing it strength. If these grains are nanoscale in size, the interface arc (grain boundary) within the material greatly increases, which again enhances its strength. For example, mechanical properties such as hardness and elastic modulus, fracture toughness, scratch resistance, fatigue strength, and hardness can be modified.

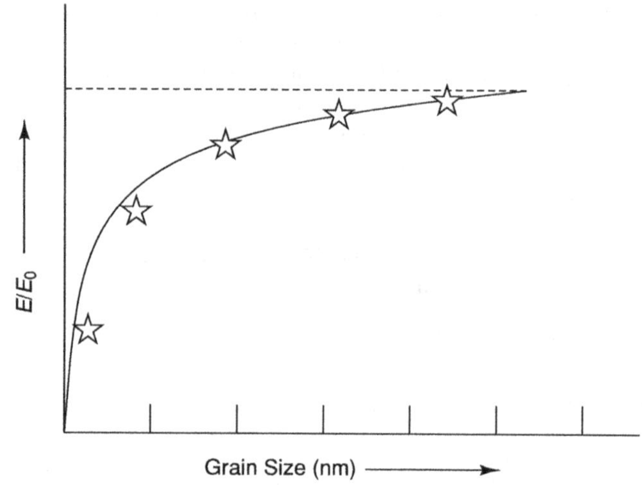

Fig 11.16 *Ratio of Young's modulus (E) or nanocrystalline materials to (E_0) of conventional grain-size materials as a function of grain size*

The mechanical properties of nanomaterials prepared by gas condensation method are of great interest. A few findings are as given below.

- 30–50% lower elastic moduli than in conventional grain-size materials
- 2–7 times higher hardness and strength—hardness values for nanocrystalline pure metals (~ 10 nm grain size) than those of large grained (> 1 μm) metal
- Hardness decreases with decreasing grain size in the nanoscale (~ 10 nm)
- Super plastic behaviour in brittle ceramics or inter-metallics with nanoscale grain sizes at low homologous temperatures

Superplasticity has been observed at somewhat lower temperatures and at higher strain rates in nanocrystalline materials. Much differing results obtained so far suggest that a very different mechanism may be causing superplastic behaviour in nanocrystalline materials.

Some properties of nanomaterials are listed in Table 11.4.

Table 11.4 *Properties of nanomaterials*

Properties	*Examples*
Catalytic	High catalytic efficiency through higher surface-to-volume ratio
Electrical	Improved electrical conductivity in ceramics and magnetic nanocomposites, increased electric resistance in metals
Magnetic	Increased magnetic coercivity up to a critical grain size, Super-paramagnetic behaviour
Mechanical	Improved hardness and toughness of metals and alloys, ductility and super-plasticity of ceramic
Optical	Spectral shift of optical absorption and fluorescence properties, increased quantum efficiency of semiconductor crystals
Biological	Increased permeability through biological barriers (members, Blood-brain barrier, etc.), improved biocompatibility.

11.25 PRODUCTION OF NANOMATERIALS

Materials that are nanoscale in one dimension (for example, very thin surface coatings), in two dimensions (for example, nanowires and nanotubes) or in all three dimensions (for example, nanoparticles) can be produced.

Nanomaterials can be synthesised by mechanical crushing of a solid into fine nanopowder (ball milling).

11.26 PREPARATION OF NANOMATERIALS

Some methods to produce nanomaterials are given below.

(i) Plasma Arcing

Plasma is an ionised gas. Ionised gas (plasma) conducts electricity. A typical plasma-arcing device consists of two electrodes. An arc passes from one electrode to the other. From the first electrode (anode)—due to the potential difference—electrons are emitted. Positively charged ions pass to the other electrode (cathode), pick up the electrons and are deposited to form nanoparticles. As a surface deposit, the depth of the coating must be only a few atoms. The interaction among them must be by hydrogen bonding or van der Waals forces. Plasma arcing is used to produce carbon nanotubes.

(ii) Chemical Vapour Deposition (CVD)

In this method, nanoparticles are deposited from the gas phase. The material is heated to form a gas and then allowed to deposit on a solid surface, usually under a vacuum condition. The deposition may be either physical or chemical. A new product is formed in deposition by chemical reaction. Nanopowders of oxides and carbides of metals can be formed if vapours of carbon or oxygen are present with the metal.

Chemical vapour deposition can also be used to grow surfaces. Surfaces with unique characteristics can be grown with this technique.

(iii) Sol-Gels

A colloid that is suspended in a liquid is called as sol. A suspension that keeps its shape is called a gel. Thus sol-gels are suspensions of colloids in liquids that keep their shape. Sol-gel formation occurs in different stages:

- Hydrolysis
- Condensation and polymerisation of monomers to form particles
- Agglomeration of particles—this is followed by the formation of networks which extends throughout the liquid medium and forms a gel

The rate of hydrolysis and condensation reactions is governed by various factors such as pH, temperature, H_2O/Si molar ratio, nature and concentration of catalyst and process of drying. Under proper conditions spherical nanoparticles are produced. Nanoparticles prepared by sol-gel method are Si $(OR)_3$, where R is alkyl groups of various types.

(iv) Electrodeposition

Using electrodeposition technique it is possible of deposit even a single layer of atoms. Nanostructured films of copper, platinum nickel, gold, etc., can be produced by electro-deposition. The films thus obtained are mechanically robust, highly flat and uniform. Since these films have larger surface areas, they exhibit quite different and favourable electrical properties. They have very wide range of applications. These include batteries, fuel, cells, solar cells, magnetic read heads, etc.

(v) Ball Milling

Ball milling can be used to prepare a wide range of elemental and oxide powders having grain sizes of 10 to 30 nm. Ball milling is the preferred method for preparing metal oxides. Nanocrystals of cerium (CeO_2) and zinc oxide (ZnO) are shown in Fig. 11.17.

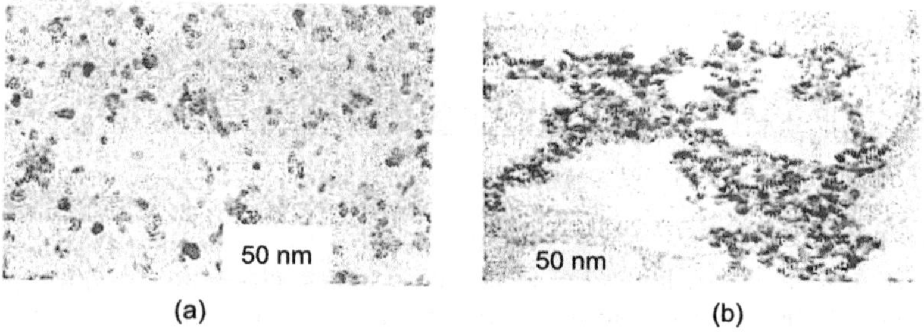

(a) (b)

Fig. 11.17 *Nanocrystals of (a) Cerium oxide, and (b) Zinc oxide*

(vi) Carbon Nanotubes (CNTs)

Carbon nanotubes were first observed by Sumio Iijima in 1991. So far we had known only three forms of carbon, namely diamond, graphite, and amorphous carbon. Now we came to know that there is a whole family of other forms of carbons known as carbon nanotubes which are related to graphite. The molecular structure of graphite is a one-atom-thick a planar network of interconnected hexagonal rings of carbon atoms. In conventional graphite, sheets of carbon are stacked on top of one another. They can easily slide over each other. That is why graphite is not hard and can be used as a lubricant. When graphite sheets are rolled into a cylinder and their edges joined, they form carbon nanotubes, i.e. carbon nanotubes are extended tubes of rolled graphite sheets.

(vii) Types of CNTs

A nanotube may consist of one tube of graphite, a one-atom thick single-walled nanotubes, or a number of concentric tubes called multiwalled nanotubes. Both of these are typically a few nanometres in diameter. Their length may vary from several micrometres to centimetres.

There are different types of CNTs because the graphitic sheets can be rolled in different ways. The three types of CNTs are zigzag. Examples are armchair and chiral. It is possible to recognise types by analysing their cross-sectional structure.

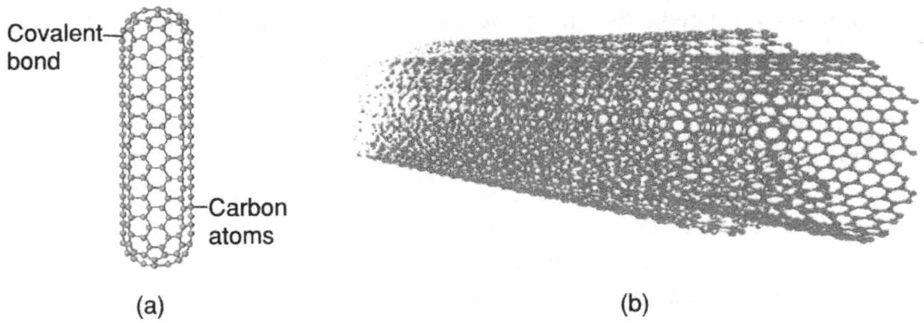

Fig. 11.18 *(a) Different types of single walled nanotubes (b) Multiwalled nanotubes*

Multiwalled nanotubes can come in an even more complex array of forms. Each concentric single-walled nanotube can have different structures, and hence, there are a variety of sequential arrangements. The simplest sequence is when concentric layers are identical, but differ in diameter. However, mixed variants are possible, consisting of two or more types of concentric CNTs arranged in different orders. These can have either regular including or random layering. The structure of the nanotube influences its properties—including electrical and thermal conductivity, density and lattice structure. Both type and diameter are important. The wider the diameter of the nanotube, the more it behaves like graphite. The narrower the diameter of the nanotube, the more its intrinsic properties depend upon its specific tube.

Carbon nanotubes have assumed an important role because of their excellent chemical and physical properties. They are

- mechanically very strong (their Young's modulus is over 1 terapascal, making CNTs as stiff as diamond), and
- flexible (about their axis) and can conduct electricity extremely well.

The helicity of the graphite sheet determines whether the CNT is a semiconductor or metal. All of these remarkable properties give CNTs a range of potential applications; for example, in reinforced composites, sensors, nanoelectronics and display devices.

11.27 APPLICATIONS OF NANOMATERIALS

Nanaoparticles are the 'small particles with a big future'. Because of their extremely large specific surface area, they are chemically very active. They are stronger and more ductile. They have electronic states quite different from those of bulk.

In *dispersed state*, nanoparticles are used as fillers, paints, magnetic recording media, ferro fluids, drugs, phosphors, rocket propellant, fuel additives, etc.

In *consolidate state*, nanoparticles are used as catalysts, electrodes of solar cells and fuel cells, sensors, adsorbents, synthetic bone, self-cleaning glass, etc.

In *ordered assembly form*, nanoparticles are used as quantum electronic devices, photonic crystals, DNA chips, biosensors, etc.

In *very dense phase*, nanoparticles are used in synthesis of flexible/ dense ceramics and insulators, harder metals, etc.

(i) Materials–Technology

- Magnets made of nanocrystalline yttrium samarium-cobalt grains exhibit unusual magnetic properties. This could lead to applications in motors and analytical instruments like Magnetic Resonance Imaging (MRI).
- Ceramics with a reduction in grain size to the nanoscale exhibit large ductility. Zironia, normally a hard, brittle ceramic, has even been rendered super plastic.
- Nancrystalline ceramics, such as silicon nitride and silicon carbide, have been used in such automotive applications as high-strength springs, ball bearings and valve lifters. They are also used as components in high-temperature furnaces.
- Nanosized titanium diozide and zinc oxide are currently used in sunscreens. They absorb and reflect ultraviolet (UV) rays and yet are transparent to visible light.
- Nanosized iron oxide is present in some lipsticks as a pigment.
- Recently developed applications include self-cleaning windows. Nanocoating of highly activated titanium dioxide is highly hydrophobic (water-repellent) and antibacterial. Coatings based on nanoparticulate oxides catalytically destroy chemical agents.
- Cutting tools made of nanocyrstalline materials, are finding applications in drills used to bore holes in circuit boards.
- Other novel and more long-term applications for nanoparticles lie in paints that change colour in response to change in temperature or chemical environment, or paints that have reduced infrared absorbtivity and so reduce heat loss.
- In general, nanoparticles have a high surface area, and hence, provide higher catalytic activity. It may also be possible to engineer specific or selective activity.
- It is now possible to synthesise harder metals having hardness five times higher than normal metals using nanoparticles.
- Flexible/dense ceramics and insulators have started replacing metals.

(ii) Information Technology

- Nanoscale-farbicated magnetic materials also have applications in data storage.
- Coatings with thickness controlled at the nano or atomic scale have been used in optoelectronic devices.
- Nanocrystalline zinc selenide, zinc sulphide, cadmium, sulphide and lead telluride synthesised by sol-gel techniques are candidates for the next generation of light-emitting phosphors. They will have huge market for large area, high brightness, flat-panel displays (as used in television screens and computer monitors).
- Quantum electronic devices have started replacing bulky conventional devices.
- Nanodimensional photonic crystals are used in chemical/ optical computers.

(iii) Biomedicals

- Nanocrystalline silicon carbide is a candidate material for artificial heart valves primarily because of its low weight, high strength and inertness.
- Biosensitive nanoparticles are used for tagging of DNA and DNA chips.
- Controlled drug delivery is possible using nanotechnology.

- Nanostructrued ceramics readily interact with bone cells, and hence, find application as an implant material.

(iv) Energy Storage

- Addition of nanoparticulate ceria (cerium oxide) to diesel fuel improves fuel economy by reducing the degradation of fuel consumption over time.
- Since the hydrogen absorbing capability increases with decrease in size of the nanoparticles, nanoparicles of Ni, Pd and Pt are useful in hydrogen storage devices.
- They are useful in magnetic refrigeration.
- Metal nanoparticles are very useful in fabrication of ionic batteries.

11.28 MODERN TECHNIQUES OF MATERIAL STUDIES

1. Differential Scanning Calorimetry

Differential Scanning Calorimetry (DSC) is a thermo-analytical technique in which the difference in the amount of heat required to increase the temperature of a sample and reference is measured as a function of temperature. Both the material and reference sample are maintained at nearly the same temperature throughout the experiment. In general, the temperature programme for a DSC analysis is designed in such a way that the sample-holder temperature increases linearly with time. A reference sample is taken, which has a well-defined heat capacity, over the range of temperatures to be scanned.

This technique was developed by E S Watson and M J O'Neill in 1962. The DSC instrument measures energy absorbed or released by the material directly and allows precise measurements of heat capacity. A differential scanning calorimeter is shown in Fig. 11.19.

Fig. 11.19 *Differential scanning calorimeter*

(i) Basic Principle

The basic principle upon which this technique works is that when the sample undergoes a physical transformation such as phase transitions, more or less heat will need to flow to it than the reference to maintain both at the same temperature. Whether less or more heat must flow to the sample depends on the process—exothermic or endothermic. For

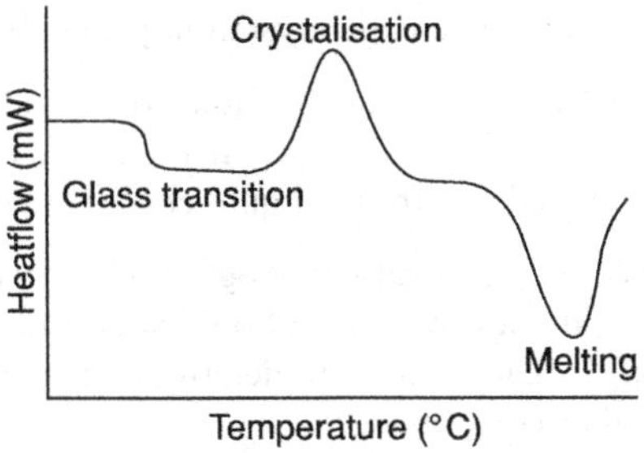

Fig. 11.20 *A DSC thermogram*

example, when a solid sample melts to a liquid state, it will require more heat flowing to the sample to increase its temperature at the same rate as the reference. This is due to the absorption of heat by the sample as it undergoes the endothermic phase transition from solid to liquid state. Likewise, when the sample undergoes exothermic processes (such as crystallisation) less heat is required to raise the temperature of the sample. By knowing the difference in heat flow between the sample and reference, differential scanning calorimeter can be used to measure the amount of heat absorbed or released during such transitions.

A DSC curve (thermogram) exhibiting endothermic and exothermic peaks in given in Fig. 11.20 which explains the phase transition well.

(ii) DTA: Differential Thermal Analysis

A similar technique, which is much in common with DSC, is differential thermal analysis (DTA). In this technique, it is the heat flow to the sample and reference that remains the same rather than the temperature. When the sample and reference are heated identically, phase changes and other thermal processes cause a difference in temperature between the sample and reference. Both DSC and DTA provide similar information. Many modern commercial DTAs are called heat flux DSCs.

(iii) Applications

DSC can be used to measure a number of characteristic properties of a sample. Using this technique, it is possible to observe fusion and crystallisation events as well as glass transition temperatures T_g. DSC can also be used to study oxidation as well as other chemical reactions.

Glass transitions may occur when the temperature of an amorphous solid is increased. These transitions appear as a step in the baseline of the recorded DSC signal. This happens because the sample undergoes a change in heat capacity and no formal phase change occurs.

As the temperature increases, an amorphous solid will become less viscous. Therefore, at some point the molecules may obtain enough freedom of motion to spontaneously arrange themselves into a crystalline form. This is known as the crystallisation temperature (T_c). This transition from amorphous solid to crystalline solid is an endothermic peak in the DSC curve. The ability to determine transition temperatures and enthalpies makes DSC a valuable tool in producing phase diagrams for various chemical systems.

- **Study of Polymers** DSC is used to examine polymers to check their composition. Melting points and glass transition temperatures for most polymers are available from standard compilations, and the method can show possible polymer degradation by the lowering of the expected melting point, T_m. Impurities in polymers can be estimated by examining thermograms and percentage crystallinity can also be obtained.

- **Liquid Crystals** DSC is used in the study of liquid crystals. Using DSC, it is possible to observe the small energy changes that occur as matter undergoes a transition from a solid to a liquid crystal and from a liquid crystal to an isotropic liquid.

- **Oxidative Stability** Differential scanning calorimetry can also be used to study the stability to oxidation of samples. Any oxidation that occurs is observed as a deviation in the baseline. Such analysis can be used to determine the stability and optimum storage conditions for a material or compound.

- **Drug Analysis** DSC is widely used in the pharmaceutical and polymer industries. For the polymer chemist, DSC is a handy tool for studying curing processes, which allows the fine tuning of polymer properties. The cross-linking of polymer molecules that occurs in the curing process is exothermic, resulting in a positive peak in the DSC curve that usually appears soon after the glass transition.

In the pharmaceutical industry, it is always advisable to have well-characterised drug compounds in order to define constituent parameters. For instance, if it is necessary to deliver a drug in the amorphours form, it is desirable to process the drug at temperatures below those at which crystallisation can occur.

• **General Chemical Analysis** It is also used in general chemical analysis and food science.

2. Transmission Electron Microscopy (TEM)—A Brief Introduction

This is a technique that utilises a beam of electrons, which is transmitted through an ultra-thin specimen. These electrons interact with the specimen as it passes through. Therefore, an image is formed from the interaction of the electrons transmitted through the specimen. This image is then magnified and focussed onto an imaging device, such as a fluorescent screen, a layer of photographic film, or to be detected by a sensor.

TEMs are capable of providing higher resolution than optical microscopes, owing to the small de Broglie wavelength of electrons. This provide the user to examine the finer details—even as small as a single column of atoms, which is tens of thousand times smaller than the smallest resolvable object in an optical microscope. Therefore, TEM is one of the major analysis methods in both physical and biological sciences. TEMs also find application in many areas including, materials science as well as pollution, nanotechnology, and semiconductor research. Figure 11.21 shows the TEM image of the polio virus.

The first TEM was built by Max Knoll and Ernst Ruska in 1931, with resolving power greater than that of light wavelength in 1933 and the first commercial TEM in 1939.

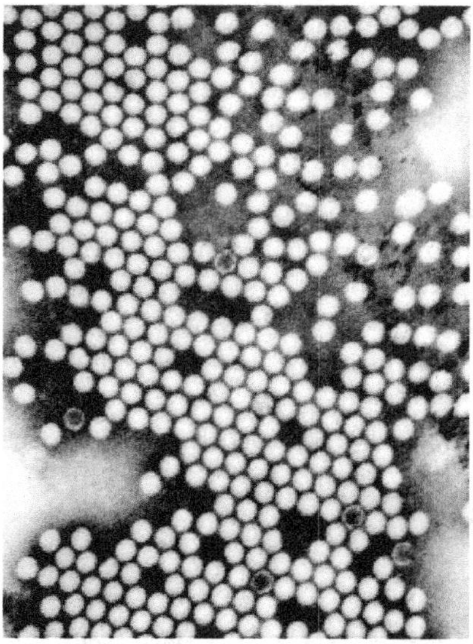

Fig. 11.21 *A TEM image of the polio virus*

(i) Principle of TEM

The accelerated electron beam presents itself as wavelike with a very small wavelength associated nearly 1 Å, so that it can undergo diffraction like a normal light wave. This fact, therefore, allowed to image microscopic objects theoretically, and hence proved to be a boon in the studing of materials.

(ii) Resolution of TEM

The resolution of an optical microscope is theoretically given by

$$d = \frac{\lambda}{2\mu \sin\alpha} \approx \frac{\lambda}{2NA} \qquad (11.3)$$

where $NA \to$ numerical aperture of the system and $\lambda \to$ optical wavelength used in the instrument Since in a TEM an electron's velocity approaches the speed of light, the wavelength of de Broglie wave can be given by

$$\lambda_e = \frac{h}{\sqrt{2m_0E\left(1+\frac{E}{2m_0c^2}\right)}} \tag{11.4}$$

where h is Planck's constant, m_0 is the rest mass of an electron and E is the energy of the accelerated electron.

(iii) Main Components of TEM

A TEM is composed of several components, which include a an ultra high (typically 10^{-7} to 10^{-9} Pa to prevent generation of an electrical arc) vacuum system in which the electrons travel, a source for generation of the electron stream (known as electron gun), a series of electromagnetic lenses and electrostatic plates. The latter two are required to manipulate the beam as required. Also required is a device to allow the insertion into motion within and removal of specimens from the beam path. Besides, imaging devices are also used to create an image from the electrons that exit the system. The detailed description of main components is given below:

- Electron gun
- Electronic lenses
- Sample holder

• **Electron Gun** This is a device to generate the electrons that uses a tungsten filament, or a lanthanum hexaborde (LaB6) source, in either hairpin style or a small and spike-shaped. LaB6 sources utilise small single crystals. By connecting this gun to a high voltage source (typically ~ 100–300 kV), the gun will begin to emit electrons either by thermionic or field emission into the vacuum. The cross-section diagram is given in Fig. 11.22.

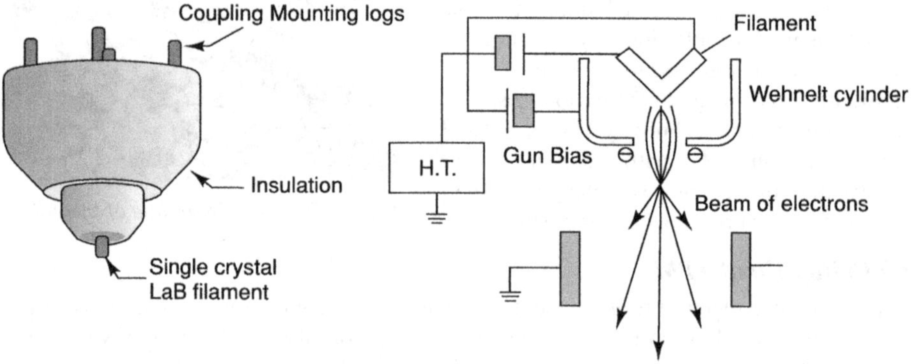

Fig. 11.22 *Electron gun and its cross-section view*

The adjustment of the electron beam is preformed using the electromagnets to focus the electron beam. The use of a magnetic field allows for the formation of a magnetic lens of variable focusing power. Apart from the magnetic field, electrostatic fields can cause the electrons to be deflected through a constant angle. The optical configuration of a TEM can be rapidly changed, similar to that in an optical microscope.

• **Electron Lens** Electron lenses are designed to act like an optical lens by focusing parallel rays at some constant focal length. Lenses may operate electrostatically or magnetically. The majority of electron lenses for TEM utilise electromagnetic coils to generate a convex lens manufactured from iron, iron-cobalt or nickel cobalt alloys, such as permalloy.

The components include the yoke, the magnetic coil, the poles, the pole piece, and the external control circuitry. The pole piece must be manufactured in a very symmetrical manner, as this provides the boundary conditions for the magnetic field that forms the lens.

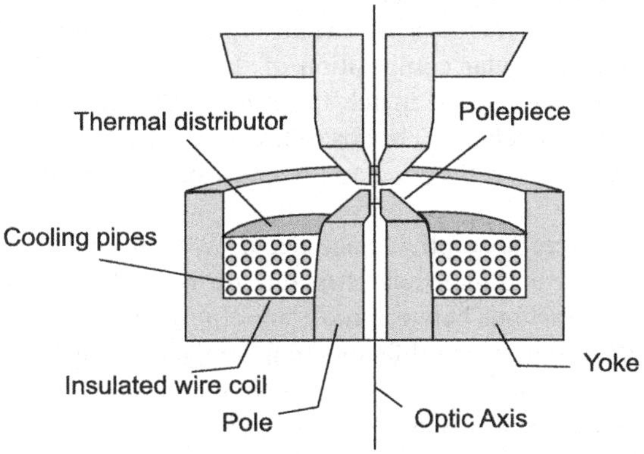

Fig. 11.23 *TEM electron lens*

Typically, a TEM consists of three stages of lensing. The stages are the condenser lenses, the objective lenses and the projector lenses. The condenser lenses are responsible for primary beam formation, whilst the objective lenses focus the beam that comes throughout the sample itself. The projector lenses are used to expand the beam onto the phosphor screen or other imaging device, such as film. The magnification of the TEM is due to the ratio of the distances between the specimen and the objective lens' image plane.

• **Sample Holder** The sample holder carries a grid of 3.05 mm diameter with a thickness and mesh size ranging from a few to 100 μm. The sample is placed on to the inner meshed area having a diameter of approximately 2.5 mm. Useful materials are copper, molybdenum, gold or platinum.

Display Screen Imaging systems in a TEM consist of phosphor screen, which may be made of fine (10–100) μm particulate zinc sulphide, for direct observation.

3. Optical Absorption Spectroscopy

This is a spectroscopic technique that measures the absorption of radiation, as a function of frequency or wavelength, due to its interaction with a material specimen. The specimen absorbs energy from the radiating field. The intensity of the absorption varies as a function of frequency and this variation is known as the *absorption spectrum*. Absorption spectroscopy is performed across the electromagnetic spectrum.

Absorption spectroscopy is an analytical tool to determine the presence of a particular substance in a sample and, in many cases, to estimate the amount of the substance present.

Infrared and ultraviolet-visible spectroscopy are particularly common in analytical applications.

In the experiment to measure absorption spectra, a beam of radiation is directed at a sample and the emerging (transmitted) intensity is detected by it. The transmitted energy can be used to calculate the absorption. The source, sample arrangement and detection technique vary significantly depending on the frequency range and the purpose of the experiment.

(i) Basic Theory of Absorption Spectrum

The absorption spectrum of a material is the fraction of incident radiation absorbed by the material over a range of frequencies. The absorption spectrum is primarily dependent upon the atomic and molecular composition of the material. Radiation is more likely to be absorbed at frequencies that match the energy difference between two quantum mechanical states of the molecules. The absorption that occurs due to a transition between two states is known as *absorption line* and a spectrum is typically composed of many such lines.

The frequencies where absorption lines occur, as well as their relative intensities, depend on the electronic and molecular structure of the molecule. The frequencies will also depend on the interactions between molecules in the sample, the crystal structure in solids, and on several environmental factors (e.g. temperature, pressure, electromagnetic field). The lines will also have a width and shape that are primarily determined by the spectral density or the density of states of the system.

The integrated intensity line is found proportional to the amount of the absorbing substance present. It is also related to the temperature of the substance and the quantum mechanical interaction between the radiation and the absorber.

Absorption and transmission spectra represent equivalent information and one can be calculated from the other through a mathematical transformation. A transmission spectrum will have its maximum intensities at wavelengths where the absorption in weakest because more light is transmitted through the sample. An absorption spectrum will have its maximum intensities at wavelengths where the absorption is strongest.

(ii) Relation to Emission Spectrum

Emission is a process by which a substance releases energy in the form of electromagnetic radiation. Emission can occur at any frequency at which absorption can occur, and this allows the absorption lines to be determined from an emission spectrum. The emission spectrum will typically have a quite different intensity pattern from the absorption spectrum, though. So the two are not equivalent.

4. Scanning Electron Microscopy—An Introduction

The Scanning Electron Microscope (SEM) is a technique that used a beam of high-energy electrons. The signals that come out from electron-sample interactions reveal information about the sample including external morphology (texture), chemical composition and crystalline structure and orientation of materials. The SEM is also capable of performing analysis of selected point locations on the sample which is especially useful in determining chemical compositions crystalline structure and crystal orientations.

(i) Basic Principles of SEM

A high-energy electron beam is allowed to interact with sample under investigation and signals are received in the form of decelerated electrons which include secondary electrons (that produce SEM images), BackScattered Electrons (BSE), diffracted backscattered electrons (EBSD that are used to determine crystal structures and orientations of minerals), photons (characteristic X-rays that are used for elemental analysis and continuum X-rays), visible light and heat. Secondary electrons and backscattered electrons are commonly used for imaging samples: secondary electrons are most valuable for showing morphology and topography on samples and backscattered electrons are most valuable for illustrating contrasts in composition in multiphase samples. X-rays are also produced by inelastic collisions of the incident electrons with electrons in discrete orbitals of atoms in the sample. As the excited electrons return to lower energy states, they yield X-rays that are of a fixed wavelength. Thus, characteristic X-rays are produced for each element in a mineral that is 'excited' by the electron beam. SEM analysis is considered to be 'non-destructive', that is X-rays generated by electron interactions do not lead to volume loss of the sample, and therefore, it is possible to analyse the same materials again and again.

Overall, SEM is a type of electron microscope that images a sample by scanning it with a high-energy beam of electrons in a raster scan pattern. Figure below illustrate the sample chamber and micrograph of SEM:

Fig. 11.24 *Micrograph of pollen grains*

Fig. 11.25 *Sample chamber of SEM*

(ii) Main Components of SEM

The main components of SEM include the following:

- Electron gun
- Electron lenses
- Sample stage
- Detectors
- Display devices
- Infrastructure components:
  - Power supply
  - Vacuum system
  - Cooling system
  - Vibration-free floor
  - Room free of ambient magnetic and electric fields

Figure. 11.26 shows a schematic diagram of SEM–EPMA.

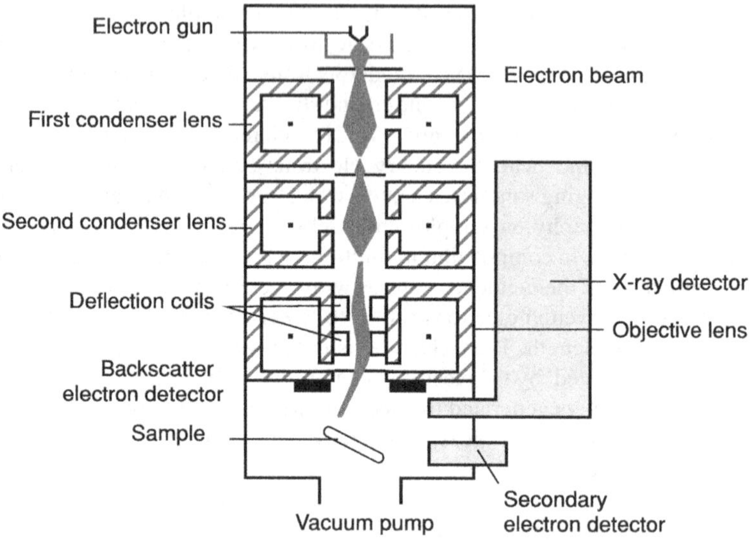

Fig. 11.26 *Schematic diagram of an SEM*

(iii) Process of Scanning and Image Formation

The emerging electrons (0.2 keV to 40 keV) are focused by one or two condenser lenses to a spot about 0.4 nm to 5 nm in diameter. The beam passes through pairs of scanning coils or pairs of deflector plates in the electrons column, typically in the final lens, which deflect the beam in the x and y axes, so that it scans in a raster fashion over a rectangular area of the sample surface.

The interaction of electron beam with the sample, can be detected and create images of the distribution of specimen current. The output signals are then amplified, image may be photographed from a high-resolution cathode ray tube, but in modern machines, it is digitally captured and displayed on a computer monitor.

(iv) Magnification

Magnification in an SEM can be controlled over a range of up to 6 orders of magnitude from about 10 to 500,000 times. Figure 11.27 shows a micrograph of a housefly's compound eye.

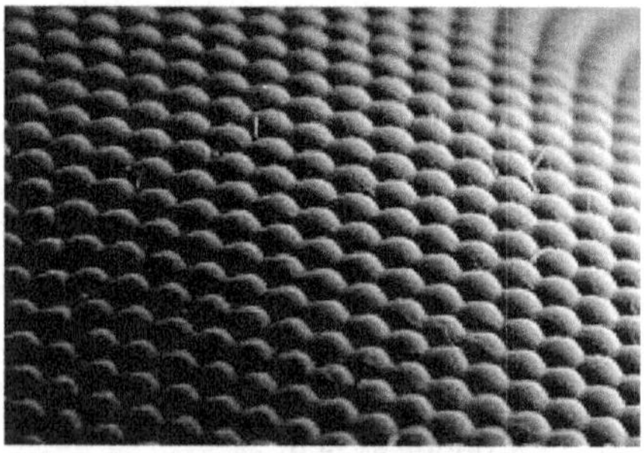

Fig. 11.27 *SEM micrograph of a housefly's compound eye*

(v) Resolution of the SEM

The spatial resolution of the SEM depends on the size of the electron spot, which in turn depends on both the wavelength of the electrons and electron-optical system that produces the scanning beam. The SEM has compensating advantages, including the ability to image a comparative large area of the specimen; the ability to image bulk materials (not just thin films or foils) and the variety of analytical modes are available for measuring the composition and properties of the specimen. Figure 11.28 shows a magnified view of a specimen:

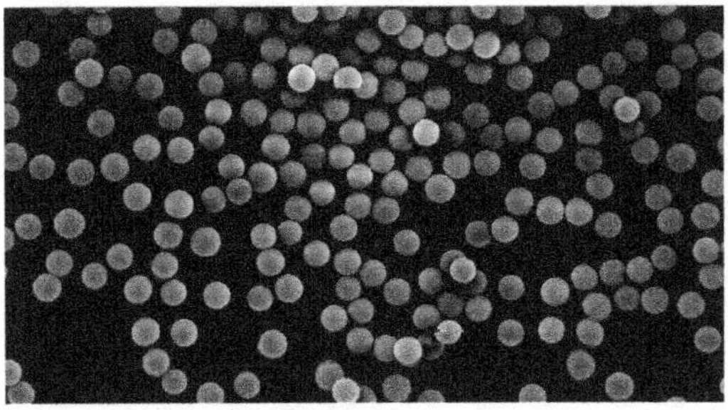

Fig. 11.28 *Magnified view of biological specimen*

(vi) Strength and Limitations of SEM

The SEM is critical in all fields that require characterisation of solid materials. Modern SEMs generate data in digital formats that are highly portable.

Samples must be solid and may be capable to fit in to the microscope chamber. Maximum size in horizontal dimensions is usually of the order of 10 cm, and vertical dimensions are approximately 40 mm. For most instruments, samples must be stable in a vacuum (nearly of the order of 10^{-5} to 10^{-6} torr). Samples likely to outgas at low pressure are not suitable for examination in SEMs.

Most SEMs use a solid state X-ray detector (EDS) and while these detectors are very fast and easy to utilise, they have relatively poor energy resolution and sensitivity to elements present in low abundances when compared to Wavelength Dispersive X-ray detectors (WDS) on most Electron Probe MicroAnalysers (EPMA). An electrically conductive coating must be applied to electrically insulating samples for study in conventional SEMs unless the instrument is capable of operation in a low vacuum mode.

(vii) Applications

It is used to generate high-resolution images of the shapes of objects (SEI) and to show spatial variation in chemical compositions. The SEM is also widely used to identify phases based on qualitative chemical analysis and/ or crystalline structure. Precise measurement of very small features and objects less than 50 nm in size is also accomplished using the SEM. Backscattered electron images (BSE) can be used for rapid discrimination of phases in multiphase samples. SEMs equipped with diffracted backscattered electron detectors (EBSD) can be used to examinemicro fabric and crystallographic orientation in many materials.

Summary

- Organic polymers are organic amorphous materials having large molecules. The molecules in polymers are in the form of long chains.
- A polymer chain is made up of large number of identical structural units called mer units. Several mer units, or mers, join together in the chain to form a molecule and several such molecular chains bond together to form the polymer.
- Polymers which contain more than one type of mer units are called co-polymers.
- The degree of polymerisation of a polymer is the average number of mer units per molecule.
- Polymers may be partially crystalline with crystalline regions dispersed within amorphous polymer region.

- Ceramics are inorganic, non-metallic materials. Mostly oxides, carbides and nitrides are classified as ceramics.
- Ceramics have structure as AX, AX_2, ABX_3 and AB_2X_4.
- The atomic bonding in ceramics is predominately iono-covalent.
- Most ceramics are brittle and have low toughness, but they have other superior mechanical characteristics such as high strength, hardness, stiffness constant, creep resistance, etc.
- Ceramics have wide application where high strength at high operating temperatures is required.
- Composites are materials formed by a combination of two or more distinct type of materials.
- The special characteristics of composites that make them unique are high strength and high elastic modulus combined with low specific gravity.
- Nanomaterials could be defined as those materials which have structured components with size less than 100 nm at least in one dimension.
- Nanoscience can be defined as the study of phenomena and manipulation of materials at atomic, molecular and macromolecular scales, where properties differ significantly from those at larger scale.
- Nanotechnology can be defined as the design, characterisation, production and application of structures, devices and systems by controlling shape and size at the nanometre scale.
- Inter-atomic spacing decreases with size.
- The melting point decreases with size.
- The ionisation potential at small sizes is higher than that for bulk and show marked fluctuations as a function of size.
- Small particles are more magnetic than bulk materials.
- Even non-magnetic solids are found to be magnetic.
- Namomaterials can be synthesised in many methods.

  (i) Plasma arcing

  (ii) Chemical vapour deposition

  (iii) Sol-gels

  (iv) Electrodeposition

  (v) Ball milling
- Carbon nanotubes (CNTs) are extended tubes of rolled graphite sheets.
- CNTs are mechanically very strong.
- CNTs are flexible (about their axis) and can conduct electricity extremely well.

Exercise

1. What do you mean by nanomaterials? Discuss the properties of metallic and semiconducting nanomaterials.
2. Describe two methods for synthesis of nanoparticles.
3. What do you mean by C60 buckyball? How was it discovered? Discuss its electric properties. How do electrical properties change if it is doped with potassium?

4. Describe two methods for fabrication of carbon nanotubes.
5. Describe the structure and properties of carbon nanotubes.
6. Describe the properties and uses of carbon nanotubes.
7. Discuss various applications of nanotechnology.
8. What are buckyballs? Discuss their properties and application.
9. What are the potential applications of carbon nanotubes?
10. What do you mean by characterisation of nanoparticles? Give one method for the characterisation and discuss important application of nanotubes.
11. What is a glass? Discuss its various kinds and important characteristics. How is it useful in electrical engineering applications?
12. What are organic polymers? Discuss some important organic polymers useful for electrical engineering purposes?
13. Discuss important properties of polyvinyl chloride. What are thermoplastics and thermosets?
14. Discuss some important properties of polymers and mention their applications.
15. What are ceramics? Discuss their structure and important applications in electrical engineering.
16. What are composite materials? How are they superior when compared with individual materials? Discuss their classification.
17. What are hybrid composites? Discuss their applications.
18. What are refractory materials? Describe their characteristic properties and applications.
19. Discuss the important properties of insulating refractory materials. Why are these important in electrical engineering?
20. What are organic semiconductors? Discuss some main applications of these materials. What is the difference between organic and inorganic semiconductors?
21. Discuss some interesting properties of super alloys and mention their applications.
22. Discuss the basic principle of differential scanning calorimetry. Mention its applications in characterising the materials.
23. What do you mean by glass transition temperature and crystallisation temperature? How do these appear in a DSC thermogram?
24. Discuss the principles of transmission electron microcopy (TEM) and mention its applications.
25. Discuss the main components of TEM and their functioning.
26. What is the principle of absorption spectroscopy? What information is obtained from absorption spectroscopy?

Objective-type Questions

1. The glass transition temperature is
   (a) at the boiling point
   (b) at melting point
   (c) between 0.5 to 0.8 times the melting point
   (d) none of these
2. The temperature at which rigid amorphous solidification takes place is known as
   (a) melting temperature
   (b) glass transition temperature
   (c) boiling temperature
   (d) all of these
3. Atomic bonding in ceramics is
   (a) ionic
   (b) covalent
   (c) van der Waals
   (d) partially ionic and partially covalent

4. Organic polymers are
   (a) metallic in character
   (b) insulators
   (c) good electrical and thermal insulators
   (d) none of these
5. Teflon exhibits a
   (a) very low melting point
   (b) high melting point
   (c) moderate melting point
   (d) none of these
6. Ceramic material exhibits
   (a) ionic bonding
   (b) covalent bonding
   (c) iono-covalent bonding
   (d) van der Waals bonding
7. Ceramics exhibit
   (a) high breakdown voltage
   (b) low breakdown voltage
   (c) moderate breakdown voltage
   (d) none of these
8. OLED is a/an
   (a) metal
   (b) insulator
   (c) refractory material
   (d) organic semiconductor
9. Melanin is a
   (a) semiconducting polymer
   (b) elemental semiconductor
   (c) metal
   (d) insulator
10. Nanoparticles have sizes between
    (a) 1 and 100 mm
    (b) 1 and 100 μm
    (c) 1 and 100 nm
    (d) 1 and 100 cm
11. Properties of materials drastically change as their size approaches
    (a) nanoscale (b) microscale
    (c) vernier scale (d) main scale
12. Magnetic nanoparticles are a class of nanoparticles which can be manipulated under the influence of
    (a) electric field
    (b) magnetic field
    (c) electrostatic field
    (d) none of these
13. Buckyball is a spheroidal fullerene consisting of
    (a) six carbon atoms
    (b) sixty carbon atoms
    (c) six hundred carbon atoms
    (d) none of these
14. Carbon nanotubes were discovered in 1991 by
    (a) R Feynman
    (b) S Iigima
    (c) Prof. Tan Zikang
    (d) Wang Ning
15. Armchair, zig-zag and chiral are the names of different types of
    (a) multiwalled nanotubes
    (b) double-walled nanotubes
    (c) single-walled nanotubes
    (d) buckyballs
16. Graphite, diamond and fullerene are the three forms of
    (a) pure boron
    (b) pure carbon
    (c) ceramic
    (d) none of these
17. C60 is actually a truncated icosahedron consisting of
    (a) 12 pentagons and 20 hexagons
    (b) 6 pentagons and 26 hexagons
    (c) 12 pentagons and 6 hexagons
    (d) none of these
18. Carbon nanotubes are about 100 times stronger than steel, but
    (a) just a sixth of weight
    (b) just sixtieth of weight
    (c) heavier than steel
    (d) none of these
19. Chemical vapour deposition is a technique for the
    (a) characterisation of nanoparticles
    (b) synthesis of carbon nanotubes
    (c) determination of the size of nanoparticles
    (d) identification of nanoparticles
20. The bonding in ceramics is
    (a) ionic
    (b) metallic
    (c) covalent
    (d) partially ionic and partially covalent

21. Vulcanised rubber is an example of
    (a) branched polymer
    (b) linear polymer
    (c) cross-linked polymer
    (d) co-polymer
22. Among the following ceramics, which one has the highest hardness?
    (a) Aluminium oxide
    (b) Boron carbide
    (c) Quartz
    (d) Silicon carbide
23. Cermets are examples of
    (a) large-particle composites
    (b) micro-composites
    (c) continuous-fibre composites
    (d) short-fibre composites
24. Among the following composites, which ones are suitable for high temperature applications?
    (a) Carbon-carbon composite
    (b) Glass-polyester composite
    (c) Carbon-resin composite
    (d) Carbon-aluminium composite

Answers

1. (c)	2. (b)	3. (d)	4. (c)	5. (b)
6. (d)	7. (a)	8. (d)	9. (a)	10. (c)
11. (a)	12. (b)	13. (b)	14. (b)	15. (c)
16. (b)	17. (a)	18. (a)	19. (b)	20. (d)
21. (c)	22. (d)	23. (a)	24. (d)	

Fill in the Blanks

1. A nanoparticle can be defined as a chunk of matter whose physical dimension lies in the range of .... to .... nm.
2. Techniques like TEM, SEM and AFM are used to .... nanoparticles.
3. The fullerenes are a class of allotropes of ....
4. High-pressure carbon monoxide deposition (HiPCO) is a technique to produce .... in bulk quantities.
5. MWNT consists of .... concentric nanotube cylinders.
6. .... and .... are two types of molecules composed of carbon atoms that have many applications in nanotechnology.
7. Bonding an atom or molecule to a .... to change the properties of buckyball is called functionalisation.
8. Single-walled carbon nanotubes can be considered to be formed by the rolling of single layer of ....
9. A one-atom-thick layer of graphite is called ....
10. The angle between the axis of the hexagonal pattern and the axis of the carbon nanotube is called nanotube's .... angle.
11. Glass is popular in electrical engineering because of its ....at elevated temperatures.
12. Glass exhibits .... dielectric constant and low power factor.
13. Polymers are composed of .... molecules.
14. The bonding in .... is covalent.
15. Replacement of H atoms by fluorine atoms gives a material known as ....
16. Teflon exhibits .... melting point.
17. Polyvinyl chloride (PVC) is a material in which alternate H atoms of its unit are replaced by .... atoms.
18. The bonding between polymer chains in thermoplastics are weak ....

19. The temperature at which a polymer becomes a rigid amorphous solid is known as ....
20. .... are good electrical and thermal insulators.
21. The atomic bonding in .... is partially ionic and partially covalent.
22. .... is one of the crystalline forms of silica.
23. Ceramics exhibit .... breakdown voltage.
24. Cermets are pressed or sintered materials consisting of ....
25. Silicon carbide is a .... material which is used to make bricks.
26. Carbon refectories are .... to sudden temperature changes.
27. Refractory materials of low thermal conductivity are used as ....
28. The organic material exhibiting .... properties is known as organic semiconductor.
29. .... are used for blades and rotors of jet engines which operate ready 10000°C.

Answers

1.	few nm, 100	2.	characterised	3.	carbon
4.	carbon nanotubes	5.	multiple		
6.	buckyball and carbon nanotubes			7.	buckyball
8.	graphite	9.	graphene	10.	chiral
11.	dielectric stability	12.	high	13.	big size
14.	polymers	15.	Teflon	16.	high
17.	chlorine	18.	secondary bonds		
19.	glass transition temperature			20.	polymers
21.	ceramics	22.	quartz	23.	high
24.	ceramics and metals	25.	Refractory	26.	very resistant
27.	heat insulators	28.	semiconducting	29.	super alloys

Reading List

1. Electron Microscopy; Methods and Protocols II Edition John Kuo, Springer, 2007.
2. Electron Microscopy Methods and Protocols MA Nasser Hajibagheri, Humana Press.
3. Scanning Electron Microscopy: Physics of Image Formation and Microanalysis Ludwig Reimer, Springer IInd Edition.
4. Scanning Electron Microscopy and X-ray microanalysis Vol-I Joseph Goldstein IIIrd Edition, Springer.
5. Transmission Electron Microscopy: A Text book for material science, Vol. I. David Bernard Williams, C. Bary Carter, Springer.
6. Differential Scanning Calorie metry: An Introduction for Practitoners Giinther Hohne, W Hemminger, HJ Flammersheim Springer, IInd Edition.
7. Optical spectroscopy of Inorganic Solids Brain Henderson and George F Im bush oxford scientific Publications.

Index

O

P

Q

R

S

Made in the USA
Monee, IL
09 June 2026

52954206R00282